A Specialist Periodical Report

# Nuclear Magnetic Resonance

Volume 4

A Review of the Literature published between June 1973 and May 1974

Senior Reporter

**R. K. Harris,** *School of Chemical Sciences, University of East Anglia*

Reporters

**P. S. Allen,** *University of Nottingham*

**W. Derbyshire,** *University of Nottingham*

**T. Drakenberg,** *Lund Institute of Technology, Lund, Sweden*

**M. I. Foreman,** *Organon Laboratories Ltd., Newhouse, Lanarkshire*

**R. Grinter,** *University of East Anglia*

**W. McFarlane,** *City of London Polytechnic*

**R. B. Mallion,** *University of Oxford*

**I. D. Robb,** *Unilever Research Ltd., Port Sunlight, Cheshire*

**D. S. Rycroft,** *University of Glasgow*

**G. J. T. Tiddy,** *Unilever Research Ltd., Port Sunlight, Cheshire*

**H. Wennerström,** *Lund Institute of Technology, Lund, Sweden*

**M. D. Zeidler,** *University of Karlsruhe, West Germany*



The Chemical Society

Burlington House, London, **W1V 0BN**

**ISBN:** 0 85186 282 9

Library of Congress Catalog Card No. 72-78527



Printed in Great Britain
at the Alden Press, Oxford

# *Foreword*

With the publication of Volume 4 of Specialist Periodical Reports on Nuclear Magnetic Resonance one feels the series is firmly established. The benefits of regular coverage are now clear. The present division into chapters is similar to that of Volume 3, but there are some changes. Thus although Chapters 1 and 2 remain as before—on Chemical Shifts and Coupling Constants, respectively—the work on Relaxation has now been split so that Chapter 3 only refers to the liquid (solution) state. Literature on relaxation in solids has been incorporated in the chapter on the Solid State, and work on gases has been omitted (it will be covered again in Volume 5). Chapters on Bandshape Phenomena, Macromolecules, and Medium Effects appear in Volume 4, as in Volume 3, except that gas-phase bandshape work is not included this year (there has scarcely been any to report!). However, it will be realized that some chapters are biennial. Thus experimental techniques, Spectral Analysis, Fourier Transform N.M.R., and CIDNP have been omitted this year, but a chapter on Multiple Resonance has been included and discusses work published over the preceding two years. The chapter on Oriented Molecules in Volume 3 has been replaced this year by one on Liquid Crystal Media, which focuses attention on a different aspect of work in this area.

The literature on N.M.R. is still increasing in volume. The reporting team have tried to avoid an undue increase in the length of the Report, but this effort necessarily involves some sacrifice in the comprehensive coverage. Thus relatively routine work on $^{13}C$ chemical shifts has been omitted from Chapter 1—a sign that studies of the $^{13}C$ nucleus are now on a par with proton work.

Reports in the current volume attempt to cover the literature from June 1973 to May 1974 inclusive, except for the biennial chapter on Multiple Resonance. However, as before, the practical coverage is from Volume 3 up to the journals received by libraries at the time the Reports were being written. This implies that coverage of Russian journals in translation will have ended well before May 1974. As before, reports of conference proceedings, or those which appear in house journals of companies or institutes are often not covered.

Coverage of papers is usually based on the U.K. Chemical Information

Service Macroprofile 'N.M.R. – Chemical Aspects', which operates from *Chemical Abstracts* magnetic tapes. It needs to be pointed out that frequently an incomplete author list is given in the Macroprofile output, and apologies are due to authors who find their names omitted. A list of recent review articles on N.M.R., compiled by the Senior Reporter, principally from the Macroprofile, is printed at the start of this Volume.

Readers will also find in the initial pages a list of the principal symbols and abbreviations (including those for chemical species) used in the present volume. It is worth noting here two innovations in notation. In the first place, as explained at the start of Chapter 2, the reports use a symbolism for coupling constants which has a consistent ordering for the nuclei, the heavier nucleus being placed first (for instance, $J_{PH}$ rather than $J_{HP}$). It is to be hoped that this practice commends itself to readers. Secondly the symbol $\chi$ and the abbreviation NQCC are used for nuclear quadrupole coupling constants. The use of $\chi$, rather than the clumsy $e^2Qq/h$, follows the practice of Lehn and Kintzinger.[1]

The International Union of Pure and Applied Chemistry have now produced Recommendations for the Presentation of N.M.R. Data relating to nuclei other than the proton. These are at a preliminary stage, and comments are invited. Copies of the Recommendations may be obtained from the IUPAC Secretariat, Bank Court Chambers, 2/3 Pound Way, Cowley Centre, Oxford OX4 3YF, U.K. The most important feature is perhaps that chemical shift scales for all nuclei are recommended to increase in the direction of increasing resonance frequency at constant applied field (*i.e.* consistent with the $\delta$ scale for $^1$H shifts). In this and other ways the recommendations agree with the practice already adopted in these Specialist Periodical Reports.

The appearance of this Volume relies on the hard work and efficiency of not only the reporting team but also many other people. We are particularly grateful to the Editorial Staff of the Chemical Society.

*October* 1974 R. K. HARRIS

[1] J. M. Lehn and J. P. Kintzinger, '$^{14}$N Nuclear Quadrupole Effects', in 'Nitrogen N.M.R.' ed. M. Witanowski and G. A. Webb, Plenum Press, London, 1973, Chapter 3.

# *Contents*

# *Symbols and Abbreviations*

These lists contain the symbols and abbreviations most frequently used in this volume, but they are not expected to be exhaustive. Some specialized notation is only defined in the relevant chapter. An attempt has been made to standardize usage throughout the volume as far as is feasible, but it must be borne in mind that the original research literature is certainly not standardized in this way, and some difficulties may arise from this fact. Trivial use of subscripts *etc.* is not always mentioned in the symbols list below. Some of the other symbols used in the text, *e.g.* for physical constants such as $h$ or $\pi$, or for thermodynamic quantities such as $H$ or $S$, are not included in the list since they are reckoned to follow completely accepted usage (see M. L. McGlashan, 'Physico-Chemical Quantities and Units', Royal Institute of Chemistry Monographs for Teachers, No. 15, second edition, 1971). The spin system notation used in the present volume of Specialist Periodical Reports on n.m.r. is that suggested by Haigh [*J. Chem. Soc.* (*A*), 1970, 1682].

# *Symbols*

| | |
|---|---|
| $a$ | hyperfine (electron–nucleus) interaction constant |
| $A$ | (i) hyperfine (electron–nucleus) interaction constant<br>(ii) parameter relating to electric field effects on nuclear shielding |
| $B$ | (i) magnetic induction field (magnetic flux density)<br>(ii) parameter relating to electric field effects on nuclear shielding |
| $B_0$ | static magnetic field of an n.m.r. or e.s.r. spectrometer |
| $B_1$, $B_2$ | r.f. magnetic fields associated with $\nu_1$, $\nu_2$ |
| $C_X$ | spin–rotation coupling constant of nucleus X (used sometimes in tensor form): $C^2 = \frac{1}{3}(C_{\parallel}^2 + C_{\perp}^2)$. |
| $C_{\parallel}$, $C_{\perp}$ | components of **C** parallel and perpendicular to a molecular symmetry axis |
| $\bar{C}$ | one third of the trace of **C**: $\bar{C} = \frac{1}{3}(C_{\parallel} + 2C_{\perp})$ |

| | |
|---|---|
| $D$ | (i) self-diffusion coefficient<br>(ii) doublet state |
| $\mathbf{D}$ | rotational diffusion tensor |
| $D_{\parallel}$, $D_{\perp}$ | components of $\mathbf{D}$ parallel and perpendicular to a molecular symmetry axis |
| $e$ | magnitude of the charge on the electron |
| $E$ | electric field |
| $E_n$ | eigenvalue of $\mathscr{H}$ (or of a contribution to $\mathscr{H}$) |
| $f$ | modulation frequency |
| $\hat{F}_G$ | nuclear spin operator for a group, $G$, of nuclei |
| $\hat{F}_{Gx}$, $\hat{F}_{Gy}$, $\hat{F}_{Gz}$ | components of $\hat{F}_G$ |
| $F_G$ | magnetic quantum number associated with $\hat{F}_G$ |
| $g$ | nuclear or electronic $g$-factor |
| $G$ | magnetic field gradient |
| $H_{ij}$ | element of matrix representation of $\mathscr{H}$ |
| $\mathscr{H}$ | Hamiltonian operator (in energy units) – subscripts indicate the nature of the operator |
| $\hat{\boldsymbol{I}}_i$ | nuclear spin operator for nucleus $i$ |
| $\hat{I}_{ix}$, $\hat{I}_{iy}$, $\hat{I}_{iz}$ | components of $\hat{\boldsymbol{I}}_i$ |
| $\hat{I}_{i+}$, $\hat{I}_{i-}$ | 'raising' and 'lowering' spin operators for nucleus $i$ |
| $I_i$ | magnetic quantum number associated with $\hat{\boldsymbol{I}}_i$ |
| $I$ | (i) ionization potential<br>(ii) moment of inertia |
| $\mathscr{S}_{mn}$ | intensity of a transition between nuclear states $m$ and $n$ |
| $^nJ$ | nuclear spin–spin coupling constant through $n$ bonds (in Hz). Further information may be given by subscripts or in brackets. Normally subscripts are only used for algebraic symbols for nuclei in spectral analysis cases, *e.g.* $J_{AX}$. Brackets are used for indicating the species of nuclei coupled, *e.g.* $J(^{13}C,^{1}H)$, or, additionally, the coupling path, *e.g.* $J$(POCF) |
| $J$ | rotational quantum number |
| $\mathscr{J}$ | electron exchange interaction |
| $^nK$ | reduced nuclear spin–spin coupling constant (see the notes concerning $^nJ$) |
| $K$ | Knight shift |
| $m_i$ | eigenvalue of $\hat{I}_{iz}$ (magnetic component quantum number) |
| $m_T$ | total magnetic quantum number for a spin system (eigenvalue of $\sum_i \hat{I}_{iz}$) |
| $M_0$ | equilibrium macroscopic magnetization of a spin system in the presence of $B_0$ |
| $M_x$, $M_y$, $M_z$ | components of macroscopic magnetization |
| $M_n$ | moment of a spectrum ($M_2$ = second moment, *etc.*) |
| $p_A$ | valence $p$ orbital of atom A |
| $P_{s_A s_B}$ | molecular orbital bond order between $s_A$ and $s_B$ |

| | |
|---|---|
| $\mathscr{P}$ | probability of a transition between two nuclear levels |
| $q$ | electric field gradient |
| $Q$ | (i) McConnell's constant relating to $a$ |
| | (ii) nuclear quadrupole moment |
| $R$ | (i) Redfield relaxation matrix |
| | (ii) magnitude of reaction electric field |
| $s_A$ | valence $s$-orbital of atom A |
| $S_A^2(0)$ | electron density in $S_A$ at nucleus A |
| $S$ | (i) singlet state |
| | (ii) electron (or, occasionally, nuclear) spin – *cf.* $I$ |
| | (iii) ordering parameter for oriented systems |
| $t$ | elapsed time |
| $t_{cp}$ | 180° pulse separation in a Carr–Purcell sequence |
| $T$ | (i) temperature |
| | (ii) triplet state |
| $T_c$ | (i) coalescence temperature for an n.m.r. spectrum |
| | (ii) critical solution temperature |
| | (iii) phase transition temperature |
| $T_1^X$ | spin–lattice relaxation time of the X nuclei (further subscripts refer to the relaxation mechanism) |
| $T_2^X$ | spin–spin relaxation time of the X nucleus (further subscripts refer to the relaxation mechanism) |
| $T_2'$ | inhomogeneity contribution to dephasing time for $M_x$ or $M_y$ |
| $T_2^*$ | total dephasing time for $M_x$ or $M_y$; $(T_2^*)^{-1} = T_2^{-1} + (T_2')^{-1}$ |
| $T_{1\rho}^X$, $T_{2\rho}^X$ | spin–lattice and spin–spin relaxation time of the X nuclei in the frame of reference rotating with $B_1$ |
| $T_{1D}$ | dipolar spin–lattice relaxation time |
| $T_d$ | pulse delay time (in FT n.m.r.) |
| $T_{ac}$ | acquisition time (in FT n.m.r.) |
| $T_p$ | period for repetitive pulses (=interpulse time $= T_{ac} + T_d$ if $\tau_p$ is negligible) |
| $X_i$ | mole fraction of compound $i$ |
| $Z_A$ | atomic number of atom A |
| $\alpha$ | (i) nuclear spin wavefunction (eigenfunction of $\hat{I}_z$) for a spin-$\frac{1}{2}$ nucleus |
| | (ii) polarizability |
| $\alpha_A^2$ | $s$-character of hybrid orbital at atom A |
| $\beta$ | nuclear spin wavefunction (eigenfunction of $\hat{I}_z$) for a spin-$\frac{1}{2}$ nucleus |
| $\gamma_X$ | (i) magnetogyric ratio of nucleus X |
| | (ii) activity coefficient of compound X |
| $\Gamma_N$, $\Gamma_M$ | polarization (net or multiplet, respectively) in Kaptein's rules for CIDNP |
| $\delta_X$ | chemical shift (for the resonance) of nucleus of element X (positive when the sample resonates to high frequency of |

| | |
|---|---|
| | the reference). Usually in p.p.m. Further information regarding solvent, references, or nucleus of interest may be given by superscripts or subscripts or in brackets. |
| $\delta_{ij}$ | Kronecker delta ($=1$ if $i=j$, and is zero otherwise) |
| $\Delta$ | (i) time between field gradient pulses<br>(ii) spectral width |
| $\Delta C$ | anisotropy in **C** ($\Delta C = C_{\parallel} - C_{\perp}$) |
| $\Delta n$ | population difference between nuclear states |
| $\Delta\delta$ | change or difference in $\delta$ |
| $\Delta\nu_{\frac{1}{2}}$ | full width (in Hz) of a resonance line at half-height |
| $\Delta\sigma$ | (i) anisotropy in $\sigma$ ($\Delta\sigma = \sigma_{\parallel} - \sigma_{\perp}$)<br>(ii) difference in $\sigma$ for two different situations |
| $\Delta\chi$ | (i) susceptibility anisotropy ($\Delta\chi = \chi_{\parallel} - \chi_{\perp}$)<br>(ii) difference in electronegativities |
| $\varepsilon_r$ | relative permittivity |
| $\varepsilon_0$ | permittivity of a vacuum |
| $\eta$ | (i) nuclear Overhauser effect<br>(ii) asymmetry factor (*e.g.* in $e^2qQh$)<br>(iii) refractive index<br>(iv) viscosity |
| $\mu$ | (i) magnetic dipole moment<br>(ii) electric dipole moment |
| $\mu_0$ | permeability of a vacuum |
| $\mu_B$ | Bohr magneton |
| $\mu_N$ | nuclear magneton |
| $\nu_i$ | Larmor precession frequency of nucleus $i$ (in Hz) |
| $\nu_0$ | (i) spectrometer operating frequency<br>(ii) Larmor precession frequency (general, or of bare nucleus) |
| $\nu_1$ | frequency of 'observing' r.f. magnetic field |
| $\nu_2$ | frequency of 'irradiating' r.f. magnetic field |
| $\Xi_X$ | resonance frequency for the nucleus of element X in a magnetic field such that the protons in TMS resonate at exactly 100 MHz |
| $\Pi_{s_A s_B}$ | mutual polarizability of $s_A$ and $s_B$ |
| $\rho^2_{s_A}$ | spin density in $s_A$ |
| $\sigma_i$ | (i) shielding constant of nucleus $i$ (used sometimes in tensor form). Usually in p.p.m. Subscripts may alternatively indicate contributions to $\sigma$.<br>(ii) collision cross-section<br>(iii) spin density matrix |
| $\sigma_{\parallel}, \sigma_{\perp}$ | components of $\sigma$ parallel and perpendicular to a molecular symmetry axis |
| $\tau$ | (i) pre-exchange lifetime of molecular species<br>(ii) cage lifetime (in CIDNP work)<br>(iii) time between r.f. pulses (general symbol) |

| | |
|---|---|
| $\tau_c$ | correlation time |
| $\tau_{coll}$ | mean time between molecular collisions in the liquid state |
| $\tau_1$, $\tau_2$ | correlation time for the spherical harmonics $Y_{1m}$, $Y_{2m}$ |
| $\tau_{\parallel}$, $\tau_{\perp}$ | correlation time for molecular rotation parallel and perpendicular to a molecular symmetry axis |
| $\tau_J$ | angular momentum correlation time |
| $\tau_p$ | pulse duration |
| $\tau_t$ | translational magnetic relaxation correlation time |
| $\chi$ | (i) magnetic susceptibility<br>(ii) electronegativity<br>(iii) nuclear quadrupole coupling constant ($=e^2qQ/h$) |
| $\omega_c$ | carrier frequency in rad $s^{-1}$ |
| $\omega_l$, $\omega_0$, $\omega_1$, $\omega_2$ | as for $\nu_l$, $\nu_0$, $\nu_2$, $\nu_2$ but in rad $s^{-1}$ |
| $\omega_m$ | modulation angular frequency (in rad $s^{-1}$) |

# Abbreviations

(a) *Physical properties*

| | |
|---|---|
| a.c. | alternating current |
| a.f. | audiofrequency |
| a.u. | atomic unit |
| a.m. | amplitude modulation |
| b.c.c. | body-centred cubic |
| c.d. | circular dichroism |
| c.m.c. | critical micelle concentration |
| d.c. | direct current |
| e.f.g. | electric field gradient |
| e.s.r. | electron spin resonance |
| f.c.c. | face-centred cubic |
| f.m. | frequency modulation |
| h.c.p. | hexagonal close-packed |
| h.f. | hyperfine |
| i.d. | inside diameter |
| i.r. | infrared |
| l.c. | liquid crystalline |
| mol. wt. | molecular weight |
| n.m.r. | nuclear magnetic resonance |
| n.q.r. | nuclear quadrupole resonance |
| o.d. | outside diameter |
| p.p.m. | parts per million |
| r.f. | radiofrequency |
| r.m.s. | root mean square |
| u.h.f. | ultra-high frequency |
| u.v. | ultraviolet |

| | |
|---|---|
| ADC | analog-to-digital converter |
| ARP | adiabatic rapid passage |
| ARPS | adiabatic rapid passage sampling |
| ASIS | aromatic solvent-induced shift |
| CAT | computer of average transients |
| CIDEP | chemically induced dynamic electron polarization |
| CIDNP | chemically induced dynamic nuclear polarization |
| CNDO | complete neglect of differential overlap |
| CPMG | Carr–Purcell pulse sequence, Meibom–Gill modification |
| CSA | chemical shift (shielding) anisotropy |
| CW | continuous wave |
| DAC | digital-to-analog converter |
| DD | dipole–dipole (interaction or relaxation mechanism) |
| DEFT | driven-equilibrium Fourier transform |
| DNP | dynamic nuclear polarization |
| EHT | extended Hückel theory |
| ENDOR | electron–nucleus double resonance |
| FC | Fermi contact |
| FET | field-effect transistor |
| FID | free induction decay |
| FPT | finite perturbation theory |
| FT | Fourier transform |
| INDO | intermediate neglect of differential overlap |
| INDOR | internuclear double resonance |
| LCAO | linear combination of atomic orbitals |
| LIS | lanthanide-induced shift |
| LSR | lanthanide shift reagent |
| MO | molecular orbital |
| NOE | nuclear Overhauser effect |
| NQCC | nuclear quadrupole coupling constant |
| OB | orbital (contribution to scalar coupling) |
| PRE | proton relaxation enhancement |
| PRFT | partially relaxed Fourier transform |
| QF | quadrupole moment/field gradient (interaction or relaxation mechanism) |
| QFT | quadra Fourier transform |
| RPFT | repetitively pulsed Fourier transform |
| SC | scalar (interaction or relaxation mechanism) |
| SCF | self-consistent field |
| SD | spin-dipolar (contribution to scalar coupling) |
| SEFT | spin-echo Fourier transform |
| S/N | signal-to-noise ratio |
| SOS | sum over states |
| SR | spin-rotation (interaction or relaxation mechanism) |
| VB | valence bond |

| | |
|---|---|
| WAHUHA | Waugh, Huber, and Haeberlen (cycle of pulses) |
| WEFT | water-eliminated Fourier transform |

(b) *Chemical species*†

| | |
|---|---|
| acac | acetyl acetonato |
| AMP | adenosine monophosphate |
| ATP | adenosine triphosphate |
| bhmc | 3-t-butylhydroxymethylene-*d*-camphorato |
| BSA | bovine serum albumin |
| DAP | dodecylammonium propionate |
| DME | 1,2-dimethoxyethane |
| DMF | dimethylformamide |
| DMSO | dimethylsulphoxide |
| DNA | deoxyribonucleic acid |
| DSS | 2,2-dimethyl-2-silapentane-5-sulphonate (usually as the sodium salt) |
| EBBA | *N*-(*p*-ethoxybenzylidene)-*p*-butylaniline |
| EDTA | ethylenediaminetetra-acetic acid |
| EMC | ethyl [*p*-(*p*-methoxybenzylidene)amino]cinnamate |
| facam | 3-trifluoroacetyl-*d*-camphorato |
| fod | 1,1,1,2,2,3,3,-heptafluoro-7,7-dimethyl-4,6-octanedionato |
| HAB | 4,4′-bis(heptyl)azoxybenzene |
| hfbc | 3-heptafluorobutyryl-*d*-camphorato |
| HOAB | *p*-n-heptyloxyazoxybenzene |
| LUNO | 2,6-dimethylpyridine-1-oxide |
| MBBA | *N*-(*p*-methoxybenzylidene)-*p*-butylaniline |
| NADH | nicotinamide adenine dinucleotide |
| PAA | *p*-azoxyanisole |
| PBLG | poly (α-benzyl L-glutamate) |
| PMAA | poly(methacrylic acid) |
| PMMA | poly(methyl methacrylate) |
| PTFE | polytetrafluoroethylene |
| PVC | poly(vinyl chloride) |
| RNA | ribonucleic acid |
| SDS | sodium dodecyl sulphate |
| TBBA | terephthal-bis-4-(4-n-butylaniline) |
| TCNQ | tetracyanoquinodimethane |
| TFA | trifluoroacetic acid |
| THF | tetrahydrofuran |
| tmhd* | 2,2,6,6-tetramethylheptane-3,5-dionato |
| TMS | tetramethylsilane |
| tRNA | transfer ribonucleic acid |

† Lower case initials are used when the species is a ligand.
* This ligand is sometimes referred to as dpm, dipivaloylmethanato.

# N.M.R. Books and Reviews

This section lists all books and reviews with n.m.r. as the principal theme, published during 1973 and 1974 (following the list given in Volume 3), which were known to the Senior Reporter at the time of going to press. Some articles dating from 1971 and 1972, omitted from the lists in Volumes 2 and 3, are also included here. Titles and numbers of pages are given where appropriate, and in the present list, for the first time, the number of references in each review and its *Chemical Abstracts* number are also listed. In general, texts of individual lectures from symposia are not listed: likewise references to technical bulletins, company or institute house journals, special reports, *etc.*, are omitted, since they are judged to be of little value to the scientific community in view of their inaccessibility and ephemeral nature. N.m.r. sections of general physical chemistry or spectroscopy textbooks have not been included. The only foreign-language articles that are normally included are those in French, German, Russian, Spanish, or Portuguese; moreover, the coverage of Russian-language publications is not likely to be comprehensive. Usually, for such foreign-language reviews the title is translated in the present list but the original language is indicated in parentheses. The source of the references is in most cases the computer-based listing of n.m.r. publications provided by the U.K. Chemical Information Service Macroprofile 'N.M.R. – Chemical Aspects'; this may sometimes result in the list of authors being abbreviated. The compilation is sectionalized for the reader's convenience.

(a) *Books*

R1. N. F. Chamberlain and J. J. R. Reed, 'N.M.R. Data of Sulphur Compounds' (Pt. 3 of 'The Analytical Chemistry of Sulphur and its Compounds', ed. J. H. Karchmer), Wiley-Interscience, 1971, 308 pages, CA 76:8823.
R2. J. T. Clerc, E. Pretsch, and S. Sternhell, 'Methods of Analysis in Chemistry', Vol. 16, '$^{13}$C NMR Spectroscopy', Akad. Verlagsges, Frankfurt, 1973, 143 pages, CA 81:9448.
R3. R. A. Dwek, 'N.M.R. in Biochemistry', Oxford University Press, 1973, 395 pages, CA 81:101 293.

R4. H. Günther, 'N.M.R. Spectroscopy: An Introduction', G. Thieme Verlag, Stuttgart, 1973, 423 pages (in German) CA 79:120 397.

R5. E. R. Haws, R. R. Hill, and D. J. Mowthorpe, 'The Interpretation of Proton Magnetic Resonance Spectra: A Programmed Introduction', Heyden & Son Ltd., London, 1973, 180 pages, CA 81:8300.

R6. J. C. Jochims, Constance University Lectures. No. 56. 'N.M.R. as a Tool of Stereochemists', Druckerei Verlagsanstalt Konstanz (Constance), 1973 (in German), CA 81:43982.

R7. P. Joseph-Nathan, 'Resonancia Magnetica Nuclear de Hidrogeno', Monograph No. 9 of Chemical Series, published by Department of Scientific Affairs, Organization of American States, Washington D.C., 1973, 109 pages (in Spanish), CA 81:129 722.

R8. M. L. Martin and G. J. Martin, 'Manual of N.M.R.', Azoulay, Paris, 1971, 214 pages (in French), CA 76:8814.

R9. I. Pop and V. Niculescu, 'N.M.R. in Metals', Acad. Repub. Socialiste Rom., Bucharest, 1973, 456 pages (in Roumanian), CA 81:31 730.

(b) *Edited Books, Books of Partial Relevance, Chapters of Books, Symposia* etc.

R10. 'Liquid Crystals and Plastic Crystals', ed. G. W. Grey and P. A. Winsor, Ellis Harwood Ltd., Chichester, 1974, Vol. 2. Relevant articles are (a) G. R. Luckhurst, 'Magnetic Resonance Spectroscopy of Liquid Crystals – Non-amphiphilic Systems', pp. 144—191 (51 refs); (b) Å. Johansson and B. Lindman, 'NMR Spectroscopy of Liquid Crystals – Amphiphilic Systems', pp. 192—230 (132 refs).

R11. 'Modern Methods of Steroid Analysis', ed. E. Heftmann, Academic Press, 1973. Relevant chapters are (a) Chapter 10, pp. 245—263, 'Introduction to NMR', D. P. Hollis (16 refs), CA 80:138 708; (b) Chapter 11, pp. 265—279, 'Applications of Lanthanide Shift Reagents', C. C. Hinckley (46 refs) CA 80:150 431.

R12. 'Resonance Techniques: Conformation and Interactions', Section VII of 'Methods in Enzymology', Vol. 26, 'Enzyme Structure', Part C, ed. C. H. W. Hirs and S. N. Timasheff, Academic Press, 1972. Relevant articles are: (a) 'High-resolution proton magnetic resonance spectroscopy of selectively deuterated enzymes', J. L. Markley, pp. 605—627 (48 refs) CA 78:94 055; (b) 'Measurement of Histidine Residue Titration by N.M.R. Spectroscopy', D. H. Meadows, pp. 638—653 (31 refs) CA 78:94 056; (c) Nuclear Relaxation Measurements of Water Protons and Other Ligands', A. S. Mildvan and J. I. Engle, pp. 654—682 (51 refs) CA 78:94 057.

R13. 'Organic Selenium Compounds: Their Chemistry and Biology', ed. D. L. Klayman, Wiley, New York, 1973. Relevant articles are (a) Section D: 'Proton N.M.R. Spectroscopy', U. Svanholm, pp. 903—932 (99 refs), CA 80:42 351; (b) Section E: '$^{77}$Se N.M.R.', M. A. Lardon, pp. 933—939 (19 refs) CA 80:42 350.

R14. 'N.M.R. of Paramagnetic Molecules: Principles and Applications', ed.

G. N. La Mar, W. D. Horrocks, and R. H. Holm, Academic Press, New York, 1973, 678 pages.

R15. 'Chemical Induced Magnetic Polarization', ed. A. R. Lepley and G. L. Closs, John Wiley & Sons, 1973, 416 pages, eight chapters (incl. one on CIDEP).

R16. 'Determination of Organic Structures by Physical Methods', ed. F. C. Nachod and J. J. Zuckerman, Academic Press, 1973, Vol. 5. Relevant articles are: (a) 'Spin Saturation Labeling', J. W. Faller, pp. 75—97 (43 refs); (b) 'Chemically and Electromagnetically Induced Dynamic Nuclear Polarization', R. G. Lawler and H. R. Ward, pp. 99—150 (94 refs).

R17 'Internal Rotation in Molecules', ed. W. J. Orville-Thomas, John Wiley & Sons, 1974. Relevant articles are: (a) 'N.M.R. and E.S.R. Studies on Simple Rotamers', J. A. Ladd and H. W. Wardale, pp. 115—155 (52 refs); (b) 'Ring Inversion in Some Six-membered Heterocyclic Compounds', M. V. Gittins, E. Wyn-Jones, and R. F. M. White, pp. 425—480 (123 refs); (c) 'Medium Effects on Rotational and Conformational Equilibria', pp. 481—584 (196 refs).

R18. 'N.M.R. Shift Reagents', ed. R. E. Sievers, Academic Press, 1973, 410 pages.

R19. 'Atlas of Spectra of Aromatic and Heterocyclic Compounds', No. 3, 'The I.R., U.V. and $^{1}$H N.M.R. Spectra of Aromatic and Some Alicyclic Polyalkyl-substituted Compounds', T. F. Ardyukova, I. K. Korobeinicheva, and A. I. Rezvukhin, Nauka Sib. Otd., Novosibirsk, U.S.S.R., 1973, 213 pages, CA 80: 8925, in Russian.

R20. 'The N.M.R. Spectra of Acridines', R. M. Acheson, in 'Acridines', ed. R. M. Acheson, pp. 687—707 (25 refs), Interscience, New York, 1973, CA 78:135 110.

R21. 'N.M.R. Spectra of Sulfur Compounds', C. Brown (CA 80: 20 868), in 'Sulfur in Organic and Inorganic Chemistry', ed. A. Senning, Dekker, New York, 1972, Vol. 3, Chapter 25, pp. 275—310 (186 refs).

R22. 'Fourier Transform Spectrometry in Chemical Research', S. T. Dunn, C. T. Foskett, R. Curbelo, and P. R. Griffiths, in 'Computers in Chemical and Biochemical Research', ed. C. E. Klopfenstein and C. L. Wilkins, Academic Press, 1972, Vol. 1, pp. 107—173 (82 refs) CA 78: 9532.

R23. 'N.M.R. and E.S.R.' B. C. Flockhart, in 'Comprehensive Analytical Chemistry', ed. C. L. Wilson and D. W. Wilson, Elsevier, Baltimore, 1971, Vol. 2C, pp. 206—317, CA 76: 51 990 (170 refs).

R24. 'N.M.R. Spectroscopy of Lipids', K. P. Henrikson, in 'Biochemistry and Methodology of Lipids', ed. A. R. Johnson and J. B. Davenport, Interscience, New York, 1971, pp. 243—249 (12 refs) CA 76: 109 810.

R25. 'Applications of N.M.R. to Investigations of Structures and Bonding in Coordination Compounds', R. H. Holm and E. H. Abbott, Amer.

Chem. Soc. Monograph, No. 168, 1971, pp. 84—133 (115 refs) CA 77:26 576.

R26. 'Dynamic Polarization of Nuclei', C. D. Jeffries, in 'Electron Paramagnetic Resonance', ed. S. Geschwind, Plenum Press, New York, 1972, pp. 217—260 (56 refs) CA 78:77 396.

R27. 'Stereochemistry and Stereochemical Non-rigidity in Transition Metal Hydrides', J. P. Jesson, in 'Transition Metal Hydrides', ed. E. L. Muetterties, Dekker, New York, 1971, Chapter 4, pp. 75—201 (317 refs) CA 77:94 791.

R28. 'N.M.R. Spectroscopy of Organometallic Compounds', R. G. Kidd, in 'Characterization of Organometallic Compounds', ed. M. Tsutsui, Wiley-Interscience, 1971, Vol. 2, pp. 374—437 (164 refs) CA 74:124 352.

R29. 'Spectroscopic Measurements', Part 4, 'N.M.R. Spectroscopy', I. R. Lantzke, in 'Physical Chemistry of Organic Solvent Systems', ed. A. K. Covington, pp. 483—523, Plenum Press, 1973 (153 refs) CA 80:88 948.

R30. 'N.M.R. Spectroscopy', B. E. Mann, in 'Spectroscopic Properties of Inorganic and Organometallic Compounds', ed. N. N. Greenwood, Chem. Soc. (Specialist Periodical Reports), The Chemical Society, 1973, Vol. 6, pp. 1—153, CA 80:88 943 (2132 refs).

R31. 'Application to Magnetic Resonance Spectroscopy: The Use of the Large Computer in Chemical Instrumentation', L. Newman (A 80: 65 088), in 'Spectroscopy and Kinetics' ('Computers in Chemistry and Instrumentation', Volume 3), ed. J. S. Mattson, H. B. Mark, jun., and H.C. MacDonald, jun., Dekker, New York, 1973, Chapter 1, pp. 3—41 (10 refs).

R32. 'Wideline N.M.R. Spectroscopy (of Cellulose)', R. A. Pittman and V. W. Tripp, in 'Instrumental Analysis of Cotton Cellulose and Modified Cotton Cellulose', ed. R. T. O'Connor, Dekker, New York, 1972, pp. 379—399 (20 refs) CA 78:112 827.

R33. 'Nuclear Magnetic Double Resonance. Principles and Applications in Solid-State Physics', D. Schmid, in 'Solid-State Physics', ed. G. Höhler, Springer Tracts on Modern Physics, 1973 Vol. 68, pp. 1—75 (110 refs) CA 80:75 967.

R34. 'Lectures and Papers from the International Conference on Molecular Spectroscopy', Wrocław, Poland (Sept. 1972), 'Advances in Molecular Relaxation Processes', 1973, Vol. 5, pp. 1—284 (a proportion of the papers relate to n.m.r.; the papers published are only those relevant to dynamic molecular properties).

R35. 'E.S.R. and N.M.R. in Biology and Medicine (Magnetic Resonance in Biological Systems)', papers from the International Conference, 1972, ed. S. E. Lasker and P. Milvy, *Ann. New York Acad. Sci.*, 1973, **222,** 1—1124.

R36. 'Modern Instrumental Methods in Natural Products Research' (symposium at the 13th Annual Meeting of the American Society of Pharmacognosy, Columbus, Ohio, 1972), *Lloydia*, 1972, **35**. Relevant articles are (a) 'Use of Solvent Effects and Rare Earth Shift Reagents in Structure Analysis', P. V. de Marco, pp. 362—374, CA 78:135 427 (13 refs); (b) '$^{13}$C N.M.R. Spectroscopy and its Application to Biological Systems', J. B. Grutzner, pp. 375—398, CA 78:96 683 (94 refs); and (c) 'Instrumental Methods in Biosynthetic Studies', H. G. Floss, pp. 399—417, CA 78:132 994 (21 refs).

R37. 'Tenth Microsymposium on Macromolecules: Conformational Structure of Polymers', Prague, Czechoslovakia, August–September, 1972 – main lectures, *Pure Appl. Chem.*, 1973, **36,** Nos. 1 and 2. Relevant articles are: (a) 'Some aspects of n.m.r. techniques for the conformational analysis of peptides', V. F. Bystrov, S. L. Portnova, T. A. Balashova, S. A. Koz'min, Yu. D. Gavrilov, and V. A. Afanas'ev, pp. 19—34 (40 refs), CA 81:78 197; (b) 'N.M.R. of synthetic polypeptides', E. M. Bradbury, P. D. Cary, C. Crane-Robinson, and P. G. Hartman, pp. 53—92 (56 refs).

R38. Eleventh European Congress on Molecular Spectroscopy (Tallinn, U.S.S.R., May–June 1973): Invited lectures, *Pure Appl. Chem.*, 1974, **37,** Nos. 1 and 2. Relevant articles are: (a) 'Equivalence of Fourier spectroscopy and slow passage in n.m.r.', R. R. Ernst, W. P. Aue, and E. Bartholdi, pp. 47—60 (35 refs); (b) '$^{13}$C n.m.r. in studies of molecular dynamics and structure', D. M. Grant, pp. 61—71 (no refs); (c) 'Study of the structure and complexation of organic and inorganic derivatives of metals by means of n.m.r. spectroscopy of heavy nuclei', V. S. Petrosyan and O. A. Reutov, pp. 147—159 (89 refs); (d) 'Nitrogen n.m.r. spectroscopy', M. Witanowski, pp. 225—233 (3 refs); (e) 'Studies of molecular conformations in proteins by $^{2}$H and $^{13}$C n.m.r. spectroscopy', K. Wüthrich, pp. 235—248 (27 refs).

R39. 'N.M.R. of Nuclei other than Protons', Papers from the Advanced Study Institute held in Tirrenia (Pisa) during Sept. 1972, ed. T. Axenrod and G. A. Webb, Wiley-Interscience, 1974, 407 pages.

R40. 'Relazioni del Corso Teorico-Pratico di Risonanza Magnetica Nucleare' (containing lecture material from a course held at the Pharmacology Research Institute, Milan, February 1973), ed. A. Frigerio, 543 pages, Tamburini Editore, Milan, 1973, 543 pages (in Italian).

R41. 'Pulsed n.m.r. and Spin Dynamics in Solids', Proceedings of the First Specialized Colloque Ampère (August 1973), ed. J. W. Hennel, Inst. of Nuclear Physics, Krakow, Poland, 1973, 271 pages, 47 papers (9 invited and 38 contributed).

R42. 'Magnetic Resonance and Related Phenomena', Proceedings of the XVIIth Congress Ampère, Turku, 1972, ed. V. Hovi, North-Holland Publishing Co., Amsterdam, 1973, 555 pages.

R43. 'Use of N.M.R. Spectroscopy to study Protein Structure and Function',

H. Rüterjans (in German), from 'Technical Biochemistry' (Papers from the 8th Tutzing Symposium, March, 1972), ed. H. J. Rehm, Deut. Ges. Chem. Apparatew. Monogr., 1973, Vol. 71, pp. 219—245 (38 refs) CA 80:11 449.

(c) *Regular Magnetic Resonance Review Series*

R44. J. Jonas, 'N.M.R. Studies in Liquids at High Pressure', *Adv. Magn. Resonance*, 1973, **6,** 73—139 (125 refs).

R45. R. W. Kreilick, 'N.M.R. Investigations of Organic Free Radicals', *Adv. Magn. Resonance*, 1973, **6,** 141—181 (40 refs).

R46. G. Mavel, 'N.M.R. Studies of Phosphorus Compounds (1965—1969)', *Ann. Reports N.M.R. Spectroscopy*, 1973, **5B,** 1—441 (1589 refs) CA 81:19 150.

R47. J. Jonas, 'N.M.R. Relaxation in Liquids and Liquid Crystals', *Magn. Resonance Rev.*, 1973, **2,** 203—220 (220 refs) CA 80:53 975.

R48. T. J. Burch and J. I. Budnick, 'N.M.R. in Magnetically Ordered Materials', *Magn. Resonance Rev.*, 1973, **2,** 247—272 (103 refs), CA 80:53971.

R49. L. R. Dalton and L. A. Dalton, 'Double Resonance', *Magn. Resonance Rev.*, 1973, **2,** 361—397 (361 refs).

R50. C. Richard and P. Granger, 'CIDNP and CIDEP', *N.M.R. Basic Principles and Progress*, 1974, **8,** 1—127 (217 refs), CA 81:161 989.

R51. J. Reuben, 'Paramagnetic Lanthanide Shift Reagents in N.M.R. Spectroscopy: Principles, Methodology, and Applications', *Progr. N.M.R. Spectroscopy*, 1973, **9,** 1—70 (280 refs) CA 80:54 443.

R52. N. M. Sergeyev, 'N.M.R. Spectroscopy of Cyclopentadienyl Compounds', *Progr. N.M.R. Spectroscopy*, 1973, **9,** 71—144 (208 refs) CA 80:65 445.

(d) *Review Articles in Regular Journals*

R53. J. P. Jesson and P. Meakin, 'Determination of Mechanistic Information from N.M.R. Lineshapes for Intramolecular Exchange', *Accounts Chem. Res.*, 1973, **6,** 269—275 (27 refs) CA 79:97 407.

R54. D. R. Kearns and R. G. Shulman, 'High-Resolution N.M.R. Studies of the Structure of Transfer Ribonucleic Acid and Other Polynucleotides in Solution', *Accounts Chem. Res.*, 1974, **7,** 33—39 (56 refs), CA 80:79210.

R55. E. Wenkert, J. S. Bindra, C.-J. Chang, D. W. Cochran, and F. M. Schnell, '$^{13}$C N.M.R. Spectroscopy of Naturally Occurring Substances. Alkaloids', *Accounts Chem. Res.*, 1974, **7,** 46—51 (33 refs) CA 80:83 333.

R56. B. Coxon, '$^{1}$H N.M.R. Spectroscopy, I', *Adv. Carbohydrate Chem.*, 1972, **27,** 7—83 (257 refs) CA 80:083 413.

R57. A. W. Garrison, L. H. Keith, and A. L. Alford, 'Confirmation of Pesticide Residues by Mass Spectrometry and N.M.R. Techniques', Adv. Chem. Ser., 1972, No. 111, 26—54 (45 refs) CA 77:148 313.

R58. L. C. Craig, H. C. Chen, and W. A. Gibbons, 'Use of Thin Film Dialysis and High Resolution N.M.R. to Study Conformation and Association Phenomena', Adv. Chem. Ser., 1973, No. 125; 'Polymer Molecular Weight Methods', ed. M. Erzin, 286—297 (42 refs) CA 80:83 810.

R59. M. D. Sauzade and S. K. Kan, 'High Resolution N.M.R. Spectroscopy in High Magnetic Fields', *Adv. Electron. Electron Phys.*, 1973, **34,** 1—93 (59 refs) CA 80:75 942.

R60. S. P. Hiremath and R. S. Hosmane, 'Applications of N.M.R. Spectroscopy to Heterocyclic Chemistry: Indole and its Derivatives', *Adv. Heterocyclic Chem.*, 1973, **15,** 277—324 (135 refs), CA 80:81 522.

R61. B. E. Mann '$^{13}C$ N.M.R. Chemical Shifts and Coupling Constants of Organometallic Compounds', *Adv. Organometallic Chem.*, 1974, **12,** 135—213 (230 refs) CA 81:24 547.

R62. D. Bethell and M. R. Brinkman, 'Chemically Induced Dynamic Nuclear Spin Polarization and its Applications', *Adv. Phys. Org. Chem.*, 1973, **10,** 53—128 (245 refs) CA 81:12 593.

R63. J. Mitchell, jun. and J. Chiu, 'Analysis of High Polymers', *Analyt. Chem.*, 1973, **45,** 273R—332R (includes 339 magnetic resonance refs., with a six-page special section) CA 78:148 261.

R64. J. R. Wasson and D. K. Johnson, 'N.M.R. Spectrometry', *Analyt. Chem.*, 1974, **46,** 314R—345R (2218 refs) CA 81:7783.

R65. H. Günther, 'Proton N.M.R. Spectra of the $[AX]_2$ and $[AB]_2$ type. Analysis and Classification', *Angew. Chem. Internat. Edn.*, 1972, **11,** 861—874 (58 refs) CA 77:170 633.

R66. H. G. Fitzky, D. Wendisch, and R. Holm, 'Structure Determinations of Nitrogen Compounds with the Aid of N.Q.R., N.M.R., and ESCA Spectroscopy', *Angew. Chem. Internat. Edn.*, 1972, **11,** 979—990 (89 refs) CA 78:57 155.

R67. B. D. Sykes and W. E. Hull, 'Recent Advances in Methodology: N.M.R.', *Ann. New York Acad. Sci.*, 1973, **226,** 60—68 (16 refs) CA 80:74 282.

R68. I. H. Sadler, 'Physical Methods: N.M.R.', *Ann. Reports* (*B*), 1972, **69,** 19—39 (138 refs) CA 80:59 035.

R69. V. A. Brown and E. G. Brame, jun., 'N.M.R. Chemical Shift Correlations for Chlorine-containing Groups', *Appl. Spectroscopy*, 1972, **26,** 224—238 (4 refs) CA 77:11 999.

R70. L. S. Kan and P. O. P. Ts'O, 'Introduction to N.M.R. Studies of Nucleic Acids', *Bot. Bull. Acad. Sinica*, 1973, **14,** 180—204 (56 refs) CA 81:579.

R71. H. Cheradame, 'Fondements Théorique et Intérêt Pratique de l'Étude de la Configuration des Polymères par RMN', *Bull. Soc. chim. France*, 1971, 2023—2033 (36 refs) (in French) CA 75:110 591.

R72 J.-R. Llinas, E.-J. Vincent, and G. Peiffer, 'Constantes de Couplages (C,C), (C,N) et (C,P) en RMN: Essai de Classification et d'Analyse', *Bull. Soc. chim. France*, 1973, 3209—3224 (94 refs) (in French), CA 80:75970

R73. M. Vincendon, 'N.M.R. of Oligo- and Poly-saccharides', *Bull. Soc. chim. France*, 1973, 3501—3511 (149 refs) (in French).

R74. I. S. Butler and D. F. R. Gilson, 'Recent Developments in Canadian Spectroscopy', *Canad. Res. Develop.*, 1973, **6**, 41—46 (3 refs) CA 80:8509.

R75. H. Beall and C. H. Bushweller, 'Dynamical Processes in Boranes, Borane Complexes, Carboranes, and Related Compounds', *Chem. Rev.* 1973, **73**, 465—486 (119 refs) CA 79:141 029.

R76. A. F. Cockerill, G. L. O. Davies, R. C. Harden, and D. M. Rackham, 'Lanthanide Shift Reagents for NMR Spectroscopy', *Chem. Rev.*, 1973, **73**, 553—588 (488 refs) CA 79:151 016.

R77. G. Kotowycz and R. U. Lemieux, 'N.M.R. in Carbohydrate Chemistry', *Chem. Rev.*, 1973, **73**, 669—698 (383 refs) CA 79:146 745.

R78. H. Günther, 'Physical Methods in Chemistry. $^{13}$C NMR Spectroscopy, I', *Chem. Unserer Zeit.*, 1974, **8**, 45—53 (*ca.* 20 refs) (in German) CA 81:7790.

R79 F. W. Wehrli, 'Fourier Transform Spectroscopy, Novel Technique for Increasing the Measuring Sensitivity in N.M.R.', *Chem.-Ztg.*, 1971, **95**, 58—65 (17 refs) (in German) CA 74:148 871.

R80. A. Maercker, 'N.M.R. Study of Organolithium and Organomagnesium Compounds', *Chem.-Ztg.*, 1973, **97**, 361—368 (55 refs) (in German) CA 79:114 635.

R81. R. Lenk, 'Brownian Movement and Spin Relaxation in Organic Molecular Systems', *Chimia*, 1973, **27**, 363—368 (23 refs) (in French) CA 79:71 648.

R82. L. W. Reeves, 'N.M.R. Studies of Liquid Crystals', *Cienc. Cult. (São Paulo)*, 1973, **25**, 516–523 (14 refs) CA 79:129 948.

R83. G. R. Khutsishvili, 'Nuclear Spin Diffusion', *Comments Solid State Phys.*, 1972, **5**, 23—29 (35 refs) CA 78:141 880.

R84. T. G. Appleton, H. C. Clark, and L. E. Manzer, 'The *trans*-Influence: Its Measurement and Significance', *Co-ord. Chem. Rev.*, 1973, **10**, 335—422, n.m.r. specifically on pp. 378—401 (total of 302 refs) CA 79:26 596.

R85. P. J. Quilley and G. A. Webb, 'Magnetic Resonance Studies of Metal-Enzymes', *Co-ord. Chem. Rev.*, 1974, **12**, 407—430 (101 refs) CA 81:73 778.

R86. G. A. Gray, '$^{13}$C NMR in Biochemistry', *Critical Rev. Biochem.*, 1973, **1**, 247—304 (175 refs).

R87. H. J. V. Tyrrell, 'Solvation', *Essays Struct. Chem.*, 1971, 296—326 (78 refs) CA 81:30 241.

R88. R. E. Richards, 'Molecular Motion in Solutions of Biopolymers using N.M.R.', *Faraday Symp. Chem. Soc.*, 1972, **3**, 176—181 (8 refs) CA 79:102 246.

R89. H.-G. Horn, 'N.M.R. Spectra and Characteristic Frequencies of Compounds Containing N-S-F Bonds', *Fluorine Chem. Rev.*, 1973, **6,** 135—192 (99 refs).

R90. A. R. Katritzky and A. Smith, 'Application of the N.M.R. to the Elucidation of Polymer Structure', *Gummi Asbest. Kunstst.*, 1973, **26,** 648—652 (in German) (no refs) CA 79:92 615.

R91. C. L. Boltz, 'Laboratory Technique which Spies on Molecules', *Indian Chem. J.*, 1973, **8,** 21—23 (no refs) CA 79:119 936.

R92. P. J. Smith and L. Smith, 'Applications of $^{119}$Sn Chemical Shifts to Structural Tin Chemistry', *Inorg. Chim. Acta Rev.*, 1973, **7,** 11—33 (73 refs) CA 80:81 517.

R93. J. H. Bradbury, B. E. Chapman, and N. L. R. King, 'Linewidths of $^{1}$H N.M.R. Spectra of Proteins in Solution', *Internat. J. Protein Res.*, 1971, **3,** 351—356 (35 refs) CA 76:59 995.

R94. J. Bartual, 'N.M.R. Spectrometry', *Invest. Inform. Text. Tensioactivos*, 1973, **16,** 191—193 (in Spanish) (no refs) CA 79:84 982.

R95. E. Gasull, 'Use of Mass Spectroscopy and N.M.R. for Studying Surfactants', *Invest. Inform. Text. Tensioactivos*, 1973, **16,** 201—219 (in Spanish) (no refs) CA 79:93 715.

R96. W. E. Stewart and T. H. Siddall, 'N.M.R. Studies of Organophosphorus Extractants', *Ion Exch. Solvent Extr.*, 1973, **3,** 83—110 (51 refs) CA 79:97 531.

R97. S. P. Sobol, R. P. Barron, and T. C. Kram, 'Use of N.M.R. and Mass Spectral Data in the Forensic Laboratory', *J. Ass. Off. Analyt. Chem.*, 1972, **55,** 866—873 (9 refs) CA 77:110 143.

R98. S. H. Pine, 'Chemically Induced Dynamic Nuclear Polarization', *J. Chem. Educ.*, 1972, **49,** 664—668 (25 refs) CA 77:158 038.

R99. W. D. Knight, 'Electronic Quantum Effects in Small Particles', *J. Vac. Sci. Technol.*, 1973, **10,** 705—712 (55 refs) CA 79:120 218.

R100. W. D. Phillips and M. Poe, 'Contact Shifts and Magnetic Susceptibilities in Iron–Sulfur Proteins as Determined from N.M.R. Spectra', *Methods Enzymol.*, 1972, **24,** 304—317 (25 refs) CA 78:68 254.

R101. W. D. Phillips, 'N.M.R. Spectroscopy of Proteins', *Methods Enzymol.*, 1973, **27,** 825—836 (23 refs) CA 80:142 515.

R102. F. R. N. Gurd and P. Keim, 'Prospects for $^{13}$C N.M.R. Studies in Enzymology', *Methods Enzymol.*, 1973, **27,** 836—911 (192 refs) CA 80:142 161.

R103. J. D. Memory and G. W. Parker, 'Resonance Studies', *Methods Exp. Phys.*, 1974, **3B,** 465—573 (259 refs) CA 80:150 417.

R104. G. R. Luckhurst, 'Magnetic Resonance Studies of Thermotropic Liquid Crystals', *Mol. Crystals Liquid Crystals*, 1973, **21,** 125—159 (70 refs) CA 79:108 882.

R105. K. Wüthrich, 'N.M.R. Spectroscopy in Biological Research', *Naturwiss.*, 1973, **60,** 221—230 (24 refs) (in German) CA 79:75 481.

R106. I. Ya. Slonim, Ya. G. Urman, and V. N. Klyuchnikov, 'Use of N.M.R. in the Chemistry and Technology of Polyesters', *Plast. Massy*, 1973, 68—70 (12 refs) (in Russian) CA 79:66 918.

R107. J. H. Bradbury, 'N.M.R. Spectroscopy in Australia with Particular Reference to Biological and Synthetic Macromolecules', *Proc. Roy. Austral. Chem. Inst.*, 1971, **38,** 357—361 (38 refs) CA 76:78 595.

R108. J. Paris, 'Application of N.M.R. in Biochemistry and Pharmacology', *Prod. Probl. Pharm.*, 1973, **28,** 355—359 (26 refs) (in French) CA 79:88 940.

R109. J. A. Walter and A. B. Hope, 'N.M.R. and the State of Water in Cells', *Progr. Biophys. Mol. Biol.*, 1971, **23,** 1—20 (96 refs) CA 76:136 913.

R110. J. A. Glasel, 'Lanthanide Ions as N.M.R. Chemical Shift Probes in Biological Systems', *Progr. Inorg. Chem.*, 1973, **18,** 383—413 (44 refs) CA 80:57 080.

R111. R. Sastre and J. L. Acosta, 'Determination of Copolymer Composition by Spectroscopic Methods', *Rev. Plast. Mod.*, 1973, **25,** 885—899 (112 refs) (in Spanish) CA 79:92 618.

R112. G. Weill, 'N.M.R. in the Study of High Polymers', *Rev. Gen. Caout. Plast.*, 1973, **50,** 1003—1006 (3 refs) (in French) CA 81:13 890.

R113. N. A. Kostromina, 'Use of Proton Resonance for Studying Complexones', *Russ. Chem. Rev.*, 1973, **42,** 261—277 (*Uspekhi Khim.*, pp. 585—614) (99 refs) CA 79:2588.

R114. N. M. Sergeev, 'Dynamic N.M.R.', *Russ. Chem. Rev.*, 1973, **42,** 339—354 (*Uspekhi Khim.*, pp. 769—798) (81 refs) CA 79:47 231.

R115. L. A. Fedorov, 'Stereochemical Non-rigidity (Internal Rotation and Metallotropy) of Organic Derivatives of Transition Metals Investigated by N.M.R. Spectroscopy', *Russ. Chem. Rev.*, 1973, **42,** 678—694 (*Uspekhi Khim.*, pp. 1481—1514) (256 refs) CA 79:125 347.

R116. I. Ya. Slonim and A. Kh. Bulai, 'Paramagnetic Shift Reagents in N.M.R. Spectroscopy', *Russ. Chem. Rev.*, 1973, **42,** 904—921 (*Uspekhi Khim.*, pp. 1976—2006) (348 refs) CA 80:59 030.

R117. C. L. Boltz, 'N.M.R. for Chemical Analysis', *Sci. Technol.*, 1974, **11,** 9—10 (no refs) CA 81:44 840.

R118. T. Wilczok and B. Lubas, 'N.M.R. Studies of the Helix–Coil Transition in Nucleic Acids', *Studia Biophys.*, 1973, **38,** 85—116 (81 refs), CA 80:79 180.

R119. T. J. Swift, 'N.M.R.', *Tech. Chem. (N.Y.)*, 1974, **6,** Pt. 2, 521—563 (52 refs) CA 80:101 708.

R120. I. D. Weisman, L. J. Swartzendruber, and L. H. Bennett, 'Nuclear Resonances in Metals. N.M.R. and Mössbauer Effect', *Tech. Metals Res.*, Pt. 2, 1973, 165—504 (148 refs) CA 80:75 983.

R121. R. A. Bell and J. K. Saunders, 'Some Chemical Applications of the Nuclear Overhauser Effect, *Topics Stereochem.*, 1973 **7,** 1—92 (99 refs) CA 78:96 678.

R122. N. K. Wilson and J. B. Stothers, 'Stereochemical Aspects of $^{13}$C

NMR Spectroscopy', *Topics Stereochem.*, 1974, **8,** 1—158 (241 refs) CA 80:119 718.

R123. D. E. Caşe, 'Place of N.M.R. Spectroscopy in Drug Metabolism Studies', *Xenobiotica*, 1973, **3,** 451—471 (35 refs) CA 80:22 411.

R124. R. Radeglia, 'Advances in High-resolution N.M.R. Spectroscopy. I. Physical Principles of Measurement, Particularly FT Spectroscopy', *Z. Chem.*, 1974, **14,** 82—92 (31 refs) (in German) CA 81:18 590.

# 1
# Nuclear Shielding

BY R. B. MALLION

## 1 Introduction

'The last thing one discovers in writing . . . is what to put first.' This remark of Pascal (Blaise, that is—*not* the pioneer in the field of magnetic susceptibilities!) is the one which springs most immediately to mind as I take over the chapter on 'Nuclear Shielding' from Dr. Raynes, who contributed this chapter for the first three volumes of these Reports.[1-3] I can, however, start by saying that I shall adhere, as far as possible, to the structure and general format of the Report which he very satisfactorily evolved over the past three years, *i.e.* it will be what the earlier Reports described as 'phenomenon-oriented rather than compound-oriented'. No major change has been made for two reasons: firstly, such a policy should be helpful to regular readers who, by now, have become accustomed to Dr. Raynes' presentation, and secondly, it seemed worthwhile to capitalize, as much as is proper, on the previous Reporter's very successful experimentation in this regard.

Adopting such a scheme has, however, presented a mild dilemma; whether to repeat, almost *verbatim*, the basic equations required for each of the separate phenomena discussed, or whether simply to cite the relevant equations detailed explicitly in the earlier Reports. The former policy would ensure that all the requisite information were in one volume, thus obviating the necessity for readers to refer back to the earlier Reports, and the latter policy would save space. Since this Report is meant to be a digest of the literature—in fact, the sort of compendium which (to use F. A. Cotton's quotation[4] of J. S. Waugh's description of another eminently assimilable publication) 'one can read in bed without a pencil'—I have decided, somewhat reluctantly, to trade brevity for readers' convenience, and (again following Raynes' practice) have repeated all the basic equations which are pertinent to the appropriate discussion in the various sections of the Report. There are, however, some

[1] W. T. Raynes, 'Nuclear Magnetic Resonance', ed. R. K. Harris (Specialist Periodical Reports), The Chemical Society, London, 1972, Vol. 1, p. 1.
[2] W. T. Raynes, 'Nuclear Magnetic Resonance', ed. R. K. Harris (Specialist Periodical Reports), The Chemical Society, London, 1973, Vol. 2, p. 1.
[3] W. T. Raynes, 'Nuclear Magnetic Resonance', ed. R. K. Harris (Specialist Periodical Reports), The Chemical Society, London, 1974, Vol. 3, p. 1.
[4] F. A. Cotton, 'Chemical Applications of Group Theory', Interscience, London and New York, 1963, Preface.

minor changes: I prefer to discuss chemical shift anisotropy in the section headed 'Basic Physical Aspects'; n.m.r. measurements relating to the concept of 'aromaticity' are discussed with 'ring current' effects (Section 4E), since authors frequently link these two subjects in the literature; and I have not been nearly so conscientious as Dr. Raynes was about distinguishing between the various sorts of 'transmitted effects' (*e.g.* the so-called 'inductive effects' and what are often dubbed 'resonance effects') discussed in Section 4. Finally, the section on 'Shieldings of Particular Species' (Section 5, devoted to $^{19}F$, $^{31}P$, and 'other nuclei') will tend more to be simply lists of papers reporting data, for those works concerning mechanisms or calculation of the shielding of such nuclei will have been dealt with earlier in the appropriate sections.

According to the Reviewer's 'brief', this Report should include articles on nuclear shielding which were published between June 1st 1973 and May 31st 1974. As before, only experimental and theoretical papers relating to nuclear shielding in *isolated molecules* have been considered [although, actually, this restriction has been somewhat arbitrarily relaxed in Section 2 ('Basic Aspects'), where I have wanted to discuss the effects of pressure and anisotropic media]; as was the case in the previous Reports, this policy precludes examination of purely experimental techniques of chemical shift measurement (covered in Chapter 4), the quantum-mechanical details of shielding-constant calculations, the mechanisms of intermolecular shielding effects (thus excluding consideration of all solution phenomena as well as the study of contact and pseudocontact shifts and of weak complex formation, all of which are dealt with in Chapter 10). Studies in the nematic phase have also been excluded, except when such investigations have led explicitly to information concerning chemical shift anisotropy.

Some attempt has been made to include coverage of certain foreign-language journals; I hope, in particular, that the literature in the French language has been covered as adequately as that in English; I must, however, apologize in advance to writers in the German language for a less than satisfactory review of their work; and papers in Russian (unless available in translation or summarized in *Chemical Abstracts*) have had, regrettably, to be omitted from consideration entirely. Finally, during the period of writing, some journals were unavailable in the Reporter's library because they were being bound; apologies are, therefore, offered to the authors of any papers which are, thereby, overlooked.

## 2 Basic Aspects of Nuclear Shielding

**A. General Theory.**—Ramsey's familiar 'two-term' expression for the component, $\sigma_{\alpha\beta}$, of the shielding tensor of a given nucleus,

$$\sigma_{\alpha\beta} = \sigma^{d}_{\alpha\beta} + \sigma^{p}_{\alpha\beta} \quad (1)$$

where the superscripts 'd' and 'p' denote 'diamagnetic' and 'paramagnetic', respectively, is valid only for the very special case in which the gauge origin (*i.e.* the origin of the vector potential due to the uniform, external magnetic field) and the origin of co-ordinates coincide at the nucleus whose shielding is being calculated; a four-term expression that allows for arbitrary choice of the gauge origin but still requires the position of the nucleus of interest to be the origin of co-ordinates has been detailed by Raynes.[1,2] In last year's Report,[3] Raynes gave a yet more general version of the Ramsey equation, applicable when the co-ordinate origin, the gauge origin, and the nucleus in question may be different points; for reference purposes these expressions are now repeated here.

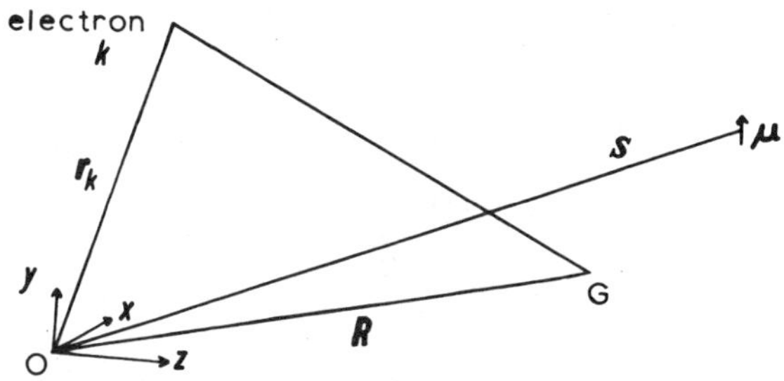


**Figure 1**

The formulae which follow should be studied with reference to Figure 1. The origin of co-ordinates is denoted O, and the gauge origin is at G, related to O by the vector $\boldsymbol{R}$. An infinitesimal 'test' dipole* of moment $\mu$ is then placed at the point† at which it is required to calculate the shielding (which point may, or may not, be the position of a nucleus[3]). The $k$th electron is, at any given time, displaced from O by the vector $\boldsymbol{r}_k$; the nuclei are assumed to be fixed. An eight-term expression then results:[3]

$$\sigma_{\alpha\beta} = \sigma_{\alpha\beta}^{\mathrm{d}} + \sigma_{\alpha\beta}^{\mathrm{dg}} + \sigma_{\alpha\beta}^{\mathrm{d}\mu} + \sigma_{\alpha\beta}^{\mathrm{dg}\mu} + \sigma_{\alpha\beta}^{\mathrm{p}} + \sigma_{\alpha\beta}^{\mathrm{pg}} + \sigma_{\alpha\beta}^{\mathrm{p}\mu} + \sigma_{\alpha\beta}^{\mathrm{pg}\mu} \quad (2)$$

In this expression, calculation of the diamagnetic terms (with superscript 'd') requires knowledge only of the unperturbed ground-state wavefunction, whereas the paramagnetic terms (superscripted 'p') are also a function of the excited-state wavefunctions (and are, thereby, that much more difficult to calculate). The terms bearing a superscript 'g' are gauge dependent, and will be zero only when the gauge origin and the origin of co-ordinates coincide. The superscript '$\mu$' refers to the point dipole, and terms containing this superscript are zero only when the dipole is located at the origin of co-ordinates. In giving explicit expressions for the terms on the right-hand side of equation (2), we again follow Raynes[3] in using SI, which is now becoming more customary:

* Dr. Raynes[3] called this dipole $m$, but it is here denoted by $\mu$ in order to avoid confusion with the symbol $m$ for the mass of the electron, used in equations (3)—(10).
† Related to O by the vector $\boldsymbol{S}$.

$$\sigma^{\rm d}_{\alpha\beta} = \frac{\mu_0}{4\pi}\frac{e^2}{2m} \langle 0|\sum_k \frac{r_k^2\delta_{\alpha\beta} - r_{k\alpha}r_{k\beta}}{|\boldsymbol{r}_k - \boldsymbol{S}|^3}|0\rangle \qquad (3)$$

$$\sigma^{\rm dg}_{\alpha\beta} = \frac{\mu_0}{4\pi}\frac{e^2}{2m} \langle 0|\sum_k \frac{R_\alpha r_{k\beta} - R_\gamma r_{k\gamma}\delta_{\alpha\beta}}{|\boldsymbol{r}_k - \boldsymbol{S}|^3}|0\rangle \qquad (4)$$

$$\sigma^{\rm d\mu}_{\alpha\beta} = \frac{\mu_0}{4\pi}\frac{e^2}{2m} \langle 0|\sum_k \frac{r_{k\alpha}S_\beta - r_{k\gamma}S_\gamma\delta_{\alpha\beta}}{|\boldsymbol{r}_k - \boldsymbol{S}|^3}|0\rangle \qquad (5)$$

$$\sigma^{\rm dg\mu}_{\alpha\beta} = \frac{\mu_0}{4\pi}\frac{e^2}{2m} \langle 0|\sum_k \frac{R_\gamma S_\gamma\delta_{\alpha\beta} - R_\alpha S_\beta}{|\boldsymbol{r}_k - \boldsymbol{S}|^3}|0\rangle \qquad (6)$$

$$\sigma^{\rm p}_{\alpha\beta} = \frac{\mu_0}{4\pi}\frac{e^2}{2m^2}\sum_n{}' (W_0 - W_n)^{-1}\left\{\langle 0|\sum_k\frac{l_{k\alpha}}{|\boldsymbol{r}_k - \boldsymbol{S}|^3}|n\rangle\langle n|\sum_k l_{k\beta}|0\rangle + \langle 0|\sum_k l_{k\beta}|n\rangle\langle n|\sum_k\frac{l_{k\alpha}}{|\boldsymbol{r}_k - \boldsymbol{S}|^3}|0\rangle\right\} \qquad (7)$$

$$\sigma^{\rm pg}_{\alpha\beta} = \frac{\mu_0}{4\pi}\frac{e^2}{2m^2}\varepsilon_{\beta\gamma\delta}R_\gamma\sum_n{}'(W_n - W_0)^{-1}\left\{\langle 0|\sum_k\frac{l_{k\alpha}}{|\boldsymbol{r}_k - \boldsymbol{S}|^3}|n\rangle\langle n|\sum_k p_{k\delta}|0\rangle + \langle 0|\sum_k p_{k\delta}|n\rangle\langle n|\sum_k\frac{l_{k\alpha}}{|\boldsymbol{r}_k - \boldsymbol{S}|^3}|0\rangle\right\} \qquad (8)$$

$$\sigma^{\rm p\mu}_{\alpha\beta} = \frac{\mu_0}{4\pi}\frac{e^2}{2m^2}\varepsilon_{\alpha\omega\psi}\, S_\omega\sum_n{}'(w_n - w_0)^{-1}\left\{\langle 0|\sum_k\frac{p_{k\psi}}{|\boldsymbol{r}_k - \boldsymbol{S}|^3}|n\rangle\langle n|\sum_k l_{k\beta}|0\rangle + \langle 0|\sum_k l_{k\beta}|n\rangle\langle n|\sum_k\frac{p_{k\psi}}{|\boldsymbol{r}_k - \boldsymbol{S}|^3}|0\rangle\right\} \qquad (9)$$

$$\sigma^{\rm pg\mu}_{\alpha\beta} = \frac{\mu_0}{4\pi}\frac{e^2}{2m^2}\varepsilon_{\alpha\omega\psi}\varepsilon_{\beta\gamma\delta}R_\omega S_\gamma\sum_n{}'(w_0 - w_n)^{-1}\left\{\langle 0|\sum_k\frac{p_{k\psi}}{|\boldsymbol{r}_k - \boldsymbol{S}|^3}|n\rangle \times \langle n|\sum_k p_{k\delta}|0\rangle + \langle 0|\sum_k p_{k\delta}|n\rangle\langle n|\sum_k\frac{p_{k\psi}}{|\boldsymbol{r}_k - \boldsymbol{S}|^3}|0\rangle\right\} \qquad (10)$$

In equations (3)—(10), $\mu_0$ is the permeability of a vacuum, $e$ and $m$ are the charge and mass, respectively, of the electron; $\boldsymbol{p}_k$ is the linear momentum operator for the $k$th electron and $l_k$ ($= \boldsymbol{r}_k \wedge \boldsymbol{p}_k$) denotes the orbital angular momentum operator for this electron. $W_n$ represents the energy of the $n$th excited state, and the prime on the summations in equations (7)—(10) is meant to indicate a summation over all values of $n$ except $n = 0$ (but including the *continuum* of states). The Greek subscripts denote Cartesian components $x$, $y$, and $z$; $\delta_{\alpha\beta}$ is 1 if $\alpha = \beta$, and zero if $\alpha \neq \beta$; $\varepsilon_{\beta\gamma\delta}$ is 1 if $(\beta\gamma\delta)$ is an even permutation of $(xyz)$, $-1$ if $(\beta\gamma\delta)$ is an odd permutation of $(xyz)$, and zero if any two of $(\beta\gamma\delta)$ are identical.

In the conventional formulation of electromagnetic theory, the external magnetic field manifests itself in the expressions obtained as an electromagnetic vector potential. These potentials are, however, defined only to within the space and time derivatives of an arbitrary function and, in two papers which were discussed in a previous Report[2] (and which have also been commented on by Moss and Perry[5]), Weisenthal and de Graaf[6,7] derived a new Hamiltonian formalism in which the coupling between the charges and the external fields is expressed in terms of multipole moments of the charge distribution and the field variables $E(x,t)$ and $B(x,t)$, rather than the usual expression involving the particle momenta and field potentials, $\phi(x,t)$ and $A(x,t)$; they made the exciting (but perhaps somewhat startling) claim that by writing the Hamiltonian explicitly in terms of the field strengths rather than their vector potentials, any question regarding gauge invariance or best choice of gauge would be obviated.[6,7] During the present review period, however, Woolley and Cordle[8] have pointed out that such gauge difficulties are not avoided by making a unitary transformation to a formalism which is independent of the electromagnetic field potentials, and that neither of the formalisms of Weisenthal and de Graaf possesses the uniqueness that has been attributed to them.[6,7] By a somewhat different reasoning, J. A. N. F. Gomes[9] (Theoretical Chemistry Department, Oxford) has also shown that, unfortunately, the new Hamiltonian, which is apparently and superficially gauge invariant, does, in fact, implicitly involve a particular choice of gauge for the vector potential, and that it has, therefore, none of the advantages which have been claimed for it.[6,7]

The general problem of 'symmetry-enforced' gauge invariance of nuclear shielding constants has been discussed in an important note by Okninski and Sadlej.[10] Since it can be shown (*e.g.* ref. 11) that a gauge transformation of the vector potential of the external magnetic field merely changes the relative contributions of the diamagnetic and paramagnetic terms of equation (1) to the total shielding tensor $\sigma_{\alpha\beta}$, it has been thought[12] that an appropriate choice of gauge might avoid completely the need for a calculation of the tiresome paramagnetic term $\sigma^{P}_{\alpha\beta}$. For the very specific gauge transformation

$$\boldsymbol{A}_B = \boldsymbol{A}^0_B - \tfrac{1}{2}\boldsymbol{B} \wedge \boldsymbol{R} \tag{11}$$

5 R. E. Moss and A. J. Perry, *Mol. Phys.*, 1972, **23**, 957.
6 L. Weisenthal and A. M. de Graaf, *Phys. Rev. Letters*, 1971, **27**, 470.
7 L. Weisenthal and A. M. de Graaf, *J. Chem. Phys.*, 1973, **58**, 249.
8 R. G. Woolley and J. E. Cordle, *Chem. Phys. Letters*, 1973, **22**, 411.
9 J. A. N. F. Gomes, personal communication, 1973.
10 A. Okninski and A. J. Sadlej, *Mol. Phys.*, 1973, **26**, 1545.
11 D. W. Davies, 'The Theory of the Electric and Magnetic Properties of Molecules', Wiley, London, 1967, Ch. 2.
12 T. K. Rebane, *Soviet Phys. JETP*, 1960, **11**, 694.

Chan and Das[13] proved the existence of such a gauge origin, which ensures a vanishing paramagnetic contribution to the shielding. In the paper under discussion,[10] Okninski and Sadlej have demonstrated that, for some symmetry species, the proof of Chan and Das[13] loses its general character and that, under certain symmetry conditions, neither term in equation (1) depends on gauge so that the *individual* terms of the whole shielding tensor become accidentally gauge invariant. Furthermore, since this feature does *not* depend on the accuracy of the wavefunctions employed, *when* such a case arises (diagnosed by two formal theorems presented in ref. 10) gauge invariance of the shielding tensor cannot be used as a criterion of completeness of the basis set. In such cases, information about completeness of the basis set can be gleaned only by a study of the origin dependence of the magnetic susceptibility, since gauge-origin independence of $\sigma_{\alpha\beta}$ does not guarantee gauge invariance of the induced currents and, hence, of the magnetic susceptibility tensor.

During the review period there has also been some discussion of relativistic and radiative interactions of many-electron atoms in an external magnetic field,[14] and the derivatives of the eigenvalues of the Hamiltonian to fifth order have been examined[15] in the context of molecular properties in general (including nuclear magnetic shielding).

**B. Basic Physical Aspects.**—This subsection will be devoted to the results of a number of experiments, and some calculations, which have a direct bearing on the more fundamental aspects of nuclear shielding.

(*i*) *Pressure and Temperature Dependences.* A particularly extensive study of the temperature and pressure dependence of $^{129}$Xe chemical shift in xenon gas has been made by Jameson *et al.*[16] Previous $^{1}$H, $^{19}$F, and $^{129}$Xe studies on gaseous systems[17,18] showed that the chemical shift has an essentially linear dependence on density up to densities of *ca.* 100 amagat; in the range 30—250 amagat, the curve is of the form

$$\sigma(\rho,T) = \sigma_0 + \sigma_1(T)\rho + \sigma_2(T)\rho^2 + \sigma_3(T)\rho^3 \qquad (12)$$

where, at room temperature,

$$\begin{aligned} \sigma_1 &= -\ 0.548 \pm 0.004 \text{ p.p.m. amagat}^{-1} \\ \sigma_2 &= (-\ 0.169 \pm 0.02) \times 10^{-3} \text{ p.p.m. amagat}^{-2} \\ \sigma_3 &= (0.163 \pm 0.01) \times 10^{-5} \text{ p.p.m. amagat}^{-3} \end{aligned} \qquad (13)$$

The work of Jameson *et al.*[16] concerns experimental determinations of $\sigma_1(T)$ for pure xenon, over the temperature range 240—440 K. An accurate tem-

13 S. I. Chan and T. P. Das, *J. Chem. Phys.*, 1962, **37**, 1527.
14 H. Grotch and R. A. Hegstrom, *Phys. Rev.* (*A*), 1973, **8**, 2771.
15 M. D. Kumanova, *Mol. Phys.*, 1973, **26**, 243.
16 C. J. Jameson, A. K. Jameson, and S. M. Cohen, *J. Chem. Phys.*, 1973, **59**, 4540.
17 A. K. Jameson, C. J. Jameson, and H. S. Gutowski, *J. Chem. Phys.*, 1970, **53**, 2310.
18 W. T. Raynes, A. D. Buckingham, and H. J. Bernstein, *J. Chem. Phys.*, 1962, **36**, 3481.

perature dependence of $\sigma_1$ is essential for determining the functional form of the shielding of two interacting molecules.[16] It is found that the experimental values can be fitted by the following fourth-degree polynomial:

$$\sigma_1(\tau) = 0.536 - 0.135 \times 10^{-2}\,\tau + 0.132 \times 10^{-4}\,\tau^2 - 0.598 \times 10^{-7}\,\tau^3 + 0.663 \times 10^{-10}\,\tau^4 \text{ p.p.m. amagat}^{-1} \qquad (14)$$

where $\tau = (T/\mathrm{K}) - 300$.

The $^1$H chemical shift of hydrogen chloride has been studied as a function of temperature and density by Trappeniers and Ahlers.[19] The quantity $\Delta\delta$ ($= \delta_{\mathrm{HCl}} - \delta_{\mathrm{CH_4}}$) was found to fit the following relation, below 293 K:

$$7.96\Delta\delta^2 + 6.32\Delta\delta - 2.83\Delta\delta t + 0.95 \times 10^{-4}\,t^2 + 5.48 \times 10^{-2} t = 2.447 \qquad (15)$$

($t$/°C). The effects of temperature on the $^{13}$C shifts of benzene, cyclohexane, pyridine, neopentane, hexane, and heptane, in the range 283—343 K, as well as the temperature shift of $^2$H in $D_2O$, have been studied by Litchman and McLaughlin.[20] All shifts were found to be linear functions of temperature, within the experimental error. The pressure dependence of the $^1$H chemical shift of chloroform has been studied by Yamada *et al.*[21] Although again slightly outside our somewhat arbitrarily restricted scope, two other papers are merely mentioned here, one concerning the temperature dependence of isotropic dipolar shieldings in oxygen and silicon,[22] and the other concerned with $^1$H and $^{19}$F shifts as a function of temperature and concentration.[23]

Raynes *et al.*[24,25] have previously reported *ab initio* calculations of bond-length dependences of nuclear shielding in the hydrogen molecule and, by means of the method of Herman and Short,[26] obtained an average over the nuclear motion which gave the shielding of $H_2$ and its isotopomers (HD, HT, $D_2$, DT, and $T_2$) in their lower rotational and vibrational states:[25] they thus deduced temperature variation of the shielding constants at low gas densities, in addition to the various isotope shifts. Unfortunately, however, no nuclear shielding data are available of sufficient accuracy to enable these calculations to be tested critically. Accordingly, in order to check their choice of wave-function, the form of perturbation theory applied,[24] and the Herman–Short

[19] N. J. Trappeniers and J. A. Ahlers, *Chem. Phys. Letters*, 1973, **21**, 226.
[20] W. Litchman and D. R. McLaughlin, *Chem. Phys. Letters*, 1973, **22**, 424.
[21] H. Yamada, T. Ishihara, and T. Kinugasa, *J. Amer. Chem. Soc.*, 1974, **96**, 1936.
[22] W. de W. Horrocks and E. S. Greenberg, *Mol. Phys.*, 1974, **27**, 993.
[23] J. Shamir and A. Netzer, *Canad. J. Chem.*, 1973, **51**, 2676.
[24] D. B. Cook, A. M. Davies, and W. T. Raynes, *Mol. Phys.*, 1971, **21**, 113.
[25] W. T. Raynes, A. M. Davies, and D. B. Cook, *Mol. Phys.*, 1971, **21**, 123.
[26] R. M. Herman and S. Short, *J. Chem. Phys.*, 1968, **48**, 1266; *ibid.*, 1969, **50**, 572.

procedure, Raynes *et al.*[27] have been forced to consider very accurate calculations of some other property of hydrogen, magnetic susceptibility being selected. They have, therefore, calculated the magnetic susceptibility of hydrogen and compared their results with those of Kolos and Wolniewicz.[28] Good agreement is obtained.[27] They also tested the Herman–Short procedure for averaging over the nuclear motion by comparing the predicted susceptibilities of $H_2$, HD, and $D_2$ with those obtained by Jain and Sahni,[29] once more with gratifying results. Meanwhile, Zeroka[30] has also independently made a theoretical study of the variation of the diamagnetic properties—both susceptibility and shielding—of the hydrogen molecule, with internuclear separation. Studying the range of internuclear separations $R$, 1.0—12.0 a.u., he has applied his previously described[31] variation–perturbation approach, using the wavefunction of Fraga and Ransil,[32] accurate over this range of $R$. The shielding at $R_e = 1.4$ a.u. is[30] 27.095 p.p.m. For normal $H_2$, the average over nuclear motion, where zero-point vibration and centrifugal distortion are taken into account, is $\langle\sigma\rangle = 26.760$ p.p.m. at 300 K and varies little in the range 20—300 K. Zeroka's results for the magnetic shielding of $H_2$ as a function of internuclear separation are shown in Table 1. The dependence of the paramagnetic shielding in diatomic molecules on the interatomic distance has also been investigated by Volodicheva and Rebane,[33] who have obtained

**Table 1** *Magnetic shielding* (p.p.m.) *in* $H_2$, *as a function of internuclear separation,* $R$/a.u. (*from Zeroka*[30])

| $R$ | $\sigma^d_{zz}$ | $\sigma^d_{xx}$ | $\sigma^p_{xx}$ | $\sigma$ | $\Delta\sigma$ |
|---|---|---|---|---|---|
| 1.0 | 33.725 | 32.130 | 0.112 | 32.736 | 1.483 |
| 1.1 | 32.056 | 30.458 | 0.133 | 31.080 | 1.465 |
| 1.2 | 30.554 | 28.972 | 0.153 | 29.601 | 1.429 |
| 1.3 | 29.200 | 27.647 | 0.173 | 28.280 | 1.381 |
| 1.4 | 27.977 | 26.465 | 0.190 | 27.095 | 1.322 |
| 1.5 | 26.867 | 25.406 | 0.205 | 26.030 | 1.255 |
| 1.6 | 25.859 | 24.457 | 0.218 | 25.070 | 1.184 |
| 1.8 | 24.105 | 22.838 | 0.236 | 23.418 | 1.030 |
| 2.0 | 22.639 | 21.523 | 0.244 | 22.057 | 0.873 |
| 2.5 | 19.904 | 19.191 | 0.212 | 19.570 | 0.501 |
| 3.0 | 18.084 | 17.766 | 0.119 | 17.952 | 0.199 |
| 4.0 | 16.039 | 16.406 | −0.191 | 16.157 | −0.176 |
| 5.0 | 15.151 | 16.036 | −0.563 | 15.366 | −0.322 |
| 6.0 | 14.825 | 16.051 | −0.889 | 15.050 | −0.336 |
| 12.0 | 14.974 | 16.106 | −1.048 | 15.030 | −0.083 |

27 W. T. Raynes, J. P. Riley, A. M. Davies, and D. B. Cook, *Chem. Phys. Letters*, 1974, **24**, 139.
28 W. Kolos and L. Wolniewicz, *J. Chem. Phys.*, 1965, **43**, 2429.
29 D. C. Jain and R. C. Sahni, *Proc. Phys. Soc.* (*London*), 1966, **89**, 409.
30 D. Zeroka, *J. Chem. Phys.*, 1973, **59**, 3835.
31 T. B. Garrett and D. Zeroka, *Internat. J. Quantum Chem.*, 1972, **6**, 651.
32 S. Fraga and B. J. Ransil, *J. Chem. Phys.*, 1970, **53**, 2992.
33 M. I. Volodicheva and T. K. Rebane, *Vestnik Leningrad. Univ.* (*Fiz.*, *Khim.*), 1973, **1**, 38 (*Chem. Abs.*, 1973, **79**, 11 723v).

vibrational–rotational corrections of shieldings calculated by the 'vector potential variation method' for HF, DF, HCl, DCl, HBr, and DBr. Dixon *et al.*,[34] in continuing their study of one-electron properties,[35] have generated LCAO–MO SCF wavefunctions for the methane molecule using a minimal basis set of SCF orbitals, and have used them to investigate the dependence of the diamagnetic shielding (and diamagnetic susceptibility) of methane on internuclear distance. The results are shown in Table 2, where the diamagnetic shielding of the proton is given in p.p.m., the C—H bond length, $R_{CH}$, is varied from 1.3 to 1.5 a.u., and $\rho_H$, the orbital exponent, ranges from 1.95 to 2.15.

**Table 2** *Diamagnetic shielding of protons in* $CH_4$ *as a function of internuclear distance* $R_{CH}$/a.u. *and orbital exponent* $\rho_H$ (*from ref.* 34)

| | $\rho_H$ | | | |
|---|---|---|---|---|
| $R_H$ | 1.30 | 1.35 | 1.40 | 1.50 |
| 1.95 | 89.66 | — | 90.55 | 91.27 |
| 2.00 | 88.10 | — | 88.97 | 89.67 |
| 2.05 | 86.59 | — | 87.44 | 88.13 |
| 2.10 | 85.14 | 85.58 | 85.98 | 86.65 |
| 2.15 | 83.74 | — | 84.56 | — |

At the equilibrium bond length, the shielding of the proton is 87.44 p.p.m. for methane, compared with 94.3 p.p.m. for the ammonium ion.[34] However, for the same bond length of 1.95 a.u., with $\rho_H = 1.5$, the difference is only 0.5 p.p.m.[34] It is pointed out[34] that the diamagnetic shielding is usually well predicted in these calculations because the function $r_H^{-1}$ occurs in the Hamiltonian operator used to generate the wavefunction.

(*ii*) *Chemical Shift Anisotropy*. The vast majority of papers on nuclear shielding (particularly those involving more routine applications of it) deal with isotropic shielding constants. It is, therefore, sometimes easy to forget that, in such cases, we are, in fact, dealing with the *trace* of a tensor quantity and that—if we could but measure it—the amount of shielding which a given nucleus in any particular molecule experiences in the presence of an external magnetic field depends on the orientation of the molecule relative to that field. Except for certain nuclei in molecules which are spherically symmetrical (for example, the carbon atoms in $CH_4$ and $CCl_4$, and the sulphur atom in $SF_6$) there will be a shielding (or 'chemical shift') anisotropy—as is, in fact, evident from Ramsey's equation (2). As Pausak, Pines, and Waugh[36] so aptly put it, even the isotropic shift is a quantity which '. . . has proven a very useful one to know, and an enormous body of lore has grown up concerning its

[34] M. Dixon, T. A. Claxton, and R. E. Overill, *J. Magn. Resonance*, 1973, **12**, 193.
[35] M. Dixon, T. A. Claxton, and J. A. S. Smith, *J.C.S. Faraday II*, 1972, **68**, 2158; *ibid.* 1973, **69**, 186.
[36] S. Pausak, A. Pines, and J. S. Waugh, *J. Chem. Phys.*, 1973, **59**, 591.

relationship to molecular structure and reactivity, substitution and solvent effects and the like. However, it is clear that a new dimension of knowledge and understanding would be added to all these matters if all six independent elements of $\sigma$ could be determined experimentally'.

There has been some noticeable increase in the number of papers reporting experimental values of chemical shift anisotropies, as compared with previous review periods;[1-3] this is mainly due to the groups of Waugh and of Dailey. $^1H$ Shielding tensors have been examined experimentally by Haeberlen and Kohlschütter,[37] Van Hecke *et al.*,[38] Raber *et al.*,[39] and Dailey *et al.*,[40,41] and theoretically by Ditchfield;[42] de Kieviet and de Lange[43] have examined the $^1H$ n.m.r. of *o*-, *m*-, and *p*-dicyanobenzene in a nematic solvent. The value of the $^1H$ shielding anisotropy in chloroform, obtained[40] from temperature studies in the nematic phase, is 9.8 $\pm$ 2.5 p.p.m.; in anhydrous $\alpha$-oxalic acid the $^1H$ chemical shielding tensor has the eigenvalues[38] (principal values) $\sigma_{11} = -17.55$ p.p.m., $\sigma_{22} = -15.41$ p.p.m., and $\sigma_{33} = 0.232$ p.p.m.; for $^1H$ chemical shift values of carboxy-groups, Raber *et al.*[39] found (with reference to TMS) $\sigma_{\parallel} = 3.5 \pm 1$ p.p.m. in the case of squaric acid and $\sigma_{\parallel} = 1.5 \pm 1$ p.p.m. for butynedioic acid, while $\sigma_{\perp} = -18.5 \pm 1$ in both cases. The anisotropies obtained for carboxylic, aliphatic, olefinic, and aromatic protons are *ca.* 1.8, 1.0, 0, and $-5$ p.p.m., respectively, with substantial variation among the various compounds studied.[37] Ditchfield[42] has made a theoretical study of $^1H$ shielding anisotropies by means of his GIAO SCF approach[44] (discussed in detail in Section 3). Calculated chemical shifts determined from the isotropic values of these tensors were found to be in excellent accord with the corresponding observed isotropic shifts for some simple polyatomic molecules. Ditchfield[42] has presented the calculated anisotropies both in a principal axis system and with respect to an X—H bond-axis framework, and has concluded that, in general, paramagnetic shielding tensors are not axially symmetric when referred to an X—H bond-axis co-ordinate system.

$^{13}C$ Shielding tensors have been examined by Pausak *et al.*,[36] Chang *et al.*,[45] Bhattacharyya and Dailey,[40,46-48] Cappelli *et al.*,[49] and Kaplan *et al.*[50] Anisotropies of $^{13}C$ shielding tensors have also been calculated (*via*

37 V. Haeberlen and V. Kohlschütter, *Chem. Phys.*, 1973, **2**, 76.
38 P. Van Hecke, J. C. Weaver, B. L. Neff, and J. S. Waugh, *J. Chem. Phys.*, 1974, **60**, 1668.
39 H. Raber, G. Brünger, and M. Mehring, *Chem. Phys. Letters*, 1973, **23**, 400.
40 P. K. Bhattacharyya and B. P. Dailey, *J. Magn. Resonance*, 1973, **12**, 36.
41 N. Zumbulyadis and B. P. Dailey, *Mol. Phys.*, 1974, **27**, 633.
42 R. Ditchfield, *Chem. Phys.*, 1973, **2**, 400.
43 W. de Kieviet and C. A. de Lange, *Chem. Phys. Letters*, 1973, **22**, 378.
44 R. Ditchfield, *J. Chem. Phys.*, 1972, **56**, 5688.
45 J. J. Chang, R. G. Griffin, and A. Pines, *J. Chem. Phys.*, 1974, **60**, 2561.
46 P. K. Bhattacharyya and B. P. Dailey, *Mol. Phys.*, 1973, **26**, 1379.
47 P. K. Bhattacharyya and B. P. Dailey, *J. Magn. Resonance*, 1974, **13**, 317.
48 P. K. Bhattacharyya and B. P. Dailey, *J. Chem. Phys.*, 1973, **59**, 3737.
49 E. Cappelli, A. Di Nola, and A. L. Segre, *Mol. Phys.*, 1974, **27**, 1385.
50 S. Kaplan, H. A. Resing, and J. S. Waugh, *J. Chem. Phys.*, 1973, **59**, 5681.

a MINDO method) by Ando *et al.*[51] (see Section 3) and the temperature dependence of such tensors has been investigated by Bonazzola and Hesse-Bezot.[52] Pausak *et al.*[36] studied the eigenvalues (and eigenvectors) of the chemical shift tensors of the various $^{13}$C nuclei in durene [1,2,4,5-tetramethylbenzene (1)]. For each type of carbon, the trace of the shielding tensor, Tr $\sigma$, agrees well with the isotropic shift, $\sigma_i$, found from the spectra of liquid solutions (Table 3).

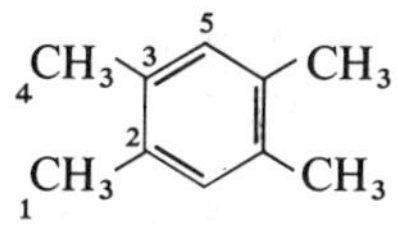

(1)

**Table 3** *Shieldings in durene* (1) (*from ref.* 36)

| *Carbon* | $\bar{\sigma}$ = Tr $\sigma$/3 | $\sigma_i$ |
|---|---|---|
| C-1 | 110.0 | 109.2 |
| C-2 | −5.8 | −5.0 |
| C-3 | −4.3 | −5.0 |
| C-4 | 109.7 | 109.2 |
| C-5 | −1.7 | −2.6 |

What Pausak *et al.*[36] describe as the 'well-known tendency' for $\sigma_i$ to be smaller for aromatic than for aliphatic carbon is ascribed largely to the very small shielding which occurs for aromatic carbons when the applied field is in the plane of the ring and tangential to it at the carbon atom in question. The shielding of $CH_3$ carbons is much less anisotropic than that of the ring carbons. What is claimed to be the first measurement of the full chemical shielding tensor for $^{13}$C in a carbonyl group (that of ammonium *d*-tartrate) yielded[37] $\sigma_{11} = -109$, $\sigma_{22} = -61$, and $\sigma_{33} = 23$ ($\pm$ 6) p.p.m. relative to liquid benzene. $^{13}$C Chemical shielding anisotropies in $^{13}CH_3I$, $^{13}CH_3Br$, and $^{13}CH_3Cl$ have been found[46] to be, respectively, $-101 \pm 15$, $-10 \pm 5$, and $22 \pm 5$ p.p.m. The $^{13}$C shielding anisotropy in chloroform obtained[40] from the temperature dependence of the nematic-phase $^{13}$C chemical shift and ordering parameter is $-39 \pm 1$ p.p.m., and the nematic–isotropic phase-difference method yields a similar value. If a comparison is made between $^{13}CHCl_3$ and $^{13}CCl_4$, which is spherically symmetric, it is found[40] that

$$\begin{aligned} \sigma_{\parallel}(^{13}CHCl_3) - \sigma_{\parallel}(^{13}CCl_4) &= -9 \text{ p.p.m.} \\ \sigma_{\perp}(^{13}CHCl_3) - \sigma_{\perp}(^{13}CCl_4) &= +30 \text{ p.p.m.} \end{aligned} \tag{16}$$

Thus, the component of the $^{13}$C shielding tensor along the molecular axis is much less sensitive to the electronegativity of the substituent than the perpendicular component in the two molecules. A study of $^{13}$C anisotropy for

[51] I. Ando, A. Nishioka, and M. Kondo, *Chem. Phys. Letters*, 1974, **25**, 212.
[52] L. Bonazzola and C. Hesse-Bezot, *J. Magn. Resonance*, 1973, **12**, 280.

absorbed benzene[50] shows that spectra do not differ significantly from those of the pure solid, as far as the principal values of the carbon shielding tensor are concerned.

$^{19}F$ chemical shift anisotropies have been reported by Griffin *et al.*,[53] den Otter and Maclean,[54] and Dailey *et al.*[47,55] Griffin *et al.*[53] studied two aromatic fluorides of potassium tetrafluorophthalate and found that the most shielded elements of the tensors lie perpendicular to the carbon ring and that, of the two in-plane components, the more shielded is along the C—F bond. Their results also show that the '*ortho* effect' of Karplus and Das[56] is not simply additive in the case of these molecules: such additivity was, in fact, assumed previously by Nehring and Saupe[57] in their study of fluorine shielding tensors in polyfluorobenzenes. Effects of temperature on $^{19}F$ tensors have been studied by Zumbulyadis and Dailey,[55,58] who have also reported work on $^{31}P$ chemical shift anisotropy[41,55,58] and on $^{15}N$ anisotropies.[59] The anisotropies in $^{15}N^{15}NO$ are $512 \pm 10$ p.p.m. (for the central N), and $369 \pm 15$ p.p.m. (for the terminal N), obtained from the temperature dependence of the nematic-phase shift; the corresponding values by the nematic–isotropic phase-difference method are $505 \pm 10$ and $366 \pm 10$ p.p.m., respectively.

The reader is also referred to three more-theoretical papers, by Meiboom *et al.*,[60-62] on n.m.r. in nematic solvents and on the formal theory of magnetic resonance parameters in anisotropic media. Although it is strictly outside our scope, mention is made here of a paper discussing the chemical shift tensor of a nucleus of spin 3/2 experiencing a large quadrupolar interaction in an applied field in a single crystal.[63]

(*iii*) *Standards and References for Chemical Shifts*. Some discussion has been devoted during the review period to the question of standards and internal references from which to measure the shieldings of various nuclei. Although arguably more appropriately dealt with in Section 5, standards as such are very much 'basic aspects' of n.m.r., and these papers are, accordingly, discussed here.

Levy *et al.*[64] have reported $^{29}Si$ chemical shifts for several organosilicons which are used as reference compounds in $^{29}Si$ n.m.r. studies, and they have

53 R. G. Griffin, H.-N. Yeung, M. D. La Prade, and J. S. Waugh, *J. Chem. Phys.*, 1973, **59**, 777.
54 G. J. den Otter and C. MacLean, *Chem. Phys. Letters*, 1973, **20**, 306.
55 N. Zumbulyadis and B. P. Dailey, *Mol. Phys.*, 1973, **26**, 777.
56 M. Karplus and T. P. Das, *J. Chem. Phys.*, 1961, **34**, 1663.
57 J. Nehring and A. Saupe, *J. Chem. Phys.*, 1970, **52**, 1307.
58 N. Zumbulyadis and B. P. Dailey, *Chem. Phys. Letters*, 1974, **26**, 273.
59 P. K. Bhattacharyya and B. P. Dailey, *J. Chem. Phys.*, 1973, **59**, 5820.
60 L. C. Snyder and S. Meiboom, *J. Chem. Phys.*, 1973, **58**, 5096.
61 R. C. Hewitt, S. Meiboom, and L. C. Snyder, *J. Chem. Phys.*, 1973, **58**, 5089.
62 Z. Luz and S. Meiboom, *J. Chem. Phys.*, 1973, **59**, 275.
63 R. B. Creel, E. von Meerwall, C. F. Griffin, and R. G. Barnes, *J. Chem. Phys.*, 1973, **58**, 4930.
64 G. C. Levy, J. D. Cargioli, G. E. Maciel, J. J. Natterstad, E. B. Whipple, and M. Ruta, *J. Magn. Resonance*, 1973, **11**, 352.

given some *desiderata* for a $^{29}Si$ standard. As they put it,[64] 'at some point in time an officially sanctioned $^{29}Si$ chemical shift reference and scale will undoubtedly emerge'. In the meantime, in the paper under discussion,[64] they present arguments which should be considered before setting $^{29}Si$ shift conventions, and give a set of accurate internal $^{29}Si$ shifts which may be used to facilitate the uniform presentation of $^{29}Si$ shift data in this interim period.

Table 4 gives their reported values of chemical shifts relative to internal tetraethyl orthosilicate, $(EtO)_4Si$; the latter is suggested as a high-field, low-frequency reference compound rather than the corresponding tetramethyl compound (which causes severe eye-damage and blindness). Solvent effects, and various *pros* and *cons*, of TMS and $(EtO)_4Si$, are discussed by Levy *et al.*,[64] and they conclude that it is possible to retain the best features of both compounds using TMS as an experimentally added reference compound, reporting shifts relative to $(EtO)_4Si$ as the primary $^{29}Si$ standard, set equal to $\delta 0.0$.

**Table 4** $^{29}Si$ *Chemical shifts relative to internal* $(EtO)_4Si$ *standard* (*in* $CCl_4$)[a] (*from ref.* 64)

| *Compound* | $\delta_{Si}$ (p.p.m.) |
|---|---|
| $(Me_3Si)_2O$ | 89.46 |
| $Me_4Si$ | 82.59 |
| $[Me_2SiO]_4$ | 63.05 |
| $Me_3SiO[Me_2SiO]_xSiMe_3$ ($\bar{x} \approx 50$) | 60.37 |
| $(MeO)_4Si$ | 3.37 |

[a] $(EtO)_4Si$ and solute concentration 15% (at *ca.* 300 K).

The usual primary reference[65] for $^{31}P$ n.m.r. is 85% phosphoric acid, $H_3PO_4$, which has the disadvantage of exhibiting a linewidth of nearly 5 Hz; $P_4O_6$ has been suggested, but has some disadvantages,[66] and Glonek and van Wazer[66] have now suggested aqueous tetrahydroxyphosphonium perchlorate as a narrow-line $^{31}P$ n.m.r. reference substance. They recommend that 85% $H_3PO_4$ at 25—30 °C be continued as the primary reference standard for $^{31}P$ n.m.r. chemical shifts, and that the tetrahydroxyphosphonium ion be used as a locking substance, as a secondary shift reference standard, and for magnetic field homogeneity adjustments.

On a somewhat less fundamental level, Live and Chan[67] have shown that two commonly used $^1H$ n.m.r. internal reference compounds, 3-(trimethylsilyl)propanesulphonic acid sodium salt and 3-(trimethylsilyl)sodium propionate, interact with purine in aqueous solution, causing significant shifts in the resonance position of their reference protons, and that tetramethylammonium chloride appears to be a more suitable reference compound for such studies. Finally, Jackowski and Kecki[68] have investigated several pos-

[65] V. Mark, C. H. Dungan, M. M. Crotchfield, and J. R. Van Wazer, in 'Topics in Phosphorus Chemistry', Interscience, New York, 1967, Vol. **5**, Ch. 4.
[66] T. Glonek and J. R. van Wazer, *J. Magn. Resonance*, 1974, **13**, 390.
[67] D. H. Live and S. I. Chan, *Org. Magn. Resonance*, 1973, **5**, 275.
[68] K. Jackowski and Z. Kecki, *Chem. analit.*, 1973, **18**, 35 (*Chem. Abs.*, 1974, **80**, 32 254e).

sible $^{1}H$ standards, and have concluded that $Bu^{t}OH$ gives a stable signal position, comparable with that of TMS.

## 3 Calculations of Nuclear Shielding

**A. Introduction.**—In this section, attention is directed to papers which present either general methods for the calculation of numerical values of nuclear shielding constants or actual calculations of shielding constants for specific molecules or classes of molecule. Papers which are devoted to the calculation of what are often considered to be particular 'transmitted effects' within molecules—such as magnetic anisotropy, electric field, and 'ring current' effects—are dealt with in Section 4.

**B. *Ab Initio* Calculations.**—The determination of shielding constants and magnetic susceptibilities by *ab initio* methods has been an important but difficult problem for even the smallest molecules,[69] a major difficulty being associated with the gauge of the vector potential describing the external magnetic field. One of the most rigorous approaches so far proposed is based on perturbed Hartree–Fock theory, as originally developed by Stevens *et al.*;[70] this method is computationally different from, but essentially similar to, that of Pople *et al.*,[71] and both approaches have been much applied and extended.[69] Another recent gauge-invariant *ab initio* SCF perturbation theory approach to magnetic shielding (already briefly mentioned in Section 2) is due to Ditchfield,[44] and preliminary applications of it[72,73] to organic molecules have given rise to calculated $^{13}C$ and $^{1}H$ chemical shifts which are in good agreement with experiment.[72,73] During the review period, there have been several contributions of extremely high quality to this field, and these will be discussed in some considerable detail in this subsection.

Appleman *et al.*[69] have presented a particularly important, extensive, and thorough investigation of 'heavy-atom' shielding in some small, polyatomic molecules. They have performed perturbed Hartree–Fock calculations on $CH_4$, $C_2H_2$, $C_2H_4$, $C_2H_6$, $CH_3F$, $CH_3OH$, $H_2O$, $NH_3$, and benzene, in order to obtain the corresponding carbon, fluorine, nitrogen, and oxygen shielding constants (as well as the magnetic susceptibilities of these molecules). They used Slater-type atomic orbitals (STOs) and, for each molecule, they examined three different basis sets: (i) a minimal* set which uses Slater's

[69] B. R. Appleman, T. Tokuhiro, G. Fraenkel, and C. W. Kern, *J. Chem. Phys.*, 1974, **60**, 2574.
[70] R. M. Stevens, R. M. Pitzer, and W. N. Lipscomb, *J. Chem. Phys.*, 1963, **38**, 550.
[71] J. A. Pople, J. W. McIver, and N. S. Ostlund, *J. Chem. Phys.*, 1968, **49**, 2960.
[72] R. Ditchfield, *Chem. Phys. Letters*, 1972, **15**, 203.
[73] R. Ditchfield and P. D. Ellis, *Chem. Phys. Letters*, 1972, **17**, 342.

* By a 'minimal basis set' is meant one which consists only of the smallest number of atomic orbitals which would be required to describe the ground state of the atom in question; (for hydrogen, this would be just 1*s*; for carbon it is 1*s*, 2*s*, and the three orthogonal 2*p*-orbitals). When more than this minimum number of functions feature in the basis set, the elements of this set constitute what is often described as an 'extended basis'.

rules for the heavy-atom exponents (SAMO)†; (ii) a minimal set with exponents optimized for the molecule (BLMO); and (iii) Clementi's double-zeta set[74] for the heavy atoms and the BLMO exponents for hydrogen (DZMO). It is found that there is generally a large dependence of $\sigma^{p}_{ii}$ (the paramagnetic part of the *i–i* shielding component), and $\sigma_{ii}$ ($= \sigma^{d}_{ii} + \sigma^{p}_{ii}$) on the basis set, the largest change occurring between the BLMO and DZMO sets; furthermore, most of the change in $\sigma_{ii}$ comes from the change in $\sigma^{p}_{ii}$. The effect of exponent optimization can, however, be substantial. Appleman *et al.*[69] also point out that when the DZMO for $H_2O$ was extended to include a third set of 2*p*-orbitals on oxygen, the effect was only *ca.* 10% of that of adding a second set of 2*p*-orbitals, suggesting that the 2*p* basis set is essentially 'saturated' at the DZMO level, although the perturbation Hartree–Fock study of Arrighini *et al.*[75] indicates that the addition of *d*-functions to oxygen does have an effect. Furthermore, all these changes are in qualitative agreement with the gauge-invariant atomic orbital (GIAO) results of Ditchfield.[44] Their estimate for the isotropic shielding of oxygen in water is 334 p.p.m., which compares favourably with Ditchfield's value[44] (328 p.p.m.) and that of Arrighini *et al.*[75] (331 p.p.m.), but poorly with the DZMO value of 249 p.p.m.; the latter is, however, comparable with that of Ditchfield *et al.*,[76] who found 267 p.p.m. using a slightly extended set of Gaussians ('4–31G') with the finite perturbation technique. The effect of Hartree–Fock coupling on $\sigma^{p}_{ii}$ is found[69] not to be significant for the SAMO and BLMO basis sets, but is, in general, quite appreciable (10—15%) for the DZMO set.

Table 5 shows the calculated[69] $^{13}C$ shieldings for several molecules which have at least one $CH_3$ group and which also possess an approximate three-fold axis of symmetry. $^{13}C$ Shieldings along the symmetry axis ($\sigma_{\parallel}$) are seen to be remarkably constant within each basis set (apart from the case of $CH_3F$, for which discrepancy reasons are advanced by Appleman *et al.*[69]). From SAMO, BLMO, and DZMO basis sets, $\sigma_{\parallel}$ average to (respectively) 192, 170, and 237 p.p.m., comparing well with similar average values from the work of Arrighini *et al.*[75] and Ditchfield *et al.*[44,76] of 190, 200, and 220 p.p.m. $^{13}C$ Shieldings for benzene are discussed,[69] as also is the importance of the use of a suitable $^{13}C$ scale when clarifying the difference between *ab initio* theories, such as the perturbed Hartree–Fock approach of Appleman *et al.*,[69] and semi-empirical models such as INDO[77,78] and MINDO.[79] Appleman *et al.*[69]

74 E. Clementi, *J. Chem. Phys.*, 1964, **40**, 1944.
75 G. P. Arrighini, M. Maestro, and R. Moccia, *J. Chem. Phys.*, 1968, **49**, 882.
76 R. Ditchfield, D. P. Miller, and J. A. Pople, *J. Chem. Phys.*, 1971, **54**, 4186.
77 P. D. Ellis, G. E. Maciel, and J. W. McIver, *J. Amer. Chem. Soc.*, 1972, **94**, 4069.
78 R. Ditchfield, in 'Molecular Structure and Properties', ed G. Allen (MTP International Review of Science), Physical Chemistry. Series One, Vol. 2, Butterworths, London, 1972.
79 A. B. Strong, D. Inkenberry, and D. M. Grant, *J. Magn. Resonance*, 1973, **9**, 145.

† This is a singularly unfortunate acronym, since the same one—S.A.M.O.—has been used recently in 'molecular fragment' calculations to denote a 'Simulated *Ab initio* Molecular Orbital' (see, for example, B. J. Duke and B. O'Leary, *Chem. Phys. Letters*, 1973, **20**, 459).

**Table 5** $^{13}C$ *Shielding*[a] *in methane and selected derivatives* (*taken from ref.* 69)

| *Component* | *Basis* | $CH_4$ | $C_2H_6$ | $CH_3F$ | $CH_3OH$[b] |
|---|---|---|---|---|---|
| $\sigma^p_\parallel$ | SAMO | −104 | −112 | −107 | −110 |
| | BLMO | −127 | −133 | −98 | −130 |
| | DZMO | −57 | −66 | −68 | −68 |
| $\sigma_\parallel$ | SAMO | 191 | 193 | 192 | 193 |
| | BLMO | 170 | 171 | 200 | 171 |
| | DZMO | 239 | 240 | 232 | 234 |
| $\sigma^p_\perp$ | SAMO | −104 | −126 | −131 | −141 |
| | BLMO | −127 | −152 | −124 | −162 |
| | DZMO | −57 | −96 | −155 | −142 |
| $\sigma_\perp$ | SAMO | 191 | 225 | 238 | 223 |
| | BLMO | 170 | 200 | 241 | 203 |
| | DZMO | 239 | 256 | 213 | 222 |

[a] In p.p.m.
[b] Average of the shielding components parallel with, or perpendicular to, the appropriate symmetry axes.

also discuss basis set dependence, the noted constancy of the $\sigma^p_\parallel$ of $^{13}C$ among $CH_3$-containing systems, the effect of coupled, 'self-coupled', and 'Dalgarno-uncoupled'[80] Hartree–Fock schemes, comparisons with other calculations, and gauge dependence. Basis set dependence is illustrated by Table 6, which shows the relative isotropic $^{13}C$ shieldings calculated by four different *ab initio* methods. A comparison of the two sets of DZMO shifts shows that the

**Table 6** *Comparison of relative* $^{13}C$ *shielding constants between various* ab initio *calculations and experiment*[a] (*taken from ref.* 69)

| *Molecule* | $\sigma$(*absolute*) − $\sigma(CH_4)$ DZMO[b] (*Gauge origin at carbon*) | DZMO[b] (*Gauge origin at centre of mass*) | *Method of Ditchfield* et al. (*Gauge origin at centre of mass*)[c] | *GIAO*[d] | *Experiment*[e] |
|---|---|---|---|---|---|
| $C_2H_2$ | −75 | −90 | −78 | −75 | −76 |
| $C_2H_4$ | −108 | −121 | −126 | −131 | −126 |
| $C_2H_6$ | +12 | +5.6 | +9.0 | −7.4 | −8.0 |
| $CH_3OH$ | −13 | −21 | −25 | −46 | −50 |
| $CH_3F$ | −25 | −31 | −41 | −64 | −78 |
| $H_2CO$ | −189 | −199 | −207 | −200 | −197 |

[a] Units of p.p.m.
[b] Double-zeta Slater basis set for which[69] $\sigma(CH_4)$ = 239 p.p.m.
[c] Slightly extended ('4–31 G') set of Gaussian orbitals, for which[76] $\sigma(CH_4)$ = 221 p.p.m.
[d] Slightly extended set of Gaussian-type, gauge-invariant orbitals, for which[44] $\sigma(CH_4)$ = 205 p.p.m.
[e] Experimental value of $\sigma(CH_4)$ taken as 196 ± 10 p.p.m.

[80] A. Dalgarno and J. M. McNamee, *J. Chem. Phys.*, 1961, **35**, 1517.

results are generally more accurate when the centre of mass, rather than the carbon nucleus, is used as gauge origin. On the question of gauge dependences, Appleman *et al.*[69] point out that the use of gauge-dependent atomic orbitals, rather than gaugeless ones, seems to have led to significant improvements in the shielding results at both *ab initio* and semi-empirical levels. They also bring to light an important difference between Ditchfield's GIAO form of perturbed Hartree–Fock theory[44] and that of Stevens *et al.*,[70] which is what Appleman *et al.*[69] are applying—that the property of 'gauge invariance' in the GIAO formulation does not refer to current conservation as such, but rather to an imposed gauge restraint on the energy. Thus, although Ditchfield's basis set forces 'gauge invariance' of the energy, current density is not necessarily a quantity which is conserved if the basis set is truncated—in fact, both theories are proper theories only when all gauge dependences disappear (*i.e.* in the idealized limit of a complete basis set). The important fundamental point to be drawn from all this is that the gauge invariance of GIAO theory is *not*, in itself, a justification of the use of such a theory (see also ref. 81). Appleman *et al.*[69] concede that Ditchfield's formulation[44] will, in general, lead to better agreement with experiment because his basis set is more complete. The major shortcoming of the perturbed Hartree–Fock theory is that the virtual orbitals do not have sufficient flexibility for adequate representation of the first-order wavefunctions and that, consequently, polarization functions, which include *d*-functions on the heavy atom and *p*-functions on hydrogen, should be used.[69]

Although it has been given considerable attention here, an adequate summary of all the interesting results and comparisons reported in this paper has not been presented, and those readers with more than a passing interest in *ab initio* calculations are strongly recommended to consult the original text.[69]

Meanwhile, Ditchfield,[82] in his quest for a self-consistent perturbation theory of diamagnetism, has made an important contribution by extending and developing his gauge-invariant LCAO method for n.m.r. chemical shifts. In contradistinction to the method of Stevens *et al.*[70] (discussed above), in which molecular orbitals, $\psi_j$, are written as linear combinations of *real* atomic orbitals,

$$\psi_j = \sum_{\nu} c_{\nu j} \phi_\nu \tag{17}$$

Ditchfield,[82] following the time-honoured procedure of London,[83] used gauge-invariant (or, better, 'gauge-dependent'[81,84]) atomic orbitals (GIAO), $\chi_\nu$, in which the real atomic basis functions $\phi_\nu$ are modified by field-dependent complex multiplicative phase factors which depend on the gauge of the vector potential $\boldsymbol{A}(\boldsymbol{r})$:

$$\chi_\nu(\boldsymbol{r}) = \phi_\nu \exp(-\,\mathrm{i}\boldsymbol{A}_\nu \cdot \boldsymbol{r}) \tag{18}$$

[81] S. T. Epstein, *J. Chem. Phys.*, 1973, **58**, 1592.
[82] R. Ditchfield, *Mol. Phys.*, 1974, **27**, 789.
[83] F. London, (*a*) *Compt. rend.*, 1937, **205**, 28; (*b*) *J. Phys. Radium*, 1937, **8**, 397; (*c*) *J. Chem. Phys.*, 1937, **5**, 837.
[84] J. A. Pople, *Discuss. Faraday Soc.*, 1962, No. 34, 7.

in which $A_v$ is the vector potential at centre $v$, due to the external field, and $\mathbf{r}$ is the position vector of an electron in $\phi_v$. Having developed his GIAO SCF perturbation theory, Ditchfield[82] shows how the final gauge-invariant magnetic shielding result can be divided up into contributions which are related to various features of electronic structure. Within this framework, several levels of theory can be defined which effectively differ only in the choice of $\phi_v$. Ditchfield considers three different basis sets: (i) a minimal basis of STOs (STO-5G); (ii) a minimal basis of Hartree–Fock atomic orbitals, in the atomic limit (LEMAO-5G); and (iii) a slightly extended basis set in which the inner shells are represented by a single sum of four Gaussians. However, the valence shells, hydrogen 1*s* and 'heavy-atom' 2*s* and 2*p*, are divided into inner and outer parts which are sums of three Gaussians and one Gaussian, respectively. There are thus two functions per hydrogen and nine per heavy atom. This basis set is denoted by 4–31G.

**Table 7** *Comparison of shieldings of various nuclei calculated (by Ditchfield's GIAO method)*[82] *from three different basis sets, with the corresponding experimental shieldings*[a]

| *Molecule* | *Nucleus of interest* | *STO-5G* | *LEMAO-5G* | *4-31G* | *Experiment*[b] |
|---|---|---|---|---|---|
| $CH_4$ | C | 0.0 | 0.0 | 0.0 | 0.0 |
| | H | 0.0 | 0.0 | 0.0 | 0.0 |
| $C_2H_6$ | C | −11.8 | 2.2 | −7.4 | −8.0 |
| | H | 0.05 | 0.03 | −0.35 | −0.75 |
| $CH_3F$ | C | −50.5 | −75.0 | −65.4 | −77.5 |
| | F | 49.2 | 128.7 | 74.0 | 76 |
| | H | −1.18 | −3.29 | −4.04 | −4.00 |
| $C_2H_4$ | C | −111.6 | −152.5 | −130.8 | −125.6 |
| | H | −5.21 | −6.20 | −5.61 | −5.18 |
| $H_2CO$ | C | −147.9 | −250.4 | −199.6 | −197 |
| | O | −757.6 | −860.3 | −858.8 | −580 to −600 |
| | H | −9.34 | −10.02 | −10.39 | −9 to −10 |
| $C_2H_2$ | C | −56.9 | −90.00 | −75.2 | −76 |
| | H | −1.41 | −2.27 | −1.25 | −1.35 |
| HCN | C | −97.5 | −147.5 | −123.5 | −120±10 |
| | N | −266.5 | −360.5 | −308.4 | −297±20 |
| | H | −3.10 | −3.79 | −2.95 | −2.83 |
| $NH_3$ | N | 0.0 | 0.0 | 0.0 | 0.0 |
| | H | 1.01 | 0.13 | 0.86 | 0.05 |
| $H_2O$ | O | 0.0 | 0.0 | 0.0 | 0.0 |
| | H | 1.37 | −1.45 | 0.06 | −0.60 |
| HF | F | 0.0 | 0.0 | 0.0 | 0.0 |
| | H | 2.20 | −4.16 | −2.12 | −2.50 |

[a] In units of p.p.m.
[b] From various sources (for details, see ref. 82). $^{13}C$ Chemical shifts are given relative to $CH_4$, $^{14}N$ shifts relative to $NH_3$, $^{17}O$ shifts relative to $H_2O$, and $^{19}F$ shifts relative to HF. A negative value indicates a high-frequency shift. The experimental $^{13}C$ shift of $CH_4$ relative to TMS is 2.1 p.p.m. All $^{1}H$ shifts are given relative to $CH_4$. The experimental $^{1}H$ shift of $CH_4$ relative to TMS is −0.23 p.p.m.

The results of Ditchfield's calculations are shown in Table 7, from which it is seen that all three SCF GIAO methods estimate shielding constants moderately well, the mean absolute differences between theory and experiment, for STO-5G, LEMAO-5G, and 4-31G, being 22.6, 22.4, and 4.1 p.p.m., respectively, for $^{13}C$ shifts and 1.31, 0.83, and 0.43 p.p.m., respectively, for $^{1}H$ shifts. Furthermore, it is immediately apparent, from the above deviations, and from Table 7, that the extended 4-31G basis gives, in general, much better agreement with experiment than do the minimal sets; this is particularly true for $^{13}C$ and $^{1}H$ chemical shifts, which are described very accurately at the 4-31G level. Ditchfield[82] also offers some comments about individual chemical shifts. The deshielding of the methane carbon when a hydrogen atom is replaced by a methyl group ('the $\alpha$ effect') is well reproduced by 4-31G and is qualitatively satisfactory but is somewhat over-estimated by STO-5G, while the LEMAO-5G set provides no interpretation at all of this effect. Some discussion is also devoted to the large deshielding effect experienced by the methane carbon when H is replaced by F, the difference in shielding at a carbon atom in C—C single and double bonds, the screening at carbon nuclei which form multiple bonds to electronegative atoms, and the limitations of the method when applied to calculate the shieldings of nitrogen, fluorine, and oxygen. Ditchfield's overall conclusion is that although the data presented in Table 7 cannot be regarded as exhaustive, the agreement between experiment and the 4-31G level of theory is 'excellent', and he promises a further detailed analysis of the various contributions to total shielding constants in a future paper. In the meantime, with Zens and Ellis, Ditchfield has further documented his earlier theory[44] by testing its predictions against experimental $^{13}C$ chemical shifts in the five possible fluoroallenes;[85] the overall agreement between theory and experiment is good.[85]

A minimum basis set of Slater orbitals has been used by Hall *et al.*[86] to form *ab initio* SCF wavefunctions from which diamagnetic properties (susceptibilities and shielding constants) of boron hydrides and their ions, including $B_8H_{12}$, $B_9H_{15}$, $[B_6H_6]^{2-}$, $[B_{10}H_{10}]^{2-}$, and $[B_{10}H_{14}]^{2-}$, were estimated. Both $^{1}H$ chemical shifts and $^{11}B$ diamagnetic shifts were calculated. Whereas a minimum basis set of STOs is known to yield values of diamagnetic susceptibilities ($\chi^d$) and diamagnetic contributions to the shielding ($\sigma^d$) which are accurate to a few per cent,[87] calculation of the paramagnetic parts, $\chi^p$ and $\sigma^p$, respectively, requires an optimized basis set at least three times the size of those used in the study under discussion.[86] Accordingly, Hall *et al.*[86] estimated values of $\chi^p$ and $\sigma^p$ from their calculated values of $\chi^d$ and $\sigma^d$, *via* the relations $\chi^p = \chi - \chi^d$ and $\sigma^p = \sigma - \sigma^d$, in which the values of $\chi$ and $\sigma$ are assumed known.[86] They find, however, that the total chemical shifts do not parallel the calculated diamagnetic shifts because of these large paramagnetic contributions. In $B_9H_{15}$ and $B_{10}H_{14}$, the $^{11}B$ diamagnetic shifts for different

85 A. P. Zens, P. D. Ellis, and R. Ditchfield, *J. Amer. Chem. Soc.*, 1974, **96**, 1309.
86 J. J. Hall, D. S. Marynick, and W. N. Lipscomb, *J. Amer. Chem. Soc.*, 1974, **96**, 770.
87 E. A. Laws, R. M. Stevens, and W. N. Lipscomb, *J. Chem. Phys.*, 1972, **56**, 2029.

boron atoms within the molecule do not follow the order of the total chemical shift.

A very detailed coupled Hartree–Fock study of molecular one-electron properties (including magnetic properties) has been presented in two important papers by Thomsen and Swanstrøm.[88,89] After outlining a general coupled Hartree–Fock scheme,[88] giving expressions for magnetic susceptibility and nuclear shielding, they have selected the water molecule and have calculated[89] various second-order properties of it, using a new Hartree–Fock Gaussian wavefunction and their coupled Hartree–Fock procedure.[88] They also have some interesting comments to make on the elusive question of gauge origin (a topic which has already been discussed at some length in Section 2). For the water molecule, there are four obvious choices of gauge origin: the oxygen atom (O), a hydrogen atom (H), the centre of mass (CM), and the centre of electronic charge (EC) (see Figure 2). As Thomsen and Swanstrøm

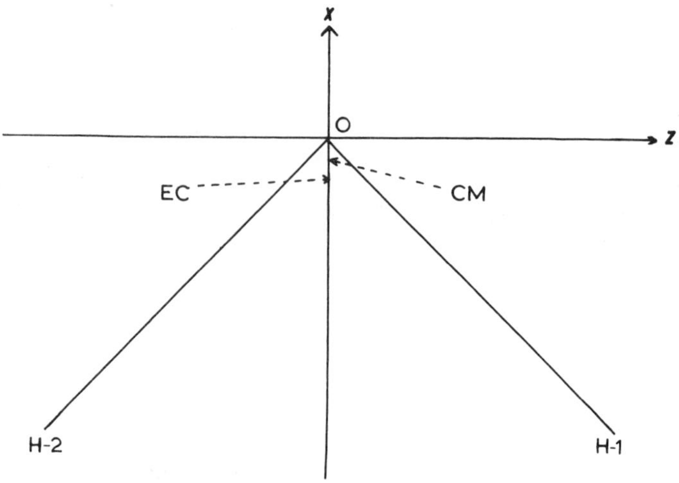


**Figure 2**
(Adapted by permission from *Mol. Phys.*, 1973, **26**, 751)

point out, and as was discussed in Section 2, the division of a magnetic property—susceptibility or shielding—into a diamagnetic part and a paramagnetic part is as arbitrary as the choice of gauge origin itself, and, in fact, only in the limit of a complete basis set will the sum of the two terms [equation (1)] be gauge invariant.[90] Of course, the paramagnetic contribution is the more troublesome one to calculate, but, according to Chan and Das,[13] magnetic properties are most accurately determined with the electronic centre (EC) as the origin. Thomsen and Swanstrøm emphasize, however, that the

[88] K. Thomsen and P. Swanstrøm, *Mol. Phys.*, 1973, **26**, 735.
[89] K. Thomsen and P. Swanstrøm, *Mol. Phys.*, 1973, **26**, 751.
[90] S. T. Epstein, *J. Chem. Phys.*, 1965, **42**, 2897.

choice of origin is not relevant for the purpose of comparisons with experimental results; the paramagnetic shielding, $\sigma_A^p$ at a nucleus A, corresponding to an origin at this nucleus, is related to the spin-rotation constant of A.[91] For the purpose of experiment *versus* theory comparisons, in the case of the paramagnetic terms, Thomsen and Swanstrøm used the following assignments:

$$\sigma_A(A) = \sigma_A^d(EC) + \sigma_A^p(EC) - \sigma_A^d(A) \tag{19}$$

in which the quantities on the right-hand side are computed from the wavefunction referred to above with the origin indicated in parentheses. This process is claimed[89] to be equivalent to calculating the magnetic properties with EC as origin and then imposing gauge invariance during a translation to the required origin. Thomsen and Swanstrøm point out that the procedure is inconsistent but is justifiable because it provides a 'flexible' basis for comparisons with experiment. Such an inconsistency is, in any case, present because of the use of a finite basis set. The results, for both oxygen and proton shielding (obtained by use of Diercksen's near Hartree–Fock wavefunction[92]) are shown in Tables 8 and 9, for various choices of origin. Proton shieldings (Table 8) agree well with the coupled Hartree–Fock calculations of Arrighini *et al.*,[93] and, when equation (19) is used, the paramagnetic terms come very close to the experimental data. The computed oxygen shielding constants (Table 9) are not significantly different from those of Arrighini *et al.*[95] or those of Ditchfield *et al.*,[76] but Thomsen and Swanstrøm[89] do not feel that they are able to assess their oxygen results critically, since no experimental data are available; however, because the first-order wavefunction is common to the calculation of the shielding constants for oxygen ($\sigma_O$) and hydrogen ($\sigma_H$) and of the magnetic susceptibility ($\chi$) of $H_2O$, and since the latter two quantities are in close agreement with experimental values, Thomsen and Swanstrøm[89] have some confidence in their results for $\sigma_O$. (This is an exactly similar line of reasoning to that advanced by Raynes *et al.* with regard to their calculations on $H_2$, detailed in ref. 27 and discussed in Section 2.)

Diamagnetic contributions to the nuclear shielding tensors (as well as to the magnetic susceptibilities) of pyridine and pyrazine calculated from 'Gaussian lobe' *ab initio* SCF molecular wavefunctions[96] have been reported by Ha and O'Konski.[97] Their results show that the hydrogens, carbons, and nitrogens in pyrazine are more diamagnetically shielded than those of pyridine; some discussion is given of the relative proton shieldings in pyridine, in comparison with those observed experimentally. In the case of carbon and nitrogen atoms, however, the (uncalculated) paramagnetic contributions are dominant, and so

91 N. F. Ramsey, 'Molecular Beams', Oxford University Press, Oxford, 1956.
92 G. H. F. Diercksen, *Theor. Chim. Acta*, 1971, **21**, 335.
93 G. P. Arrighini, M. Maestro, and R. Moccia, *J. Chem. Phys.*, 1970, **52**, 6411.
94 M. Jaszunski and A. J. Sadlej, *Theor. Chim. Acta*, 1972, **27**, 135.
95 G. P. Arrighini, M. Maestro, and R. Moccia, *Chem. Phys. Letters*, 1970, **7**, 351.
96 J. D. Petke, J. L. Whitten, and J. A. Ryan, *J. Chem. Phys.*, 1968, **48**, 953.
97 T. K. Ha and C. T. O'Konski, *Internat. J. Quantum Chem.*, 1973, **7**, 609.

**Table 8** *Various* $^1$H *shieldings in* $H_2O$ (*from ref.* 89)[e]

| | *Thomsen and Swanstrøm*[89] | | | | *Ref.* 93 | *Ref.* 94 | *Experiment* | |
|---|---|---|---|---|---|---|---|---|
| *Gauge origin*[a] | O | H | EC | H[b] | H | H | H[c] | H[d] |
| $\sigma^d_{xx}$ | 23.3 | 101.1 | 23.3 | 101.1 | 102.8 | 101.0 | — | — |
| $\sigma^d_{yy}$ | 8.9 | 130.6 | 14.3 | 130.6 | 130.2 | 130.4 | — | — |
| $\sigma^d_{zz}$ | 31.9 | 75.8 | 37.2 | 75.8 | 75.3 | 75.7 | — | — |
| $\sigma^d_{xz}(=\sigma^d_{zx})$ | −20.9 | −39.3 | −13.6 | −39.3 | — | — | — | — |
| $\bar{\sigma}^d$ | 21.4 | 102.5 | 24.9 | 102.5 | 102.8 | 102.3 | — | — |
| $\sigma^p_{xx}$ | 6.2 | −51.2 | 6.2 | −71.6 | −52.6 | −76.7 | −71.8 | −71.3 |
| $\sigma^p_{yy}$ | 13.0 | −77.0 | 9.0 | −107.3 | −79.2 | −118.4 | −107.0 | −108.3 |
| $\sigma^p_{zz}$ | 5.8 | −26.5 | 1.9 | −36.7 | −25.8 | −39.0 | −36.6 | −37.3 |
| $\sigma^p_{xz}$ | 11.2 | 33.2 | 5.8 | 31.5 | — | — | — | — |
| $\sigma^p_{zx}$ | 7.8 | 34.0 | 7.8 | 33.5 | — | — | — | — |
| $\bar{\sigma}^p$ | 8.2 | −51.6 | 5.7 | −71.9 | −52.5 | −78.0 | −71.8 | −72.2 |
| $\bar{\sigma}$ | 29.6 | 50.9 | 30.6 | 30.6 | 50.2 | 24.3 | 30.2 | |

[a] See Figure 2.
[b] Assignment of $\sigma^p$ by equation (19).
[c] $H_2O$ spectrum.
[d] $D_2O$ spectrum.
[e] After publication of ref. 89 the authors discovered some errors in the reported calculated shieldings. Tables 8 and 9 incorporate the corrected values. (P. Swanstrøm, personal communication, January 13, 1975.)

it is not meaningful to compare the calculated diamagnetic shielding values of these nuclei with the total shieldings measured experimentally.

The calculation of second-order properties by the coupled Hartree–Fock procedure has, in the past, proved 'expensive' because of the large number of electron-interaction integrals required in the MO basis; this difficulty can be

**Table 9** *Variously calculated* $^{17}$O *shieldings in* $H_2O$ (*from ref.* 89)[c]

| | *Thomsen and Swanstrøm*[89] | | | *Ref.* 95 | *Ref.* 76 |
|---|---|---|---|---|---|
| *Gauge origin*[a] | O | EC | O[b] | O | CM |
| $\sigma^{d}_{xx}$ | 415.7 | 415.7 | 415.7 | 415.2 | — |
| $\sigma^{d}_{yy}$ | 417.0 | 415.9 | 417.0 | 415.6 | — |
| $\sigma^{d}_{zz}$ | 415.3 | 414.3 | 415.3 | 414.3 | — |
| $\bar{\sigma}^{d}$ | 416.0 | 415.3 | 416.0 | 415.0 | 415.5 |
| $\sigma^{p}_{xx}$ | −122.0 | −122.0 | −122.0 | −108.2 | — |
| $\sigma^{p}_{yy}$ | −134.2 | −133.2 | −134.3 | −90.3 | — |
| $\sigma^{p}_{zz}$ | −65.3 | −60.1 | −61.1 | −53.4 | — |
| $\bar{\sigma}^{p}$ | −107.2 | −105.1 | −105.8 | −84.0 | −148.0 |
| $\bar{\sigma}$ | 308.8 | 310.2 | 310.2 | 331.1 | 267.3 |

[a] See Figure 2.
[b] Assignment of $\sigma^{p}$ by equation (19).
[c] Corrected value (see footnote to Table 8).

partially obviated by using an uncoupled Hartree–Fock method in which the relevant equations are solved approximately by neglecting all non-diagonal terms.[98] During the review period, a general appraisal has been given[99] of uncoupled SCF and Hartree–Fock perturbation theories, while Birnstock[100] has presented uncoupled Hartree–Fock calculations of $^{13}$C shielding constants by means of *ab initio* and other wavefunctions. Birnstock has decoupled the equations of the 'single excitation' scheme of Caves and Karplus;[98] in this investigation, the *ab initio* minimum basis wavefunctions of Palke and Lipscomb[101] and NEMO[102] wavefunctions have been used to perform calculations on $CH_4$, $C_2H_2$, $C_2H_4$, $C_2H_6$, and HCN. The gauge origin was taken as the centre of nuclear mass. When compared with the shielding constants of Ditchfield *et al.*, computed by the finite perturbation method with an STO-5G basis set, those of Birnstock,[100] which arise from calculations employing minimal basis set wavefunctions, revealed the same trends, though their actual sizes are quite different. Birnstock[100] has also shown that isotropic shielding constants calculated with NEMO wavefunctions are able to reflect multiple-bond effects (though deficiencies due to variation of the gauge origin are more marked here); in general, the results parallel those of Ditchfield

98 T. C. Caves and M. Karplus, *J. Chem. Phys.*, 1969, **50**, 3649.
99 A. J. Sadlej, *Mol. Phys.*, 1973, **26**, 1445.
100 F. Birnstock, *Mol. Phys.*, 1973, **26**, 343.
101 W. E. Palke and W. N. Lipscomb, *J. Amer. Chem. Soc.*, 1966, **88**, 2384.
102 J. D. Newton, F. P. Boer, and W. N. Lipscomb, *J. Amer. Chem. Soc.*, 1966, **88**, 2353, 2367.

*et al.* In a further paper, with Klöpper, Birnstock[103] has calculated similar $^{13}C$ shielding constants by means of second-order perturbation theory in order to test the ability of certain recently proposed[104] improved virtual orbitals (which are unitary transformations of the canonical MOs) to accelerate the convergence of the sum over states, as compared with the canonical MOs themselves. Although this procedure did prove to be conducive to the calculation of ($^{13}C$,H) spin coupling constants, the changes it produced in $^{13}C$ shielding constants turned out not to be significant.[100]

This subsection is concluded by the mention of an *ab initio* calculation of the magnetic susceptibility of ammonia[105] and an *erratum* by Zeroka and Hameka[106] to their 1966 paper[107] on the calculation of magnetic shielding constants of diatomic molecules, in particular to the general theory and its application to the HF molecule. The error is in certain integrals, and the use of the correct integrals leads to the following shielding constants for the HF molecule at an internuclear separation of 1.733 a.u.: $\sigma_H = 29.49$, $\sigma_F = 411.83$ p.p.m., in reasonable agreement with the values of Hindermann and Cornwell[108] for the non-vibrating HF monomer ($\sigma_H = 29.2 \pm 0.5$ p.p.m.; $\sigma_F = 419.7 \pm 0.3$ p.p.m.). Finally, general expressions have been given for all magnetic dipole-moment integrals over STOs in a form suitable or ready computation.[109]

**C. Empirical and Semi-empirical Calculations.**—Aesthetically satisfying as *ab initio* calculations may be, they are, unfortunately, not yet feasible in the case of larger molecules; accordingly, those interested in the calculation of nuclear shielding constants in such molecules must, for the present, content themselves with choosing, and applying, one of the large number of semi-empirical formalisms and parametrizations which are available, from CNDO down to simple Hückel. Since a large amount of space has (intentionally) already been devoted to *ab initio* calculations, the semi-empirical ones discussed in this subsection must necessarily be dealt with rather more cursorily. During the review period, a particularly comprehensive general discussion of the theoretical foundations of purely semi-empirical quantum chemistry has been presented by Freed.[110]

Quirt *et al.*[111] have used a modification of the standard CNDO/2 program to calculate $^{13}C$ 'titration shifts' in amino-acids; the diamagnetic contribution, $\sigma^d$, was evaluated from local charge densities and the paramagnetic contribution, $\sigma^p$, was calculated by use of the average-energy approximation of

103 F. Birnstock and D. Klöpper, *Chem. Phys. Letters*, 1973, **20**, 542.
104 S. Huzinaga and C. Arnau, *Phys. Rev.* (*A*), 1970, **1**, 1285.
105 S. S. Chang and H. F. Hameka, *J. Chem. Phys.*, 1973, **59**, 3297.
106 D. Zeroka and H. F. Hameka, *J. Chem. Phys.*, 1973, **59**, 2171.
107 D. Zeroka and H. F. Hameka, *J. Chem. Phys.*, 1966, **45**, 300.
108 D. K. Hindermann and C. D. Cornwell, *J. Chem. Phys.*, 1968, **48**, 4148.
109 H. Ichimura and A. Rauk, *J. Chem. Phys.*, 1973, **59**, 5720.
110 K. F. Freed, *J. Chem. Phys.*, 1974, **60**, 1765.
111 A. R. Quirt, J. R. Lyerla, I. R. Peat, J. S. Cohen, W. F. Reynolds, and M. H. Freedman, *J. Amer. Chem. Soc.*, 1974, **96**, 570.

Pugmire and Grant,[112] and by what Quirt *et al.*[111] call a 'sum-over-states' method. Sasaki[113] has also used a modified CNDO/2 method in which the excitation energies of the transitions were taken into account, in order to calculate $^{13}C$ and $^{1}H$ shifts of $C_2H_6$, $CH_4$, $CH_3OH$, and $CH_3NH_2$. Application of the mean excitation energy approximation was not considered legitimate in this calculation, and contributions of excited states were not limited to those of the highest-filled and lowest-vacant orbitals. Ando *et al.*[114] have corrected an earlier formula of Yonezawa *et al.*[115] for the calculation of $^{13}C$ shifts by the CNDO/2 method. Ando *et al.*[114] recommend:

$$\sigma^{d} = 4.45[3.25 - 0.35\,(q - 4)]q$$

$$\sigma^{p} = (671.1Q_{2p} - 2068.7)/\Delta E \qquad (20)$$

in which $q$ and $Q_{2p}$ are, respectively, the total electron density and the excess charge of the $2p$-electrons on the carbon atom whose shielding is being calculated, and $\Delta E$ is an average excitation energy. Ando *et al.*[114] have applied this formalism to the methylene carbons of *meso*- and *rac*-2,4-dimethoxypentanes as model compounds of poly(vinyl methyl ether). Although strictly outside the scope of this chapter, a paper by Sadlej and Sadlej[116] should be noted, in which a semi-empirical version of the single-excitation model based on the CNDO/2 approximation is used to calculate oxygen chemical shifts due to solvent–solute interactions in electrolyte solutions. There is also a study[117] by means of CNDO/2 (and modified CNDO/2 and extended Hückel) of the diamagnetic susceptibility of a number of cyclic compounds.

In 1972, Ellis *et al.*[118] reported a finite perturbation SCF–MO theory of $^{13}C$ chemical shifts (which, like Ditchfield's,[44] employed gauge-invariant atomic orbitals) at the INDO level of approximation.[119] Maciel *et al.*[120] have now extended this approach to include the effect of an electronegative atom or substituent. In their approach, the effect of such a substituent was simulated in the following two ways: (i) by finding a set of INDO atomic parameters which would qualitatively simulate the influence exerted by a fluorine substituent on the $^{13}C$ shieldings of the hydrocarbon frameworks of simple fluorocarbons, and (ii) by using their previously described[121,122] 'pseudo-atom' method to simulate electronegative substituents. The INDO atomic parameters were chosen from an experimental comparison with $^{13}C$ shifts in

112 R. J. Pugmire and D. M. Grant, *J. Amer. Chem. Soc.*, 1968, **90**, 687.
113 Y. Sasaki, *Chem. and Pharm. Bull. (Japan)*, 1973, **21**, 888 (*Chem. Abs.*, 1973, **79**, 41 732q).
114 I. Ando, I. Masahiro, and T. Ikuta, *Makromol. Chem.*, 1973, **169**, 285.
115 T. Yonezawa, I. Morishima, and H. Kato, *Bull. Chem. Soc. Japan*, 1966, **39**, 1389.
116 J. Sadlej and A. J. Sadlej, *J. Magn. Resonance*, 1974, **14**, 97.
117 J. E. Bloor and Z. B. Maksić, *Mol. Phys.*, 1973, **26**, 397.
118 P. D. Ellis, G. E. Maciel, and J. W. McIver, *J. Amer. Chem. Soc.*, 1972, **94**, 4069.
119 J. A. Pople, D. L. Beveridge, and P. A. Dobosh, *J. Amer. Chem. Soc.*, 1968, **90**, 4201.
120 G. E. Maciel, J. L. Dallas, and R. L. Elliott, *J. Amer. Chem. Soc.*, 1973, **95**, 5857.
121 G. E. Maciel and K. D. Summerhays, *J. Amer. Chem. Soc.*, 1971, **93**, 520.
122 K. D. Summerhays and G. E. Maciel, *Mol. Phys.*, 1972, **24**, 913.

$CH_2F_2$, $CH_3F$, $CH_3CH_2F$, $CH_2{=}CHF$, $CH{\equiv}CF$, and $C_6H_5F$, and the pseudo-atom method[121,122] was applied[120] to the systems $CH_3X$, $CH_3CH_2X$, $CH_2{=}CHX$, and $CH{\equiv}CX$. INDO calculations of $^{13}C$ chemical shifts, this time *via* the Pople–Karplus theory,[123] have been reported by Galasso;[124] in the Pople–Karplus approach, the total shielding constant of a given carbon atom is expressed in terms of the excess charge density, the bond orders between $2p$ atomic orbitals, and an average excitation energy. Galasso's results suggest that the approximations adopted by the Pople–Karplus theory cause the dependence of $^{13}C$ chemical shift on charge to be underestimated, and that shifts predicted by this theory correlate reasonably well with experimental values only if an empirical charge-dependent term is added to the diamagnetic and paramagnetic contributions. Two papers not available to the Reporter are merely mentioned here, *viz.* an INDO finite perturbation study of $^{13}C$ chemical shifts by Fukui[125] and a semi-empirical calculation of carbon shieldings in 12 cycloalkanes, by Solkan *et al.*[126]

INDO calculations of $^{15}N$ and $^{17}O$ shieldings have been reported by Moniz and Poranski,[127] who used the Pople theory for atoms containing *p*-electrons; they evaluated the diamagnetic shielding from atomic terms, taking account only of contributions from the atom in question and from atoms directly bonded to it. Moniz and Poranski[127] point out that the paramagnetic term in Pople's theory has 'three non-observables embodied in its evaluation'. These are (i) $\langle r^{-3}\rangle_{2p}$, the mean inverse-cube of the distance between the nucleus and its $2p$-electrons; (ii) $\Delta E$, the mean value of the singlet electronic excitation energies; and (iii) $\Sigma Q$, the charge-density bond-order matrix elements appropriately summed over the occupied molecular orbitals. After discussing some of the many assumptions involved in the Pople[128] theory, Moniz and Poranski[127] are of the opinion that '. . . all in all, despite the superposition of approximation upon approximation, the Pople MO theory of chemical shifts has often led to reasonable explanations of experimentally observed shifts, and to a grasp of the principal factors governing the shift of a nucleus in a particular functional group or molecular environment'. Encouraged by this, Moniz and Poranski[127] have devised a semi-empirical method which they regard as '. . . a synthesis of approaches used previously on a piecemeal basis'. The INDO approximation was used to evaluate $\Sigma Q$ and $\langle r^{-3}\rangle_{2p}$, $\Delta E$ was taken as the lowest observed singlet transition energy, and a local diamagnetic term was adopted, as proposed by Mason.[129] They also gave separate consideration to $\Sigma Q_{AA}$, which depends only on the charge densities of the atom in question. Better fits to observed $^{15}N$ and $^{17}O$ shifts were obtained using $\Sigma Q_{AA}$ than $\Sigma Q$.

123 J. A. Pople and M. Karplus, *J. Chem. Phys.*, 1963, **38,** 2803.
124 V. Galasso, *Mol. Phys.*, 1973, **26,** 81.
125 H. Fukui, *Bull. Chem. Soc. Japan*, 1974, **47,** 751.
126 V. N. Solkan, V. Y. Kimenko, and Yu. A. Ustynyuk, *Teor. i eksp. Khim.*, 1973, **9,** 254 (*Chem. Abs.*, 1973, **79,** 25 112g).
127 W. B. Moniz and C. F. Poranski, *J. Magn. Resonance*, 1973, **11,** 62.
128 J. A. Pople, *J. Chem. Phys.*, 1962, **37,** 53, and subsequent papers.
129 J. Mason, *J. Chem. Soc.* (*A*), 1971, 1038.

A nucleus bonded to a heavy atom such as bromine or iodine (*e.g.* proton shieldings in HX, and $^{13}C$ shieldings in $CH_3X$, X = Br or I) resonates at abnormally low frequency. In an interesting and important paper, Morishima *et al.*[130] have examined this phenomenon theoretically in terms of a third-order perturbation theory which includes the large spin–orbit interaction ($\lambda \boldsymbol{L}\cdot\boldsymbol{S}$) characteristic of the heavy atom. They present an expression for this new type of shielding constant, $\sigma_{LS}$, the spin polarization shift due to such a spin–orbit interaction, in its complete form and in a simplified form with average excitation energy approximation. Semi-empirical SCF calculations predict[130] that $\sigma_{LS}$ accounts for 51% of the total proton shielding in HI.

Yet again, reference must be made to the method of Chan and Das[13] for calculating paramagnetic contributions to the chemical shielding parameter. Their procedure, which is based on purely electrostatic considerations and does not require knowledge of excited-state wavefunctions, has been applied by Mohanty[131] to some tetrahedral and octahedral molecules. Agreement with experiment was found to be excellent for proton shieldings, and less good (though not as bad as previous proton results in some less symmetrical molecules) for fluorine shieldings.

Litchman[132] has used Ramsey's analysis (Section 2) in the context of a VB formalism to calculate $^{15}N$ shifts in methylamines and methylamine hydrochlorides. In this study, the paramagnetic contribution was discussed in terms of the average excitation energy and it was concluded that protonation shifts of the methylamines are primarily due to changes in the diamagnetic terms, whereas the chemical shifts within each series are due to changes in the paramagnetic terms. Engelhardt *et al.*[133] have used a simplified model, in which only 3*p*-orbitals of silicon and 2*p*-orbitals of carbon are considered, in order to discuss the relative paramagnetic contributions to $^{29}Si$ shifts of $(CH_3)_{4-n}SiX_n$ compounds and $^{13}C$ shifts of analogous carbon compounds. Their results are interpreted[133] as showing clearly that both $^{29}Si$ and $^{13}C$ shieldings depend primarily on the *sigma* charge of the silicon and carbon atoms, and that $(p\text{–}d)\pi$ interactions on the silicon atom are of minor importance. Tarasevich and Egorov[134] have obtained good agreement between calculated and experimental $^{31}P$ shifts in symmetrical phosphorus compounds and cyclic phosphonitrile chlorides $(Cl_2PN)_n$ ($n$ = 3, 4, or 5). These calculations offer the possibility[134] of solving some problems of Letcher and van Wazer[135] without losing their basic physical ideas. A model for the *a priori* empirical calculation of $^{1}H$ and $^{19}F$ chemical shifts in molecules has been proposed by Stahl-Larivière.[136] The method, which uses a restricted number

130 I. Morishima, K. Endo, and T. Yonezawa, *J. Chem. Phys.*, 1973, **59**, 3356.
131 S. Mohanty, *J. Chem. Phys.*, 1973, **59**, 4415.
132 W. M. Litchman, *J. Magn. Resonance*, 1973, **12**, 182.
133 G. Engelhardt, R. Radeglia, H. Jancke, E. Lippmaa, and M. Mägi, *Org. Magn. Resonance*, 1973, **5**, 561.
134 A. S. Tarasevich and Y. P. Egorov, *Teor. i eksp. Khim.*, 1973, **9**, 73 (*Chem. Abs.*, 1973, **78**, 166 673 g).
135 J. H. Letcher and J. R. van Wazer, *J. Chem. Phys.*, 1966, **45**, 2926.
136 H. Stahl-Larivière, *Org. Magn. Resonance*, 1974, **6**, 170.

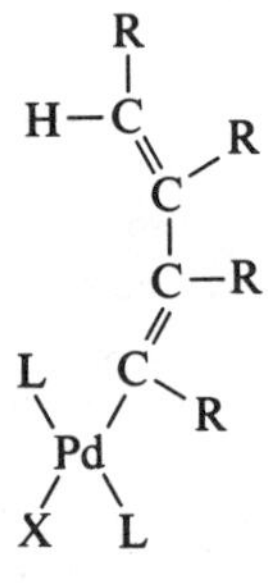

(2)

of parameters obtained from experiment, yields predicted shifts in 49 molecules with a mean accuracy of $\pm 0.32$ p.p.m. for $^1$H and $\pm 11$ p.p.m. for $^{19}$F.

Finally, Buckingham and Urland[137] have evaluated the *d*-electron contribution to the shielding of the apical proton in various planar Pd$^{II}$ complexes of the form $PdL_2(C_4R_4H)X$(2), in which L is a ligand, R = $CO_2CH_3$, and X = Br or $OCOCF_3$. Using a previously detailed formalism, and taking transition energies from the absorption spectrum of $[PdCl_4]^{2-}$, they obtained the results given in Table 10 for various orbital exponents, $\rho$. A 'near neighbour' dipole approximation yields $\sigma = -18.66$ p.p.m. In order to obtain a significant increase over this value, it is necessary to choose $\rho < 2$; Slater's rules, however, require that $\rho = 2.04$.

**Table 10** *Values of the* Pd 4*d-electron magnetic shielding* (*in* p.p.m.) *of a proton at* 0.23 nm *above the* Pd *nucleus and the plane of the complex of* $C_{4v}$ *symmetry for various values of the orbital exponent*, $\rho$ (*from ref.* 137)

| | | | | | | | | | | |
|---|---|---|---|---|---|---|---|---|---|---|
| $\rho$ | 0.5 | 0.9 | 0.93 | 0.94 | 1.0 | 1.2 | 1.6 | 2.0 | 3.0 | 4.0 |
| $\sigma$ | 11.8 | −0.7 | −1.9 | −2.3 | −4.5 | −10.8 | −16.9 | −18.3 | −18.6 | −18.6 |

## 4 Transmission of Effects within Molecules

**A. Introduction.**—In this section we leave the rigours of *ab initio* calculations on small molecules and turn our attention to what is, perhaps, the more 'real' world of the chemistry of larger molecules; it is, however, from the nuclear shielding point of view, necessarily a less fundamental and more qualitative world, mainly because present-day ignorance of excited-state wavefunctions compels us to make, at least mentally, a somewhat subjective and—perhaps one could not unfairly say—rather *ad hoc* partition of the total shielding effect into a series of contributions which can be explained (or, perhaps more

137 A. D. Buckingham and W. Urland, *Mol. Phys.*, 1973, **26**, 1571.

correctly, rationalized) in terms of well-established 'physical' and 'chemical' concepts.[1-3,138] The first such division is into a 'local' part, $\sigma_{loc}$, on the nucleus in question, and a part due to the remainder of the molecule. This local part is then further broken down into a diamagnetic part, $\sigma^{d}_{loc}$, and a paramagnetic part, $\sigma^{p}_{loc}$, approximate expressions for which are[139]

$$\sigma^{d}_{loc} = \frac{\mu_0}{4\pi} \times \frac{e^2}{3m} \langle 0|\sum_i r_i^{-1}|0\rangle \qquad (21)$$

$$\sigma^{p}_{loc} = -\frac{\mu_0}{4\pi} \frac{\hbar^2 e^2 \langle r^{-3}\rangle_{np}}{2m^2 \Delta E} \sum_B Q_{AB} \qquad (22)$$

Here, $r_i$ is the distance of the $i^{th}$ electron involved in the local electron distribution at the nucleus, $\Delta E$ is, again, a mean excitation energy, $\sum_B Q_{AB}$ represents the amount of 'imbalance' in the populations of the orbitals about the given nucleus (this is zero for spherical symmetry), and $\langle r^{-3}\rangle_{np}$ is the mean inverse-cube of the distance of the valence shell ($n$) $p$-electrons from the nucleus (the original Karplus–Pople theory[139] concerns only $p$-electrons, not $d$).

Although $\sigma^{d}_{loc}$ is usually of greater magnitude, numerically, than $\sigma^{p}_{loc}$, the latter is more sensitive to changing electronic influences which occur in a series of similar compounds.[2] As Raynes has pointed out, however,[140] this division into 'local' and 'non-local' terms is rather unsatisfactory since the local contributions have their origin, at least to some degree, in the remainder of the molecule, whereas some of the non-local contributions alter the shielding indirectly by changing the local contributions. It is, therefore, more convenient to classify the *mechanisms* by which contributions are made to $\sigma^{d}_{loc}$ and $\sigma^{p}_{loc}$ into 'through-bond' or 'chemical' effects ('inductive effects', 'resonance effects', and intramolecular hydrogen-bond effects) and 'through-space' or 'physical' effects (electric field effects, van der Waals interactions, and steric effects). Contributions to shieldings from distant electrons are also, of course, necessarily 'through-space' effects, and arise from the magnetic anisotropy of particular groups, 'ring current' effects, and pseudo-contact (or 'dipolar') shifts (the latter of which are not considered here but are reviewed in Chapter 10). All these considerations have been gone into in great depth by Raynes, both in earlier Reports,[1-3] and in an extensive paper on the subject,[140] and so no further details are given here. As Raynes has also pointed out (and as will certainly be evident from this section) it is often very difficult to distinguish these various effects from each other—and, indeed, attempts to unravel the numerous combinations of contributions considered in many papers in subsection B, in particular, will not be made here. The major 'through-bond' effects will be discussed first, followed by 'through-

[138] A. Saika and C. P. Slichter, *J. Chem. Phys.*, 1954, **22**, 26.
[139] M. Karplus and J. A. Pople, *J. Chem. Phys.*, 1963, **38**, 2803.
[140] W. T. Raynes, *Mol. Phys.*, 1971, **20**, 321.

space' effects. Isotope effects are considered by Raynes[1-3] to be largely a 'through-bond' effect, and they are, accordingly, dealt with at the end of this section.

**B. 'Resonance Effects', 'Inductive Effects', and 'Substituent Effects' in General.**—The number of papers which have appeared in the review period on 'inductive effects', 'resonance effects', and 'substituent effects' in general (including correlations between chemical shifts and charge densities) has been quite overwhelming, and the present Reporter will, somewhat inadequately, do little more than cite the various papers here, with intermittent comment. In addition to being a vehicle for the discussion of 'resonance' and 'inductive' effects specifically, this subsection will be used as a repository for papers which claim to deal with 'substituent effects' in general.

Beistel *et al.*[141] have studied proton shift behaviour in nine families of disubstituted benzenes, and have provided quantitative correlations of internal and *meta* shifts among all families. Their studies demonstrate shift additivity for the 4-substituted halogenobenzenes and suggest a chain-rule relationship for substituent effects in those compounds. For this reason, Beistel *et al.*[141] reach the important conclusion that previous theories of substituent interactions appear to be incorrect when applied to proton shift data. As Reynolds *et al.*[142] point out, the nature of $\pi$-inductive effects is controversial, but they regard the question as being largely a semantic one, since, they claim,[142] the term '$\pi$-inductive effect' carries different connotations for different groups of investigators; their CNDO/2 calculations and $^{13}C$ chemical shifts for $\alpha$- and $\beta$-substituted phenylalkanes are said to provide evidence for a '$\pi$-polarization effect', and they conclude that this is a significant mechanism for transmission of electronic substituent effects in phenyl derivatives and may be the dominant effect in systems in which the substituent is insulated from direct conjugative interaction with the phenyl group. Deady and Willett[143] have studied the chemical shifts of the *N*-methyl group in a series of 2-, 3-, and 4-substituted *N*-methylpyridinium salts, and have concluded that the shift is affected primarily by resonance interaction between the substituent and the aza-group; their results also suggest that the resonance effect from a 2-substituent is at least as great as that from the corresponding 4-substituent; Beeby *et al.*[144] have noted a regular relationship between the influence of substituents on the chemical shifts of the *ortho* protons in monosubstituted benzenes and the influence of the same substituents on the protons in the *cis* relationship in monosubstituted ethylenes.

Inductive and resonance effects on $^{13}C$ shieldings have been reported by

[141] D. W. Beistel, H. E. Chen, and P. J. Fryatt, *J. Amer. Chem. Soc.*, 1973, **95**, 5455.
[142] W. F. Reynolds, I. R. Peat, M. H. Freedman, and J. R. Lyerla, *Canad. J. Chem.*, 1973, **51**, 1857.
[143] L. W. Deady and G. D. Willett, *Org. Magn. Resonance*, 1974, **6**, 53.
[144] J. Beeby, L. Drake, R. Duffiin, S. Sternhell, E. Pretsch, and W. Simon, *Org. Magn. Resonance*, 1973, **5**, 163.

several workers.[145-147] Clark and Ward[145] find that only minor changes in $\sigma$-inductive, $\pi$-inductive, and $\sigma$–$\pi$-conjugative effects occur between the platinum atom and the phenyl ring in certain platinum organometallic compounds; Olah and Westerman[146] have investigated the relative order of charge delocalization by phenyl, cyclopropyl, and methyl groups in carbonium ions, by means of $^{13}C$ chemical shifts; and Stephany *et al.*[147] have discussed $^{13}C$ shifts in some aliphatic, unsaturated, and aromatic isocyanides, and their complexes, in terms of the inductive and mesomeric substituent effects on the polarization and charge density of the C—N≡C bonds.

As for $^{19}F$ shifts of substituted fluorobenzenes, it was recognized over 20 years ago[148] that such shifts correlate well with the classical Hammett substituent constants.[149] This correlation was greatly extended by Taft *et al.*[150]

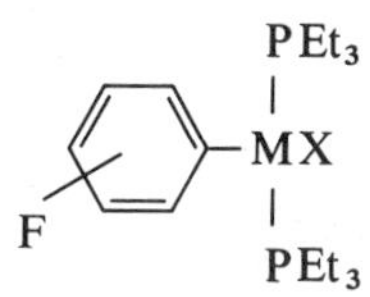


(3)

and, in the review period, has been used by Parshall[149] for the study of aryl complexes of nickel, palladium, and platinum, such as (3), where X represents various ligands. The inductive ($\sigma_I$) and resonance ($\sigma_R$) substituent parameters are derived from the equation of Taft *et al.*:[150]

$$\delta_m = -7.10\sigma_I + 0.60 \quad (23)$$

$$\delta_p - \delta_m = -29.5\sigma_R \quad (24)$$

It is found[149] that the chemical shifts and Taft substituent constants, $\sigma_R$, for *p*-fluorophenyl compounds of palladium and platinum are quite similar, but that those for nickel complexes are generally larger, indicating greater transfer of electron density to the carbon. Parshall's results[149] confirm that halides give higher $\sigma_R$ values than do other ligands, and that an 'n.m.r. *trans*-influence series' can be defined: I, Br, Cl $<$ CN $<$ Ph, Me. 'At the risk of provoking further controversy', Parshall is 'tempted' to speculate [149] that nickel is more effective in $\pi$-interactions with ligands than are palladium and platinum.

In an important paper, Dawson *et al.*[151] have reconsidered the validity of correlating solvent-induced proton shielding constants in aromatic derivatives with reactivity parameters; they conclude that comparison of $^1H$ shifts and

145 H. C. Clark and J. E. H. Ward, *J. Amer. Chem. Soc.*, 1974, **96**, 1741.
146 G. A. Olah and R. W. Westerman, *J. Amer. Chem. Soc.*, 1973, **95**, 7530.
147 R. W. Stephany, M. J. A. de Bie, and W. Drenth, *Org. Magn. Resonance*, 1974, **6**, 45.
148 H. S. Gutowsky, P. W. McCall, B. R. McGarvey, and L. H. Meyer, *J. Amer. Chem. Soc.*, 1952, **74**, 4809.
149 G. W. Parshall, *J. Amer. Chem. Soc.*, 1974, **96**, 2360.
150 S. Ehrenson, R. T. C. Brownlee, and R. W. Taft, *Progr. Phys. Org. Chem.*, 1973, **10**, 1.
151 D. A. Dawson, G. K. Hamer, and W. F. Reynolds, *Canad. J. Chem.*, 1974, **52**, 39.

CNDO/2 charge densities in 4-substituted derivatives of styrene, toluene, benzyl chloride, and *NNN*-trimethylphenylammonium indicate that such shifts primarily reflect intermolecular electronic effects. This was deduced from a correlation with the '*F*' and '*R*' ('field' and 'resonance') parameters of Swain and Lupton.[152] Dawson *et al.*[151] conclude that it *is* valid to correlate substituent-induced chemical shifts for aromatic derivatives with these 'divided' reactivity parameters, provided that (i) shifts are measured at infinite dilution in a non-polar solvent, and (ii) at least ten derivatives of each family are investigated, and they dismiss the suggestions of Wiley and Miller[153] that such correlations are invalid.

'Hammett', 'Taft', and 'Swain–Lupton' types of correlation have, in fact, been very popular during the review period amongst a number of workers who have applied them, with varying degrees of success, to several nuclei. There is insufficient space to go into the details here, but these procedures have been nicely summarized by Phillips *et al.*,[154,155] who have investigated the transmission of substituent effects across two aromatic rings connected by an N—N link. Briggs and Randall[156] have sought correlations in a series of 17 monosubstituted derivatives of hexafluorobenzenes with (i) $^{13}$C shifts of monosubstituted benzenes and *para*-substituted fluorobenzenes; (ii) $^{19}$F chemical shifts of monosubstituted hexafluorobenzenes; and (iii) Taft's $\sigma_I$ and $\sigma_R$ parameters. They conclude that the dependence on $\sigma_I$ suggests that calculations based on $\pi$ density alone should not be expected to rationalize the $^{13}$C shifts of $C_6F_5X$ compounds. $^{19}$F Shifts of $[TiF_{6-x}X_x]^{2-}$ ($X = NCS^-$ or $NCO^-$) and some related mixed complexes have been found by Dean and Ferguson[157] to be reasonably represented by an equation of the form:

$$\delta = pc + qt \qquad (25)$$

where *c* and *t* are characteristic of X, and *p* and *q* are the number of ligands, X, *cis* and *trans*, respectively, to the resonating fluorine. They plotted the *t* constants against Hammett–Taft parameters $\sigma_R^0$, which are thought[158] to be a measure of the $\pi$-donor abilities of the various substituents. Other Hammett $\sigma$ investigations have included a study of N-substituted azoles,[159] exocyclic protons in *para*-substituted benzene derivatives,[160] and oxirans.[161] Caplin[162] has reported a very thorough investigation of substituent effects in the $^1$H n.m.r. of monosubstituted acetophenones, by means of the Swain–Lupton[152]

152 G. C. Swain and E. C. Lupton, *J. Amer. Chem. Soc.*, 1968, **90**, 4328.
153 G. R. Wiley and S. I. Miller, *J. Org. Chem.*, 1972, **37**, 767.
154 P. J. Mitchell and L. Phillips, *J.C.S. Perkin II*, 1974, 109.
155 M. G. Belsham, A. R. Muir, M. Kinns, L. Phillips, and L.-M. Twanmoh, *J.C.S. Perkin II*, 1974, 119.
156 J. M. Briggs and E. W. Randall, *J.C.S. Perkin II*, 1973, 1789.
157 P. A. W. Dean and B. J. Ferguson, *Canad. J. Chem.*, 1974, **52**, 667.
158 R. W. Taft, E. Price, F. R. Fox, I. C. Lewis, K. K. Anderson, and G. T. Davies, *J. Amer. Chem. Soc.*, 1973, **85**, 3146.
159 P. Bouchet, C. Coquelet, and J. Elguero, *J.C.S. Perkin II*, 1974, 449.
160 B. M. Lynch, *Org. Magn. Resonance*, 1974, **6**, 190.
161 L. Villa, L. Schenetti, and F. Taddei, *Org. Magn. Resonance*, 1973, **5**, 593.
162 G. A. Caplin, *Org. Magn. Resonance*, 1974, **6**, 99.

multiple-correlation analysis, and has also examined substituent effects in 1,2,3-selenadiazoles.[163,164] Okuyama *et al.*[165] have investigated $^{13}C$ and $^{1}H$ in 1-phenyl-1,3-butadiene and phenylallene, *via* Hammett-type relations, while, in studying the transmission of substituent effects through oxygen and sulphur atoms, Kagimoto *et al.*[166] have found that Hammett plots of the $^{13}C$ shifts of the terminal carbon in ring-substituted phenyl vinyl ethers and sulphides confirm the greater transmission efficiency of the sulphur atom. Although actually published before the review period, two reviews on the Hammett equation and linear correlations of substituent effects in $^{1}H$, $^{19}F$, and $^{13}C$ shielding[167,168] have recently come to the Reporter's attention.

Papers on $^{1}H$ substituent effects for the following cases may be noted: 74 alkyl-substituted 1,3-dioxans by Taverner and Anteunis,[169] 51 methyl- and t-butyl-substituted cyclohexanes by Werner *et al.*,[170] 1,3-alkoxybutadienes,[171] acetals,[172] amino-group proton shifts in the amino-*sym*-triazine series[173] and phenylhydrazines,[174] esters,[175,176] 1,3-dioxans,[177] methylborazines,[178] and nitrophenylhydrazone $\alpha$-carbonyls.[179] Substituent effects on the $\alpha$, $\beta$, $\gamma$, and $\delta$ atoms of 2-substituted tetrahydropyrans have been discussed by de Hoog[180] and inductive and mesomeric contributions to the $^{1}H$ shifts of thiophens have been discussed by Bugge.[181] Maciel and Rausch[182] have reported various proton effects on $\pi$-cyclopentadienyl and $\pi$-indenyl complexes of titanium, zirconium, and hafnium containing $\sigma$-bonded organic substituents, Schönberg and Anderson[183] have examined electronic (and steric) effects in the ligands of

[163] G. A. Caplin, *J.C.S. Perkin I*, 1974, 30.
[164] G. A. Caplin, *Org. Magn. Resonance*, 1973, **5**, 169.
[165] T. Okuyama, K. Izawa, and T. Fueno, *Bull. Chem. Soc. Japan*, 1974, **47**, 410.
[166] O. Kajimoto, M. Kobayashi, and T. Fueno, *Bull. Chem. Soc. Japan*, 1973, **46**, 1422.
[167] M. T. Tribble and J. G. Traynham, *Adv. Linear Free Energy Relat.*, 1972, 143–201. (*Chem. Abs.*, 1973, **78**, 135 092n).
[168] O. Exner, *Adv. Linear Free Energy Relat.*, 1972, 1–69 (*Chem. Abs.*, 1973, **78**, 135 091m).
[169] D. Tavernier and M. Anteunis, *J. Magn. Resonance*, 1974, **13**, 181.
[170] H. Werner, H. J. Köhler, M. Mühlstädt, A. Zschunke, and G. Mann, *Org. Magn. Resonance*, 1973, **5**, 119.
[171] J. P. Dorie, M. L. Martin, S. Odiot, and F. Tonnard, *Org. Magn. Resonance*, 1973, **5**, 265.
[172] F. Borremans, M. Anteunis, and F. Anteunis-De Ketelaere, *Org. Magn. Resonance*, 1973, **5**, 299.
[173] O. P. Shkurko, L. R. Roitman, and V. P. Mamaev, *Zhur. org. Khim.*, 1973, **9**, 1012 (*Chem. Abs.*, 1973, **79**, 41 735t).
[174] E. V. Titov, L. M. Kapkan, V. I. Rybachenko and A. Y. Chervinskii, *Ukrain khim. Zhur.*, 1973, **39**, 927 (*Chem. Abs.*, 1973, **79**, 145 830m).
[175] J. Dabrowski, A. Ejchart, and K. Kamienska-Trela, *Org. Magn. Resonance*, 1973, **5**, 483.
[176] K. D. Bartle, G. Hallas, and J. D. Hepworth, *Org. Magn. Resonance*, 1973, **5**, 479.
[177] P. Maroni, L. Cazaux, J.-P. Gorrichon, P. Tisnes, and J.-G. Wolf, *Org. Magn. Resonance*, 1973, **5**, 517.
[178] G. Baumelou, M. Pasdeloup, and J.-P. Laurent, *Org. Magn. Resonance*, 1973, **5**, 585.
[179] F. Venien and M. Lefeuvre, *Org. Magn. Resonance*, 1973, **5**, 113.
[180] A. J. de Hoog, *Org. Magn. Resonance*, 1974, **6**, 233.
[181] A. Bugge, *Chemica. Scripta* 1973, **3**, 190 (*Chem. Abs.*, 1973, **78**, 158 748v).
[182] E. Samuel and M. D. Rausch, *J. Amer. Chem. Soc.*, 1973, **95**, 6263.
[183] A. R. Schönberg and W. P. Anderson, *Inorg. Chem.*, 1974, **13**, 465.

certain organometallic compounds, and Solo *et al.*[184] have discussed the effects of D-ring substituents on the chemical shifts of the angular protons in certain steroids. There has been an *erratum*, by Hamer *et al.*,[185] to their earlier paper[186] on substituent effects in substituted styrenes.

In an important paper, Emsley *et al.*,[187] after giving a very useful summary of simple theories of the origin of proton shielding (magnetic anisotropy, electric field effects, *etc.*) have continued their examination of the proton shifts of a number of 1-substituted naphthalenes in the light of some of these simple theories of substituent chemical shifts which have previously been claimed to be successful for benzene derivatives. In fact, none of these methods gives the correct results for all positions in naphthalene, and they conclude[187] that attempts to use proton shielding as an indication of molecular structure must be treated with caution, and that reliable information on such structure can be obtained only when proton shielding can be ascribed to changes in the electron density on the proton (which they calculate by a CNDO/2 method.)

As for substituent effects on $^{13}C$ chemical shifts, Yonemoto[188] has given a thorough analysis of these in nitriles. He finds that the cyano-carbon resonance in substituted acetonitriles shifts to lower frequency with increasing electronegativity of the substituent and explains this in terms of the $\pi$-bond polarization which is caused by the electronic field of the substituent. The $\alpha$-carbon shifts do not show correlation with cyano-carbon shifts, nor with the electronegativity of the substituents. Gray[189] has reported a very convincing correlation between $\delta(^{13}CH_3X)$ and $\delta(Ph_3\overset{+}{P}{}^{13}CH_2X)$ (4). Variation of $^{13}C$ shifts (including those in the phenyl rings) with the substituent X in (4) is illustrated

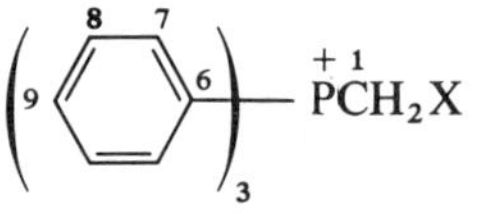


(4)

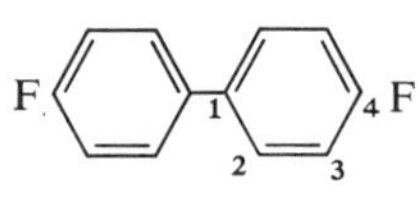


(5)

in Table 11. Imanari *et al.*[190] have studied $^{13}C$ shieldings in halogenobiphenyls, and for *para*-derivatives [such as (5)] they find that the observed chemical shifts of the C-1, C-2, and C-3 positions give an excellent linear dependence on the electronegativity of the substituent; C-4, however, shows a deviation, which may be due to 'heavy atom' effects. Other substituent effects on $^{13}C$ shifts which have been discussed are the $\alpha$, $\beta$, $\gamma$, and $\delta$ effects in

[184] A. J. Solo, S. Eng, and B. Singh, *J. Pharm. Sci.*, 1973, **62**, 1471 (*Chem. Abs.*, 1973, **79**, 126 698h).
[185] G. K. Hamer, I. R. Peat, and W. F. Reynolds, *Canad. J. Chem.*, 1973, **51**, 2596.
[186] G. K. Hamer, I. R. Peat, and W. F. Reynolds, *Canad. J. Chem.*, 1973, **51**, 915.
[187] J. W. Emsley, J. C. Linden, S. R. Salman, and D. T. Clark, *J.C.S. Perkin II*, 1973, 611.
[188] T. Yonemoto, *J. Magn. Resonance*, 1974, **13**, 153.
[189] G. A. Gray, *J. Amer. Chem. Soc.*, 1973, **95**, 7736.
[190] M. Imanari, M. Kohno, M. Ohuchi, and K. Ishizu, *Bull. Chem. Soc. Japan*, 1974, **47**, 708.

**Table 11** $^{13}$C *Chemical shifts* (p.p.m.) *in* (4) (*from ref.* 189)

| X | C-6 | C-7 | C-8 | C-9 | C-1 |
|---|---|---|---|---|---|
| H | 119.30 | 133.51 | 130.66 | 135.36 | 10.56 |
| Me | 119.00 | 134.42 | 131.52 | 136.27 | 16.39 |
| Et | 118.27 | 133.76 | 130.73 | 135.23 | 24.20 |
| Pr$^n$ | 118.34 | 133.79 | 130.76 | 135.30 | 22.50 |
| CH=CH$_2$ | 117.95 | 134.09 | 130.57 | 135.33 | 28.53 |
| Ph | 118.57 | 134.99 | 131.35 | 136.50 | 30.32 |
| CH$_2$OH | 119.02 | 133.14 | 130.55 | 135.13 | 27.10 |
| COMe | 119.05 | 134.15 | 130.30 | 134.90 | 40.14 |
| CO$_2$Et | 117.96 | 134.11 | 130.65 | 135.63 | 32.89 |
| OMe | 117.02 | 134.58 | 130.98 | 135.93 | 66.19 |
| Cl | 116.48 | 134.62 | 130.89 | 136.01 | 33.79 |
| COPh | 119.04 | 134.27 | 130.40 | 135.06 | 38.66 |

steroids,[190,191] alkyl substituent effects,[192,193] and the transmission of the electronic effects of substituents through C—C triple and double bonds.[194] Simeral and Maciel[195] have reported substituent parameters for the $^{13}$C shifts of 1,2-disubstituted ethanes, while Kamazaki *et al.*[196] have studied the additivity of substituent effects on $^{13}$C shifts in substituted benzenes; $^{13}$C substituent effects have also been considered by Günther *et al.*[197] and by Gray.[198] Voelter and Breitmaier[199] have studied $^{13}$C shifts of aldopyranose carbons in a series of pyranosyl-adenines and have found that anomeric carbons with an axially oriented heterocycle always resonate at lower frequency than the corresponding ones with an equatorial position. Finally, Lippmaa *et al.*[200] have interpreted the $^{13}$C spectra of a series of substituted nortricyclenes (6) in terms of $\alpha$, $\beta$, $\gamma$, and $\delta$ effects.

With regard to substituent effects on $^{19}$F chemical shifts, the comprehensive papers of Phillips *et al.*[154,155] have already been mentioned; Adcock

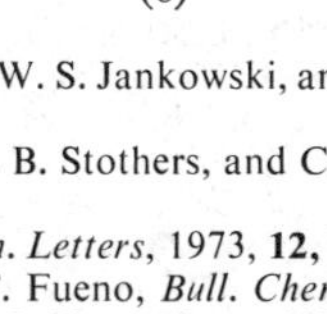


(6)

[191] N. S. Bhacca, D. D. Giannini, W. S. Jankowski, and M. E. Wolff, *J. Amer. Chem. Soc.*, 1973, **95**, 8421.
[192] S. H. Grover, J. P. Guthrie, J. B. Stothers, and C. T. Tan, *J. Magn. Resonance*, 1973, **10**, 227.
[193] T. Ibusuki and Y. Saito, *Chem. Letters*, 1973, **12**, 1255 (*Chem. Abs.*, 1974, **80**, 81 559r.)
[194] K. Izana, T. Okuyama, and T. Fueno, *Bull. Chem. Soc. Japan*, 1973, **46**, 2881.
[195] L. Simeral and G. E. Maciel, *J. Phys. Chem.*, 1973, **77**, 1590.
[196] M. Yamazaki, T. Usami, and T. Takeuchi, *Nippon Kagaku Kaishi*, 1973, **11**, 2135 (*Chem. Abs.*, 1974, **80**, 47 030b).
[197] H. Günther, W. Peters, and R. Wehner, *Chem. Ber.*, 1973, **106**, 3683.
[198] G. A. Gray, *J. Amer. Chem. Soc.*, 1973, **95**, 5092.
[199] W. Voelter and E. Breitmaier, *Org. Magn. Resonance*, 1973, **5**, 311.
[200] E. Lippmaa, T. Pehk, and J. Paasivirta, *Org. Magn. Resonance*, 1973, **5**, 277.

*et al.*[201] have studied a number of substituted aryl fluorides, claiming that the results provide 'conclusive experimental proof' that polar effects play a major role in determining $^{19}F$ shifts in these molecules, as well as evidence for 'hyperconjugative' electron withdrawal by substituent methyl groups in the ground state of neutral molecules. Yolles and Woodland[202] have reported substituent $^{19}F$ shifts in fluorodiphenylsilanes and their carbon analogues (7; Y = Si or C, X = H or a donor atom or group). Appealing, again, to the

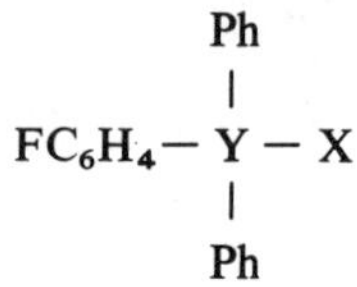

(7)

Taft procedure, they find small $\sigma_I$ parameters from the *meta* shifts for both the silicon and the carbon series. A $^{19}F$ investigation of the transmission of electronic effects through tin–aromatic carbon bonds has been reported,[203] as also have substituent effects of (transition-metallo) methyl groups.[204] Because it is strictly outside the present scope, an important paper by Dayd and Taft,[205] concerning the effects of aprotic solvents on $^{19}F$ substituent shifts, is merely mentioned here.

Correlation of chemical shifts of individual nuclei with the variously calculated charge densities on particular atoms has been a very popular activity during the review period, and some of the numerous papers on this subject are conveniently dealt with here. Sterk and Holzer,[206] in studying a large number of aryl protons, found the relationship

$$\delta_H = 51.5 - 22.7q_H - 5.3q_C \quad (26)$$

(where $q_C$ is the charge density at the corresponding carbon atom) when the various charges were calculated by a CNDO method; they found, in fact, that the smallest standard error for the prediction of $\delta_H$ values is obtained when the charges are calculated by CNDO or INDO methods. Correlations with CNDO/2 charges have also been documented by Tonnard *et al.*,[207] by Lazzeretti *et al.*[208] (in aryl epoxides), and by Fournier *et al.*[209,210] Sharpless

201 W. Adcock, M. J. S. Dewar, and B. D. Gupta, *J. Amer. Chem. Soc.*, 1973, **95**, 7353.
202 S. Yolles and J. H. R. Woodland, *J. Organometallic Chem.*, 1973, **54**, 95.
203 D. N. Kravtsov, B. A. Kvasov, T. S. Khazanova, and E. I. Fedin, *J. Organometallic Chem.*, 1973, **61**, 219.
204 C. W. Fong and M. D. Johnson, *J.C.S. Perkin II*, 1973, 986.
205 S. K. Dayal and R. W. Taft, *J. Amer. Chem. Soc.*, 1973, **95**, 5595.
206 H. Sterk and H. Holzer, *Org. Magn. Resonance*, 1974, **6**, 133.
207 F. Tonnard, S. Odiot, J. P. Dorie, and M. L. Martin, *Org. Magn. Resonance*, 1973, **5**, 271.
208 P. Lazzeretti, I. Moretti, F. Taddei, and G. Torre, *Org. Magn. Resonance*, 1973, **5**, 385.
209 J. Fournier, E. J. Vincent, A. M. Chauvet, and A. Crevat, *Org. Magn. Resonance*, 1973, **5**, 573.
210 J. Fournier, *J. Mol. Structure*, 1973, **18**, 391.

*et al.*[211] have examined the $^1H$ n.m.r. of a series of heterocyclic analogues of anthracene and have concluded (somewhat surprisingly, perhaps) that although, for planar molecules, the shift of the proton *ortho* to the heteroatom is linear with respect to the electron density on that heteroatom, $^1H$ chemical shifts are otherwise independent of electron density. As Sharpless *et al.*[211] point out, however, the correlation of chemical shifts and electron densities is often a frustrating problem; it seems that a high correlation between these variables often results from a fortuitous balance between other factors, such as the diamagnetic anisotropy of the heteroatom or the presence of neighbouring currents or of lone-pair electrons. Lazzeretti *et al.*,[208,212,213] in fact, subtract a calculated estimate of 'ring current' effects before attempting CNDO charge-density *versus* proton chemical shift correlations in certain heterocyclic molecules.

Charge densities have also been correlated with $^{13}C$ shieldings. Boicelli *et al.*[214] have examined the four symmetrical naphthyridines (8)—(11) and

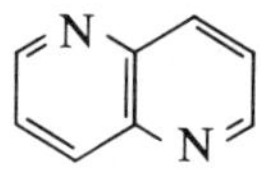


(8)

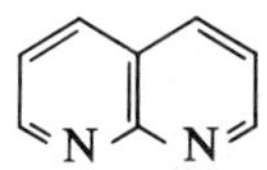


(9)

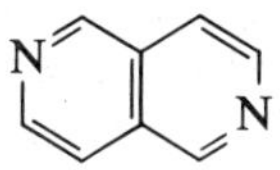


(10)

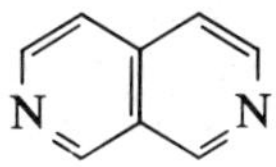


(11)

have found that the $^{13}C$ shifts of their constituent carbons depend linearly on the total charge density at these atoms, as calculated by the INDO and CNDO approximations. The work of Reynolds *et al.*[142] on $\pi$-inductive effects in phenylalkanes with polar or charged substituents has already been discussed; here the use of CNDO/2 calculations of electron density in that investigation should be noted. Van Dongen *et al.*[215] have observed a close relation between the variations of $\delta(C_\alpha)$, $\delta(C_\beta)$, and $\delta(C_\gamma)$ in acetylenes and cumulenes (12) and the total charge densities, calculated by the CNDO/2 method, on the corresponding atoms of the unsubstituted systems with

$$R_2C(=C)_n=CR_2$$

(12) $R = H, n = 0$—$2$
$R = Ph, n = 0$—$4$

211 N. E. Sharpless, R. B. Bradley, and J. A. Ferretti, *Org. Magn. Resonance*, 1974, **6**, 115.
212 L. Marchetti, L. Pentimalli, P. Lazzeretti, L. Schenetti, and F. Taddei, *J.C.S. Perkin II*, 1973, 1926.
213 P. Lazzeretti, G. Monari, and F. Taddei, *Mol. Phys.*, 1973, **26**, 311.
214 A. C. Boicelli, R. Danieli, A. Mangini, L. Lunazzi, and G. Placucci, *J.C.S. Perkin II*, 1973, 1024.
215 J. P. C. M. Van Dongen, and M. J. A. De Bie, *Tetrahedron Letters*, 1973, **16**, 1371.

$n = 0$—5. A low-frequency shift of the $C_\alpha$ carbon, on going to the next-highest homologue, is always accompanied by a high-frequency shift of the $C_\beta$ carbon, and a low-frequency shift of the $C_\gamma$ carbon (if present), and *vice versa*, so that $\delta(C_\beta) - \delta(C_\alpha)$ is much larger for the cumulenes with $n$ odd than with $n$ even. Other discussions involving electron densities are those of Vögeli and von Phillipsborn,[216] Dabrowski *et al.*,[217] and Hogeveen and Kwant.[218]

This subsection is concluded by citing papers on the effect of spiroconjugation on $^1H$ shifts,[219] electronegativities in pentaborane derivatives from ethyl-group $^{13}C$ shifts,[220] a comparison of $\sigma$–$\pi$ delocalization in phenylnorbornyl and related cations,[221] and a re-examination of the controversial spectrum of cyclopentadienyltitanium.[222] For want of a more appropriate context, two papers correlating coupling constants and chemical shifts are also mentioned here.[223,224]

**C. Magnetic Anisotropy Effects.**—Fields due to the anisotropy of magnetic susceptibility of distant groups or bonds affect the shielding of a given nucleus directly and may be considered to act independently of, and in addition to, any effects on shielding caused by 'resonance' and 'inductive' mechanisms *etc.*[1] The fields which such effects provide, 'from afar', so-to-speak, are, of course, independent of the nucleus which is being shielded by them and they are, therefore, most evident in proton shielding, because of the relatively small range of shielding which this nucleus may undergo. The well-known classical equation of this effect, due to McConnell,[225] is

$$\sigma_m = \frac{\mu_0}{4\pi}\frac{\Delta\chi}{3R^3}(1 - 3\cos^2\theta) \qquad (27)$$

in which $\sigma_m$ is the shielding due to a bond with anisotropy $\Delta\chi$ at a point related to the site of the assumed point dipole in that bond *via* the quantities $R$ and $\theta$ (see Figure 3). This equation, the refinements to it due to ApSimon *et al.*,[226] the difficulties associated with the legitimacy of the concept of bond anisotropies,[140] and the practical difficulties concerning the point which should be taken as the 'electronic centre' of a given bond[140,227] have all been well documented in the references given, and are not further discussed here.

216 U. Vögeli and W. von Philipsborn, *Org. Magn. Resonance*, 1973, **5**, 551.
217 J. Dabrowski, K. Kamienska-Trela, and L. Kozerski, *Org. Magn. Resonance*, 1974, **6**, 43.
218 H. Hogeveen and P. W. Kwant, *J. Amer. Chem. Soc.*, 1973, **95**, 7315.
219 M. F. Semmelhack, J. S. Foos, and S. Katz, *J. Amer. Chem. Soc.*, 1973, **95**, 7325.
220 T. Onak and E. Wan, *J. Magn. Resonance*, 1974, **14**, 66.
221 G. A. Olah and G. Liang, *J. Amer. Chem. Soc.*, 1974, **96**, 195.
222 A. Glivicky and J. D. McCowan, *Canad. J. Chem.*, 1973, **51**, 2609.
223 K. P. Shrestha and K. L. Henold, *J. Amer. Chem. Soc.*, 1973, **95**, 6699.
224 T. Pehk, M. Alla, and E. Lippmaa, *Org. Magn. Resonance*, 1973, **5**, 351.
225 H. M. McConnell, *J. Chem. Phys.*, 1957, **27**, 226.
226 J. W. ApSimon, W. G. Craig, P. V. Demarco, D. W. Mathieson, L. Saunders, and W. B. Whalley, *Chem. Comm.*, 1966, 359.
227 C. W. Haigh, R. B. Mallion, and E. A. G. Armour, *Mol. Phys.*, 1970, **18**, 751.

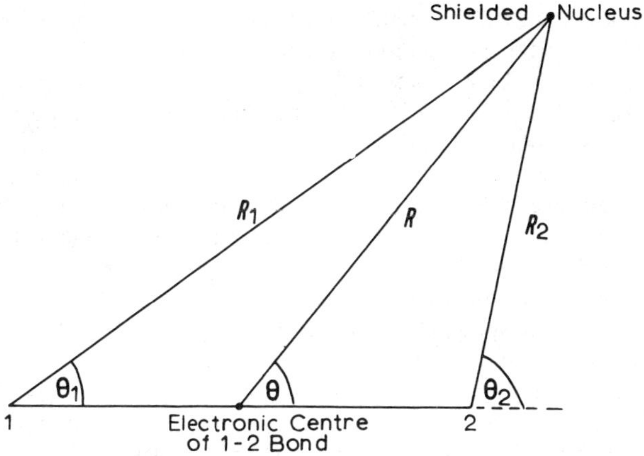


**Figure 3**

One of the most general papers dealing with this effect to appear during the review period is by Schmalz *et al.*,[228] on localized magnetic susceptibility anisotropies. The paper is concerned primarily with the magnetic susceptibility tensor, and a theoretical framework is given for partitioning this tensor into local contributions. However, Schmalz *et al.* do have some comments to make on chemical shift measurements, which are the most common source of magnetic susceptibility anisotropies. None of the values arising from numerous attempts in the literature to derive group anisotropies from chemical shift measurements is in agreement with the corresponding values reported by Schmalz *et al.*[228] They speculate that the poor agreement between magnetic susceptibility anisotropies derived from chemical shifts and those measured directly would seem to indicate that some other contribution to the chemical shift has not been correctly evaluated. Schmalz *et al.*[228] also state that the uncertainty surrounding the appropriate origin[140,227] to be assigned to a bond susceptibility when relating it to magnetic shielding has led to a general practice of ignoring susceptibility contributions in treatments of the chemical shift. However, the derivation which they present, of atom as well as bond susceptibilities, makes possible an unambiguous origin for such a calculation, since, manifestly, the origin of an *atom* susceptibility is the nucleus itself.

A very comprehensive account of magnetic anisotropy effects in $^{13}C$ spectra of cyclic alkynes, allenes, and alkenes has been detailed by Charrier *et al.*[229] In this paper, polar plots are given of contours of constant $(1 - 3\cos^2\theta)/R^3$ from the McConnell equation [equation (27)]. The authors point out, however, that the 'single point dipole' approximation has obvious shortcomings for calculating magnetic shielding produced by triple bonds in that it does not give a good reflection of the overall electron distributions extant in

[228] T. G. Schmalz, C. L. Norris, and W. H. Flygare, *J. Amer. Chem. Soc.*, 1973, **95,** 7961.
[229] C. Charrier, D. E. Dorman, and J. D. Roberts, *J. Org. Chem.*, 1973, **38,** 2644.

such bonds. Better results would, they claim, be expected from Pople's procedure,[230] in which two point dipoles (denoted '1' and '2') are centred on each end of the bond (0.12 nm apart), and Charrier *et al.* also give a polar plot of $[(1 - 3\cos^2\theta_1)/R_1^3 + (1 - 3\cos^2\theta_2)/R_2^3]$ (see Figure 3 for the variables $R_1$, $R_2$, $\theta_1$, and $\theta_2$). The shielding and deshielding regions of these two plots have approximately the same angular dependencies at distances greater than 0.15 nm from the centre of the bond, and both shielding and deshielding effects die away more rapidly with distance from the bond centre for the 'split dipole' model. Charrier *et al.*[229] also discuss the adaptation of these models to accommodate two or more triple bonds in conjugation.

Recently there have also been discussions on anisotropy effects of the carboxy-group in the $^1H$ n.m.r. of L-hydroxyproline[231] and on long-range magnetic anisotropic effects on olefinic protons by *endo*-carbonyl groups;[232] anisotropic effects have been invoked in the interpretation of $^1H$ and $^{13}C$ n.m.r. of complexes[233] and in hydrogen-bonding studies;[234] distant shielding effects have also been considered in trinucleoside diphosphates.[235]

**D. 'Ring Current' Effects and Considerations of 'Aromaticity'.**—The concept of $\pi$-electron 'ring currents' in conjugated systems (proposed originally by Pauling,[235] Lonsdale,[236] and London[83] to account for magnetic anisotropies, and later applied to n.m.r. by Pople,[237,238] McWeeny,[239] and others) continues to hold the interest of many workers (not least that of the present Reporter), and some forty or fifty papers directly relating to this phenomenon have been published during the review period. The 'ring current' picture is not without its conceptual difficulties, however, which perhaps arise from an incomplete emancipation from Pauling's[235] original proposal that calculating the magnetic anisotropies of conjugated systems is equivalent to 'calculating the magnetic effect of the currents induced in a conducting network.' The 'ring current' idea is, nevertheless, certainly very useful conceptually, even though (like most of the 'transmitted effects' being considered in this section) it manifestly represents a grossly incomplete physical picture[240]—though not necessarily because of the arguments put forward some years ago by Musher;[241] the separation between 'local' and 'non-local' effects seems largely arbitrary, and it would evidently be theoretically more sound to

230 J. A. Pople, *J. Amer. Chem. Soc.*, 1966, **68,** 4811.
231 L. Pogliani and M. Ellenberger, *J. Amer. Chem. Soc.*, 1974, **96,** 1621.
232 K. Fujita, K. Hata, and R. Oda, *J. Org. Chem.*, 1973, **38,** 2640.
233 T. B. Brill and A. J. Kotlar, *Inorg. Chem.*, 1974, **13,** 470.
243 L. S. Kan, J. C. Barrett, and P. O. P. T'So, *Biopolymers*, 1973, **12,** 2409 (*Chem. Abs.*, 1974, **80,** 37 420q).
235 L. Pauling, *J. Chem. Phys.*, 1936, **4,** 673.
236 K. Lonsdale, *Proc. Roy. Soc.*, 1937, **A159,** 149.
237 J. A. Pople, *J. Chem. Phys.*, 1956, **24,** 1111.
238 J. A. Pople, *Mol. Phys.*, 1958, **1,** 175.
239 R. McWeeny, *Mol. Phys.*, 1958, **1,** 311.
240 C. A. Coulson, J. A. N. F. Gomes, and R. B. Mallion, 1974, *Mol. Phys.*, in press.
241 J. I. Musher, *J. Chem. Phys.*, 1965, **43,** 4081.

consider the total $\pi$-electron contribution to the susceptibility perpendicular to the plane of a conjugated system, or to the chemical shift of its constituent protons (of which that due to 'ring currents' is only a part) on the well-understood assumption of a $\sigma$–$\pi$ separation.[240]

In recent years, Roberts has made several contributions to the literature on uncoupled Hartree–Fock 'ring current' calculations (*e.g.* refs. 242—244) and, in the present review period, he has given[245] an upper bound for the error involved in applying the so-called 'geometric approximation'[242] (by means of which a 'coupled' Hartree–Fock calculation—such as that of Hall and Hardisson[246]—is simulated from an 'uncoupled' procedure) when calculating the 'ring current' properties of conjugated molecules. The error resulting from the use of such an approximation was found[245] to vary from some 2%, in the case of 11,12-benzofluoranthene, to zero in the case of benzene (where the first-order corrections to the bond-order matrix—due to the presence of the external magnetic field—are identically zero when the origin is chosen at the ring centre, thereby making the coupled and uncoupled procedures exactly equivalent when applied to this molecule). Roberts[247] has also extended his previous investigations[244] into the numerical repercussions of the London integral approximation. The London approximation consisted essentially in replacing $\boldsymbol{r}$, a vector relating the position of an *electron* in a given orbital to an arbitrary origin, by the quantity $\frac{1}{2}(\boldsymbol{R}_s + \boldsymbol{R}_t)$ in the vicinity of the bond $s$—$t$ of the conjugated system; $\boldsymbol{R}_s$ and $\boldsymbol{R}_t$ are the vectors relating the *vertices s* and *t*, respectively, of the conjugated system, to this same origin. Since $\boldsymbol{r}$ occurs in certain integrals which arise in the magnetic theory, these integrals are thus much simplified by this procedure. The London integral approximation appears to serve well in the calculation of anisotropies;[83,247] when it comes to calculating chemical shifts, however, terms in $1/r^3$ occur, and then there arises the question of whether to use the substitution

$$1/r^3 \rightarrow 1/\left|\tfrac{1}{2}(\boldsymbol{R}_s + \boldsymbol{R}_t)\right|^3 \qquad (28)$$

(which Roberts[247] designates 'LA I') or whether the substitution

$$1/r^3 \rightarrow \tfrac{1}{2}(1/R_s^3 + 1/R_t^3) \qquad (29)$$

(called 'LA II' by Roberts[247]) is more appropriate. As is, of course, intuitively reasonable, both approximations LA I and LA II improve considerably as the distance of the $s$—$t$ bond from the proton in question increases. Table 12 illustrates this point for the $\beta$-proton (Martin's terminology[248]) of anthracene (Figure 4), from which it is clear that no significant loss of accuracy would result if 'exact' evaluation of the integrals concerned (*i.e.* evaluation without

242 H. G. Ff. Roberts, *Theor. Chim. Acta*, 1969, **15**, 63.
243 A. T. Amos and H. G. Ff. Roberts, *Mol. Phys.*, 1971, **20**, 1073.
244 H. G. Ff. Roberts, *Chem. Phys. Letters*, 1971, **11**, 259.
245 H. G. Ff. Roberts, *Theor. Chim. Acta*, 1974, **33**, 269.
246 G. G. Hall and A. Hardisson, *Proc. Roy. Soc.*, 1962, **A268**, 328.
247 H. G. Ff. Roberts, *Mol. Phys.*, 1974, **27**, 843.
248 R. H. Martin, *Tetrahedron*, 1964, **20**, 897.

**Table 12** *Contributions to the chemical shift of the β-proton in anthracene from the several bonds in the molecule, calculated by various methods (from ref.* 247).

| *Bond*[a] | *Contribution to shift* (p.p.m.) *LA I* | *LA II* | *Exact* |
|---|---|---|---|
| 1—2 | −1.298 | −2.174 | −1.076 |
| 2—3 | −1.113 | −1.864 | −0.923 |
| 3—4 | 0.300 | 0.396 | 0.289 |
| 4—12 | 0.359 | 0.350 | 0.343 |
| 11—12 | −0.061 | −0.060 | −0.059 |
| 1—11 | 0.283 | 0.373 | 0.273 |
| 10—12 | 0.000 | 0.000 | 0.000 |
| 9—11 | −0.183 | −0.206 | −0.181 |
| 9—14 | 0.093 | 0.101 | 0.090 |
| - - - - - - | - - - - - - | - - - - - - | - - - - - -[b] |
| 10—13 | 0.135 | 0.137 | 0.132 |
| 13—14 | 0.023 | 0.023 | 0.023 |
| 5—13 | −0.019 | −0.020 | −0.019 |
| 5—6 | 0.060 | 0.061 | 0.060 |
| 6—7 | 0.064 | 0.063 | 0.063 |
| 7—8 | 0.043 | 0.045 | 0.043 |
| 8—14 | −0.063 | −0.066 | −0.063 |

[a] Numbering as in Figure 4.
[b] This broken line divides those bonds (above) which fall within the 'four-bond' criterion (see text) and those (below) which fall outside it.

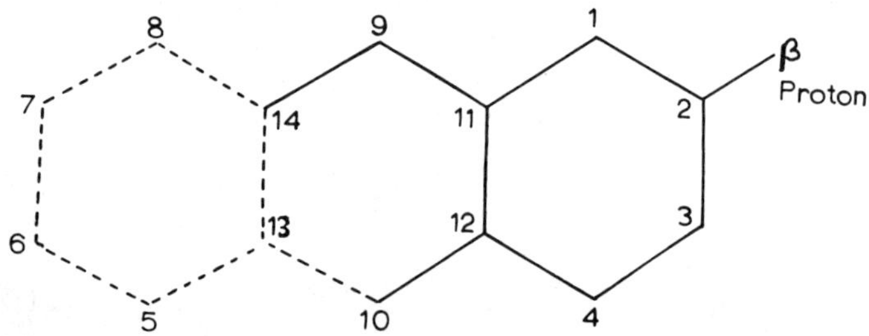


**Figure 4**
(Adapted by permission from *Mol. Phys.*, 1974, **27**, 843)

use of either LA I or LA II) were to be restricted to bonds 'near' the proton. Roberts[247] finds that the following definition of 'nearness' is suitable in practice: let a given proton, *P*, be bonded to an atom, *a*, on the (hexagonal) conjugated network, and let *a′* be the atom nearest to *P* on this network; then, an integral on the *s*—*t* bond is evaluated exactly only if this bond can be reached from either *a* or *a′* by a walk of four bond lengths (or less) along the network. In the diagram of anthracene shown in Figure 4, those bonds which satisfy this criterion for the β-proton are drawn as continuous lines; those whose contribution to the shielding of this proton can 'safely' be estimated by use of the London approximation are depicted as broken lines. In Table 13, shieldings calculated by means of LA I, LA II, and LA I + 'four-bond' policy are compared with the corresponding experimental[249] values (obtained from iterative, second-order analyses, and extrapolated to infinite dilution in $CCl_4$). Overall, the results are quite satisfactory, although, curiously, over a wide range of benzenoid hydrocarbons, the experiment–theory deviations

**Table 13** *Comparison of theoretical and experimental shifts* (p.p.m.), *those for naphthalene and anthracene being given relative to the benzene value* (*from ref.* 247)

| *Molecule* | *Proton*[a] | *LA I* | *LA II* | *'Four-bond'*[c] + *LA I* | *Experiment*[b] |
|---|---|---|---|---|---|
| Benzene | | −1.06 | −2.41 | −0.72 | — |
| Naphthalene | α | −0.59 | −0.74 | −0.48 | −0.46 |
| | β | −0.25 | −0.35 | −0.21 | −0.11 |
| Anthracene | α | −0.85 | −1.08 | −0.70 | −0.66 |
| | β | −0.32 | −0.43 | −0.28 | −0.12 |
| | γ | −1.17 | −1.47 | −0.93 | −1.09 |

[a] Martin's nomenclature.[248] [b] Ref. 249. [c] See text.

follow very similar patterns to those encountered when the simple Hückel–McWeeny method[239] was applied to the same molecules[227] (β-protons deviating markedly *etc.*). Furthermore, no 'ring current' theory (except the very crude, semi-classical one of Johnson and Bovey![250,251]) has yet predicted correctly the relative order of the three distinct proton chemical shifts in pyrene. The absolute value for benzene, from Roberts' work (Table 13), is −0.72 p.p.m., which is in good agreement with Pople's 1964 experimental estimate,[252] though different from a recent semi-empirical estimate[227] of −1.56 p.p.m. and Vysotskii's theoretical value[253] of −1.20 p.p.m., discussed by Raynes last year[3] (see also later in this subsection).

Uncoupled and coupled Hartree–Fock 'ring current' calculations have also been reported by Corradi *et al.*[254] in an attempt to elucidate the relative 'aromatic characters' of certain five-membered heterocycles. They also report some relevant proton chemical shifts, extrapolated to infinite dilution in cyclohexane. Over a wide range of molecules, the following relation is obtained between $\delta_r^{CHF}$, the chemical shift of the $r$th proton, corrected for the 'ring

[249] C. W. Haigh and R. B. Mallion, *Mol. Phys.*, 1970, **18**, 737.
[250] C. E. Johnson and F. A. Bovey, *J. Chem. Phys.*, 1958, **29**, 1012; the tables are available, for example, in J. W. Emsley, J. Feeney, and L. H. Sutcliffe 'High Resolution Nuclear Magnetic Resonance Spectroscopy', Pergamon Press, Oxford, 1965, Appendix B, and in F. A. Bovey, 'Nuclear Magnetic Resonance Spectroscopy', Academic Press, New York, 1969, Appendix.
[251] R. B. Mallion, *J. Chem. Soc.* (*B*), 1971, 681.
[252] J. A. Pople, *J. Chem. Phys.*, 1964, **41**, 2559.
[253] Y. B. Vysotskii, *J. Struct. Chem.*, 1972, **13**, 462.
[254] E. Corradi, P. Lazzeretti, and F. Taddei, *Mol. Phys.*, 1973, **26**, 41.

current' contribution calculated by the coupled procedure, and $q_r$, the charge on the carbon atom to which the $r$th resonating proton is attached:

$$\delta_r^{\mathrm{CHF}} = (28.73 \pm 2.28)\, q_r + (7.57 \pm 0.18) \qquad (30)$$

When the 'ring current' corrections are made *via* an uncoupled Hartree–Fock method, the corresponding correlation is

$$\delta_r^{\mathrm{UCHF}} = (28.63 \pm 2.22)\, q_r + (7.51 \pm 0.18) \qquad (31)$$

Both these lines have a correlation coefficient of 0.96. Corradi *et al.* claim that the calculated 'ring current' contributions to proton shielding constants show that pyrrole must be regarded as being more 'aromatic' than furan and thiophen, and that the same conclusions can be inferred by examining the calculated 'ring current' contributions to the magnetic susceptibility normal to the molecular planes of such molecules. (The reader will gather, from the judicious use of quotation marks adopted here, that the present Reporter is sceptical about *direct* connections between the idea of 'ring currents' and the concept of 'aromaticity'[255]). The 'aromaticity scale' for benzene, pyrrole, furan, and thiophen is, according to Corradi *et al.*, 1:0.55:0.50:0.45, which is different from the order obtained previously by Davies.[256,257] This discrepancy is, perhaps, not surprising, since the two investigations used different methods of calculation and different parametrizations; nevertheless, the remarks of Corradi *et al.*[254] have provoked critical comment from Davies[258] and, in turn, a reply from Lazzeretti and Taddei.[259] The controversy seems to be centred mainly around the question of parametrizations, though, on a more fundamental level, it should not be overlooked that the two methods of calculation employed *are* different; in particular, in Davies' 'finite perturbation' method,[256] the charge–bond-order matrix is regarded as being unchanged in the presence of a magnetic field, whereas in the coupled Hartree–Fock method of Hall and Hardisson[246] (and also in the uncoupled versions of it,[242,243] when the 'geometric' correction is applied) this is, in general, not so. There has also been a further study[260] of the relative 'aromaticities' of furan and thiophen—as well as those of selenophen and tellurophen—in which n.m.r. measurements were one of the criteria applied.

The present Reporter[261] has examined recent criticisms[262] of the gauge factors usually employed in the 'test-dipole' method[239,242,246] for the calculation of 'ring current' chemical shifts in conjugated molecules. It was shown that, in a simple, semi-empirical scheme of the London–Pople–McWeeny type,[83,238,239] insertion of a dipole contribution into the vector potential appearing in the gauge factor, although having no effect on the calculated

255 C. A. Coulson and R. B. Mallion, submitted for publication in *J. Amer. Chem. Soc.*
256 D. W. Davies, *Trans. Faraday Soc.*, 1961, **57**, 2081.
257 D. W. Davies, *Chem. Comm.*, 1965, 258.
258 D. W. Davies, *Mol. Phys.*, 1974, **27**, 1437.
259 P. Lazzeretti and F. Taddei, *Mol. Phys.*, 1974, **27**, 1439.
260 F. Fringuelli, G. Marino, A. Taticchi, and G. Grandolini, *J.C.S. Perkin II*, 1974, 332.
261 R. B. Mallion, *Mol. Phys.*, 1973, **25**, 1415.
262 R. McWeeny, *Chem. Phys. Letters*, 1971, **9**, 341.

'ring current' intensities, is algebraically analogous (and, at large distances from ring centres, numerically equivalent) to estimating the secondary field at the origin due to a set of classical line currents, as discussed originally by Longuet-Higgins and Salem.[263] The 'ring current' secondary field, $\boldsymbol{B}'$, at a given proton in a polycyclic molecule is still of the form

$$\boldsymbol{B}' \propto \sum_{\substack{\text{rings}\\ l}} J_l K(\boldsymbol{r}_l) \tag{32}$$

in which $J_l$ is the 'McWeeny' 'ring current' intensity in the $l$th ring of the molecule (an explicit expression for which is given in ref. 261) and $K(\boldsymbol{r}_l)$ is a geometric factor which depends merely on the physical position in space of the proton in question, relative to the $l$th ring; however, whereas, on the McWeeny theory, this geometric factor (in terms of the quantities shown in Figure 5 for the bond $i$—$j$), is given by

$$\left(\frac{S}{a^2}\right) K^{\text{McW}}(\boldsymbol{r}_l) = \sum_{\substack{\text{Bonds}\\ i\to j\\ \text{around}\\ \text{ring } l}} -\frac{1}{r_{ij}}\left[\frac{(x_i - x_j)}{r_{ij}} - \frac{3}{4}\frac{(x_i - x_j)(x_i{}^2 + x_j{}^2)}{r_{ij}^3} + \ldots\right] \tag{33}$$

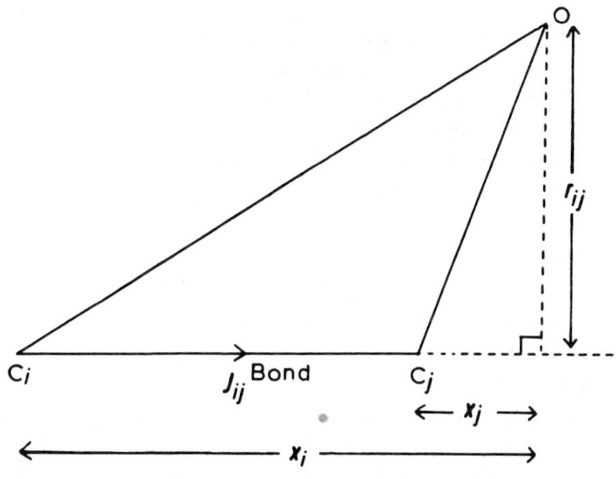


**Figure 5**
(Adapted by permission from *Mol. Phys.*, 1973, **25**, 1415)

($a$ being the length of a carbon–carbon bond in benzene and $S$ the area of a standard benzene hexagon), when this Biot–Savart Law approach is used, the quantity on the left-hand side of equation (33) is replaced by

$$K^{\text{BS}}(\boldsymbol{r}_l) = \sum_{\substack{\text{Bonds}\\ i\to j\\ \text{around}\\ \text{ring } l}} -\frac{1}{r_{ij}}\left[\frac{(x_i - x_j)}{r_{ij}} - \frac{1}{2}\frac{(x_i^3 - x_j^3)}{r_{ij}^3} + \ldots\ldots\right] \tag{34}$$

[263] H. C. Longuet-Higgins and L. Salem, *Proc. Roy. Soc.*, 1960, **A257**, 445.

It is seen that these two quantities are equal, to first order; the second-order term vanishes and the two expressions thus differ only in the third-order term, important for small values of $r_{ij}$ (such as occur, for example, in the benzene geometric factor, and that relating the α-proton in naphthalene to the more distant ring in that molecule) but not important for more distant points from ring centres (as the sample results in Table 14 indicate). The situation is

**Table 14** *Geometric factors for benzene, naphthalene, and anthracene (from ref.* 261)

| *Geometric situation* | | *Geometric factor* $(S/a^2)K^{McW}(\mathbf{r}_l)^a$ | $K^{BS}(\mathbf{r}_l)^b$ |
|---|---|---|---|
| Benzene | | −1.0436 | −0.6583 |
| Naphthalene | α proton | −0.2052 | −0.2002 |
| | β proton | −0.0740 | −0.0726 |
| Anthracene | α proton | −0.0486 | −0.0469 |
| | β proton | −0.0206 | −0.0205 |

$^a$ From equation (33). $^b$ From equation (34).

entirely analogous to the one concerning the various London approximations encountered when reviewing Roberts' work[247] above. It was concluded[261] that any errors involved in the procedure of inserting a dipole contribution into the gauge factor (as well as errors due to the London approximation) can be compensated for, numerically, by an appropriate empirical parametrization, and that such parametrizations should, therefore, be treated with more scepticism than has previously been thought necessary. For example, application of equations (32) and (34) to calculate the secondary fields, $B_r'$ at the various protons in 16 different, condensed, benzenoid hydrocarbons produced the following correlation with the corresponding experimental chemical shifts[249,264] (sterically unhindered protons only):

$$\delta_r^{exp} = 1.21(B_r'/B_{benzene}') + 5.96 \tag{35}$$

(correlation coefficient: 0.98). This implies that the 'ring current' contribution to the chemical shift of a benzene proton is 1.21 p.p.m., in good agreement with Vysotskii's recent value[253] (1.20 p.p.m.) but quite different from the value of 1.56 p.p.m. previously obtained[227] *via* the unmodified McWeeny method.[239]

Haley and Hameka[265] have used London's method, and extensions of it, to calculate the magnetic susceptibilities of some nitrogen heterocycles, while Labzowsky and Lozovic[266] have studied conjugated molecules in strong magnetic fields by means of Rebane's free-electron model.[267]

In the area of 'organic' applications of 'ring current' effects, there are two

[264] C. W. Haigh and R. B. Mallion, *J. Mol. Spectroscopy*, 1969, **29**, 478.
[265] L. V. Haley and H. F. Hameka, *J. Amer. Chem. Soc.*, 1974, **96**, 2020.
[266] L. N. Labzowsky and Y. E. Lozovik, *Internat. J. Quantum Chem.*, 1973, **7**, 985.
[267] T. K. Rebane, *Vestnik Leningrad. Gosodarst. Univ.*, 1957, 16, 19.

interesting papers by Vos *et al.*[268,269] on the subject of 'paratropism', *i.e.* the ability of certain rings in a polycyclic system to sustain paramagnetic 'ring currents' such as are encountered, for example, in the so-called 'anti-aromatic' 4*n*-annulenes.[270] Vos *et al.*[268] examined the $^1H$ n.m.r. spectra of the carbanions of dihydroanthracene [$AH^-$], xanthene [$XH^-$], and thioxanthene [$TxH^-$] [(13)—(15)]. The central rings of these carbanions may be regarded as

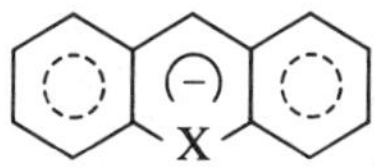

(13) X = $CH_2$ ([$AH]^-$)
(14) X = O ([$XH]^-$)
(15) X = S ([$TxH]^-$)

possessing 4*n* π-electrons ($n = 2$), since the oxygen and sulphur atoms each contribute two electrons to the π-system; they are thus expected to bear paramagnetic 'ring currents'. The side rings have (formally) six π-electrons and are thus expected to be 'diatropic'.[268] Using, essentially, equation (32), Vos *et al.*[268] expressed the 'ring current' contribution to the shielding ($\sigma^r_{RC}$) of the *r*th proton, in any one of the above three carbanions, as

$$\sigma^r_{RC} = J_1[K^r_{JB}(I) + K^r_{JB}(III)] + J_2K^r_{JB}(II) \qquad (36)$$

in which 'I' and 'III' denote the side rings of the carbanion in question and 'II' its central ring. $K^r_{JB}$ (I, II, III) are the 'ring current' contributions due to a unit ('benzene') diamagnetic current in rings I, II, and III, respectively, obtained from the tables of Johnson and Bovey,[250] and $J_1$ and $J_2$ are the 'ring current' intensities (as yet undetermined) in the outer and inner rings, respectively, of the relevant carbanion. By subtracting from the observed shieldings those shifts due to electric field and non-uniform charge effects, Vos *et al.*[268] estimated an experimental 'ring current' contribution to the shifts of the five non-equivalent protons in each of the carbanions considered and, from a least-squares fit of the remainder, with equation (36), obtained values for $J_1$ and $J_2$. The results point to diamagnetic 'ring currents' in the side rings and to paramagnetic ones in the central rings (as anticipated), the latter increasing in the order [$AH^-$], [$TxH^-$], [$XH^-$]. When the $K^r_{JB}$ (I), $K^r_{JB}$ (II), *etc.* factors of equation (36) are estimated by other methods[261,271,272] more recent than that of Johnson and Bovey,[250] different $J_i$ values are obtained, though the general trends exhibited are not seriously affected. An exactly

268 H. W. Vos, Y. W. Bakker, C. MacLean, and N. H. Velthorst, *Org. Magn. Resonance*, 1974, **6**, 245.
269 H. W. Vos, Y. W. Bakker, C. MacLean, and N. H. Velthorst, *Chem. Phys. Letters*, 1974, **25**, 80.
270 J. A. Pople and K. G. Untch, *J. Amer. Chem. Soc.*, 1966, **88**, 4811.
271 C. W. Haigh and R. B. Mallion, *Org. Magn. Resonance*, 1972, **4**, 203.
272 R. C. Haddon, *Tetrahedron*, 1972, **28**, 3613.

similar argument, given in ref. 269, indicates the existence of a strong paramagnetic 'ring current' in the seven-membered rings of the carbanion of 5*H*-dibenzo[*a*,*d*]cycloheptene (16) ($-2.24$, as a ratio to the benzene value) and of the nitranion of 5*H*-dibenz[*b*,*f*]azepine (17) ($-1.50$). Evidence for a

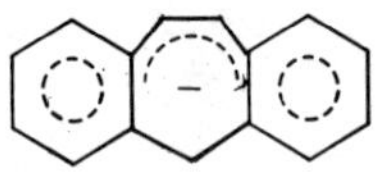

(16)

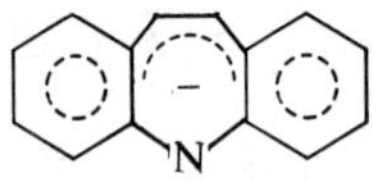

(17)

paramagnetic 'ring current' in the dianion of [18]annulene has been presented by Oth *et al.*[273] The $^{1}$H n.m.r. spectrum at $-110$ °C shows that the dianion exists in solution as an equilibrium between two species (18) and (19) in the ratio 7:3. Both these species have six intra-annular and 12 extra-annular protons. The inference that they sustain a paramagnetic 'ring current' is made from the fact that the inner protons resonate at $\delta$ 29.5 and 28.1, whereas the signal due to the outer protons appears at $\delta$ $-1.13$.

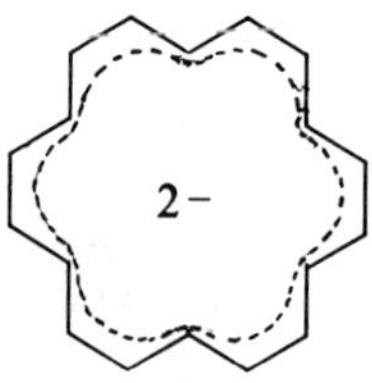

(18)

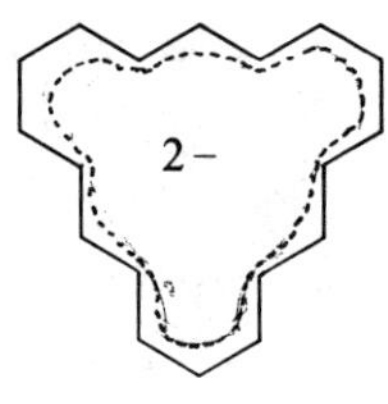

(19)

The present review period witnessed the first demonstration of the chemical shift of a proton directly attached to a [4]annulene ring; Masamune *et al.*[274] have examined the $^{1}$H n.m.r. spectrum of tri-t-butyl[4]annulene (20) in [$^{2}H_{8}$]-tetrahydrofuran, thereby providing important information concerning the induced 'ring current' in the cyclic 4$\pi$-electron system. The spectrum consists of three singlets, and the shifts are as follows:

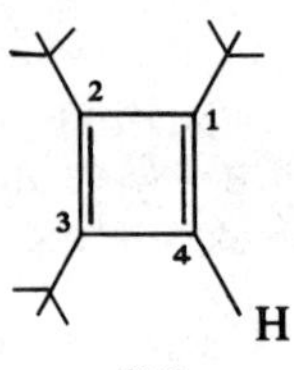

(20)

[273] J. F. M. Oth, E. P. Woo, and F. Sondheimer, *J. Amer. Chem. Soc.*, 1973, **95**, 7337.
[274] S. Masamune, N. Nakamura, M. Suda, and H. Ona, *J. Amer. Chem. Soc.*, 1973, **95**, 8481.

| | |
|---|---|
| Protons of t-butyl groups at positions 1 and 3 | $\delta$ 1.22 |
| Protons of t-butyl group at position 2 | $\delta$ 1.05 |
| Proton attached directly to the ring at position 4 | $\delta$ 5.38 |

Using the C-2 proton ($\delta = 6.42$) of cyclopentadiene as reference, Masamune *et al.*[274] assumed that the difference, $\Delta\delta = 6.42 - 5.38 = 1.04$, may be taken as the paramagnetic contribution to the shift of proton 4 from the induced 'ring current' in (20). By use of the analysis of Pople and Untch,[270] and certain assumptions about geometry, they calculated $\Delta\delta$ to be about 1.18 p.p.m.—an excellent accord, but subject, of course, to an arbitrary choice of reference compound.[274]

The thought of paramagnetic 'ring currents' inevitably leads us back to the vexed question of 'aromaticity', a subject conveniently dealt with here. Those who have reservations about direct connections between these two concepts will have enjoyed considerable *Schadenfreude* on reading a paper by Staley and Kingsley[275] which showed that the two most commonly employed criteria for 'aromaticity'—diamagnetic 'ring currents' and $\pi$-resonance energies—lead to opposite conclusions when the spiro[2,7]decatrienyl anion (21) is compared with the closely related methylenecyclo-octatrienyl anion (22). The n.m.r. of (21) is explained on the basis of a substantial diamagnetic 'ring

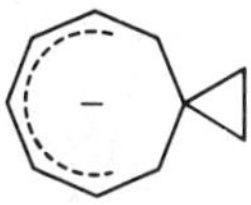

(21)

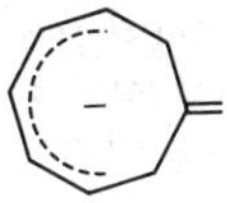

(22)

current' in its $\pi$-system—the first case in which a 'ring current' has apparently resulted from spirocyclopropyl conjugation. (22), on the other hand, has been shown in an earlier paper by Staley and Kingsley[276] to be essentially 'atropic', *i.e.* it has *no* 'ring current', of either sign. They also suggest[276] that it may be a nearly ideal '$4n + 1$' closed-shell $\pi$-system for study. Their investigation shows that '. . . there is no evidence for substantial diamagnetic or paramagnetic ring currents in this anion, thereby indicating that '$4n + 1$' $\pi$-systems are atropic'. The 'aromaticity' and $^1$H n.m.r. spectra of bicyclo-[3,3,2]decatrienyl dianion have also been discussed.[277] A paramagnetic 'ring current' has been confirmed in [16]annulene 1,4:7,10-dioxide[278] and in monosubstituted [16]annulenes,[279] and Vogel *et al.*[280] have examined semibullvalenes such as (23) (which are said[280] to be 'potential homoaromatic compounds') in terms of 'ring current' effects on their n.m.r. spectra. Vogel

[275] S. W. Staley and W. G. Kingsley, *J. Amer. Chem. Soc.*, 1973, **95**, 5805.
[276] S. W. Staley and G. M. Cramer, *J. Amer. Chem. Soc.*, 1973, **95**, 5052.
[277] M. J. Goldstein, S. Tomoda, and G. Whittaker, *J. Amer. Chem. Soc.*, 1973, **96**, 3677.
[278] H. Ogawa, M. Kubo, and I. Tabushi, *Tetrahedron Letters*, 1973, 361.
[279] G. Schröder, G. Kirsch, and J. F. M. Oth, *Chem. Ber.*, 1974, **107**, 460.
[280] E. Vogel, U. H. Brinkler, K. Nachtkamp, J. Wassen, and K. Müllen, *Angew. Chem. Internat. Edn.*, 1973, **12**, 758.

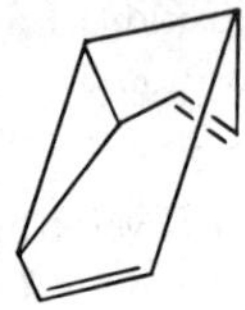

(23)

*et al.*[281] have also provided an extremely elegant demonstration of a diamagnetic 'ring current' in the 'bridged' [$(4n + 2)\pi$-electron] annulene 1,6-methano[10]annulene (24) and of a relatively strong paramagnetic one in the bridged ($4n$ $\pi$-electron) annulene 1,7-methano[12]annulene (25). The 'bridge' protons of (24) resonate at $\delta$ −0.5, whereas the signals from those of (25) appear at $\delta$ 6.1. Weavers and Sondheimer[282] have examined the effect of

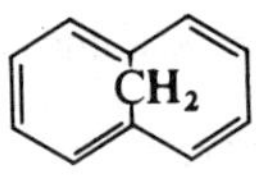

(24)

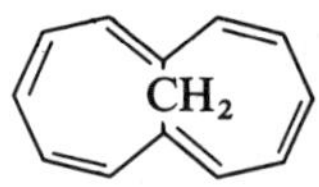

(25)

benzoannelation on the diamagnetic 'ring current' of $(4n + 2)$ $\pi$-electron systems, showing that annelation of the benzene ring reduces, but does not completely suppress, this current.

Conveniently mentioned here also are papers reporting $^1$H chemical shift data on heptahelicene,[283] bridged biphenyls,[284] porphyrins,[285—288] cyclophanes[289—293] ferrocenes,[294] bridged annulenes,[295] dianions of bridged [12]annulenes[296] (being novel, 'aromatic' 14 $\pi$-electron systems), tetradehydro-[14]annulenes[297,298] [18$\pi$-electron analogues of isobenzofuran (26)], aza[13]-

281 E. Vogel, H. Königshafen, K. Müllen, and J. F. M. Oth, *Angew. Chem. Internat. Edn.* 1974, **13**, 281.
282 R. T. Weavers and F. Sondheimer, *Angew. Chem. Internat. Edn.*, 1974, **13**, 141.
283 N. Defay, *Org. Magn. Resonance*, 1974, **6**, 221.
284 D. M. Hall, H. Hwang-Yong, and B. Bhanthumnavin, *J.C.S. Perkin II*, 1973, 2131.
285 C. D. Barry, H. A. O. Hill, B. E. Mann, P. J. Sadler, and R. J. P. Williams, *J. Amer. Chem. Soc.*, 1973, **95**, 4545.
286 M. Tsutsui and C. P. Hrung, *J. Amer. Chem. Soc.*, 1974, **96**, 2638.
287 F. Walker, A. la Mar, and N. Gerd, *Ann. New York Acad. Sci.*, 1973, **206**, 328.
288 H. Ogoshi, W. Watanabe, and Z. Yoshida, *Tetrahedron*, 1973, **29**, 3241.
289 E. A. Truesdale and D. J. Cram, *J. Amer. Chem. Soc.*, 1973, **95**, 5825.
290 K. Nishiyama, H. Kazuo, and T. Sato, *J.C.S. Perkin II*, 1974, 577.
291 T. Otsubo and S. Mizogami, *Bull. Chem. Soc. Japan*, 1973, **46**, 3831.
292 T. Umemoto, T. Otsubo, and Y. Socata, *Tetrahedron Letters*, 1973, **3**, 593.
293 S. Rosenfeld and P. M. Keehn, *Tetrahedron Letters*, 1973, **41**, 4021.
294 D. W. Slocum and C. R. Ernst, *J. Org. Chem.*, 1973, **38**, 1620.
295 E. Vogel, M. Mann, Y. Sakata, K. Müllen, and J. F. M. Oth, *Angew. Chem. Internat. Edn.*, 1974, **13**, 283.
296 J. F. M. Oth, K. Müllen, H. Königshafen, M. Mann, Y. Sakata, and E. Vogel, *Angew. Chem. Internat. Edn.*, 1974, **13**, 284.
297 R. T. Weavers and F. Sondheimer, *Angew. Chem. Internat. Edn.*, 1974, **13**, 139.
298 P. J. Beebey, R. T. Weavers, and F. Sondheimer, *Angew. Chem. Internat. Edn.*, 1974, **13**, 138.

(26)

and aza[17]-bridged tetrahydro-oxa- and thia-[17]annulenes[299] ('diatropic' homologues of furan and thiophen), monosubstituted [16]annulenes,[300] *ortho*-hexaphenylene,[301] benzofluoranthenes and thiophens,[302] cyclopenta[*c*,*d*]-azulene,[303] substituted acenaphthenes,[304] pleiadienes,[305] and dianions of polycyclic hydrocarbons.[306] There has also been some discussion of '$\pi$-electron delocalization' in borazine.[307]

As for classical (as opposed to quantum-mechanical) 'ring current' models, the crude point-dipole approach has been applied, and the much-neglected Mayo–Goldstein[308] 'current-loop' method (which, as classical treatments go, is quite sophisticated) has been used and extended by Kamezawa,[309] and applied successfully to ferrocenes. A perennial favourite is, of course, the Johnson–Bovey model,[250] and their tables have again been much used during the review period, quite often routinely (*e.g.* refs. 310—313) and sometimes as the cornerstone of a conformational determination, *e.g.* ref. 314. In analysing the conformations of bridged biphenyls, Benjamins and Chandler[315] have used the 'double-difference' method,[316] which, they claim, permits estimation of that part of the chemical shift of ring protons which is due solely to 'conformationally dependent anisotropic effects'. (Since the Johnson–Bovey paper[250] of 1958 is mentioned, on average, more than 50 times each year—over the past four years, at least, this has been the case—the latter are somewhat random examples of the many which could have been chosen.) Some

299 J. M. Brown and F. Sondheimer, *Angew. Chem. Internat. Edn.*, 1974, **13,** 337, 339.
300 G. Schröder, G. Frank, H. Röttele, and J. F. M. Oth, *Angew. Chem. Internat. Edn.*, 1974, **13,** 205.
301 L. Ernst, A. Mannschreck, and K.-D. Rümpler, *Org. Magn. Resonance*, 1973, **5,** 125.
302 D. W. Jones, R. S. Matthews, and K. D. Bartle, *Spectrochim Acta*, 1974, **30A,** 489.
303 K. Hafner, K.-P. Meinhardt, and W. Richarz, *Angew. Chem. Internat. Edn.*, 1974, **13,** 204.
304 C. K. Fay, J. B. Grutzner, L. F. Johnson, S. Sternhell, and P. W. Westerman, *J. Org. Chem.*, 1973, **38,** 3122.
305 J. Kolc and J. Michl, *J. Amer. Chem. Soc.*, 1973, **95,** 7391.
306 R. G. Harvey and H. Cho, *J. Amer. Chem. Soc.*, 1974, **96,** 2434.
307 J. B. Rowbotham and T. Schaefer, *Canad. J. Chem.*, 1974, **52,** 489.
308 R. E. Mayo and J. H. Goldstein, *Mol. Phys.*, 1966, **10,** 301.
309 N. Kamezawa, *J. Magn. Resonance*, 1973, **11,** 88.
310 R. Benassi, P. Lazzeretti, I. Moretti, F. Faddei, and G. Torre, *Org. Magn. Resonance*, 1973, **5,** 391.
311 K. W. Blake and B. Jaques, *J.C.S. Perkin II*, 1973, 1660.
312 F. J. Koer and C. Altona, *Rec. Trav. chim.*, 1974, **93,** 147.
313 R. Korenstein, K. A. Muszkat, and S. Sharafy-Ozeri, *J. Amer. Chem. Soc.*, 1973, **95,** 6177.
314 C. F. G. C. Geraldes, V. M. S. Gil, and D. H. Eargle, *J. Chem. Phys.*, 1973, **59,** 1166, 1171.
315 H. Benjamins and W. D. Chandler, *Canad. J. Chem.*, 1974, **52,** 597.
316 J. J. Bergman, W. D. Chandler, and R. Y. Moir, *Canad. J. Chem.*, 1971, **49,** 225.

investigators (*e.g.* refs. 315, 317—319) have sought to obtain a 'second opinion' on the predictions of the Johnson–Bovey tables[250] by also using the more recent semi-empirical quantum-mechanical tables due to Haigh and the present Reporter.[271] Although Crespi *et al.*[317] have observed that, in their study of a flavin mononucleotide, an extrinsic 'ring current' effect alone '. . . will not give even the crudest fit to iso-shielding lines calculated by Bovey and Johnson or Haigh and Mallion without placing the tyrosyl group much closer to the iso-alloxazine moiety than van der Waals radii will allow . . .', this sort of critical comment appears to be rare, and the present Reporter feels that there is, in general, a little too much 'blind faith' displayed in the use of tables of this sort. Neither of these tables is a *panacea* for the problem they both try to solve; the Johnson–Bovey tables[250] overestimate 'ring current' deshielding in the 'deshielding zone' in and near the plane of a benzenoid ring, while the quantum-mechanical tables,[207] although quite good in the deshielding region,[227] underestimate shielding in the 'shielding cone' above the plane of a benzene ring[320,321] (the latter because of particularly serious breakdown of the London approximation[320]). Until such time as a set of quantum-mechanical tables is available which makes no appeal to the London approximation, perhaps judicious exploitation of the (fortuitous) complementary nature of the two tables, referred to above, should be advised. Lazzeretti *et al.*[208] have used the Johnson–Bovey tables[250] to calculate shielding curves as a function of phenyl-ring rotation for the three non-equivalent, non-aromatic protons in styrene oxide, and have compared these with similar curves obtained *via* the much more sophisticated Hall and Hardisson method[246] (which, however, still uses the London integral approximation). Although one can question their use of an empirical factor[251] of *ca.* 0.6 (to 'scale' entries in the Johnson–Bovey tables) for 'correcting' shieldings at points *outside* that region of space in which the factor was empirically determined (*i.e.* outside the plane of the particular aromatic ring whose anisotropy is causing the shielding being estimated), it can be said, with reservations, that this result does provide at least some reassurance about the Johnson–Bovey tables.[250]

'Ring current' effects on $^{13}$C chemical shifts are small, and will normally be almost wholly overwhelmed by other influences on such shifts.[322,323,275] In the review period, further attention has been given to these effects.[324] Extensive $^{13}$C data have also been reported for porphyrins by Abraham *et*

317 H. L. Crespi, J. R. Norvis, J. P. Bays, and J. J. Katz, *Ann. New York Acad. Sci.*, 1973, **222**, 800.
318 A. G. Gonzalez, *J.C.S. Perkin I*, 1973, 2637.
319 T. L. James, M. L. Ludwig, and M. Cohn, *Proc. Nat. Acad. Sci., U.S.A.*, 1973, **70**, 3292.
320 C. W. Haigh and R. B. Mallion, *Mol. Phys.*, 1971, **22**, 955.
321 P. I. Rose, *Org. Magn. Resonance*, 1973, **5**, 187.
322 R. H. Levin and J. D. Roberts, *Tetrahedron Letters*, 1973, 135.
323 H. Günther and H. Schmickler, *Angew. Chem. Internat. Edn.*, 1973, **12**, 243.
324 V. M. Mamaev, Y. K. Grishin, and F. M. Smirnova, *Doklady Akad. Nauk S.S.S.R.*, 1973, **213**, 386 (*Chem. Abs.*, 1974, **80**, 59 315m).

*al.*,[325,326] who give a discussion which illustrates the dangers of using $^{13}C$ chemical shifts, without further supporting evidence, to determine such a 'subtle'[326] problem as aromatic delocalization. Günther *et al.*[327] also point out that attempts to use $^{13}C$ data as criteria of 'aromaticity' in comparative studies of olefinic and potentially 'aromatic' systems '. . . are not without risk',[327] as they show[327] for several molecules.

There are also two papers[328,329] dealing with intermolecular 'ring current' effects in dilution studies (and thus strictly outside the present scope), an 'atlas' of spectra of aromatic and heterocyclic compounds,[330] and an (unsuccessful!) attempt to correlate relative 'ring current' intensities in several carcinogenic, polycyclic, condensed, benzenoid hydrocarbons and their heterocyclic analogues with their respective biological activities.[331]

**E. Electric Field Effects.**—An electric field due to a polar substituent causes a distortion of the electronic distribution giving rise to changes in the nuclear shielding as compared with the same compound without the substituent in question; these changes, $\sigma_E$, are given by Buckingham's equation:

$$\sigma_E = - AE_z - BE^2 \tag{37}$$

in which $E$ is a uniform electric field acting in the vicinity of a conically symmetric bond X—Y (where Y is the nucleus being shielded), $A$ and $B$ are bond parameters dependent only on X, and $E_z$ is the component of $E$ along the X—Y bond. The various approximations in this equation have been detailed by Raynes in earlier Reports.[1-3]

A particularly comprehensive study of electric field effects in $^{13}C$ spectra of unsubstituted fatty acids has been given by Batchelor *et al.*[332] as a potential aid to conformational analysis. Writing Buckingham's equation in the equivalent form

$$\sigma_E = - E_i A_{ij}\, dR_j + E_i E_j B_{ijk}\, dR_k \tag{38}$$

(in which $E_i$ are components of the field vector, $dR_k$ are components of a unit vector, fixed in the molecular framework, $A_{ij}$ are components of a second-rank tensor, and $B_{ijk}$ are components of a third-rank tensor), they have applied the equation to determine the source of non-equivalence in the $^{13}C$ shifts of unsaturated hydrocarbons and fatty acids. They also investigate the dependence of the shift on double bond position and further find that the electric

325 R. J. Abraham and G. H. Barnett, *Tetrahedron*, 1973, **29,** 553.
326 R. J. Abraham, G. E. Hawkes, and K. M. Smith, *J.C.S. Perkin II*, 1974, 627.
327 H. Günther, H. Schmickler, U. H. Brinker, K. Nachtkamp, J. Wassen, and E. Vogel, *Angew. Chem. Internat. Edn.*, 1973, **12,** 760.
328 D. F. Ewing, *Org. Magn. Resonance*, 1973, **5,** 321.
329 M. Vida, J. Cvengros, and V. Kello, *Org. Magn. Resonance*, 1973, **5,** 549.
330 T. F. Ardyukova, I. K. Korobeinicheva, and A. I. Rezvukhin, 'Atlas of Spectra of Aromatic and Heterocyclic Compounds', No. 3 (*Chem. Abs.*, 1974, **80,** 8925c).
331 R. B. Mallion, *Biochimie*, 1974, **56,** 187.
332 J. G. Batchelor, J. H. Prestegard, R. J. Cushley, and S. R. Lipsky, *J. Amer. Chem. Soc.*, 1973, **95,** 6358.

field effect is larger than previously reported for protons and fluorine and is discernible at least eight C—C bonds away from the head group of the dipole. In fact, because of electric field effects, the $^{13}C$ chemical shift parameter is one which will be largely dependent on molecular geometry, and Batchelor *et al.*[332] predict that such effects will play an important part in future $^{13}C$ n.m.r. conformational studies.

The Buckingham equation has also been applied and extended by Slejko and Drago[333] in an investigation of Lewis acid–base interactions, while Lazzeretti *et al.*[213] have used the equation in estimating the anisotropic effects of the carbonyl group; Fournier *et al.*[209] have used it in their investigations of thiamine and its polyphosphoric esters.

There is also a paper by Pazzoli *et al.*[334] which establishes, by means of INDO calculations, a relationship between proton chemical shift and hybridization at carbon in open-chain alkanes, suggesting that, in these compounds, proton shielding depends mainly on C—H bond electrons. The most interesting feature of this correlation is that the well-documented decrease in proton shielding on going from $CH_3$— to C—$CH_2$— to (C—)$_2$CH— is fully accounted for by changes in a local property of the C—H bond without the necessity of invoking contributions due to distant atoms or bonds.

**F. van der Waals and Steric Effects.**—That intramolecular dispersion forces contribute to nuclear shielding is now well established, owing mainly to the work of Buckingham *et al.*,[335,336] Marshall and Pople,[337] and Yonemoto,[338] amongst others. The problem has also been studied by Homer and Callaghan,[339] and has been discussed in detail by Raynes in earlier Reports.[1—3] The effect is of the form[2]

$$\sigma_w = -3B\alpha I/R^6 \qquad (39)$$

in which $B$ is a constant often equated with the coefficient of the quadratic term in the expression for the electric field contribution [equation (37)] and $\alpha$ and $I$ are, respectively, the polarizability and ionization potential of the perturbing group or atom; $R$ is the distance from this perturbing group or atom to the nucleus in question.

Chemical shifts due to long-range dispersion interactions resulting from changes from the gaseous to the liquid state have been examined experimentally, for a variety of solute–solvent systems, by Senthilnathan and Singh,[340] *via* three different methods. They point out, however, that they have no basis on which to favour any particular one of the three methods used and that they are, therefore, unable to state which of the various values for dispersion

333 F. L. Slejko and R. S. Drago, *J. Amer. Chem. Soc.*, 1973, **95**, 6935.
334 S. A. Pozzoli, P. Lazzeretti, and F. Taddei, *J.C.S. Chem. Comm.*, 1973, 536.
335 A. D. Buckingham, *Canad. J. Chem.*, 1960, **38**, 300.
336 W. T. Raynes, A. D. Buckingham, and H. J. Bernstein, *J. Chem. Phys.*, 1962, **36**, 3481.
337 T. W. Marshall and J. A. Pople, *Mol. Phys.*, 1958, **1**, 199.
338 T. Yonemoto, *Canad. J. Chem.*, 1966, **44**, 223.
339 J. Homer and D. Callaghan, *J. Chem. Soc.* (*B*), 1971, 2430.
340 V. P. Senthilnathan and S. Singh, *Chem. Phys. Letters*, 1973, **21**, 147.

shifts are the most reliable. Buckingham *et al.*[341] have examined the polarizability of a pair of hydrogen atoms at long range and have calculated the coefficients of $R^{-6}$ and $R^{-8}$ in the long-range expression of the parallel and perpendicular components of a pair of such atoms at separation $R$. The coefficients of $R^{-6}$ (2558.59 a.u. for the parallel component, and 1268.25 a.u. for the perpendicular one) are somewhat larger than previous theoretical estimates, and much larger than the classical values of 729 and 182.25 a.u., respectively. The terms in $R^{-8}$ arise from the field of the quadrupole moment induced by the field gradient at each atom.

Steric effects on proton chemical shifts have been discussed by Peeling *et al.*[342] in their study of the rotation of dichloromethyl groups. In this investigation, they emphasize an obvious, but nevertheless major, difficulty (see, for example, ref. 227) in applying equation (39) quantitatively to any given interaction—namely that since steric deshielding is proportional to $R^{-6}$, only very slight changes in $R$ ($< 0.01$ nm) in the region of the van der Waals contact distance can cause marked changes in the chemical shift of the proton in question.

Stothers *et al.*[343,344] have studied long-range shielding effects (including steric effects) on $^{13}C$ shifts in methyl-substituted cyclohexanones and methylcyclopentanones, and in alicyclic systems. In particular, they noted marked deshielding trends associated with sterically crowded $\delta$-nuclei in *syn*-axial arrangements in such molecules. Differential $^{13}C$ shieldings in disubstituted ethylenes have been found by Haan and van de Ven[345] to be larger than those in disubstituted ethylenes, and this observation has been attributed by them to the consequences of steric interaction between *geminal* substituents; Dabrowski *et al.*,[217] in studying the $^{13}C$ spectra of quaternary ammonium salts derived from enamino-ketones, have accounted for shift differences of the C-3

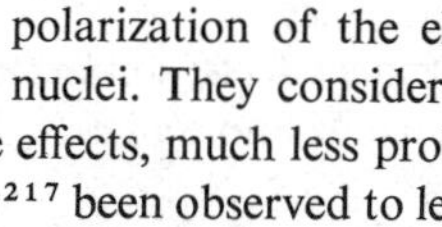


(27)

resonance in (27) of *ca.* 3.0 p.p.m. in terms of what they call a 'steric compression shift'—*i.e.* by the polarization of the electron clouds of the C—H bonds towards the carbon nuclei. They consider an alternative explanation, based on hyperconjugative effects, much less probable, since the same structural change had previously[217] been observed to lead to a shift in the opposite direction of the nitrogen resonance signal of *ca.* 2.0 p.p.m. Steric effects in

[341] A. D. Buckingham, P. H. Martin, and R. S. Watts, *Chem. Phys. Letters*, 1973, **21**, 186.
[342] J. Peeling, B. W. Goodwin, T. Schaefer, and J. B. Rowbotham, *Canad. J. Chem.*, 1973, **51**, 2110.
[343] J. B. Stothers and C. T. Tan, *Canad. J. Chem.*, 1974, **52**, 308.
[344] S. H. Grover and J. B. Stothers, *Canad. J. Chem.*, 1974, **52**, 870.
[345] J. W. de Haan and L. J. M. van de Ven, *Org. Magn. Resonance*, 1973, **5**, 147.

$^{13}C$ shielding have also been invoked by Simeral and Maciel,[346] while Farnum and Patton[347] have discussed steric and inductive effects in the $^{19}F$ n.m.r. of *p*-fluorophenyl carbonium ions.

Finally, Sharrod *et al.*[348] have reported an extensive study of steric effects in [2,2]metacyclophanes, including a steric isotope effect, remote substituent effects on a steric barrier, and other steric phenomena; Boekelheide *et al.*[349] have examined [2,2](2,6)-pyridinoparacyclophane-1,9-diene, and, continuing their investigation of restricted rotations involving tetrahedral carbon, Nakamura *et al.*[350] have made a direct observation of the hindered rotation of a methyl group.

**G. Hydrogen Bonding Effects.**—When a proton forms a hydrogen bond with an electron donor atom, a strong electric field is created in the vicinity of the proton which causes it to be substantially deshielded. Kielmann *et al.*[351] have observed such phenomena in 1,1,1-trifluoro-2-hydroxy-4-alkanones; hydrogen-bonding studies, based on $^{1}H$ and $^{19}F$ n.m.r., have revealed strong intramolecular $\rangle$C=O---H—O bonding (with relatively weak intermolecular interactions in certain of these molecules [*e.g.* (28)], and intermolecular F...H—O bonding in some others [*e.g.* (29)]. From the $^{1}H$ spectra of some *o*-amido-benzanilides, Prasad and Venkataratram[352] have deduced that secondary amido-substituents in the 2-position form a strong hydrogen bond with

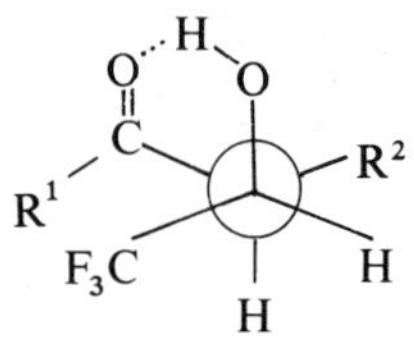


(28)

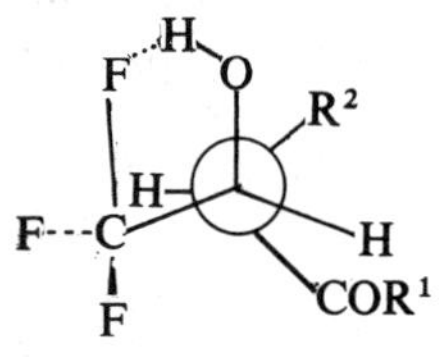


(29)

the anilide proton. Langer and Hyne[353] have discussed the $^{1}H$ n.m.r. of variously substituted aryl sulphanes of the general formula $RC_6H_4S_xH$, in terms of intramolecular electronic effects, conformational factors, and hydrogen-bonding. Evidence for an intramolecular hydrogen-bond shift in these

346 L. Simeral and G. E. Maciel, *Org. Magn. Resonance*, 1974, **6**, 226.
347 D. G. Farnum and D. S. Patton, *J. Amer. Chem. Soc.*, 1973, **95**, 7728.
348 S. A. Sherrod, R. L. da Costa, R. A. Barnes, and V. Boekelheide, *J. Amer. Chem. Soc.*, 1974, **96**, 1565.
349 V. Boekelheide, K. Galuszko, and K. S. Szeto, *J. Amer. Chem. Soc.*, 1974, **96**, 1578.
350 M. Nakamura, M. Ōki, and H. Nakanishi, *J. Amer. Chem. Soc.*, 1973, **95**, 7169.
351 E. Kiehlmann, B. C. Menon, and N. McGillivray, *Canad. J. Chem.*, 1973, **51**, 3177.
352 K. K. Prasad and R. V. Venkataratnam, *Austral. J. Chem.*, 1973, **26**, 1263.
353 H. J. Langer and J. B. Hyne, *Canad. J. Chem.*, 1973, **51**, 3403.

molecules is by no means unambiguous since an 'anomalous' low-frequency shift which is observed for the terminal S—H proton, though unlikely to be inductive or 'electronic' in nature, may possibly be rationalized, at least in phenylsulphanes, by effects of the diamagnetic anisotropy of the aromatic ring (*cf.* Johnson and Bovey[250]). Intermolecular hydrogen-bonding is, however, thought to be responsible for the concentration dependence of chemical shifts in these molecules. The effect of hydrogen-bonding on the $^{13}C$ shifts in the enamino carbonyl groups of compounds of the general formula $O{=}CR^1{-}CR^2{=}CR^3NR^4_2$ has been discussed by Kozerski and Dabrowski.[354] Finally, Yablonskii and Lastochkina[355] have made a $^1H$ study of hydrogen-bonding in hydroperoxides of hydrocarbons.

Although strictly outside the present scope, several papers dealing with intermolecular hydrogen-bonding effects are mentioned here: in concentrated aqueous electrolyte systems,[356] in the interaction of cyclopropane and $H^+$,[357] in the study of Lewis acid–base interactions,[333] and in the association of BuOH in benzene, hexane, cyclohexane, and chloroform;[358] Stephany *et al.*[147] have stated that the hydrogen-bond shift of the carbon in isocyanides and some of their complexes is much smaller than previous data would suggest. The reader's attention is also drawn to a deuterium magnetic resonance study[359] of hydrogen-bonding in crystals of sodium hydrogen oxalate monohydrate ($NaHC_2O_4,H_2O$), and to an n.m.r. investigation of hydrogen-bonding in hormones[360] and in nucleobases with 'Watson–Crick' base pairs.[361]

**H. Isotope Shifts.**—When an atom in a molecule is substituted by a heavier isotope of the same element, there almost always follows an observed increase in shielding in the molecule in question.[1—3,362] This is because such a substitution usually leads to lower rotational and vibrational levels in the molecule which, in turn, leads to smaller amplitudes of bond stretchings; since stretching of a chemical bond generally produces a reduction of shielding, the ultimate consequence of this substitution (when a suitable average has been taken over all available vibrational and rotational states) is an increase in shielding over that which obtained in the given molecule before the introduction of the heavier isotope.[25] During the review period, there have been about 10 papers on such 'isotope effects' (as they are called) which have come to the Reporter's notice.

354 L. Kozerski and J. Dabrowski, *Org. Magn. Resonance*, 1973, **5**, 459.
355 O. P. Yablonskii, N. S. Lastochkina, and V. A. Belyaev, *Neftekhimiya*, 1973, **13**, 851 (*Chem. Abs.*, 1974, **80**, 81 796r).
356 E. J. Sare, C. I. Moynihan, and C. A. Angell, *J. Phys. Chem.*, 1973, **77**, 1869.
357 J. L. Pierre and R. Perraud, *Tetrahedron Letters*, 1973, 1935.
358 L. M. Imanov and F. G. Mirzoev, *Izvest. Akad. Nauk Azerb. S.S.R., Ser. Fiz. Tekh. Mat. Nauk*, 1972, **2**, 48 (*Chem. Abs.*, 1973, **78**, 123 436q).
359 B. Berglund and J. Tegenfeldt, *Mol. Phys.*, 1973, **26**, 633.
360 W. Schimmack and W. Lohmann, *Z. Naturforsch.*, 1973, **28c**, 494.
361 D. M. Crothers, C. W. Hilbers, and R. G. Shulman, *Proc. Nat. Acad. Sci. U.S.A.*, 1973, **70**, 2899.
362 H. Batitz-Hernandez and R. A. Bernheim in 'Progress in N.M.R. Spectroscopy', ed. J. W. Emsley, J. Feeney, and L. H. Sutcliffe, Pergamon Press, Oxford, 1967, vol. 3, p. 63.

If the isotope effect in question concerns directly the n.m.r. parameters of the actual substituted nucleus, it is often referred to as a 'primary isotope effect'. Much more commonly encountered, however, are 'secondary' isotope effects; deuterium isotope effects have been particularly exploited in the review period. Baird[363] has studied such effects on proton shifts in deuteriated styrenes [*e.g.* (30)], finding (from all isomers of [$^2H_1$]- and [$^2H_2$]-styrenes deuteriated on the vinyl positions) that the isotope shifts are (at 60 MHz) 1.0—1.1 Hz (H *gem* to D), 0.3—0.4 Hz (H *cis* to D), and 0.7—0.8 Hz (H *trans* to D); furthermore, the overall effects are approximately additive in the spectra of the dideuteriated compounds. Baird concludes[363] that, although 'inductive' effects *may* contribute to isotope shifts, it seems likely that the explanation lies rather in the different zero-point vibrations for the isotopically substituted species. Deuterium isotope effects on proton shifts have also been studied by Haddon and Jackman,[364] who, using data for t-butylcyclohexane and cyclohexane itself, deduce the following five isotope effects: geminal ($\Delta\delta_{gem}$), vicinal equatorial D on axial H ($\Delta\delta_{eq\ ax}$), vicinal equatorial D on equatorial H($\Delta\delta_{eq\ eq}$), vicinal axial D on axial H ($\Delta\delta_{ax,ax}$), and a long-range effect ($\Delta\delta_{lr}$) associated with the deuterium four bonds removed. A regression analysis, with correlation coefficient 0.996, yielded the parameters given in Table 15.

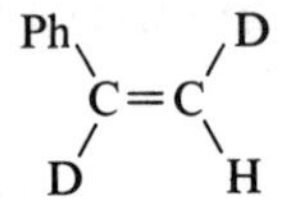

(30)

**Table 15** *Deuterium isotope effects on proton chemical shifts*[a] (*from ref.* 364)

| | |
|---|---|
| $\Delta\delta_{gem}$ | 0.019 ±0.002 |
| $\Delta\delta_{eq,ax}$ | 0.007 ±0.002 |
| $\Delta\delta_{eq,eq}$ | 0.007 ±0.002 |
| $\Delta\delta_{ax,ax}$ | 0.014 ±0.001 |
| $\Delta\delta_{lr}$ | $0.003_5$ ±0.001 |

[a] In p.p.m.; the presence of the deuterium causes *deshielding*.

The dependence of the deuterium isotope effect on $^{19}F$ resonance in mono-deuteriated cyclohexyl fluoride [(31)—(38)] on stereochemical factors has been investigated in an elegant piece of work by Lambert and Greifenstein.[365] In (31)—(34) the vicinal isotope effect on the $^{19}F$ resonance was measured by comparison with the analogous undeuteriated compound, cyclohexyl fluoride (37),(38), and the geminal effect was similarly evaluated with compounds (35)

363 M. C. Baird, *J. Magn. Resonance*, 1974, **14**, 117.
364 V. R. Haddon and L. M. Jackman, *Org. Magn. Resonance*, 1973, **5**, 333.
365 J. B. Lambert and L. G. Greifenstein, *J. Amer. Chem. Soc.*, 1973, **95**, 6150.

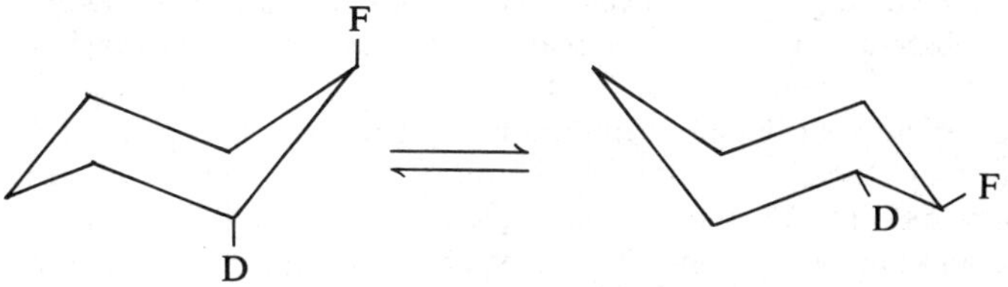


(31) (32)

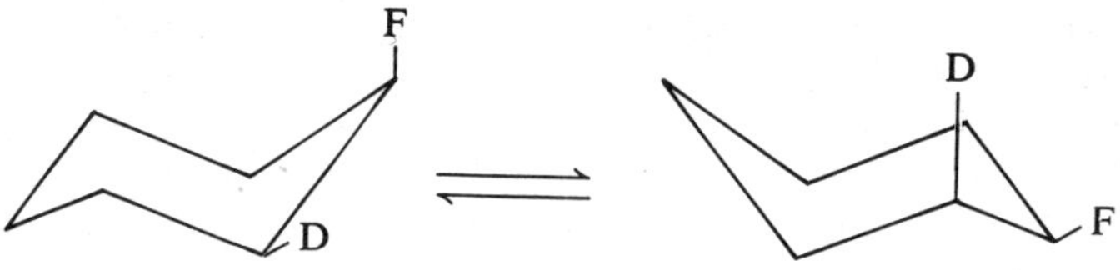


(33) (34)

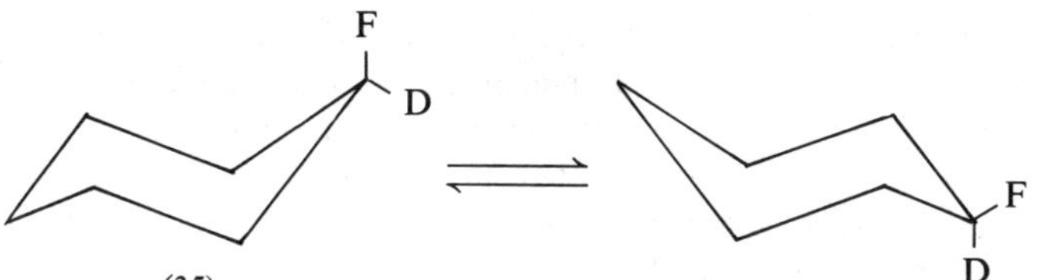


(35) (36)

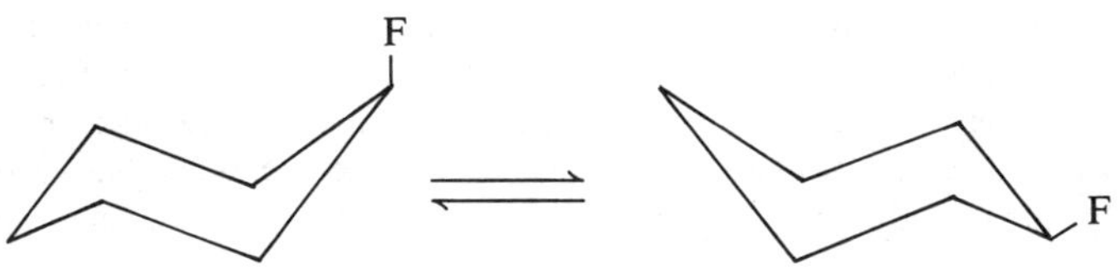


(37) (38)

and (36). The vicinal isotope effect has a strong stereochemical dependence: the effect is twice as large when the dihedral angle is 180° as when it is 60°. These observations place certain constraints on the various theories of isotope effects (a thorough discussion of which—not reproduced here because of space limitations—is given by Lambert and Greifenstein[365]). The observed vicinal isotope shifts are additive. The shift of the axial fluorine in the [$^2H_4$]-compound (1.06 $\pm$ 0.03 p.p.m.) compares well with the sum of the two axial–axial and two axial–equatorial shifts (0.98 $\pm$ 0.11). The shift of the equatorial fluorine in another [$^2H_4$] compound (0.60 $\pm$ 0.03) is close to the sum of two equatorial–equatorial and two equatorial–axial shifts (0.66 $\pm$ 0.07).

Deuterium isotopic effects on $^{13}C$ chemical shifts have been studied by several workers.[191,366–368] Feeney *et al.*[366] have used the small low-frequency shifts observed in the $^{13}$C resonances of the neighbouring carbonyl and α-carbons when the proton on the nitrogen atom in a peptide bond is replaced by deuterium to distinguish peptide carbonyl signals from amide and carboxylic carbonyl signals; Tulloch and Mazurek[367] have found measurable low-frequency shifts of β- and γ-carbon signals after introduction of one or more deuterium atoms into long-chain aliphatic compounds; Breitmaier *et al.*[368] find that isotope effects correlate with hybridization and electron withdrawal at the coupling carbon within a series of comparable compounds; and Bhacca *et al.*[191] have found that monodeuteriation of the steroid cortisol causes $^{13}$C chemical shifts of carbons adjacent to a deuteriated carbon to change by *ca.* 0.1 p.p.m. Other papers dealing with isotope effects to appear during the review period include a study of deuteriated ethanols,[369] solvent isotope shifts of thallium ions in 'light' and 'heavy' water,[370] isotope shifts in $^{19}$F resonance,[371] and the effect of (H,D) isotopic substitution on the parameters of the $^{1}$H n.m.r. spectrum of 1,1,1,3,3-pentachloropropane[372] and other polyhalogenoalkanes.[373]

## 5 Shielding of Particular Nuclear Species

**A. Introduction.**—In the first Report in this series,[1] Raynes listed in this section papers giving chemical shift data of specific nuclei, including all relevant papers on $^{1}$H and $^{13}$C shielding. By only the second Report,[2] it became evident that limitations of space and the abundance of new data on the shieldings of other nuclei necessitated the exclusion of further discussion of any of the very large number of papers presenting new $^{1}$H chemical shift data. The situation has now been reached whereby it is no longer possible to give further discussion of papers reporting $^{13}$C shielding data either, this nucleus now having pride of place as the 'most studied' of all those nuclei which lend themselves to investigation by n.m.r. techniques. Furthermore, as discussed in Section 1, space considerations and copy-deadlines dictate that even those nuclear species that *are* dealt with are treated very cursorily. In the succeeding subsections, the shielding of fluorine, phosphorus, nitrogen, and 'other' nuclei will be considered.

**B. Fluorine Chemical Shifts.**—The origins of shielding changes in the $^{19}$F nucleus (regarded as being almost entirely caused by changes of $\sigma^{p}_{loc}$) have

366 J. Feeney, P. Partington, and G. C. K. Roberts, *J. Magn. Resonance*, 1974, **13**, 268.
367 A. P. Tulloch and M. Mazurek, *J.C.S. Chem. Comm.*, 1973, 692.
368 E. Breitmaier, G. Jung, W. Völter, and L. Pohl, *Tetrahedron*, 1973, **29**, 2485.
369 E. Bengsch, M. Corval, G. L. Martin, and M. Marchetti, *Org. Magn. Resonance*, 1974, **6**, 195.
370 S. O. Chan and L. W. Reeves, *J. Amer. Chem. Soc.*, 1974, **96**, 404.
371 V. A. Shcherbakov, B. N. Chernyshov and R. L. Davidovich, *Zhur. strukt. Khim.*, 1973, **14**, 1109 (*Chem. Abs.*, 1974, **80**, 76 372v).
372 B. A. Englin, Y. V. Belov, A. E. Zolotarev, T. A. Onishchenko, I. K. Shmyrev, and R. K. Freidlina, *Doklady Akad. Nauk S.S.S.R.*, 1973, **210**, 880 (*Chem. Abs.*, 1973, **79**, 65 335p).
373 B. A. Englin, Y. V. Belov, A. Zolotarev and T. A. Onishchenko, *Izvest. Akad. Nauk S.S.S.R., Ser. khim.*, 1973, **8**, 1734 (*Chem. Abs.*, 1973, **79**, 145 846w).

been discussed in some detail by Raynes;[2] we merely list papers reporting data here which have not been referred to in earlier sections of this Report. There are some 40 such papers. Data have been reported on alkanes and alkenes,[374] hexafluorophosphates,[375] fluorodiphenylsilanes,[202] oxypentafluoro-selenium and -tellurium groups,[376] alkyl trifluoroacetates,[377] fluoroethylenes,[378] polyfluorinated naphthalenes,[379] trifluoromethylphenyl acetate derivatives,[380] polybutadienes,[381] (pentafluorophenyl)-zinc and -cadmium halides,[382] complexes of di-(t-arsine) chelates,[383] fluorophenyl carbonium ions,[347] 2-fluoroethyl trichloroacetate,[384] sulphur, phosphorus, and silicon fluorides,[385] —$CF_2CF_2CF_2$—systems in fluorocyclopentenes,[386] 1,3-difluoroacetone and 1,1,3,3-tetrafluoroacetone,[387] trifluoromethyl complexes of platinum, palladium, nickel, and iridium,[388] tungstyl fluoro-complexes,[389] polymeric fluorides of tin(IV) and titanium(IV) and other tin(IV) compounds,[390,391] beryllium fluoride,[392] fluorophenyl derivatives of silicon–platinum complexes,[393] peroxyfluorotantalates,[394] and cyclopentadienylperfluoroalkyl-iron and -nickel complexes.[395] $^{19}F$ Data have also arisen from a study of XF addition to acenaphthylene,[396] the transfer of electron effects across bismuth–aromatic-carbon bonds,[397] in the determination of phenols,[398] and in the study of tetrafluoroethylene–propylene copolymers.[399] Data have also been

374 J. F. Hinton and L. W. Jaques, *J. Magn. Resonance*, 1973, **11**, 229.
375 R. E. J. Sears, *J. Chem. Phys.*, 1973, **59**, 973.
376 K. Seppelt, *Z. anorg. Chem.*, 1973, **399**, 65 (*Chem. Abs.*, 1973, **79**, 59 702q).
377 I. B. Golovanov, V. N. Gagloev, V. M. Sobolev, I. A. Soboleva, and V. V. Smolyaninov, *Zhur. obshchei Khim.*, 1973, **43**, 2666 (*Chem. Abs.*, 1974, **80**, 107 474m).
378 A. V. Fokin, M. A. Landau, and A. S. Kabankin, *Zhur. fiz. Khim.*, 1973, **47**, 2688 (*Chem. Abs.*, 1974, **80**, 114 546q).
379 L. S. Kobrina, V. D. Shteingarts, and L. N. Shchegoleva, *Izvest. sibirsk. Otdel. Akad. Nauk, Ser. khim. Nauk*, 1974, 68 (*Chem. Abs.*, 1974, **80**, 126 541x).
380 G. R. Sullivan and J. A. Dale, *J. Org. Chem.*, 1973, **38**, 2143.
381 K. C. Ramey, M. W. Hayes, and A. G. Altenau, *Macromolecules*, 1973, **6**, 795 (*Chem. Abs.*, 1973, **79**, 146 935m).
382 D. F. Evans and R. F. Phillips, *J.C.S. Dalton*, 1973, 978.
383 W. R. Cullen, L. D. Hall, and J. E. H. Ward, *J. Amer. Chem. Soc.*, 1973, **95**, 3431.
384 R. J. Abraham and J. R. Monasterios, *Org. Magn. Resonance*, 1973, **5**, 305.
385 J. A. Gibson, D. G. Ibbott, and A. F. Janzen, *Canad. J. Chem.*, 1973, **51**, 3203.
386 L. Cavalli and R. K. Harris, *J. Magn. Resonance*, 1973, **10**, 355.
387 B. L. Shapiro, W. A. Thomas, J. L. McClanahan, and M. D. Johnston, *J. Magn. Resonance*, 1973, **11**, 355.
388 K. R. Dixon and K. C. Moss, *J.C.S. Dalton*, 1973, 1528.
389 B. N. Chernyshov, V. A. Shcherbakov, R. L. Davidovich and Y. V. Gagarinskii, *Doklady Akad. Nauk S.S.S.R.*, 1973, **210**, 903 (*Chem. Abs.*, 1973, **79**, 72 043v).
390 P. A. W. Dean, *Canad. J. Chem.*, 1973, **51**, 4024.
391 J. Sala-Pala and J. E. Guerchais, *Bull. Soc. chim. France*, 1973, 2913.
392 R. E. J. Sears, *J. Chem. Phys.*, 1973, **59**, 5213.
393 C. Eaborn, M. R. Harrison, P. N. Kapoor, and D. R. M. Walton, *J. Organometallic Chem.*, 1973, **63**, 99.
394 J. Y. Calves and J. E. Guerchais, *J. Fluorine Chem.*, 1974, **4**, 47.
395 K. Stanley, R. A. Zelonka, J. Thomson, P. Fiess, and M. C. Baird, *Canad. J. Chem.*, 1974, **52**, 1781.
396 L. D. Hall and D. L. Jones, *Canad. J. Chem.*, 1973, **51**, 2902.
397 D. N. Kravtsov, B. A. Kvasov, S. I. Pombrik, and E. I. Fedin, *Izvest. Akad. Nauk S.S.S.R., Ser. khim.*, 1974, 234 (*Chem. Abs.*, 1974, **80**, 107 794r).
398 F. F. L. Ho, *Analyt. Chem.*, 1974, **46**, 496.
399 K. Ishigure, Y. Tabata, and K. Oshima, *Macromolecules*, 1973, **6**, 584.

reported on cadmium and erbium fluorides,[400] octahedral fluorothiocyanato- and fluorocyanato-*N*-titanates(IV),[157] and in hexafluoroacetone adducts in polymers;[401] there have been two biological studies involving $^{19}F$ n.m.r.—a study of *cis*- and *trans*-4-fluoro-L-proline[402] and one of angiotensin II analogues containing fluorine.[403] Other papers involving $^{19}F$ are refs. 404—406 Finally, Radeglia *et al.*[407] have studied $^{19}F$ shielding (as well as $^{13}C$ and $^{14}N$) in *para*-aromatic nitriles, in the context of substituent constants and $\pi$-electron densities, while Bildsøe and Schaumburg[408] have made a CNDO/2 and INDO study of $^{19}F$ (and $^{1}H$) shielding in aromatic carbonyl fluorides.[408]

**C. Phosphorus Chemical Shifts.**—As discussed at some length in previous Reports,[2,3] the theory of phosphorus chemical shifts has been treated much more systematically than that of any other comparable nucleus, the review of Letcher and van Wazer being the standard reference on the subject.[409] During the review period, in addition to *ca.* 30 papers reporting data on $^{31}P$ shielding, a 400-page *magnum opus* by Mavel has appeared;[410] it is a thoroughly comprehensive account of n.m.r. studies of phosphorus compounds, covering the period 1965—1969. Of the regular papers, there have been $^{31}P$ studies of phosphine complexes,[411—414] including those of triarylphosphine[413] and pentamethylcyclopentaphosphine.[414] There have also been studies of phosphoranes,[415—419] including fluorophosphoranes,[415] trimethylenephosphorane,[416] dimethylaminomethylenephosphoranes,[417] tetraoxyazaphosphorane,[418] and 1,2,3-oxa-azaphosphoranes.[419] There have been many $^{31}P$ studies of phosphorus complexes, *e.g.* rhodium(I) complexes containing


400 E. Banks, M. Greenblatt, and B. R. McGarvey, *J. Chem. Phys.*, 1973, **58**, 4787.
401 F. F. L. Ho, *Analyt. Chem.*, 1973, **45**, 603.
402 J. T. Gerig and R. S. McLeod, *J. Amer. Chem. Soc.*, 1973, **95**, 5725.
403 W. H. Vine and D. A. Brueckner, *Biochemistry*, 1973, **12**, 1630.
404 R. E. Schuster and R. D. Bennett, *J. Org. Chem.*, 1973, **38**, 2904.
405 J. Y. Calves and J. E. Guerchais, *Bull. Soc. chim. France*, 1973, 1222.
406 G. A. Olah, J. Nishimura, and P. Kreienbühl, *J. Amer. Chem. Soc.*, 1973, **95**, 7672.
407 R. Radeglia, W. Storek, G. Engelhardt, F. Ritschl, E. Lippmaa, T. Pehk, M. Magi, and D. Martin, *Org. Magn. Resonance*, 1973, **5**, 419.
408 H. Bildsøe and K. Schaumburg, *J. Magn. Resonance*, 1974, **14**, 223.
409 J. H. Letcher and J. R. van Wazer, 'Topics in Phosphorus Chemistry, $^{31}P$ Nuclear Magnetic Resonance', Interscience, New York, 1967, Vol. 5, Ch. 2 and 3.
410 G. Mavel, 'N.M.R. Studies of Phosphorus Compounds (1965–1969)', in *Ann. Rev. N.M.R. Spectroscopy*, 1973, **5B**, 1–441.
411 R. M. Lynden-Bell and G. B. Mather, *J.C.S. Dalton*, 1973, 715.
412 M. A. Bennett, R. N. Johnson, and I. B. Tomkins, *J. Amer. Chem. Soc.*, 1974, **96**, 61.
413 C. A. Tolman, P. Z. Meakin, D. L. Linder, and J. P. Jesson, *J. Amer. Chem. Soc.*, 1974, **96**, 2762.
414 J. P. Albrand, D. Gagnaire, and J. B. Robert, *J. Amer. Chem. Soc.*, 1973, **95**, 6498.
415 M. A. Landau, A. S. Kabankin, and A. V. Fokin, *Zhur. fiz. Khim.*, 1973, **47**, 2916 (*Chem. Abs.*, 1974, **80**, 36 595p).
416 H. Schmidbauer and W. Buchner, *Chem. Ber.*, 1973, **106**, 1251.
417 K. Issleib, M. Lischewski, and A. Zschunke, *Org. Magn. Resonance*, 1973, **5**, 401.
418 W. C. Hamilton, J. S. Ricci, F. Ramirez, L. Kramer, and P. Stern, *J. Amer. Chem. Soc.*, 1973, **95**, 6335.
419 Y. Y. Samitov, M. A. Pudovik, A. I. Khayarov, and L. K. Kibardina, *Zhur. obshchei Khim.*, 1973, **43**, 46 (*Chem. Abs.*, 1973, **78**, 123 882g).

phosphorus,[420] triphenylphosphine complexes of ruthenium(II),[421] three-co-ordinate nickel complexes,[422,423] $[FeNO(diphos)_2]^+[PF_6]^-$,[424] t-phosphine–mercury(II) halide complexes,[425] and complexes of $BF_3$[426] (the latter paper also reporting $^{19}F$ studies). In the review period the Reporter has also encountered papers reporting $^{31}P$ studies of compounds containing alkyl groups,[427] of four- and five-co-ordinate phosphorus,[428] of arylphosphonic acid chlorides,[429] and of perfluoroalkyl-substituted phosphorus(V) hydrides.[430] $^{31}P$ Shifts in $RP(X)NR_2$ (X = Cl or F) have been reported,[431] as have $^{31}P$ (and $^{19}F$) studies on fluorophosphate anions containing a P—P bond.[432]

Some biological studies involving $^{31}P$ shifts are: the determination of intercellular pH by $^{31}P$,[433] the observation of a change of $^{31}P$ shift during protonation of phosphoryl and thiophosphoryl compounds,[434] the study of phospholipids[435,436] and oligonucleotides,[437] and of the binding of 3′-cytidine monophosphate to ribonuclease A.[438] Finally, although published before the review period, a review article on $^{31}P$ has recently been mentioned in *Chemical Abstracts*.[439]

**D. Boron Chemical Shifts.**—Here, about a dozen papers published in 1974 are listed: there have been $^{11}B$ studies of new metallo-carbaboranes,[440—442] of potassium tetra-arylborates,[443] of pentaboranes,[444] of 12-carbaborane


420 P. Meakin and J. P. Jesson, *J. Amer. Chem. Soc.*, 1973, **95**, 7272.
421 K. G. Caulton, *J. Amer. Chem. Soc.*, 1974, **96**, 3007.
422 P. Meakin, R. A. Schunn, and J. P. Jesson, *J. Amer. Chem. Soc.*, 1974, **96**, 279.
423 C. A. Tolman, W. C. Seidel, and L. W. Gosser, *J. Amer. Chem. Soc.*, 1974, **96**, 53.
424 P. R. Hoffman, J. S. Miller, C. B. Ungermann, and K. G. Caulton, *J. Amer. Chem. Soc.*, 1973, **95**, 7902.
425 S. O. Grim, P. J. Lui, and R. L. Keiter, *Inorg. Chem.*, 1974, **13**, 342.
426 R. Bravo, M. Durand, and J.-P. Laurent, *Org. Magn. Resonance*, 1973, **5**, 357.
427 P. W. Clark, J. L. S. Curtis, P. E. Garrow, and G. E. Hartwell, *Canad. J. Chem.*, 1974, **52**, 1714.
428 M. A. Landau, A. S. Kabankin, and A. V. Fokin, *Izvest. Akad. Nauk S.S.S.R., Ser. khim.*, 1973, **10**, 2244 (*Chem. Abs.*, 1974, **80**, 42 655z).
429 F. M. Kharrasova, T. V. Zykova, R. A. Salakhutdinov, and G. I. Rakhimova, *Zhur. obshchei Khim.*, 1973, **43**, 2642 (*Chem. Abs.*, 1974, **80**, 107 473k).
430 J. W. Gilje, R. W. Braun, and A. H. Cowley, *J.C.S. Chem. Comm.*, 1973, 813.
431 S. Distefano, H. Goldwhite, and E. Mazzola, *Org. Magn. Resonance*, 1974, **6**, 1.
432 H. Falius and M. Murray, *J. Magn. Resonance*, 1973, **10**, 127.
433 R. B. Moon and J. H. Richards, *J. Biol. Chem.*, 1973, **248**, 7276.
434 N. K. Skvortsov, B. I. Ionin, and A. A. Petrov, *Zhur. obshchei Khim.*, 1974, **44**, 220 (*Chem. Abs.*, 1974, **80**, 107 536h).
435 G. S. Assmann, E. A. Sokoloski, and H. B. Brewer, *Proc. Nat. Acad. Sci. U.S.A.*, 1974, **71**, 549.
436 T. O. Henderson, T. Glonek, and T. C. Myers, *Biochemistry*, 1974, **13**, 623.
437 V. F. Zarytova, D. G. Knorre, A. V. Lebedev, A. S. Levina, and A. I. Rezvukhin, *Doklady Akad. Nauk S.S.S.R.*, 1973, **212**, 630 (*Chem. Abs.* 1973, **79**, 146 782j).
438 D. G. Gorenstev and A. Wyrwicz, *Biochem. Biophys. Res. Comm.*, 1973, **54**, 976.
439 J. R. Van Wazer, *Determination Org. Structure Phys. Methods*, 1971, **4**, 323 (*Chem. Abs.*, 1974, **80**, 107 375e).
440 P. C. Keller, *J. Amer. Chem. Soc.*, 1974, **96**, 3078.
441 D. F. Dustin and M. F. Hawthorne, *J. Amer. Chem. Soc.*, 1974, **96**, 3462.
442 R. J. Wiersema and M. F. Hawthorne, *J. Amer. Chem. Soc.*, 1974, **96**, 761.
443 J. T. Vandeberg and C. E. Moore, *Org. Magn. Resonance*, 1973, **5**, 57.
444 R. R. Rietz and R. Schaeffer, *J. Amer. Chem. Soc.*, 1973, **95**, 4580.

isomers,[445] and of $B_6H_{10}$.[446] There have also been $^{11}B$ investigations of $C_2B_6H_{10}$, a new *nido*-carbaborane,[447] *nido*-dicarbanonaboranes and *nido*-dicarbadecaboranes,[448] and *nido* 11-atom metallocarbaboranes and their Lewis acid adducts.[449] Two other papers involving $^{11}B$ shifts are also listed here.[450,451]

**E. Silicon Chemical Shifts.**—The chemical shift scale proposed by Levy *et al.*[64] has already been mentioned in Section 2. An interpretation of some $^{29}Si$ n.m.r. shifts has been attempted by Engelhardt *et al.*[133], and unusual $^{29}Si$ chemical shift trends in substituted phenyltrifluorosilane systems have been investigated by Ernst *et al.*[452] Other papers include a study of organosilicons containing chloromethyl groups,[453] phenyl- and benzyl-substituted silanes,[454] methyl- and methylphenyl-siloxanes,[455] and silicon–nitrogen compounds.[456] Other $^{29}Si$ studies include refs 457—459.

**F. Nitrogen Chemical Shifts.**—Nitrogen chemical shifts are usually rationalized in terms of changes in $\sigma_{loc}^{p}$, as Raynes has previously discussed.[2] New nitrogen shift data have been published for monosubstituted pyridines,[460] edta complexes,[461] concentrated aqueous solutions of ammonium salts,[462] pyrrole,[463] substituted pyrazoles,[464] and organometallic azides.[465] 'Unexpected' $^{15}N$ (and $^{13}C$) parameters have been derived from the spectra of 2,4,6-trimethylbenzonitrile oxide.[466] There has been a $^{14}N$ study of pyridine exchange in bis-(2,4-pentanedionato)dipyridine-cobalt(II) and -nickel(II) complexes,[467]

[445] T. A. Babushkina, V. V. Khrapov, V. A. Brattsev, Y. V. Golk'tyapin, and V. I. Stanko, *Zhur. strukt. Khim.*, 1973, **14**, 1018 (*Chem. Abs.*, 1974, **80**, 69 889r).
[446] V. T. Brice, H. J. Johnson, and S. G. Shore, *J. Amer. Chem. Soc.*, 1973, **95**, 6629.
[447] A. J. Gotcher, J. F. Ditter, and R. E. Williams, *J. Amer. Chem. Soc.*, 1973, **95**, 7514.
[448] R. R. Rietz and R. Schaeffer, *J. Amer. Chem. Soc.*, 1973, **95**, 6254.
[449] C. J. Jones, J. N. Francis, and M. F. Hawthorne, *J. Amer. Chem. Soc.*, 1973, **95**, 7633.
[450] W. G. Henderson, M. J. How, G. R. Kennedy, and E. F. Mooney, *Carbohydrate Res.*, 1973, **28**, 1 (*Chem. Abs.*, 1973, **79**, 65 363w).
[451] A. K. Covington and K. E. Newman, *J. Inorg. Nuclear Chem.*, 1973, **35**, 3257.
[452] C. R. Ernst, L. Spialter, G. R. Buell, and D. L. Wilhite, *J. Organometallic Chem.*, 1973, **59**, C13.
[453] J. Schraml, J. Vcelak, and V. Chvalovsky, *Coll. Czech. Chem. Comm.*, 1974, **39**, 267.
[454] J. Schraml, D. C. Nguyen, V. Chvalovsky, M. Mägi, and E. Lippmaa, *J. Organometallic Chem.*, 1973, **51**, C5. (*Chem. Abs.*, 1973, **78**, 166 726b).
[455] G. Engelhardt, M. Mägi, and E. Lippmaa, *J. Organometallic Chem.*, 1973, **54**, 115.
[456] H. Jancke, G. Engelhardt, and M. Mägi, *Z. Chem.*, 1973, **13**, 435.
[457] H. C. Marsmann, *Chem.-Ztg.*, 1973, **97**, 128 (*Chem. Abs.*, 1973, **78**, 166 213g).
[458] H. Saji, T. Yamadaya, and M. Asanuma, *Phys. Letters*, 1973, **45**, 109.
[459] G. C. Levy and J. D. Cargioli, *J. Amer. Soc.*, 1973, **95**, 3445.
[460] M. Witanowski, L. Stefaniak, H. Januszewski, T. Saluvere, and G. A. Webb, *Adv. Mol. Relaxation Processes*, 1973, **5**, 169 (*Chem. Abs.*, 1974, **80**, 42 624p).
[461] R. Hagen, J. P. Warren, D. H. Hunter, and J. D. Roberts, *J. Amer. Chem. Soc.*, 1973, **95**, 5712.
[462] J. M. Briggs and E. W. Randall, *Mol. Phys.*, 1973, **26**, 699.
[463] J. M. Briggs, E. Rahkama, and E. W. Randall, *J. Magn. Resonance*, 1973, **11**, 416.
[464] M. Witanowski, L. Stefaniak, H. Januszewski, and J. Elguero, *J. Chim. phys.*, 1973, **70**, 697.
[465] J. Mueller, *J. Organometallic Chem.*, 1973, **51**, 119.
[466] M. Christl and J. P. Warren, *J. Amer. Chem. Soc.*, 1973, **95**, 4392.
[467] S. F. Lincoln, *J.C.S. Dalton*, 1973, 1896.

and of $^{15}N$ shieldings in dinitrogen complexes of permethyltitanocene.[468] A review of $^{15}N$ shielding has appeared recently.[469] Also mentioned here are three papers which have included $^{14}N$ as well as other nuclear shifts: a $^{14}N$ (as well as $^{11}B$) study of silylamines and silylaminoboranes[470] and of alkoxy-, alkylthio-, and 1-pyrrolyl-boranes,[471] and a $^{14}N$ (as well as $^{19}F$) study of 2,4,6-trifluoronitrobenzene.[472]

**G. Chemical Shifts of other Nuclei.**—The following have been reported during the review period: $^{2}H$, refs. 473—475; $^{17}O$, ref. 476; $^{7}Li$ and $^{23}Na$, refs. 477—484; $^{27}Al$, ref. 485; $^{33}S$, ref. 486; $^{35}Cl$, ref. 487 (also includes some $^{14}N$ data), ref. 488 (also includes $^{1}H$ and $^{2}H$ studies), and refs. 489 and 490; $^{43}Ca$, ref. 491; $^{59}Co$, refs. 492—495; $^{73}Ge$, ref. 496; $^{77}Se$, ref. 497; $^{81}Br$, ref. 498; $^{93}Nb$, ref. 499 (also reports a $^{19}F$ study), ref. 500 (also $^{2}H$), and ref. 501; $^{107}Ag$ and

468 J. E. Bercaw and E. Rosenberg, *J. Amer. Chem. Soc.*, 1974, **96**, 612.
469 R. L. Lichter, *Determination Org. Struct. Phys. Methods*, 1971, **4**, 195 (*Chem. Abs.*, 1974, **80**, 107 374d).
470 H. Noeth, W. Tinhof, and B. Wrackmeyer, *Chem. Ber.*, 1974, **107**, 518.
471 H. Noeth and B. Wrackmeyer, *Chem. Ber.*, 1973, **106**, 1145.
472 J. Biemond, B. J. M. Neijzen, and C. MacLean, *Chem. Phys.*, 1973, **1**, 335.
473 L. S. Bresler, A. S. Khachaturov, and I. Y. Poddubnyi, *J. Organometallic Chem.*, 1974, **64**, 335.
474 H. H. Mantsch, H. Saito, and L. C. Leitch, *J. Amer. Chem. Soc.*, 1974, **96**, 256.
475 J. Seelig and W. Niederberger, *J. Amer. Chem. Soc.*, 1974, **96**, 2069.
476 L. O. Anderson and J. Mason, *J.C.S. Dalton*, 1974, 202.
477 M. S. Greenberg and R. L. Bodner, *J. Phys. Chem.*, 1973, **77**, 2449.
478 H. Lechert, *Ber. Bunsengesellschaft phys. Chem.*, 1973, **77**, 697 (*Chem. Abs.*, 1974, **80**, 32 235y).
479 F. Kraemer, W. Müller-Warmuth, J. Scheerer, and H. Dutz, *Z. Naturforsch.*, 1973, **28a**, 1338.
480 J. P. Kintzinger and J. M. Lehn, *J. Amer. Chem. Soc.*, 1974, **96**, 3313.
481 N. S. Magnuson and J. A. Magnuson, *Biophys. J.*, 1973, **13**, 1117.
482 T. L. James and J. H. Noggle, *Bioinorg. Chem.*, 1973, **2**, 69 (*Chem. Abs.*, 1973, **79**, 118 971h).
483 J. B. Smart, R. Hogan, P. A. Scherr, M. T. Emerson, and J. P. Oliver, *J. Organometallic Chem.*, 1974, **64**, 1.
484 M. M. Exner, R. Waack, and E. C. Steiner, *J. Amer. Chem. Soc.*, 1973, **95**, 7009.
485 A. D. Toy, T. D. Smith, and J. R. Pilbrow, *Austral. J. Chem.*, 1973, **26**, 1889.
486 O. Lutz, A. Nolle, and A. Schwenk, *Z. Naturforsch.*, 1973, **28a**, 1370.
487 A. P. Gulya, V. A. Shcher'akkov, and A. V. Ablov, *Doklady Akad. Nauk S.S.S.R.*, 1973, **209**, 854 (*Chem. Abs.*, 1973, **79**, 120 276x).
488 D. E. O'Reilly, E. M. Peterson, and C. E. Scheie, *J. Chem. Phys.*, 1973, **29**, 3576.
489 D. E. Carlstrom and W. G. Miller, *J. Phys. Chem.*, 1973, **77**, 2759.
490 J. A. Happe, *J. Amer. Chem. Soc.*, 1973, **95**, 6232.
491 O. Lutz, A. Schwenk, and A. Uhl, *Z. Naturforsch.*, 1973, **28a**, 1534.
492 J. Wilinski and R. J. Kurland, *Inorg. Chem.*, 1973, **12**, 2202.
493 K. L. Craighead, *J. Amer. Chem. Soc.*, 1973, **95**, 4434.
494 R. A. LaRossa and T. L. Brown, *J. Amer. Chem. Soc.*, 1974, **96**, 2072.
495 A. Johnson and G. W. Everett, *Inorg. Chem.*, 1973, **12**, 2801.
496 H. Saji. *Phys. Letters* (*A*), 1973, **45**, 469.
497 M. A. Lardon, 'Organic Selenium Compounds: Their Biology and Chemistry', ed. D. L. Klayman, Wiley, New York, 1973, p. 933 (*Chem. Abs.*, 1974, **80**, 42 350w).
498 R. L. Ward and P. L. Whitney, *Biochem. Biophys. Res. Comm.*, 1973, **51**, 343.
499 Yu. A. Buslaev, V. D. Kopanv, S. M. Sinitsyna, and V. G. Khlebodarov, *Zhur. neorg. Khim.*, 1973, **18**, 2567 (*Chem. Abs.*, 1973, **79**, 152 472z).
500 K. Nakamura, *Bull. Chem. Soc. Japan*, 1973, **46**, 2028.
501 R. G. Kidd and H. G. Spinney, *Inorg. Chem.*, 1973, **12**, 1967.

$^{109}$Ag, ref. 502; $^{113}$Cd, refs. 503—505; $^{119}$Sn, refs. 506—508; $^{181}$Ta, ref. 509; $^{199}$Hg, ref. 510; $^{205}$Tl, refs. 511, 370; $^{207}$Pb, refs. 512, 513.

Finally, the following papers each involve the shifts of different combinations of nuclei: one by Canters and de Boer[514] ($^{6}$Li, $^{7}$Li, $^{23}$Na, $^{39}$K, $^{85}$Rb, $^{87}$Rb, and $^{133}$Cs), a study of phosphine–borane derivatives *via* $^{11}$B, $^{19}$F, and $^{31}$P n.m.r. by Rapp and Drake,[515] and a comparative study[516] of the $^{29}$Si and $^{119}$Ti n.m.r. of molecules of the type $Me_nMX_{4-n}$, where M = Si or Sn, and X = Cl, Br, or I.

## 6 Conclusions

Any overall account of a full year's work in such a diverse and intensively studied area as nuclear shielding is bound to be subjective and to reflect, to a greater or lesser degree, the tastes and prejudices of the reviewer. With these reservations in mind, it can be concluded that this has been a most active and productive year, particularly valuable contributions and advances having been made in basic theory (Section 2A), in *ab initio* calculations (Section 3B), and in the study of certain 'transmitted effects' (Section 4).

Finally, I should like to thank my predecessor on this Report, Dr. W. T. Raynes, for helpful discussion and advice. I am also grateful to the Dean, Canons, and Students of Christ Church, Oxford, for my tenure of a *Research Lecturership of the House*.

502 C. W. Burges, R. Koschmieder, W. Sahm, and A. Schwenk, *Z. Naturforsch.*, 1973, **23a**, 1753.
503 G. E. Maciel and M. Borzo, *J.C.S. Chem. Comm.*, 1973, 394.
504 R. J. Kostelnick and A. A. Bothner-By, *J. Magn. Resonance*, 1974, **14**, 141.
505 V. V. Zhukov, I. D. Weisman, and L. H. Bennett, *J. Res. Nat. Bur. Stand., Sect. A*, 1973, **77**, 713. (*Chem. Abs.*, 1974, **80**, 42 609m).
506 R. R. Sharp, *J. Chem. Phys.*, 1974, **60**, 1149.
507 E. V. van den Berghe and G. P. van der Kelen, *J. Organometallic Chem.*, 1973, **61**, 197.
508 P. J. Smith and L. Smith, *Inorg. Chim. Acta, Rev.*, 1973, **7**, 11 (*Chem. Abs.*, **1974, 80,** 81 517a).
509 L. C. Erich, A. C. Gossard, and R. L. Hartless, *J. Chem. Phys.*, 1973, **59**, 3911.
510 G. E. Maciel and M. Borzo, *J. Magn. Resonance*, 1973, **10**, 388.
511 L. A. Baidakov and A. N. Kudryavtsev, *Izvest. Akad. Nauk S.S.S.R., Neorg. Materialy*, 1973, **9**, 1055 (*Chem. Abs.*, 1973, **79**, 85 342s).
512 M. J. Cooper, A. K. Holliday, P. H. Makin, and R. J. Puddephatt, *J. Organometallic Chem.*, 1974, **65**, 377.
513 R. M. Hawk and R. R. Sharp, *J. Chem. Phys.*, 1974, **60**, 1522.
514 G. W. Canters and E. de Boer, *Mol. Phys.*, 1973, **26**, 1185.
515 B. Rapp and J. E. Drake, *Inorg. Chem.*, 1973, **12**, 2868.
516 E. V. van den Berghe and G. P. van der Kelen, *J. Organometallic Chem.*, 1973, **59**, 175.

# 2

# Nuclear Spin–Spin Coupling

BY R. GRINTER

## 1. Introduction

The general outline of this chapter of the Report again follows the pattern set in earlier volumes quite closely. Only minor changes have been made, but it is hoped that these changes will help to make the information in the chapter more accessible, particularly to the reader wishing to make a rapid search for particular types of coupling constant. These innovations arose as the result of an attempt to choose a particular order in which to write coupling constants, *i.e.* $^{3}J$(PCCH) or $^{3}J$(HCCP)?, $^{4}J$(HNCCH) or $^{4}J$(HCCNH)? The solution proposed is that, following the almost universal use of $^{1}J$(CH), $^{1}J$(PH), *etc.*, the nucleus of greatest atomic mass should be written first so that we write $^{3}J$(PCCH) rather than $^{3}J$(HCCP), or $^{2}J$(DCH) rather than $^{2}J$(HCD). In the event that the coupled nuclei are identical then the atoms in the coupling path determine the order, that of greatest atomic mass coming as early in the order as possible. Thus we write $^{6}J$(HCOCCCH) and not $^{6}J$(HCCCOCH).

A problem of much rarer occurrence, though one which is likely to increase in importance, is the question of indicating in the coupling symbol the nucleus whose resonance was observed to obtain the result quoted, *i.e.* has $^{3}J$(PCCH) been obtained from the $^{31}$P or the $^{1}$H spectrum? It has been proposed by Dr. B. J. Kimber (personal communication) that this be indicated by placing the nucleus whose resonance was observed in bold type, *e.g.* $^{3}J$(**P**CCH) indicates that the coupling quoted is taken from the $^{31}$P spectrum. The average of $^{3}J$(**P**CCH) and $^{3}J$(PCC**H**) would be indicated simply by the absence of a bold **P** or **H**, *i.e.* $^{3}J$(PCCH). This would also be the form used in situations where the nucleus observed is not specified.

It is hoped that these suggestions will commend themselves to n.m.r. spectroscopists and we (the Senior Reporter and the present Reporter) would be pleased to receive comments on them. The first, but not the second, has been strictly observed in the following pages.

With regard to the ordering of couplings within a particular section of the Report, an attempt has been made to follow the Periodic Table more closely than in previous years. Though no rigorous rules can be laid down, the following guiding principles have been adhered to as closely as seems reasonable. For couplings of the general form $^{n}J$(X,Y) we start with Y = $^{1}$H and vary X across the main groups of the Periodic Table from I to VIII, including all the elements in a particular group before proceeding to the next. The transition metals, and also occasionally some main-group metals, are taken last. Y is then moved along the table in the same manner, each step in Y being followed by a variation of X across the remaining groups.

It also appears appropriate here to mention a small matter concerning the use of the phrase 'reduced coupling constant'. This term is normally used for the quantity $K$ defined by equation (1), but recently[1] it has also been used to

$$K_{AB} = (2\pi/\hbar\gamma_A\gamma_B)J_{AB} \tag{1}$$

describe the splittings of reduced magnitude which are observed in continuous-wave spin-decoupled spectra. Although it seems unlikely that serious confusion will arise from these two different uses of 'reduced coupling constant' it might be advisable if authors were to retain this terminology exclusively for $K$, and adopt an expression such as 'diminished coupling constant' for the other phenomenon.

Finally, as in previous volumes, considerable use is made of common abbreviations such as CNDO/2, FPT, *etc.* Readers are referred to page xiv for a list of these abbreviations and of the other symbols used.

## 2. Theoretical Work

**A. *Ab Initio* Calculations**—Voitländer[2] has continued his earlier work on the variational calculation of $^1J$(DH). The essential feature of the new calculation is the use of a contact operator which corresponds to a nuclear magnetic moment which is spread out over a finite region of space, rather than a delta function. The self-coupling energy is then finite and the variational treatment can be based upon a minimization of the complete second-order energy. Kowalewski *et al.*[3] have also calculated $^1J$(DH) for the hydrogen molecule using SCF LCAO MO wave functions and a large amount of configuration interaction; all singly and doubly excited triplet states were included. The result, after vibrational averaging, of $^1J$(DH) = 43.48 Hz is in very good agreement with experiment (42.94 ± 0.1 Hz). The authors compare and discuss a number of other calculations of this coupling.

Birnstock and Klöpper[4] have shown that use of improved virtual orbitals with LCAO MO wavefunctions can make a significant difference to the agreement of calculated $^1J$(CH) values with experiment.

**B. Semi-empirical Calculations**—There seems to have been rather less activity in this field than of late. Blizzard and Santry[5] have noted small but significant errors in their formulation of the INDO FPT method, and these require

[1] R. A. Newmark and J. R. Hill, *J. Amer. Chem. Soc.*, 1973, **95**, 4435.
[2] W. Sänger and J. Voitländer, *Z. Naturforsch.*, 1973, **28a**, 1866.
[3] J. Kowalewski, B. Roos, P. Siegbahn, and R. Vestin, *Chem. Phys.*, 1974, **3**, 70.
[4] F. Birnstock and D. Klöpper, *Chem. Phys, Letters*, 1973, **20**, 542.
[5] A. C. Blizzard and D. P. Santry, *J. Chem. Phys.*, 1973, **58**, 4714.

changes in the published values of their calculated[6] coupling constants. Hilton and Sutcliffe[7] have questioned the calculations of $^4J$(FCCCF) by Hirao *et al.*[8] with regard to both the signs and the magnitudes of the calculated couplings.

den Otter *et al.*[9] have shown experimentally that for the *trans*-(F,F) coupling in *trans*-difluoroethylene there is an anisotropic contribution, some 5% of which arises from the indirect (electron-mediated) coupling. This contribution is quantitatively predicted by INDO FPT calculations but the authors warn that such excellent agreement of theory with experiment cannot be regarded as the norm for this type of calculation. INDO FPT calculations of (H,H) and (F,F) couplings in halogenotoluenes have been reported.[10]

INDO FPT calculations of $^1J$($^{15}$NH) and $^2J$(HNH) have been described by Wasylishen and Schaefer.[11] They find that both couplings are very sensitive to the HNH angle in the range 110—120°. $^1J$($^{15}$NH) was found to be almost unaffected by changes of nuclear charge but $^2J$(HNH) changed by as much as 2 Hz.

Solkan, Sergeyev, and their co-workers have reported a series of INDO FPT calculations of (H,H) and (C,H) coupling.[12,13] Their results for cycloalkenes[13] may be summarized as follows. The calculated values of $^2J$(HCH) are very poor. $^3J$(HCCH) depends upon C—C bond length and increases by 0.6—0.7 Hz for a decrease in bond length of 0.02 Å (2 pm), but this is not sufficient to explain the observed variations. For a diene (see Figure), the effect

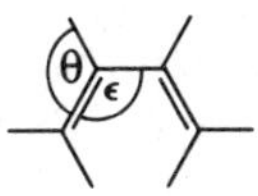


of changes in $\theta$ and $\varepsilon$ and in the dihedral angle $\phi$ are investigated.[13] For the planar *cis* configuration ($\phi$ = 180°) there is a dependence of $^3J$(HCCH) on both $\varepsilon$ and $\theta$, that upon $\theta$ being much more pronounced. For $^4J$(HCCH) calculated couplings are about twice as large as those observed but the experimental dependence upon $\varepsilon$ is confirmed theoretically and the contributions of $\sigma$- and $\pi$- electrons are discussed. In cyclopentadiene the value of $^4J$(HCCH) between $CH_2$-group and CH-group protons depends upon the angle between the planes of the CH—$CH_2$—CH and CH=CH—CH=CH moieties. Finally, $^5J$(HCCCCH) depends upon $\varepsilon$ (Figure), becoming larger as $\varepsilon$ decreases.

6 A. C. Blizzard and D. P. Santry, *J. Chem. Phys.*, 1971, **55**, 950.
7 J. Hilton and L. H. Sutcliffe, *J. Magn. Resonance*, 1974, **14**, 241.
8 K. Hirao, H. Nakatsuji, H. Kato, and T. Yonezawa, *J. Amer. Chem. Soc.*, 1972, **94**, 4078; K. Hirao, H. Nakatsuji, and H. Kato, *ibid.*, 1973, **95**, 31.
9 G. J. den Otter, C. MacLean, C. W. Haigh, and S. Sykes, *J.C.S. Chem. Comm.* 1974, 24.
10 J. B. Rowbotham, A. F. Janzen, J. Peeling, and T. Schaefer, *Canad. J. Chem.*, 1974, **52**, 481.
11 R. E. Wasylishen and T. Schaefer, *Canad. J. Chem.*, 1973, **51**, 3087.
12 V. N. Solkan, N. M. Sergeyev, Yu. V. Klimenko, and Yu. A. Ustynyuk, *Teor. i eksp. Khim.*, 1973, **9**, 258.
13 V. N. Solkan and N. M. Sergeyev, *Org. Magn. Resonance*, 1974, **6**, 200.

Rowbotham and Schaefer[14] have discussed methyl–methyl proton coupling constants in methyl-substituted aromatic systems, including borazine. They have analysed the couplings in terms of $\sigma$- and $\pi$-electron contributions, and INDO FPT calculations are reported. The Canadian workers have also investigated (H,H) coupling in vinylpyridines both experimentally and theoretically (INDO FPT).[15] The couplings over six and seven bonds are found to be $\pi$-electron-dominated and there is some evidence that $^4J(H_\alpha,H_o)$ in 2-methyl-5-vinylpyridine is not stereospecific and therefore is also $\pi$-electron-dominated.

Bildsøe and Schaumburg[16] have reported SOS calculations (using both CNDO/2 and INDO MO's) of (H,H) and (F,H) couplings in aromatic carbonyl fluorides. The results are not encouraging, though they do show that INDO is superior to CNDO/2 in all cases except $^3J$(HCCH). Similarly disappointing were the results of Galasso[17,18] who has described INDO FPT calculations of (H,H) coupling in thiophens. Although the signs of the couplings are correctly predicted[17] the relative magnitudes of $^4J$(HCCCH) and $^4J$(HCSCH) are reversed. Attempts to improve the results by including sulphur *d*-orbitals[18] in a variety of parameterizations produced only a slight improvement in one case. Generally the results were inferior. The same author has used the INDO FPT method to calculate (H,H) couplings in pyridine aldehydes.[19] The calculated values are somewhat larger than the observed and it is suggested that this discrepancy may stem from an overestimate of the $\pi$ contribution. However, the calculated couplings reflect the experimental trends and, in particular, the stereospecificity of $^5J$(HCCCCH) between a ring proton and the aldehydic proton is correctly reproduced. Galasso[20] has also reported INDO FPT calculations of (H,H), (N,H), and (N,C) couplings in pyridine. The (H,H) and (N,H) values are quite good, except for an incorrect sign for $^4J$(HCNCH), but the (N,C) results are poor, probably owing to the neglect of orbital and spin-dipolar terms.

Sasaki[21] has carried out calculations of $^1J$(CH) and $^2J$(HCH) in methane and three simple derivatives using a modified CNDO/2 method and SOS. He has particularly discussed the large number of excited states which make significant contributions to these couplings and concludes that the application of an average-energy approximation in these situations is quite meaningless. CNDO/2 SOS calculations of (C,H) coupling in protonated benzene have been reported by Zhidomirov and Abronin.[22]

Radeglia and Gey[23] have discussed the prediction of (C,H) and (H,H)

14 J. B. Rowbotham and T. Schaefer, *Canad. J. Chem.*, 1974, **52**, 489.
15 J. B. Rowbotham and T. Schaefer, *Canad. J. Chem.*, 1974, **52**, 136.
16 H. Bildsøe and K. Schaumburg, *J. Magn. Resonance*, 1974, **14**, 223.
17 V. Galasso, *Chem. Phys. Letters*, 1973, **21**, 54.
18 V. Galasso, *Z. Naturforsch.*, 1973, **28a**, 1951.
19 V. Galasso, *Mol. Phys.*, 1973, **26**, 81.
20 V. Galasso, *Org. Magn. Resonance*, 1974, **6**, 5.
21 Y. Sasaki, *Chem. and Pharm. Bull.* (*Japan*), 1973, **21**, 888.
22 G. M. Zhidomirov and I. A. Abronin, *Chem. Phys. Letters*, 1973, **21**, 409.
23 R. Radeglia and E. Gey, *Z. phys. Chem.* (*Leipzig*), 1973, **252**, 345.

coupling constants by means of equation (2), in which $\pi'_{s_A s_B}$ is the mutual

$$J_{AB} = a\pi'_{s_A s_B} + b \tag{2}$$

polarizability of the valence *s*-orbitals of the coupled nuclei calculated with the addition of 0.5 atomic units to the energy denominator and *a* and *b* are constants. Using equation (2) $^1J$(CH) and *cis*, *trans*, and *gem* (H,H) couplings in monosubstituted ethylenes can be predicted to within 10 Hz in the case of $^1J$-(CH) and 2 Hz in the case of (H,H). No explanation of the various *a* and *b* values can be offered by the authors at the present time. Radeglia and his co-workers[24] have also calculated $^3J$(HCCH) values for merocyanines, cyanines, and oxanoles *via* H—H bond orders, *P*, calculated from CNDO/2 MO's. There is a fair linear relationship between *J* and *P*, but there are marked deviations from it. The latter occur in ethylenic fragments which are directly bound to oxygen, and the authors suggest that the discrepancy is due to the presence of *cis* structures—the calculations considered only all *trans* conformations. Peat and Reynolds[25] have continued their studies of long-range (H,H) coupling. Application of INDO and CNDO/2 FPT to coupling in conjugated dienes and diynes has again enabled them to form estimates of the $\pi$- and $\sigma$-electron contributions to the coupling constants. Earlier results concerning the reliability of the INDO FPT method for the calculations of long-range $^nJ$(H,H) are confirmed; in all cases signs are given correctly and in most cases magnitudes are accurately predicted.

Lozac'h and Braillon[26] have measured the signs and magnitudes of (H,H) and (C,H) couplings in 2,5-dihydrofuran, 2,5-dihydrothiophen, and butadiene sulphone, paying particular attention to the $^4J$(HCXCH) values in these and similar small molecules. They interpret these 'cross-ring' couplings by means of a novel semi-empirical theory of (H,H) coupling which uses the sum-over-states formalism (with the average-energy approximation) but places heavy emphasis on the particular molecular orbitals which contribute to the coupling. In this the symmetry of the MO's in question plays an important role. This is an interesting approach which throws light upon the values of coupling constants in a way which more elaborate MO methods do not. It would be nice to see it applied to other problems and to have a detailed comparison with the results of more conventional techniques. One important conclusion is that where two identical coupling paths exist the measured coupling should be *four* rather than *two* times that observed for one path.

Theoretical prediction of the effect of a change of dihedral angle upon three-bond coupling has attracted much less attention lately. However, the dependence upon geometry of $^3J$(HNC$^\alpha$H) in peptides is a subject of special interest and a detailed theoretical study (INDO FPT) has been made by Solkan and Bystrov.[27] There is a slight problem in notation here. The standard usage in

[24] R. Radeglia, H. Gey, H. Hennig, W. Kühnel, D. Labes, and H.-G. Sieg, *Z. Chem.*, 1973, **13**, 182.
[25] I. R. Peat and W. F. Reynolds, *Canad. J. Chem.*, 1973, **51**, 2968.
[26] R. Lozac'h and B. Braillon, *J. Magn. Resonance*, 1973, **12**, 244.
[27] V. N. Solkan and V. F. Bystrov, *Tetrahedron Letters*, 1973, 2261.

peptide chemistry is to designate the HNC$^{\alpha}$H dihedral angle by $\phi$, but to measure $\phi$ in such a way that it is equal to $+60°$ and $-120°$ when the C$^{\alpha}$—H bond is *cis* and *trans* respectively to the N—H bond. In the conventional n.m.r. usage these conformations would correspond to dihedral angles of 0° and 180° respectively. To avoid confusion Solkan and Bystrov discuss their results in terms of an HNC$^{\alpha}$H dihedral angle,[27] $\theta$, which is zero when N—H and C$^{\alpha}$—H are eclipsed. We also adopt this notation in describing their work, in which they take *N*-formylaminoacetaldehyde (OCH—NH—C$^{\alpha}$H$_2$CHO) as their model and, using the INDO FPT method, determine the dependence of $^3J$(HNC$^{\alpha}$H) upon $\theta$ and upon $\psi$, the C$^{\alpha}$C torsion (or dihedral) angle. They find the coupling to be positive over almost the whole range of $\theta$ values, the general shape of the dependence being that of a Karplus curve. For all angles $^3J$(HNC$^{\alpha}$H) is smaller for *cis*-peptide-bond than for *trans*-peptide-bond isomers. The value of $\psi$ also influences the (H,H) coupling, but not by more than $\pm$ 1 Hz so that the $J$–$\theta$ relationship remains useful for the study of peptide conformations. The calculated curve is in quite good agreement with the experimental, though the latter shows a smaller difference between the coupling at the two maxima ($\theta = 0°$ and $\theta = 180°$) than does the former.

Solkan and Bystrov[28] have also carried out similar analyses of $^3J$($^{15}$NCCH) and $^3J$($^{15}$NCC$^{13}$C) in peptides and the dependence of these couplings on the rotational state of the central C—C bond.

Giessner-Prettre and Pullman[29] have used a variety of semi-empirical methods to investigate the effect of conformation upon $^3J$(HCCH) in a number of nucleosides, and Giessner-Prettre[30] has made a similar investigation of $^3J$(HNC$^{\alpha}$H) in *N*-methylacetamide. The length of the N—C$^{\alpha}$ bond is an important factor in the latter. *N*-Methylacetamide was also the object of Barfield and Gearhardt's study[31] of $^3J$($^{15}$NCCH) using the INDO FPT technique. They found that their theoretical curve of *J versus* dihedral angle $\phi$ was well reproduced by equation (3) for the *cis* (la) and equation (4) for the *trans* (1b) arrangements of the amide bond. Since both couplings show a change of sign, sign determination is essential if $^3J$(NCCH) is to be used for structural assignments.

$$^3J(cis) = -1.60 + 2.30\cos\phi - 2.58\cos 2\phi \qquad (3)$$
$$^3J(trans) = -1.80 + 2.15\cos\phi - 2.67\cos 2\phi \qquad (4)$$

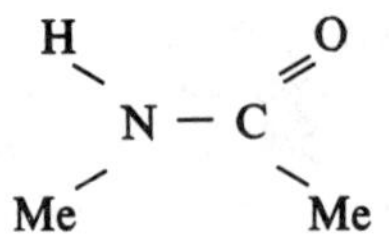

(1a)

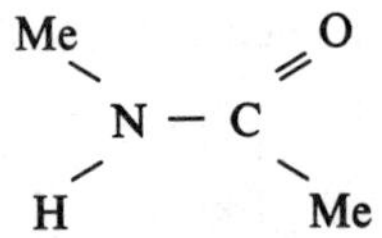

(1b)

[28] V. N. Solkan and V. F. Bystrov, *Izvest. Akad. Nauk S.S.S.R. Ser. khim.*, 1974, 102.
[29] C. Giessner-Prettre and B. Pullman, *J. Theor. Biol.*, 1973, **40**, 441.
[30] C. Giessner-Prettre, *Compt. rend.*, 1973, **276**, *D*, 1045.
[31] M. Barfield and H. L. Gearhart, *Mol. Phys.*, 1974, **27**, 899.

An experimental and theoretical (INDO FPT) analysis of $^3J$(CCCC) has been made by Doddrell *et al.*[32] The calculated curve of $^3J$(C,C) against $\phi$ for butan-2-ol is almost symmetrical about $\phi = 180°$ with maxima at 180° (~4 Hz), 0° and 360° (~5.5 Hz) and minima at 90° and 270° (~0.5 Hz). These figures are in good accord with the experimentally determined couplings for some alcohols of known conformation.

Pomerantz and Hillenbrand[33] have noted an interesting aspect of the $^1J(C^1C^2)$ value in 1-cyanobicyclo[1,1,0]butane (2). Whereas simple calculations (FC term only) based upon hybridization or maximum overlap concepts

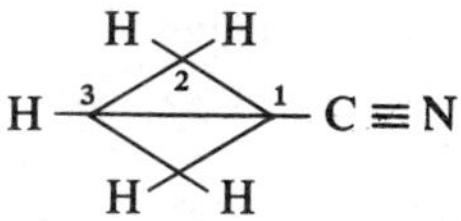

(2)

give values of the correct order of magnitude and positive sign for both $^1J(C^1C^2)$ and $^1J(C^1C^3)$, more sophisticated INDO calculations give 23.4 Hz for the former but − 8.0 Hz for the latter. The experimental values are ±22 Hz and ± 16 Hz respectively and the authors show that the orbital and spin-dipolar terms make significant contributions to the calculated coupling. The authors hope to determine the sign of $^1J(C^1C^3)$ experimentally.

$^1J$(SiC) values have been calculated using values of $\alpha_C^2$ and $\alpha_{Si}^2$ obtained from maximum orbital overlap wavefunctions.[34] The experimental data are found to fit equation (5), in which $\alpha_C^2$ and $\alpha_{Si}^2$ are respectively the carbon and

$$^1J(\text{SiC}) = 555.4\alpha_C^2\alpha_{Si}^2 + 18.2 \quad (5)$$

silicon *s*-characters of the Si—C bond. The authors conclude that the variation in hybridization of the relevant carbon atom accounts for the changes in $^1J$(SiC) observed experimentally, but all the compounds quoted have only carbon atoms directly bonded to the silicon. The factors involved in simple relationships for calculating $^1J$(CH), particularly $\alpha_C^2$, the average energy and the co-ordination number of carbon, have been discussed by Gil and Geraldes.[35]

Seeking to emphasize the simple but clear view of nuclear spin–spin coupling *via* spin propagation and the pairing of electrons in localized bonds, Purcell and Martin[36] have adopted a novel approach. They have focused their attention upon $^2J$(HCH) and their method is to take INDO MO's for the molecules in question, calculated in the presence of a finite perturbation, and then to localize these separately for the $\alpha$ and $\beta$ spin orbitals. The result is a set of localized orbitals which are essentially combinations of *s*-orbitals from H

[32] D. Doddrell, I. Burfitt, J. B. Grutzner, and M. Barfield, *J. Amer. Chem. Soc.*, 1974, **96**, 1241.
[33] M. Pomerantz and D. F. Hillendrand, *J. Amer. Chem. Soc.*, 1973, **95**, 5809.
[34] K. Kovačevič and Z. B. Maksič, *J. Mol. Structure*, 1973, **17**, 203.
[35] V. M. S. Gil and C. F. G. C. Geraldes, *J. Magn. Resonance*, 1973, **11**, 268.
[36] K. F. Purcell and R. L. Martin, *J. Amer. Chem. Soc.*, 1973, **95**, 7956.

and hybrids from the heavier atoms. The differences in the $\alpha$ and $\beta$ orbital populations show how spin density is propagated through the molecule, and diagrams illustrating this can be interpreted in the usual manner and lead to predictions concerning the sign of the coupling between any two nuclei. Using this form of analysis the authors are able to show that the sign of $^{2}J$(HCH) is determined by a balance between one-centre exchange energies and the percentage *s*-character of the carbon hybrids. Some conclusions concerning the effect of hyperconjugation on $^{2}J$(HCH) can also be drawn.

Mallory[37] has proposed a theory of 'through-space' (F,F) coupling based upon the idea that lone-pair orbitals from the coupling fluorine atoms overlap directly to form bonding and antibonding MO's delocalized over both fluorines. The fact that these couplings are observed to be positive is satisfactorily explained in terms of spin correlation between the outer 'overlapping' electrons and spin polarization of the fluorine core electrons. However, the paper lacks a quantitative estimate of the coupling to be expected *via* such a mechanism, and of its dependence upon F—F distance. This would appear to be essential if the theory is to be generally accepted.

**C. Other Topics of Theoretical Interest.**—Observations of CIDNP have been used to determine the absolute signs of $^{19}$F hyperfine splitting constants for substituted benzyl radicals.[38,39] The results are in agreement with theory. Raynes and Riley[40] have used the theoretical results of Schulman and Kaufman[41] to obtain estimates of the bond-length dependence of the one-bond coupling in hydrogen and its isotopomers. They show that the observed couplings should increase with increasing temperature and estimate that $^{1}J$(TH) should rise from *ca.* 336 Hz at 100 to 337.5 Hz at 450 K.

## 3 Coupling of Directly Bonded Nuclei

**A. Coupling to Hydrogen.**—(*i*) *Comparisons of* $^{1}J$(XH) *and* $^{1}J$(XD). James *et al.*[42] have measured $^{1}J(^{11}$BH) and $^{1}J(^{11}$BD) in $LiBH_nD_{n-4}$. The absolute value of $^{1}J(^{11}$BH) decreases as *n* decreases. Two detailed studies of $^{1}J$(CH) and $^{1}J$(CD) have been published.[43,44] Both find that the deviations from the relationship $J(CH)/J(CD) = \gamma_H/\gamma_D$ are very small and $^{1}J$(CD) can be predicted from $^{1}J$(CH) to within 1 Hz. Additional deuterium atoms attached to the same carbon do not appear to affect the coupling to the remaining hydrogens.[43] A few other values of $^{1}J$(CD) have been reported.[45,46]

37 F. B. Mallory, *J. Amer. Chem. Soc.*, 1973, **95**, 7747.
38 M. R. Brinkman, D. Bethell, and J. Hayes, *J. Chem. Phys.*, 1973, **59**, 3431.
39 J. Bargon and K. G. Seifert, *J. Phys. Chem.*, 1973, **77**, 2877.
40 W. T. Raynes and J. P. Riley, *Mol. Phys.*, 1974, **27**, 337.
41 J. M. Schulman and D. N. Kaufman, *J. Chem. Phys.*, 1972, **57**, 2328.
42 B. D. James, B. E. Smith, and R. H. Newman, *J.C.S. Chem. Comm.*, 1974, 294.
43 H. N. Colli, V. Gold, and J. E. Pearson, *J.C.S. Chem. Comm.*, 1973, 408.
44 E. Breitmaier, G. Jung, W. Voelter, and L. Pohl, *Tetrahedron*, 1973, **29**, 2485.
45 H. Günther and G. Jikeli, *Chem. Ber.*, 1973, **106**, 1863.
46 R. H. Martin, J. Moriau, and N. Defay, *Tetrahedron*, 1974, **30**, 179.

(*ii*) *Elements of Group IVB*. Theoretical work on $^1J$(CH) has already been described in Sections 2A and 2B.

Iordache[47] has made a careful statistical analysis of the many relationships between $^1J$(CH) and structural parameters such as C–H bond length, CCC angle, *etc*. He concludes that all such correlations are both valid and useful in relating *J* to structural features, and *vice versa*, but he emphasizes the necessity for careful choice of experimental data and the correct use of the correlations obtained.

Masuda and Imamoto[48] have attempted to separate the influence of effective nuclear charge and hybridization upon $^1J$(CH) by investigating values of this coupling in aromatic methyl compounds of general formula Ph—Y—Me. They have correlated the coupling with substituent-group electronegativities $\chi_g$(PhY), which have been related to effective nuclear charge, and with orbital electronegativities $\chi_o$(Y), which are linearly related to the *s*-character of the carbon hybrid orbitals. They find good linear correlations of $^1J$(CH) with both $\chi_g$(PhY) and $\chi_o$(Y), demonstrating that the coupling depends on both effective nuclear charge and *s*-character. A linear relationship was also found between $^1J$(CH) and $(1 + n^*)$ where $n^*$, the number of valence electrons on the atom Y, was related to the electronegativity of Y by Gordy[49] for atoms, and modified for radicals by Wilmshurst.[50] Gil and Geraldes[35] have considered the relationship between $^1J$(CH) and carbon *s*-character, $\alpha_C^2$, average energy, $\Delta E$, and co-ordination number of carbon, *n*. With reference to equation (6), which was suggested by Maksić *et al.*[51] they propose that $k_1$

$$J(\mathrm{CH}) = \frac{k_1}{1 + S_{\mathrm{CH}}^2}\,\alpha_C^2 - k_2 \qquad (6)$$

depends upon $\Delta E$ and should change with *n*. If this were taken into account $k_2$ would appear to be an artifact of one of the basic assumptions made in the theory, namely the neglect of delocalization in almost localized MOs. The effect of substituents on $\alpha_C^2$ and hence upon $^1J$(CH) in cyclopropyl ethers has been discussed by Galy, Sarrazin, and Vincent.[52] A similar analysis has been applied by Barboiu[53] to the signs and magnitudes of $^1J$(CH) for acetic acid, ethanol, acetaldehyde, and some of their derivatives. Kennedy and McFarlane[54,55] have reported numerous $^1J$(CH) values for the methyl groups in a variety of organotins. As *n* increases in the series $Me_{4-n}Sn(SR)_n$ (R = alkyl or phenyl) $^1J$(CH) in the tin methyl groups increases[55] and this is attributed to the effect of nuclear charge and/or the diversion of *s*-character from the C—

[47] D. Iordache, *Rev. Roumaine Phys.*, 1973, **18**, 1165.
[48] S. Masuda and N. Inamoto, *Chem. Letters*, 1973, 655.
[49] W. Gordy, *Phys. Rev.*, 1946, **69**, 604.
[50] J. K. Wilmshurst, *J. Chem. Phys.*, 1957, **27**, 1129.
[51] Z. B. Maksič, M. Eckert-Maksič, and M. Randič, *Theor. Chim. Acta*, 1971, **22**, 70.
[52] J. P. Galy, M. Sarrazin, and E. J. Vincent, *Compt. rend.*, 1973, **276**, *C*, 791.
[53] V. Barboiu, *Rev. Roumaine Chim.*, 1974, **19**, 363.
[54] J. D. Kennedy and W. McFarlane, *J.C.S. Dalton*, 1973, 2134.
[55] J. D. Kennedy and W. McFarlane, *J.C.S. Perkin II*, 1974, 146.

Sn bond as a result of electron withdrawal from Sn by the electronegative thiolate groups. The replacement of tin methyl groups by selenium has exactly the same effect on the (C,H) coupling in the remaining methyl groups.[54] For the compounds $Me_{4-n}MX_n$ (M = Si or Sn, X = halogen) the variation of $^1J$(CH) with $n$ also shows an increase of the coupling as $n$ increases.[56] However, for the methylsilicon halides variation of the halogen reveals that the coupling decreases in the series iodide > bromide > chloride over the whole range except for $MeSiI_3$ and $MeSiBr_3$. This is also true of the tin monohalides; with the di- and tri-halides the effect is diminished and the values approach each other. The authors suggest an explanation in terms of the contributions of various valence bond structures to the coupling.[56]

Linear correlations of $^1J$(CH) with the Swain–Lipton $F$ parameter,[57] Brown $\sigma^+$ substituent constants,[58] and Hammett's $\sigma_p$[59] have been reported. Ewing[60] has investigated the $^1J$(CH) values of a series of vinyl derivatives and related compounds (3). $^1J$(CH) for the hydrogen attached to the same carbon

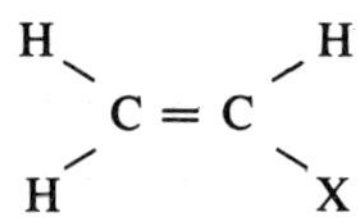

(3)

as X correlates very badly with substituent electronegativity, but very well with $\sigma_I$ if conjugating substituents are excluded. In the case of halogen substituents the marked stereospecificity of the $^1J$(CH) values of the protons *cis* and *trans* to X is rationalized in terms of an intrinsic dependence of the atom–atom polarizability on the dihedral angle between the coupling atoms and the substituent. An additional positive increment to the coupling of the *cis* proton, due to direct interaction of the substituent non-bonding electrons or an orbital effect, is also suggested.

Ewing[61] has also examined the $^1H$ spectra of *ortho*-substituted benzaldehydes for evidence of the influence of the benzene-ring $\pi$-system on the $^1J$(CH) and $^2J$(CCH) couplings. Unfortunately, the expected effects, if indeed they are present, are obscured by larger ones due to the changing mesomeric interaction in the different molecules. An increase in $^1J$(CH) values for benzene and anisole is found on complexing with $Cr(CO)_3$.[62] Values of the increase for ring protons lay in the range 13—18 Hz, the only low value (1.5 Hz) being that for the OMe group of anisole.

[56] E. V. van den Berghe and G. P. van der Kelen, *J. Organometallic Chem.*, 1973, **59,** 175.
[57] N. H. Werstiuk, R. Taillefer, R. A. Bell, and B. Sayer, *Canad. J. Chem.*, 1973, **51,** 3010.
[58] A. Grinvald and M. Rabinovitz, *J.C.S. Perkin II*, 1974, 94.
[59] Vo-Kim-Yen, Z. Papoušková, J. Schraml, and V. Chvalovský, *Coll. Czech. Chem. Comm.*, 1973, **38,** 3167.
[60] D. F. Ewing, *Org. Magn. Resonance*, 1973, **5,** 567.
[61] D. F. Ewing, *Org. Magn. Resonance*, 1974, **6,** 298.
[62] G. M. Bodmer and L. J. Todd, *Inorg. Chem.*, 1974, **13,** 360.

Considerable attention has been paid to the effect on $^1J$(CH) of the lone-pair electrons on near-by nitrogen or oxygen atoms. Thus Ban Binst and Tourwe[63] found that $^1J$(CH) for carbons adjacent to a bridgehead nitrogen in quinolizidines was 6—12 Hz larger for conformations in which the lone pair was *cis* to the hydrogen than for those in which it was *trans*, while Takeuchi, Chivers, and Crabb[64] have noted that $^1J$(CH) for a C—H bond *trans* coplanar to the lone pair of an adjacent nitrogen is of the order of 10 Hz larger than it is in the case of a *gauche* relationship between C—H bond and lone pair. The effects of oxygen lone pairs have been noted by Albriktsen who, in the case of cyclic sulphites, finds that the $^1J$(CH) values for carbons adjacent to oxygen differ by about 8 Hz.[65] He assigns the larger coupling to the equatorial proton, *i.e.* the proton having a *gauche* relationship to both oxygen lone pairs. Similarly, in the 1,3-dioxans $^1J$(CH) for the carbon between the two oxygen atoms is *ca.* 8 Hz larger for the coupling to the equatorial than to the axial proton,[66] and Albriktsen attributes this difference to the fact that the equatorial proton is nearer to the oxygen lone pairs than the axial proton. The dioxans do not appear to show $^1J$(CH) values which are sensitive to the orientation of more remote lone pairs, though there is some evidence of this in the cyclic sulphites.[65] There are rather interesting differences between the effects of oxygen and nitrogen lone pairs on $^1J$(CH), and the results described in the two papers dealing with nitrogen lone pairs do not show an obvious relationship to each other. These problems deserve further theoretical and experimental study.

Werstiuk *et al.*[57] have compared $^1J$(CH) values in bicyclic and tricyclic ketones with those in acyclic and monocyclic compounds. The coupling constant at the bridgehead $\alpha$ to the carbonyl group is increased in the former as compared with the latter group of molecules. A detailed study of $^1J$(CH) in hexapyranoses has been reported.[67] In almost all cases a difference of *ca.* 10 Hz was found between the $\alpha$- and $\beta$-anomers, the former being *ca.* 170 Hz, the latter *ca.* 160 Hz. The dependence of $^1J$(CH) on temperature for butadiene and some deuteriated derivatives has been described.[68]

Pehk, Alla, and Lippmaa[69] have noted a linear relationship between $^1J$(CH) and $\delta_C$ for the methyl groups of some rigid hydrocarbons. Bhattacharyya and Dailey have shown that $^1J$(CH) is positive in methyl alcohol[70] and also in methyl chloride, bromide, and iodide.[71]

One-bond (C,H) coupling in the allylic portion of $\pi$-allylic palladium complexes has been discussed by Mann *et al.*,[72] and $^1J$(CH) values have been

[63] G. Ban Binst and D. Tourwe, *Heterocycles*, 1973, **1**, 257.
[64] Y. Takeuchi, P. J. Chivers, and T. A. Crabb, *J.C.S. Chem. Comm.*, 1974, 210.
[65] P. Albriktsen, *Acta Chem. Scand.*, 1973, **27**, 3889.
[66] K. Bock and L. Wiebe, *Acta Chem. Scand.*, 1973, **27**, 2676.
[67] K. Bock and C. Pedersen, *J.C.S. Perkin II*, 1974, 293.
[68] R. L. Lipnick and E. W. Garbisch, *J. Amer. Chem. Soc.*, 1973, **95**, 6370, 6375.
[69] T. Pehk, M. Alla, and E. Lippmaa, *Org. Magn. Resonance*, 1973, **5**, 351.
[70] P. K. Bhattacharyya and B. P. Dailey, *J. Chem. Phys.*, 1973, **59**, 3737.
[71] P. K. Bhattacharyya and B. P. Dailey, *Mol. Phys.*, 1973, **26**, 1379.
[72] B. E. Mann, R. Pietropaolo, and B. L. Shaw, *J.C.S. Dalton*, 1973, 2390.

reported for ferrocenylmethyl carbonium ions,[73] keten,[74] 1,5-cyclo-octadivne,[75] and the trishomocyclopropenyl cation.[76]

Rastelli and Pozzoli[77] have applied the results of a novel calculation of hybridization in methyl and phenyl derivatives of silane to the estimation of $^1J(\text{SiH})$ values. They find that the experimental results lie on lines parallel to that given by equation (7) and that the points can be brought on to this line by

$$^1J(\text{SiH}) = 810\alpha_{\text{Si}}^2 \qquad (7)$$

means of additive contributions which are negative for the methyl group and positive for phenyl. The experimental points are then given well by equation (8), in which $N_{\text{Ph}}$ and $M_{\text{Me}}$ are respectively the numbers of phenyl and methyl groups attached to silicon.

$$^1J(\text{SiH}) = 810\alpha_{\text{Si}}^2 + 4.30\ N_{\text{Ph}} - 1.40\ N_{\text{Me}} \qquad (8)$$

(*iii*) *Elements of Group VB.* Theoretical work on $^1J(^{15}\text{NH})$ by Wasylishen and Schaefer[11] has been described in Section 2B. Cowley and Schweiger[78] have measured this coupling in some $^{15}$N-labelled aminophosphines, aminophosphoranes, aminoarsines, sulphenamides, and aminosilanes. They find very little dependence of the coupling on solvent. There is a dependence upon electronegativity which makes itself felt in the failure of Binsch's relationship (9), which is based on a series of organonitrogen compounds and does not

$$\alpha_{\text{N}}^2 = 0.43\ ^1J(\text{NH}) - 6 \qquad (9)$$

hold in cases where the electronegativity of the heteroatom is markedly different from that of carbon. Indeed, in the fragment X—$NH_2$ there is a good correlation between $^1J(\text{NH})$ and the electronegativity of the atom or group X. There is not a linear relationship between $^1J(\text{NH})$ and $\delta_{\text{H}}$.[78]

Measurements in nematic solvents have shown that $^1J(\text{PH})$ is positive in phosphine.[79] In dialkylhypophosphites of Al, Ga, and In $^1J(\text{PH})$ decreases as we go down the group and this has been ascribed to the increasing of the polar character of the P—H bond due to the increasing electronegativity of the metal.[80] Stec[81] has reported studies of the effect of conformation on $^1J(\text{PH})$ in dioxaphosphorinans.

Further reports of one-bond coupling to hydrogen may be found in the

[73] G. H. Williams, D. D. Traficante, and D. Seyferth, *J. Organometallic Chem.*, 1973, **60**, C53.
[74] J. Firl and W. Runge, *Angew. Chem. Internat. Edn.*, 1973, **12**, 668.
[75] E. Kloster-Jensen and J. Wirz, *Angew. Chem. Internat. Edn.*, 1973, **12**, 671.
[76] S. Masamune, M. Saki, A. V. Kemp-Jones, and T. Nakashima, *Canad. J. Chem.*, 1974, **52**, 855.
[77] A. Rastelli and S. A. Pozzoli, *J. Mol. Structure*, 1973, **18**, 463.
[78] A. H. Cowley and J. R. Schweiger, *J. Amer. Chem. Soc.*, 1973, **95**, 4179.
[79] N. Zumbulyadis and B. P. Dailey, *Mol. Phys.*, 1974, **27**, 633.
[80] B. Schaible, W. Haubold, and J. Weidlein, *Z. anorg. Chem.*, 1974, **403**, 289.
[81] W. Stec, *Z. Naturforsch.*, 1974, **29b**, 109.

following papers: $^1J$(BH) (refs. 82—90), $^1J$(CH) (refs. 26 and 90—112), $^1J$(SiH) (refs. 113—115), $^1J$(NH) (refs. 96, 113, and 116—119), $^1J$(PH) (refs. 86, 120, and 121) and $^1J$(PtH) (refs. 115 and 122—126).

[82] R. K. Hertz, H. D. Johnson, and S. G. Shore, *Inorg. Chem.*, 1973, **12,** 1875.
[83] R. Schaeffer and E. Walter, *Inorg. Chem.*, 1973, **12,** 2209.
[84] G. N. Boiko, Yu. I. Malov, and K. N. Semenenko, *Izvest. Akad. Nauk S.S.S.R., Ser. khim.*, 1973, 1143.
[85] J. P. Laussac, G. Jugie, J. P. Laurent, and F. Gallais, *Compt. rend.*, 1973, **276** *C*, 1497.
[86] B. Rapp and J. E. Drake, *Inorg. Chem.*, 1973, **12,** 2868.
[87] T. Onak, B. Lockman, and G. Haran, *J.C.S. Dalton*, 1973, 2115.
[88] T. Onak, K. Gross, J. Tse, and J. Howard, *J.C.S. Dalton*, 1973, 2633.
[89] J. Rathke and R. Schaeffer, *Inorg. Chem.*, 1974, **13,** 760.
[90] T. Onak and E. Wan, *J.C.S. Dalton*, 1974, 665.
[91] L. A. Fedorov, Z. Stumbreviciute, A. K. Prokof'ev, and E. I. Fedin, *Doklady Akad. Nauk S.S.S.R.*, 1973, **209,** 134.
[92] R. J. Abraham and P. Loftus, *J.C.S. Chem. Comm.*, 1973, 454.
[93] K. Hildenbrand and H. Dreeskamp, *Z. Naturforsch*, 1973, **28b,** 226.
[94] J. F. Hinton and L. W. Jaques, *J. Magn. Resonance*, 1973, **11,** 229.
[95] L. F. Williams and A. A. Bothner-By, *J. Magn. Resonance*, 1973, **11,** 314.
[96] J. M. Briggs, E. Rahkamaa, and E. W. Randall, *J. Magn. Resonance*, 1973, **12,** 40.
[97] B. L. Shapiro and M. D. Johnston, *J. Magn. Resonance*, 1973, **12,** 199.
[98] H. A. Brune, G. Horlbeck, H. Röttele, and U. Tanger, *Z. Naturforsch.*, 1973, **28b,** 68.
[99] J. W. Anderson, G. K. Barker, J. E. Drake, and M. Rodger, *J.C.S. Dalton*, 1973, 1716.
[100] H. J. C. Yeh, H. Ziffer, D. M. Jerina, and D. R. Boyd, *J. Amer. Chem. Soc.*, 1973, **95,** 2741.
[101] H. A. Brune and G. Horlbeck, *Z. Naturforsch.*, 1973, **28b,** 656.
[102] R. P. Hughes and J. Powell, *J. Organometallic Chem.*, 1973, **60,** 427.
[103] H. Günther, G. Jikeli, H. Schmickler, and J. Prestien, *Angew. Chem. Internat. Edn.*, 1973, **12,** 762.
[104] M. Begtrup, *Acta Chem. Scand.*, 1973, **27,** 3101.
[105] B. Birdsall, J. Feeney, and P. Partington, *J.C.S. Perkin II*, 1973, 2145.
[106] J. Runsink, J. de Wit, and W. D. Weringa, *Tetrahedron Letters*, 1974, 55.
[107] G. Fronza, A. Gamba, R. Mondelli, and G. Pagani, *J. Magn. Resonance*, 1973, **12,** 231.
[108] M. L. Sheer, *Org. Magn. Resonance*, 1974, **6,** 85.
[109] T. Onak and E. Wan, *J. Magn. Resonance*, 1974, **14,** 66.
[110] G. Müller and W. von Philipsborn, *Helv. Chim. Acta*, 1973, **56,** 2680.
[111] H. C. Clark, L. E. Manzer, and J. E. H. Ward, *Canad. J. Chem.*, 1974, **52,** 1165.
[112] P. Granger and M. Maugras, *Chem. Phys. Letters*, 1974, **24,** 331.
[113] D. W. W. Anderson, J. E. Bentham, and D. W. H. Rankin, *J.C.S. Dalton*, 1973, 1215.
[114] G. Fritz and N. Braunagel, *Z. anorg. Chem.*, 1973, **399,** 280.
[115] D. W. W. Anderson, E. A. V. Ebsworth, and D. W. H. Rankin, *J.C.S. Dalton*, 1973, 2370.
[116] J. A. Sogn, W. A. Gibbons, and E. W. Randall, *Biochemistry*, 1973, **12,** 2100.
[117] J. M. Briggs, E. Rahkamaa, and E. W. Randall, *J. Magn. Resonance*, 1973, **11,** 416.
[118] W. Walter, M. F. Sieveking, and E. Schaumann, *Tetrahedron Letters*, 1974, 839.
[119] A. Maquestiau, Y. Van Haverbeke, and R. N. Muller, *Org. Magn. Resonance*, 1974, **6,** 224.
[120] J. W. Gilje, R. W. Braun, and A. H. Cowley, *J.C.S. Chem. Comm.*, 1973, 813.
[121] R. Burgada and C. Laurenço, *J. Organometallic Chem.*, 1974, **66,** 255.
[122] D. W. W. Anderson, E. A. V. Ebsworth, J. K. Macdougall, and D. W. H. Rankin, *J. Inorg. Nuclear Chem.*, 1973, **35,** 2259.
[123] K. R. Dixon, K. C. Moss, and M. A. R. Smith, *J.C.S. Dalton*, 1973, 1528.
[124] K. S. Murray, B. E. Reichert, and B. O. West, *J. Organometallic Chem.*, 1973, **63,** 461.
[125] T. W. Dingle and K. R. Dixon, *Inorg. Chem.*, 1974, **13,** 846.
[126] W. McFarlane, *J.C.S. Dalton*, 1974, 324.

**B. Coupling to Elements of Groups IIIB and IVB.**—(*i*) *Coupling between Carbon and Boron*. Previous results[127] on (C,B) coupling in sodium tetraphenylboron have been amended and values of $^{1}J(C^{10}B)$ and $^{3}J(CCC^{10}B)$, 16.4 Hz and *ca.* 0.9 Hz respectively, added.[128]

(*ii*) *Elements of Group IVB and Mercury*. The interesting theoretical problems associated with the value of $^{1}J(C^{1}3^{3})$ in 1-cyanobicyclo[1,1,0]butane (2)[33] have been discussed in Section 2B. Günther and Herrig[129] have made an extensive investigation of $^{1}J(CC)$ in methylenecycloalkanes (4) and in 1,1′-di-t-butylethylene (5). $^{1}J(C^{2}C^{3})$ drops from 95.2 Hz for $n = 1$ to 73.2 Hz for

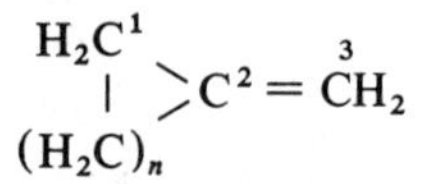

(4) $n = 1$—6

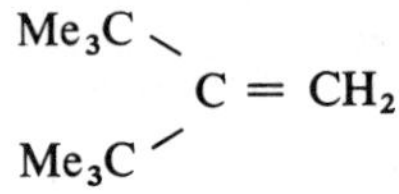

(5)

$n = 2$ and to 72.0 Hz for $n = 6$. $^{1}J(C^{1}C^{2})$ changes more slowly from 23.2 Hz for $n = 1$ to 40.3 Hz for $n = 4$. It then remains effectively constant. The authors conclude that these results can be satisfactorily interpreted in terms of carbon *s*-character, $\alpha_{C}^{2}$, and propose the equation (10).

$$^{1}J(C^{i}C^{j}) = (550 \pm 50)\alpha_{Ci}^{2}\,\alpha_{Cj}^{2} \qquad (10)$$

Where carbon atoms with similar hybridizations are involved good linear correlations of $^{1}J(CC)$ with carbon–carbon distance (in Å), $R_{CC}$, are found:[129] for $sp^{3}$–$sp^{3}$, $sp^{3}$–$sp^{2}$, and $sp^{2}$–$sp^{2}$ equation (11) and for $sp$–$sp$, $sp$–$sp^{2}$, and $sp$–$sp^{3}$ equation (12).

$$^{1}J(CC) = (-\,167.2 \pm 9.4)R_{CC} + (294.4 \pm 13.5) \qquad (11)$$

$$^{1}J(CC) = (-\,570.5 \pm 24.8)R_{CC} + (854.8 \pm 33.7) \qquad (12)$$

Olah and Westerman[130] have reported values of $^{1}J(CC)$ for a number of carbocations. They find an approximate correlation with the hybridization at each carbon atom, and the effect of charge seems to be quite small. The effect of pH on (C,C) coupling in amino-acids has been studied,[131] and here also the effect of charge on the coupling seems rather small. The authors estimate that for leucine (6), $^{1}J(C_{\alpha}C[O_{2}H]) = 59.5 \pm 0.5$ Hz and $^{1}J(C_{\alpha}C[O_{2}{}^{-}]) = 53.5 \pm 0.5$ Hz. Marshall and Miller[132] have noted that $^{1}J(CC)$ does not depend simply on ring strain.

Theoretical work on $^{1}J(SiC)$ has already been described in Section 2B. A large amount of data on $^{1}J(^{119}SnC)$ has been given by Mitchell,[133] who has a

127 D. E. Axelson, A. J. Oliver, and C. E. Holloway, *Org. Magn. Resonance*, 1973, **5**, 255.
128 D. E. Axelson, A. J. Oliver, and C. E. Holloway, *Org. Magn. Resonance*, 1974, **6**, 64.
129 H. Günther and W. Herrig, *Chem. Ber.*, 1973, **106**, 3938.
130 G. A. Olah and P. W. Westerman, *J. Amer. Chem. Soc.*, 1974, **96**, 2229.
131 Son Tran-Dinh, S. Fermandjian, E. Sala, R. Mermet-Bouvier, M. Cohen, and P. Fromageot, *J. Amer. Chem. Soc.*, 1974, **96**, 1484.
132 J. L. Marshall and D. E. Miller, *J. Amer. Chem. Soc.*, 1973, **95**, 8305.
133 T. N. Mitchell, *J. Organometallic Chem.*, 1973, **59**, 189.

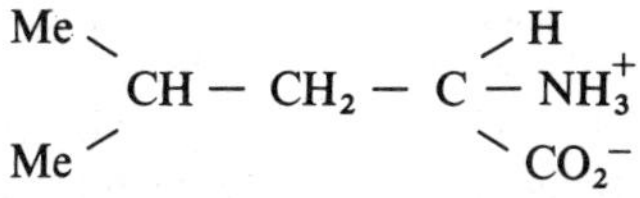

(6)

useful table of ranges of this coupling in alkytins, *viz.* 300–340 Hz, tetra-alkytins; 330—390 and 450—480 Hz, 4- and 5-co-ordinated trialkyltins respectively; and 370—480 and 900—970 Hz, 4- and 6-co-ordinated dialkyltins respectively. Schaeffer and Zuckerman[134] have found that $^1J$(SnC) for a number of substituted aryltrimethyltin derivatives correlates very well with the Hammett $\sigma$ values for the substituents. A plot of $^1J$(SnC) for the methyl carbon against that for the phenyl carbon gives a poor straight line of negative slope which does not pass through the origin. This could indicate contributions to the coupling other than FC.

Doddrell *et al.*[135] have investigated $^1J$(MC) (M = $^{119}$Sn, $^{207}$Pb, or $^{199}$Hg) in furyl and thienyl organometallic compounds (7). They find large increases in (C,M) coupling as methyl groups are replaced by furyl and thienyl groups which are not reasonably explained by changes in *s*-character. Similar conclusions have been drawn by others.[136] They[135] suggest that effective nuclear charge and $\Delta E$ also play an important role.

(7) X = O or S
M = Sn, Pb, or Hg

Mitchell[137] has found that $^1J$(SnSn) in hexaorganotins shows a marked dependence upon the effective nuclear charge at tin.

(*iii*) *Elements of Group VB.* The idea that there may be a simple connection between $^1J$(NC) and *s*-character at nitrogen or carbon has been increasingly questioned. Axelson *et al.*[138] have discussed this problem in some detail and can find no interpretation of $^1J$(NC) in terms of these factors plus the effect of nitrogen lone pairs. They conclude that further efforts in this direction will be unfruitful until more information concerning the signs of the couplings becomes available, and that separate consideration of lone-pair, substituent, and $\pi$-bond effects is necessary. A value of $^1J(^{15}$NC) of $\pm$ 77.5 Hz, the largest

[134] C. D. Schaeffer and J. J. Zuckerman, *J. Organometallic Chem.*, 1973, **55**, 97.
[135] D. Doddrell, K. G. Lewis, C. E. Mulquiney, W. Adcock, W. Kitching, and M. Bullpitt, *Austral. J. Chem.*, 1974, **27**, 417.
[136] H. G. Kuivila, J. L. Considine, R. J. Mynott, and R. H. Sarma, *J. Organometallic Chem.*, 1973, **55**, C11.
[137] T. N. Mitchell, *J. Organometallic Chem.*, 1974, **70**, C1.
[138] D. E. Axelson, I. M. Walker, A. J. Oliver, and C. E. Holloway, *Spectroscopy Letters*, 1973, **6**, 475.

yet reported, has been measured in 2,4,6-trimethylbenzonitrile $^{15}N$-oxide.[139] The very high value of the coupling is again at odds with the relationship between $^1J$(NC) and $\alpha_N^2$ and $\alpha_C^2$, on the basis of which it should be *ca.* 30 Hz. This may well indicate the importance of terms other than FC, in which case such contributions to the coupling appear to be more important for $^1J$(NC) than for $^1J$(CC). The idea that OB and SD terms make important contributions to $^1J$(NC) receives some support from the poor correlation between that coupling and $^2J$(NCC) which was observed in a series of saturated amides.[140] The $^1J(^{14}NC)$ coupling corresponding to the large $^1J(^{15}NC)$ above was also measured;[139] its value is 52 ± 2 Hz ($77.5 \times \gamma_{^{15}N}/\gamma_{^{14}N} = 55.3$ Hz).

Lichter, Dorman, and Wasylishen[141] have also discussed the effect of N lone-pair orientation on $^1J$(NC). The effect is quite marked. In acetaldoxime (8), for example, $^1J$(NC) is 4.0 Hz for the *E*-isomer and 2.3 Hz for the *Z*, but

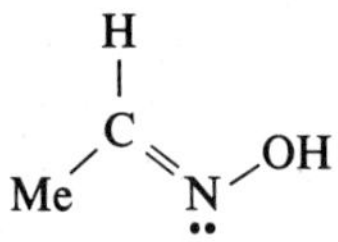


*E*-isomer
(8a)

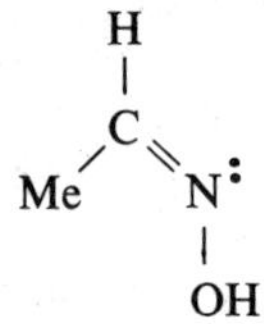


*Z*-isomer
(8b)

in geometrically isomeric oximes in general substantial differences in $^1J$(NC) arise only when one substituent is a proton. INDO FPT calculations suggest[141] that the much larger $^1J$(NC) values in pyridinium and quinolinium ions are negative whereas the small oxime values are positive. Again, the oxime values fall far below the 13 Hz expected on the basis of the *s*-character at C and N.

$^1K(^{15}N^{13}C)$ has been shown to be negative,[100] and for formamide in aqueous solutions its variation with concentration has been measured.[142] [For $^1J$(SiN) see Section 3C.]

Gray has reported values of $^1J$(PC) for ylides[143] and has also compared the coupling with $^1J$(CC) and $^1J$(CH) in systems of the form Y—$^{13}CH_2$—X, where X may be one of a variety of substituents and Y = $^1$H, $^{13}$C, or $^{31}$P.[144] There is a fairly good correlation of $^1J$(PC) with $^1J$(CC) but not with $^1J$(CH). $^1J$(PC) is large in ylides[144] and varies with the stereochemistry of the organic part of the molecule. No obvious explanation of this is apparent since we have $P^+$, and hence no lone pairs, and the conformation in the immediate region of the C—P bond is the same. In the case of phenyl phosphorus ylides the

[139] M. Christl, J. P. Warren, B. L. Hawkins, and J. D. Roberts, *J. Amer. Chem. Soc.*, 1973, **95**, 4392.
[140] R. L. Lichter, C. G. Fehder, P. H. Patton, J. Combes, and D. E. Dorman, *J.C.S. Chem. Comm.*, 1974, 114.
[141] R. L. Lichter, D. E. Dorman, and R. Wasylishen, *J. Amer. Chem. Soc.*, 1974, **96**, 930.
[142] J. F. Hinton, K. H. Ladner, and W. E. Stewart, *J. Magn. Resonance*, 1973, **12**, 90.
[143] G. A. Gray, *J. Amer. Chem. Soc.*, 1973, **95**, 5092.
[144] G. A. Gray, *J. Amer. Chem. Soc.*, 1973, **95**, 7736.

one-bond coupling to the ylide carbon is much greater than that to the phenyl carbon.

(*iv*) *Fluorine*. Hinto and Jaques[94] have found a fairly good linear relationship between $^1J$(FC) and the F chemical shift for a variety of fluorinated hydrocarbons.

(*v*) *Metals*. For carbonyl complexes of Mo, W, V, Fe, and Co Mann[145] has shown that the value of $n \times {}^1K(\mathrm{M}^{13}\mathrm{C})/s^2_{\mathrm{M}}(0)$ (where $n$ = the number of carbonyl groups attached to M) is a constant in the range 40—50 or 44—62 (dimensions in the units used by Mann[145]) depending upon the theoretical values chosen for $s^2_{\mathrm{M}}(0)$. This result is consistent with the proposition that these couplings are FC-dominated. The $^1J$(WC) values seem to give a measure of the *trans* effect of the ligand. Braterman *et al.*[146] have reported $^1J$(WC) values for a number of tungsten carbonyl derivatives. The fact that the ratio of $^1J$(MC) in the molybdenum and tungsten hexacarbonyls is approximately equal to the ratio of $^2J$(PMC) in analogous Mo and W complexes is, they suggest, further evidence for the importance of *s*-electron density at the metal nucleus in these couplings. $^1J(^{59}\mathrm{CoC})$ in $[\mathrm{Co(CN)_6}]^{3-}$ = 126 ± 4 Hz.[147]

Values of $^1J(^{103}\mathrm{RhC})$ for rhodium olefin and carbonyl complexes have been reported by Axelson, Holloway, and Oliver,[148] who have discussed their results in relation to the corresponding couplings to Pt. Good linear correlations between $^1J$(PtC) and $^2J$(PtCH),[149] $^2J$(PtCC),[150] and $^3J$(PtCCC)[150] have been reported. These results must indicate a dependence of all the couplings on similar factors, and in view of the relationship to $^2J$(PtCH) a dominant FC term is indicated. A similar conclusion might be drawn from the work of Cardin *et al.*[151] who find that trends in $^1J$(PtC) do not support the view that the coupling is inversely proportional to the charge-donor ability of the *trans* ligand, but may reflect changes in the hybridization in the Pt—C bond. Cooper *et al.*[152] have measured $^1J$(PtC) in complexes of styrene derivatives with $\mathrm{Pt^{II}}$. The coupling is found to correlate linearly with Brown's $\sigma_p^+$ value for the substituent on styrene. The coupling to the C-1 ethylenic carbon, which is considerably greater than that to C-2, is rationalized in VB terms. $^1J$(PtC) in *cis*-methylplatinum(II) derivatives has been tentatively related to the steric bulk of the ligands.[153]

145 B. E. Mann, *J.C.S. Dalton*, 1973, 2012.
146 P. S. Braterman, D. W. Milne, E. W. Randall, and E. Rosenberg, *J.C.S. Dalton*, 1973, 1027.
147 M. Hirota, Y. Koike, H. Ishizuka, A. Yamasaki, and S. Fujiwara, *Chem. Letters*, 1973, 853.
148 D. E. Axelson, C. E. Holloway, and A. J. Oliver, *Inorg. Nuclear Chem. Letters*, 1973, **9**, 885.
149 M. H. Chisholm, H. C. Clark, L. E. Manzer, J. B. Stothers, and J. E. H. Ward, *J. Amer. Chem. Soc.*, 1973, **95**, 8574.
150 H. C. Clark and J. E. H. Ward, *J. Amer. Chem. Soc.*, 1974, **96**, 1741.
151 D. J. Cardin, B. Cetinkaya, E. Cetinkaya, M. F. Lappert, E. W. Randall, and E. Rosenberg, *J.C.S. Dalton*, 1973, 1982.
152 D. G. Cooper, G. K. Hamer, J. Powell, and W. F. Reynolds, *J.C.S. Chem. Comm.*, 1973, 449.
153 H. C. Clark and J. E. H. Ward, *Canad. J. Chem.*, 1974, **52**, 570.

Ibusuki and Saito[154] have noted a linear correlation between $^1J$(HgC) and $^2J$(HgCH), and Fedorov *et al.*[91] have found a linear correlation with hybridization at C. Both results suggest a dominant FC term in this coupling and, indeed, the evidence seems to be that the FC term is dominant in all one-bond metal–carbon couplings for which reasonable quantities of data are available. $^1J$(HgC) values have also been reported by Doddrell and his co-workers,[135] and Hagen *et al.*[155] have measured $^1J(^{113/111}CdC)$ in the $Cd^{II}$ edta complex.

Other one-bond couplings to elements of Groups IIIB and IVB may be found in the following papers: $^1J$(CB) (ref. 109); $^1J$(PB) (refs. 85 and 86); $^1J$(FB) (refs. 86 and 156); $^1J$(CC) (refs. 90 and 157—159); $^1J$(PC) (refs. 93, 145, 146, 160, and 161); $^1J$(FC) (refs. 53, 92, 101, 102, and 162—164); $^1J$(MetalC) (refs. 102, 111, and 165—167); and $^1J$(PtSi) (ref. 115).

**C. Coupling to Elements of Group VB.**—(*i*) *Silicon and Elements of Group VB.* Anderson, Bentham, and Rankin[113] failed to observe $^1J$(SiN) in $PF_2$-$^{15}NHSiH_3$ and $(SiH_3)_2{}^{15}NH$, but for $(SiH)_3{}^{15}N$ they found a value of + 6 Hz. They also determined the signs and magnitudes of $^1J$(PN) in $PF_2{}^{15}NH_2$ ( + 73 Hz) and $PF_2{}^{15}NHSiH_3$ ( + 69 Hz). Bhattacharyya and Dailey[168] have used measurements in the nematic phase to determine the sign of $^1J(^{15}N^{15}N)$ in $^{15}N^{15}NO$. A negative coupling constant ( $-$ 9.16 $\pm$ 0.30 Hz) is more consistent with the experimental data than a positive coupling. Bulusu *et al.*[169] have determined the value of $^1J(^{15}N^{15}N)$ in eight doubly labelled compounds. These results, together with two from the literature, show some evidence of a relationship between $^1J$(NN) and hybridization at N. Thus phenylhydrazine ($sp^3$–$sp^3$) has $^1J$(NN) = 6.7 Hz, *p*-nitrobenzaldehyde ($sp^3$–$sp^2$) 10.7 Hz, and *trans*-azoxybenzene, *p*-hydroxyazobenzene, and dibenzylnitrosamine ($sp^2$–$sp^2$) show couplings of 13.7, 15.0, and 19.0 Hz respectively. However, the authors fail to rationalize the couplings in

154 T. Ibusuki and Y. Saito, *Chem. Letters*, 1974, 311.
155 R. Hagen, J. P. Warren, D. H. Hunter, and J. D. Roberts, *J. Amer. Chem. Soc.*, 1973, **95**, 5712.
156 M. J. Bula and J. S. Hartman, *J.C.S. Dalton*, 1973, 1047.
157 H. Seto, T. Sato, and H. Yonehara, *J. Amer. Chem. Soc.*, 1973, **95**, 8461.
158 H. Seto, *Tetrahedron Letters*, 1974, 651.
159 H. Seto, L. W. Cary, and M. Tanabe, *J.C.S. Chem. Comm.*, 1973, 867.
160 M. Fild and W. Althoff, *J.C.S. Chem. Comm.*, 1973, 933.
161 S. Aime and R. K. Harris, *J. Magn. Resonance*, 1974, **13**, 236.
162 D. H. Lemmon and J. A. Jackson, *Spectrochim. Acta*, 1973, **29A**, 1899.
163 R. G. Green, L. H. Sutcliffe, B. Taylor, and S. M. Walker, *Spectrochim. Acta*, 1974, **30A**, 703.
164 M. Imanari, M. Kohno, M. Ohuchi, and K. Ishizu, *Bull. Chem. Soc. Japan*, 1974, **47**, 708.
165 T. Yamamoto, A. R. Garber, G. M. Bodner, L. J. Todd, M. D. Rausch, and S. A Gardner, *J. Organometallic Chem.*, 1973, **56**, C23.
166 J. Evans, B. F. G. Johnson, J. Lewis, J. R. Norton, and F. A. Cotton, *J.C.S. Chem. Comm.*, 1973, 807.
167 P. Chini, S. Martinengo, D. J. A. McCaffrey, and B. T. Heaton, *J.C.S. Chem. Comm.*, 1974, 310.
168 P. K. Bhattacharyya and B. P. Dailey, *J. Chem. Phys.*, 1973, **59**, 5820.
169 S. Bulusu, J. R. Autera, and T. Axenrod, *J.C.S. Chem. Comm.*, 1973, 602.

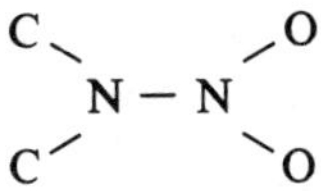

(9)

various nitramines, *i.e.* in the group (9), for which $^1J(^{15}N^{15}N)$ tends to be either ~9 or ~5 Hz. The result for NNO quoted above and the value of 7 ± 2 Hz found in a titanium dinitrogen complex[170] also fail to fit in with the scheme relating the coupling to hybridization at N. Extraordinary differences in $^1J$(PP) values for pentamethylcyclopentaphosphine, for which (10) is the assumed structure, suggest a large stereochemical dependence for this coupling.[171] The values found are − 248.6, − 236.4, and − 310.3 Hz for $J_{12}$, $J_{23}$, and $J_{34}$ respectively.

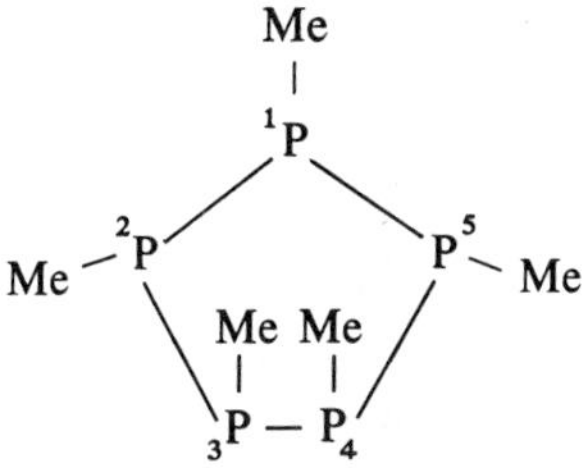

(10)

(*ii*) *Elements of Group VIB.* $^1J$(SeP) values for a series of organophosphorus selenides have been reported by McFarlane and Rycroft.[172] The values range from − 684 to − 1046 Hz and increase as the electronegativity of the substituents on phosphorus increases. This is consistent with Bent's rules, and similar observations have been made for $^1J$(TeP).[173] Values of $^1J$(SeP) can be predicted to within ± 10 Hz by use of additive substituent parameters.[172] Stec[81] has discussed the effect of conformation on $^1J$(SeP).

(*iii*) *Fluorine.* An n.m.r. study of phosphoryl fluoride orientated in a nematic phase has been compared with an electron-diffraction study.[174] There is a significant discrepancy between the structures determined by these two methods, and an orientation-dependent geometry could be the explanation of this fact. However, a large anisotropy in the indirect (P,F) coupling, which would also explain the experimental result, cannot be ruled out. Phosphorus trifluoride has been examined in a nematic solvent by Zumbulyadis and Dailey.[175] From these measurements the authors find some support for a

170 J. E. Bercaw, E. Rosenberg, and J. D. Roberts, *J. Amer. Chem. Soc.*, 1974, **96**, 612.
171 J. P. Albrand, D. Gagnaire, and J. B. Robert, *J. Amer. Chem. Soc.*, 1973, **95**, 6498.
172 W. McFarlane and D. S. Rycroft, *J.C.S. Dalton*, 1973, 2162.
173 I. A. Nuretdinov and E. I. Loginova, *Izvest. Akad. Nauk S.S.S.R., Ser. khim.*, 1973, 2827.
174 J. Bulthuis and C. A. de Lange, *J. Magn. Resonance*, 1974, **14**, 13.
175 N. Zumbulyadis and B. P. Dailey, *Mol. Phys.*, 1973, **26**, 777.

negative sign for $^{1}J$(PF); other workers, however,[113] take this to be well established. Values of $^{1}J$(PF) have been reported[176] for compounds of the form $RP(Se)F_2$, RP(Se) FCl, and $RP(Se)Cl_2$ where R may be Me, Et, Ph, $NMe_2$, $NEt_2$ or $NHSiMe_3$.

(*iv*) *Metals*. The value of $^{1}J(^{113/111}Cd^{15}N)$ in the edta complex of $Cd^{II}$ is 81 Hz.[155] Goggin and his co-workers[177] have compared $^{1}J$(PtN) ($Pt^{II}$ and $Pt^{IV}$) with $^{1}J(Pt^{II}P)$ and $^{1}J(Pt^{IV}P)$. The values of $^{1}J$(PtN) measured, 157, 166, and 209 Hz, are much smaller than the only other value of this coupling known, 430 Hz.

Mather, Pidcock, and Rapsey[178] have discussed $^{1}J$(PtP) in the complexes $[PtCl_{4-n}(R_3P)_n]^{(n-2)+}$ and $[PtCl_{6-n}(R_3P)_n]^{(n-2)+}$ ($n = 1$, 2, or 3) in terms of the MO expression for the FC term only. Support for FC-dominated coupling is found in the correlation of $^{1}J$(PtP) with the Pt—P bond length. This relationship of coupling to bond length is also found in *mer*-tris(phosphine) complexes of $W^{IV}$, $Rh^{I}$, $Rh^{III}$, $Pt^{II}$, and $Pt^{IV}$ and within one and the same complex where $^{1}J$(MP) *trans* to Cl is greater than $^{1}J$(MP) *trans* to P, and P—M bonds are shorter when *trans* to Cl than when *trans* to P. The ratios of $^{1}J$(PtP) to $^{2}J$(PtCH) for related $Pt^{II}$ and $Pt^{IV}$ complexes also reflect bond-length differences in the two oxidation states. Mather, McLaughlin, and Pidcock have also discussed the dependence of $^{1}J$(MP) upon M—P bond length[179] and the application to coupling constant estimation of equations of the form (13). The effect on $^{1}J$(PtP) of the electronegativity of the groups attached to P has been noted.[126]

$$^{1}J(\mathrm{SnP}) = [^{2}J(\mathrm{SnCH})/^{2}J(\mathrm{PtCH})]\ ^{1}J(\mathrm{PtP}) \qquad (13)$$

Grim, Pui Jun Lui, and Keiter[180] have measured values of $^{1}J$(HgP) in tertiary phosphine–mercury(II) halide complexes. Their results may be summarized as follows: (*a*) for a particular phosphine $^{1}J$(HgP) increases in the order of the electronegativity of the halogen ligand, I < Br < Cl; (*b*) the coupling is larger for dimers $L_2Hg_2X_4$ than for monomers $L_2Hg X_2$, when L and X are the same in each; (*c*) the absolute value of the co-ordination chemical shift is larger for those compounds with the larger $^{1}J$(HgP) values; (*d*) the most basic phosphine has the largest coupling constant. These trends are interpreted as following the electron deficiency of the mercury and the consequent electron donation of the phosphine groups, greater $\sigma$ donation leading to greater coupling.

Other values of one-bond couplings to Group VB elements include

176 H. W. Roesky and W. Kloker, *Z. Naturforsch.*, 1973, **28b**, 697.
177 P. L. Goggin, R. J. Goodfellow, S. R. Haddock, J. R. Knight, F. J. S. Reed, and B. F. Taylor, *J.C.S. Dalton*, 1974, 523.
178 G. G. Mather, A. Pidcock, and G. J. N. Rapsey, *J.C.S. Dalton*, 1973, 2095.
179 G. G. Mather, G. M. McLaughlin, and A. Pidcock, *J.C.S. Dalton*, 1973, 1823.
180 S. O. Grim, Pui Jun Lui, and R. L. Keiter, *Inorg. Chem.*, 1974, **13**, 342.

$^{1}J$(SeP) (ref. 176) and $^{1}J$(MetalP) (refs. 115, 122, 125, 146, 153, and 181—186).

**D. Coupling to Elements of Group VIB.**—Seppelt[187–189] has published very extensive collections of data on $^{1}J$(SeF) and $^{1}J$(TeF). For octahedrally co-ordinated complexes the coupling of selenium to an axial fluorine is usually smaller than that to an equatorial fluorine. The data on tellurium is insufficient for a similar comparison to be made.

**E. Coupling to Elements of Group VIIB.**—(*i*) *Elements of Group VIIB.* $^{1}J$(BrF) in $[BrF_6]^+$ has been measured by Gillespie and Schrobilgen,[190,191] who find values of 1575 and 1697 Hz for $^{79}$Br and $^{81}$Br respectively. They note[191] that the hexafluoro cations of Group VII do not follow the parallelogram rule proposed for Groups IV—VI by Reeves *et al.*,[192] which correlates the square root of the reduced coupling constant with the atomic number for an iso-electronic series of hexafluorides. However, it does seem safe to use the rule to predict that both *reduced* (Br,F) couplings are negative.

(*ii*) *Elements of Group VIII.* The same workers have also investigated one-bond coupling of fluorine to xenon.[190,193] They find [193] that $^{1}J(^{129}$XeF) shows a smooth correlation with $\delta_F$ for a wide range of compounds. This provides a rationale for the small couplings to equatorial F in $[XeF_5]^+$ and $[XeO_2F]^+$ and leads to the conclusion that the axial couplings in these ions are of the opposite sign to that of $^{1}J$(XeF) in other molecules. By comparing reduced couplings in the hexafluorides of $Sn^{IV}$ to $I^{VII}$, which are isoelectronic with the hypothetical $Xe^{VIII}$ cation $[XeF_6]^{2+}$, the authors arrive at the conclusion that $^{1}K$(XeF) of $Xe^{VIII}$ and $Xe^{VI}$, with the exceptions of $XeOF_4$, $XeO_2F_2$, and possibly $XeF_6$ and $[XeO_2F]^+$, are negative whereas $^{1}K$(XeF) for $Xe^{IV}$ and $Xe^{II}$ are positive. For $[XeF_5]^+$ a double-resonance experiment showed that the signs of both (Xe,F) couplings are the same, *i.e.* negative. $^{1}J$(XeF) becomes more negative the higher the positive charge on xenon.

181 P. L. Goggin, R. J. Goodfellow, J. R. Knight, M. G. Norton, and B. F. Taylor, *J.C.S. Dalton*, 1973, 2220.
182 N. Ahmad, E. W. Ainscough, T. A. James, and S. D. Robinson, *J.C.S. Dalton*, 1973, 1148.
183 N. Ahmad, E. W. Ainscough, T. A. James and S. D. Robinson, *J.C.S. Dalton*, 1973, 1151.
184 R. J. Goodfellow, M. J. Hardy, and B. F. Taylor, *J.C.S. Dalton*, 1973, 2450.
185 W. McFarlane and D. S. Rycroft, *J.C.S. Faraday II*, 1974, **70**, 377.
186 S. O. Grim, W. L. Briggs, R. C. Barth, C. A. Tolman, and J. P. Jesson, *Inorg. Chem.*, 1974, **13**, 1095.
187 K. Seppelt, *Z. anorg. Chem.*, 1973, **399**, 65.
188 K. Seppelt, *Angew. Chem. Internat. Edn.*, 1972, **11**, 723.
189 K. Seppelt, *Angew. Chem. Internat. Edn.*, 1974, **13**, 92.
190 R. J. Gillespie and G. J. Schrobilgen, *J.C.S. Chem. Comm.*, 1974, 90.
191 R. J. Gillespie and G. J. Schrobilgen, *Inorg. Chem.*, 1974, **13**, 1230.
192 L. W. Reeves, *J. Chem. Phys.*, 1964, **40**, 2423; J. Feeney, R. Haque, L. W. Reeves, and C. P. Yue, *Canad. J. Chem.*, 1968, **46**, 1389.
193 R. J. Gillespie and G. J. Schrobilgen, *Inorg. Chem.*, 1974, **13**, 765.

Further values of one-bond coupling to Group VIIB elements may be found in the following publications: $^{1}J$(SiF) (refs. 194 and 195); $^{1}J$(SnF) (ref. 196); $^{1}J$(PF) (refs. 80, 120, 194, and 197—202); $^{1}J$(SeF) (refs. 203—205); $^{1}J$(XeF) (ref. 188); and $^{1}J$(WF) (refs. 206—208).

**F. Miscellaneous One-bond Couplings.**—Kennedy and McFarlane[54] have shown that $^{1}J(^{119}Sn^{77}Se)$ in some organotin selenides is positive, *i.e.* $^{1}K$(SnSe) is negative. The coupling increases in magnitude as the groups attached to Sn become more electron-withdrawing. McFarlane and Rycroft[209] have found a value of +15 Hz for $^{1}J(^{13}C^{7}Li)$ in the methyl-lithium tetramer. They believe this to be in accord with a dominant FC mechanism.

A number of $^{1}J(^{65}Cu^{63}Cu)$ values for copper metal have been reported.[210] The theoretical values are in quite good agreement with experiment. The experimental result for nearest-neighbour atoms is +230 Hz, a value in strange contrast with a measurement of $^{1}J(^{63}Cu^{63}Cu)$ in copper acetate monohydrate which gave a result of 10 940 Hz.[211]

## 4 Coupling between Nuclei Separated by Two Chemical Bonds, $^{2}J$

**A. Geminal Proton–Proton Coupling.**—Theoretical work on $^{2}J$(HCH) and $^{2}J$(HNH) has already been noted in Section 2B.

Fay *et al.*[212] have carried out a detailed investigation of the effect of substituents on geminal and vicinal (H,H) coupling in systems of well-defined geometry. They find that, for the common range of functional groups, the dependence of these couplings on substituents is complex and not simply a question of electronegativity. They suggest that empirical 'substituent effects' might be used predictively. Curry and Sardella[213] have given an interesting

194 J. A. Gibson, D. G. Ibbott, and A. F. Janzen, *Canad. J. Chem.*, 1973, **51**, 3203.
195 H. C. Marsmann and R. Loewer, *Chem.-Ztg.*, 1973, **97**, 660.
196 P. A. W. Dean, *Canad. J. Chem.*, 1973, **51**, 4024.
197 R. G. Cavell and A. R. Sanger, *Inorg. Nuclear Chem. Letters*, 1973, **9**, 461.
198 J. F. Nixon, *J. Fluorine Chem.*, 1973, **3**, 179.
199 V. I. Zakharov, Yu. V. Belov, B. I. Ionin, and A. A. Petrov, *Doklady Akad. Nauk S.S.S.R.*, 1973, **209**, 1343.
200 O. Stelzer and E. Unger, *J.C.S. Dalton*, 1973, 1783.
201 R. Schmutzler, *J.C.S. Dalton*, 1973, 2687.
202 W. M. Douglas, R. B. Johannesen, and J. K. Ruff, *Inorg. Chem.*, 1974, **13**, 371.
203 K. Seppelt, *Angew Chem. Internat. Edn.*, 1972, **11**, 630.
204 K. Seppelt, *Chem. Ber.*, 1973, **105**, 2431.
205 K. Seppelt, *Chem. Ber.*, 1973, **106**, 157.
206 J. Y. Calves and J. E. Guerchais, *Bull. Soc. chim. France*, 1973, 1220.
207 M. Harman, D. W. A. Sharp, and J. M. Winfield, *Inorg. Nuclear Chem. Letters*, 1974, **10**, 183.
208 Yu. A. Buslaev, Yu. V. Kokunov, and Yu. D. Chubar, *Doklady Akad. Nauk S.S.S.R.*, 1973, **213**, 1083.
209 W. McFarlane and D. S. Rycroft, *J. Organometallic Chem.*, 1974, **64**, 303.
210 E. R. Andrew and W. S. Hinshaw, *Phys. Letters (A)*, 1973, **43**, 113.
211 A. Kawamori, R. Takagi, and G. Soda, *Mol. Phys.*, 1973, **25**, 489.
212 C. K. Fay, J. B. Grutzner, L. F. Johnson, S. Sternhell, and P. W. Westerman, *J. Org. Chem.*, 1973, **38**, 3122.
213 E. R. Curry and D. J. Sardella, *J. Amer. Chem. Soc.*, 1974, **96**, 1822.

discussion of the relationship between $^2J$(HCH) and $^4J$(HCCCH) in molecules of the form $H_2C{=}X$ and $Me_2C{=}X$ respectively. An excellent linear correlation between these two couplings (correlation coefficient 0.997) suggests that the same mechanism of spin information transfer is operative in both series of molecules.

The effect of solvent upon geminal coupling between non-equivalent methylenic protons has been examined by Brink.[214] $^2J$(HCH) in formaldehyde is +42.2 Hz[213] and $^2J$(TCH) in tritiated penicillic acid is 15.0 Hz.[215]

Other values of geminal proton–proton coupling may be found in the following papers: $^2J$(HBH) (ref. 82); $^2J$(HCH) (refs. 13, 107, 109, and 216–245); and $^2J$(HNH) (ref. 118).

**B. Two-bond Coupling of Other Nuclei to Protons.**—(*i*) *Elements of Group IVB.* Barboiu[53] has measured the signs and magnitudes of $^2J$(CCH) in acetic acid, ethanol, acetaldehyde, and some of their derivatives. The values range from −8.7 to +31 Hz and the following generalizations can be made: (*a*) $^2J(C_{sp^3}$-

214 M. Brink, *Monatsh.*, 1973, **104**, 619.
215 J. M. A. Al-Rawi, J. A. Elvidge, D. K. Jaiswal, J. R. Jones, and R. Thomas, *J.C.S. Chem. Comm.*, 1974, 220.
216 K. Bergesen and M. Bjorøy, *Acta Chem. Scand.*, 1973, **27**, 1103.
217 D. W. Aksnes and O. Vikane, *Acta Chem. Scand.*, 1973, **27**, 1337.
218 T. A. Crabb and M. J. Hall, *J.C.S. Perkin II*, 1973, 1379.
219 G. M. Sharma and O. A. Roels, *J. Org. Chem.*, 1973, **38**, 3648.
220 J. D. Remijnse, B. G. M. Vandeginste, and B. M. Wepster, *Rec. Trav. chim.*, 1973, **92**, 804.
221 G. Desimoni, L. Astolfi, M. Cambieri, A. Gamba, and G. Tacconi, *Tetrahedron*, 1973, **29**, 2627.
222 V. R. Haddon and L. M. Jackman, *Org. Magn. Resonance*, 1973, **5**, 333.
223 Y. P. Wong and H. F. White, *J. Magn. Resonance*, 1973, **12**, 1.
224 B. A. Englin, T. A. Onishchenko, Yu. V. Belov, V. I. Dostovalova, I. K. Shmyrev and U. Utebaev, *Izvest. Akad. Nauk S.S.S.R., Ser. khim.*, 1973, 982.
225 J. Fouché and S. Combrisson, *Bull. Soc. chim. France*, 1973, 1693.
226 P. Äyräs and K. Pihlaja, *Tetrahedron*, 1973, **29**, 1311.
227 J. T. Gerig and R. S. McLeod, *J. Amer. Chem. Soc.*, 1973, **95**, 5725.
228 J. B. Lambert, C. E. Mixan, and D. H. Johnson, *J. Amer. Chem. Soc.*, 1973, **95**, 4634.
229 R. M. Lequan and M. P. Simonnin, *Tetrahedron*, 1973, **29**, 3363.
230 P. Albriktsen and R. K. Harris, *Acta Chem. Scand.*, 1973, **27**, 1875.
231 M. Anteunis, R. Van Cauwenberghe, and C. Becu, *Bull. Soc. chim. belges*, 1973, **82**, 591.
232 C. H. Green and D. G. Hellier, *J.C.S. Perkin II*, 1973, 1966.
233 R. G. Green, P. Shaw, L. H. Sutcliffe, B. Taylor, and S. M. Walker, *Spectrochim. Acta*, 1974, **30A**, 529.
234 K. F. Wong and S. Ng, *J. Magn. Resonance*, 1974, **13**, 239.
235 A. De Smet and M. Anteunis, *Org. Magn. Resonance*, 1973, **5**, 589.
236 E. M. Gaydou and J. R. Llinas, *Org. Magn. Resonance*, 1974, **6**, 23.
237 A. V. Cunliffe and R. K. Harris, *Org. Magn. Resonance*, 1974, **6**, 121.
238 D. W. Aksnes and J. Støgard, *Acta Chem. Scand.*, 1973, **27**, 3277.
239 K. Bergesen and M. Bjorøy, *Acta Chem. Scand.*, 1973, **27**, 3477.
240 T. A. Crabb, M. J. Hall, and R. O. Williams, *Tetrahedron*, 1973, **29**, 3389.
241 W. Ried and H. J. Herrmann, *Annalen*, 1974, 201.
242 J. Jokisaari and E. Rahkamaa, *Z. Naturforsch.*, 1973, **28a**, 30.
243 K. Takahashi, T. Akasaka, S. Kondo, and K. Tsuda, *Org. Magn. Resonance*, 1974, **6**, 193.
244 R. Gassend, Y. Limouzin, and J. C. Maire, *Org. Magn. Resonance*, 1974, **6**, 259.
245 T. W. Dolzine and J. P. Oliver, *J. Amer. Chem. Soc.*, 1974, **96**, 1737.

$C_{sp^3}H$) is negative; (*b*) $^2J(C_{sp^2}C_{sp^3}H)$ is also negative, but difluoroacetic acid (+2.2 Hz) is an exception; (*c*) $^2J(C_{sp^3}C_{sp^2}H)$ can be of either sign; and (*d*) the magnitudes of the negative couplings are less than 3 Hz but the positive can be very large (*e.g.* +46.3 Hz for chloral). The author suggests that (*d*) indicates an important positive $\pi$ contribution to the coupling, and a comparison of $^2J$(CCH) in acetaldehyde (+26.6 Hz) and crotonaldehyde (+4.5 Hz), where the intervening carbon atom is doubly bonded to oxygen in the first case and to carbon in the second, indicates that this $\pi$ contribution increases with increasing polarization of the $\pi$-bonds.

Runsink, de Wit, and Weringa[106] have reassigned the long-range (C,H) couplings in furan and find that $|^2J(CCH)| > |^3J(CCCH)|$, which is not often the case in conjugated heterocycles. However, the same observation has been made by Sørensen *et al.*[246] for substituted furans and thiophens, and they point out that $|^2J(CCH)/^3J(CCCH)|$ lies between 1.1 and 2.1 and that this can be useful in $^{13}C$ assignment problems. Ewing[61] and Williams and Bothner-By[95] have also reported $^2J$(CCH) values.

van den Berghe and van der Kelen[56,247,248] have investigated $^2J$(SiCH) and $^2J$(SnCH) in methyl-silanes and -stannanes. $^2J(^{29}SiCH)$ in the trimethylsilicon halides shows an inverse dependence on the electronegativity of the halogen atom, *i.e.* $|J_I| > |J_{Cl}|$.[56] In the trihalides, however, the situation is completely reversed and the coupling follows Bent's rules. $^2J$(SnCH) always has $|J_I| < |J_{Cl}|$.[56] The authors suggest an explanation of these facts based upon a change in the *s*-character of the Si—C bond brought about by *structurally* induced bond-angle changes rather than the *electronegativity* effects invoked by Bent. There is some support for this view in the $^1J$(CH) values of $CH_3X$ which increase as the angle HCX decreases from X = F to X = I. Although 'normal' behaviour of $^2J$(SnCH) was found in the halogen-substituted methylstannanes[56] the same authors have noted that in the series of compounds $Me_3SnX$ (X = $NMe_2$, Cl, Br, or I) the dimethylamino-compound has the smallest $^2J$(SnCH) value although nitrogen is the most electronegative substituent.[247] In this case the authors propose an explanation based upon the effect of intermolecular association which is possible in the case of $Me_3Sn$-$NMe_2$ but not with more bulky halogen substituents. Finally, the same authors invoke polymerization to explain some unusual results for $^2J$(SnCH) in molecules of general formulae $Me_{4-n}Sn(OR)_n$ (R = Me or Et).[248]

Kennedy and McFarlane[54,55] have also reported some $^2J$(SnCH) values. In the series of compounds $Me_{4-n}Sn(SR)_n$ (R = alkyl or phenyl), there appears to be a dependence of $^2J$(SnCH) upon hybridization at Sn, as well as upon charge distribution, since the changes in $^2J$ as *n* varies are affected by the bulk of R.[55] Some rather low values of $^2J$(SnCH) have been recorded for four-co-ordinate trimethyltin derivatives.[249]

246 S. Sørensen, M. Hansen, and H. J. Jakobsen, *J. Magn. Resonance*, 1973, **12**, 340.
247 E. V. van den Berghe and G. P. van der Kelen, *J. Organometallic Chem.*, 1973, **61**, 197.
248 E. V. van den Berghe and G. P. van der Kelen, *J. Mol. Structure*, 1974, **20**, 147.
249 E. B. Bradley, R. H. Herber, P. J. Busse, and K. Niedenzu, *J. Organometallic Chem.*, 1973, **52**, 297.

(*ii*) *Elements of Group VB*. Yeh *et al.*[100] have determined $^2J(^{15}NCH)$ values in four pairs of molecules of general formulae (11). The absolute signs were determined for $R^1$ = 9-anthryl, $R^2$ = butyl and the others were assumed to be analogous. For the *E*-isomers $^2J(NCH)$ lay between +3.4 and +3.7 Hz

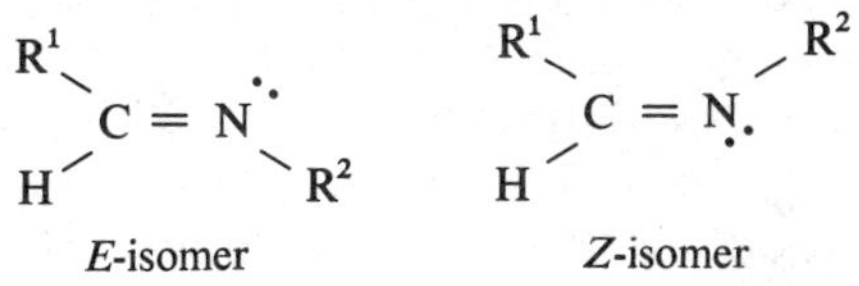

(11) $R^1$ = 9-anthryl, naphthyl, or Ph; $R^2$ = Me or $Bu^t$

while for the *Z*-isomers the range was −9.8 to −10.4 Hz. Thus this *cis* relationship of proton and nitrogen lone pair in the *Z*-isomer leads to a large negative contribution to the coupling. A $^2J(^{14}NCH)$ value of 28 Hz has been measured in $WF_4(NMe)(NCMe)$[207] and $^2J(^{15}NCH) = -4.53 \pm 0.05$ Hz in [$^{15}$N]pyrrole.[96,117] Anderson, Bentham, and Rankin[113] have found values of $^2J(^{15}NSiH)$ of −4.1, −4.3, and −4.2 Hz in $PF_2{}^{15}NHSiH_3$, $(SiH_3)_2{}^{15}NH$, and $(SiH_3)_3{}^{15}N$ respectively. The signs were determined by double-resonance methods.

It has been reported from two laboratories that $^2J(PCH)$ depends upon hybridization at C.[93,250] When the intervening carbon is $sp^2$-hybridized the coupling is positive and of the order of 6 Hz; an $sp^3$-hybridized C results in negative couplings of the order of 13 Hz. In the case of ethylenic phosphines, Lequan and Simonnin have found that $^2J(PCH)$ is large and positive when the C—H bond and the P lone pair lie in a parallel relationship (12a) and small, perhaps even negative, when they are antiparallel (12b).[229] Furthermore, they have also noted[250] that quaternization of phosphorus diminishes $^2J(PCH)$ in

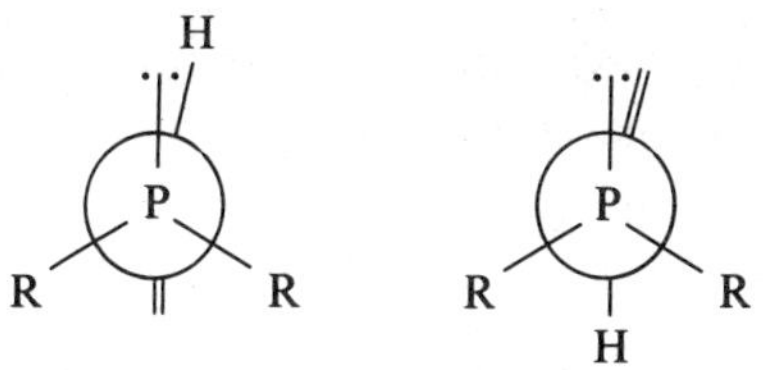

(12a) (12b)

the former case, the α-conformer, and increases the coupling in the latter case, the β-conformer. Since the α- and β-conformers described above correspond closely to the *Z*- and *E*-isomers of structure (11) it is tempting here to seek an analogy between $^2J(PCH)$ and $^2J(NCH)$ on the one hand and $^2J(PCC)$ and $^2J(NCC)$ on the other. For the last two couplings it seems to be well established that a nearby lone pair makes a positive contribution to the *reduced*

[250] R. M. Lequan and M. P. Simonnin, *Bull. Soc. chim. France*, 1973, 2365.

coupling constant in both phosphorus-[251] and nitrogen-containing[252] molecules, and, bearing in mind the negative magnetogyric ratio of $^{15}N$, this would appear to be also the situation in $^2J$(PCH) and $^2J$(NCH). Further confirmation of the positive sign of $^2J$(PCH) when the intervening carbon is $sp^2$-hybridized has been obtained by Zakharov *et al.*[199] $^2J$(PCH) is found to be much larger in phosphine oxides than in the corresponding phosphines.[253] $^2J$(PNH) is positive and *ca.* 19 Hz in $PF_2NH_2$ and $PF_2NHSiH_3$.[113]

(*iii*) *Elements of Group VIB.* Several values of $^2J$(SeCH) have been reported. They are normally positive.[54,254] Some measurements of $^2J$(TeCH) have been made by McFarlane and McFarlane.[255] They find that in general $^2J$(TeCH) is approximately twice as large as the corresponding $^2J$(SeCH); the ratio $\gamma(^{125}\text{Te})S^2_{\text{Te}}(0)/\gamma(^{77}\text{Se})S^2_{\text{Se}}(0)$ is *ca.* 2.

(*iv*) *Metals.* Anderson, Ebsworth, and Rankin[115] have reported a large number of coupling constants, and their signs, for triethylphosphine complexes of platinum. Almost all the reduced couplings were found to be positive over odd numbers of bonds and negative over even, *i.e.* consistent with a dominant FC term. However, $^2J$(PtSiH) and $^2J$(PtGeH) are positive where the signs are known even though the range of values observed is very large, 20—434 Hz. This suggests that another term outweighs the FC in these cases. Mather *et al.*[179] have discussed the use of equations (14) and (15) for the estimation

$$^2J(\text{RhCH}) \simeq [^1J(\text{RhP})/^1J(\text{PtP})].^2J(\text{PtCH}) \qquad (14)$$
$$^2J(\text{HgCH}) \simeq [^1J(\text{HgP})/^1J(\text{PtP})].^2J(\text{PtCH}) \qquad (15)$$

of coupling constants. They find that such equations have some use when a suitable range of complexes and coupling constants is available, but they are not helpful when the metal–ligand bonds are very dissimilar. A linear correlation of $^2J$(PtCH) with $^1J$(PtC) has been reported.[111] $^2J$(HgCH) has been found to correlate linearly with the p$K$ of the corresponding acid in the case of the methylmercury salts of organic acids,[256] with the p$K_a$ and p$K_m$ values of the carbanions R in the molecules MeHgR,[257] and with $^1J$(HgC) and $\delta_C$.[154]

Further two-bond couplings to hydrogen may be found in the following publications: $^2J$(BBH) (ref. 90); $^2J$(BCH) (ref. 90); $^2J$(AlBH) (ref. 84); $^2J$(TlCH) (refs. 80 and 258); $^2J$(CCH) (refs. 26, 65, 90, 92, 104, 105, 107, 110, 112, and 155); $^2J$(SiCH) (refs. 99 and 114); $^2J$(SnCH) (refs. 99 and 258—

251 M. P. Simonnin, R. M. Lequan, and F. W. Wehrli, *J.C.S. Chem. Comm.*, 1972, 1204; S. Sørenson, R. S. Hansen, and H. J. Jakobsen, *J. Amer. Chem. Soc.*, 1972, **94**, 5900.
252 W. B. Jennings, D. R. Boyd, C. G. Watson, E. D. Becker, R. B. Bradley, and D. M. Jerina, *J. Amer. Chem. Soc.*, 1972, **94**, 8501.
253 P. W. Clarke, J. L. S. Curtis, P. E. Garrou, and G. E. Hartwell, *Canad. J. Chem.*, 1974, **52**, 1714.
254 M. P. Simonnin, M. J. Pouet, and C. Paulmier, *Tetrahedron Letters*, 1973, 4487.
255 H. C. E. McFarlane and W. McFarlane, *J.C.S. Dalton*, 1973, 2416.
256 L. F. Sytsma and R. J. Kline, *J. Organometallic Chem.*, 1973, **54**, 15.
257 K. P. Butin, I. P. Beletskaya, and O. A. Reutov, *J. Organometallic Chem.*, 1974, **64**, 323.
258 A. T. Weibel and J. P. Oliver, *J. Organometallic Chem.*, 1973, **57**, 313.

260); $^2J$(NCH) (refs. 139 and 261); $^2J$(NOH) (ref. 261); $^2J$(PCH) (refs. 80, 85, 115, 122, 172, 177, 181, and 262—271); $^2J$(PPtH) (ref. 125); $^2J$(SeCH) (ref. 99); $^2J$(FCH) (refs. 10, 223, 227, 243, 272, and 273); $^2J$(FPH) (ref. 120); $^2J$(HgCH) (ref. 274); $^2J$(PtCH) (refs. 153, 178, 184, 268, and 275—278); $^2J$(RhCH) (ref. 279); and $^2J$(PtSiH) (ref. 122).

**C. Two-bond Coupling to Carbon, Excluding $^2J(^{13}CX^1H)$.**—(*i*) *Elements of Group IIIB*. Values of $^2J(CC^{11}B)$[128] and $^2J(^{205/203}TlNC)$[280] have been reported. The last (18 Hz) is approximately equal to $^4J$(TlNCC) (12 and 15 Hz), which is reminiscent of other carbon–metal couplings and of olefinic (H,H) coupling. It lends support to the comparison of thallium and hydrogen coupling.

(*ii*) *Elements of Group IVB*. Many values of $^2J$(SnCC) have been published[133,134,136] and there is a satisfactory correlation of this coupling with Hammett $\sigma$ values.[134] Some measurements of $^2J$(SnNC) and $^2J$(SnSnC) have been made.[133]

(*iii*) *Elements of Group VB*. Lichter and his co-workers have measured $^2J$(NCC) in a series of amides[140] and find only a rough correlation between this coupling and $^1J$(NC). They predict that the sign of $^2J(^{15}NC^{13}C)$ is negative on the basis of a behaviour similar to that of $^2J$(CCC). They find[141] that in oximes $^2J$(NCC) is sensitive to the orientation of the nitrogen lone pair, the *E*-isomer (8a) showing a greater coupling than the *Z* (8b) (9.0 and 1.8 Hz respectively in acetaldoxime, for example).

Burdon, Hotchkiss, and Jennings[281] have made an interesting study of the molecules (13) and (14). At −40 °C $^2J$(PNC) in (13a) = 25 and 12 Hz and

259 G.-E. Matsubayashi, M. Okunaka, and T. Tanaka, *J. Organometallic Chem.*, 1973, **56**, 215.
260 U. Folli, D. Iarossi and F. Taddei, *J.C.S. Perkin II*, 1973, 1284.
261 S. Polance, B. Vercek, B. Stanovnik, and M. Tisler, *J. Heterocyclic Chem.*, 1973, **11**, 103.
262 A. Zschunke, A. Hauser and H. Oehme, *Z. Chem.*, 1973, **13**, 310.
263 G. Hägele, K. Diemert, and W. Kuchen, *Z. Naturforsch.*, 1973, **28b**, 185.
264 L. Eberhard, J. P. Lampin, F. Mathey, and C. Badellon, *Tetrahedron*, 1973, **29**, 2909.
265 T. Osaki, J. Otera, and Y. Kawasaki, *Bull. Chem. Soc. Japan*, 1973, **46**, 1803.
266 J. Chatt, G. J. Leigh, and R. M. Slade, *J.C.S. Dalton*, 1973, 2021.
267 H. H. Falius, K. P. Giesen, and U. Wannagat, *Z. anorg. Chem.*, 1973, **402**, 139.
268 J. M. Duff, B. E. Mann, B. L. Shaw, and B. Turtle, *J.C.S. Dalton*, 1974, 139.
269 W. Wolfsberger, *Z. Naturforsch.*, 1974, **29b**, 35.
270 D. A. Allison and J. G. Verkade, *Phosphorus*, 1973, **2**, 257.
271 K. Issleib, M. Lischewski, and A. Zschunke, *Org. Magn. Resonance*, 1973, **5**, 401.
272 B. L. Shapiro, W. A. Thomas, J. L. McClanahan, and M. D. Johnston, *J. Magn. Resonance*, 1973, **11**, 355.
273 R. Guedj, R. Nabet, and T. Wade, *Tetrahedron Letters*, 1973, 907.
274 F. Glocking, S. R. Stobart, and J. J. Sweeney, *J.C.S. Dalton*, 1973, 2029.
275 J. R. Hall and G. A. Swile, *J. Organometallic Chem.*, 1973, **56**, 419.
276 D. R. McMillin and R. S. Drago, *Inorg. Chem.*, 1974, **13**, 546.
277 J. R. Hall and G. A. Swile, *J. Organometallic Chem.*, 1974, **67**, 455.
278 D. Hollings, M. Green, and D. V. Claridge, *J. Organometallic Chem.*, 1973, **54**, 399.
279 M. Green and G. J. Parker, *J.C.S. Dalton*, 1974, 333.
280 R. J. Abraham, G. E. Hawkes, and K. M. Smith, *J.C.S. Chem. Comm.*, 1973, 401.
281 J. Burdon, J. C. Hotchkiss, and W. B. Jennings, *Tetrahedron Letters*, 1973, 4919.

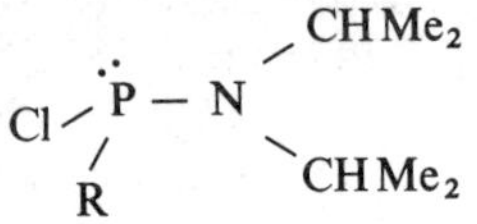 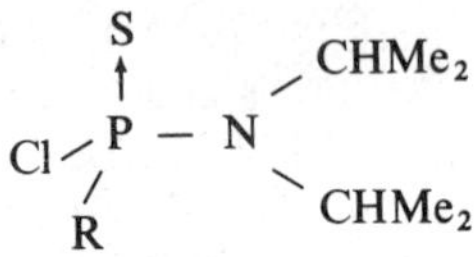

(13) a; R = Ph
b; R = Me

(14) a; R = Ph
b; R = Me

in (13b) 25 and 13 Hz. At 55 °C the time-averaged values are 7 and 6 Hz respectively. The two couplings must therefore have different signs. In (14a) and (14b) $^2J$(PNC) $\simeq$ 3 Hz at both −40 °C and 55 °C so the large difference in the values of the coupling in (13a) and (13b) at low temperature suggests an important lone-pair effect. A strong dependence of $^2J(P^{III}OC)$ on the dihedral angle between the P lone pair and the O—C bond has been observed.[282]

Two recent papers have pointed out the similarities between $^2J$(PMC) and $^2J$(PMP), where M is a transition metal.[146,151] In particular one finds that the *trans* coupling is always greater than the *cis*, and couplings in phosphite complexes are larger than those in phosphine complexes. These effects can be interpreted in terms of *s*-electron density in the relevant bonds and of the effective nuclear charge of P. There is some evidence for an angular dependence of $^2J$(PMC) (M = W or Mo).[146]

(*iv*) *Coupling of Transition Metals to Carbon*. Clark and Ward[150] have reported values of $^2J$(PtCC) and $^2J$(PtAsC) and have correlated these couplings with $^1J$(PtC) and $^3J$(PtCCC). The evidence is that these couplings are all FC-dominated.

Other two-bond couplings to carbon include $^2J$(CCC) (refs. 33, 131, and 132); $^2J$(SnCC) (refs. 135 and 283); $^2J$(PbCC) (ref. 135); $^2J$(NCC) (ref. 139); $^2J$(PCC) (refs. 93, 144, and 151); $^2J$(POC) (ref. 146); $^2J$(PPC) (ref. 161); $^2J$(PMC) (refs. 145 and 149); $^2J$(FCC) (refs. 53, 92, 94, 102, and 162—164); $^2J$(PtCC) (refs. 102 and 149); $^2J$(HgCC) (refs. 91 and 135); $^2J$(PtNC) (ref. 153); $^2J$(PtPC) (ref. 151); and $^2J$(PtAsC) (refs. 149 and 153).

**D. $^2J(^{31}P,^{29}Si)$.**—Niecke and Flick[284] have measured $^2J$(PNSi) values in bis(trimethylsilyl)aminobis(trimethylisilylmino)phosphorane (15). The couplings are very different, being 1.2 Hz for the amino- and 16.7 Hz for the imino-group, while the $^2J$(SiCH) values are effectively the same. $^2J$(PPtSi) has been measured.[115]

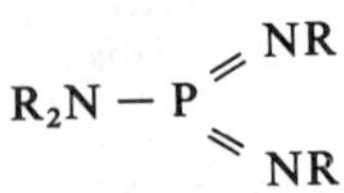

(15) R = $SiMe_3$

282 M. Haemers, R. Ottinger, D. Zimmermann, and J. Reisse, *Tetrahedron Letters*, 1973, 2241.
283 G. Matsubayashi and T. Tanaka, *Spectrochim. Acta*, 1974, **30A**, 869.
284 E. Niecke and W. Flick, *Angew. Chem. Internat. Edn.*, 1974, **13**, 134.

**E. Coupling of Group VB and VIIB Elements to Group VB.**—(*i*) *Elements of Group VB*. Schumann and Schmidpeter[285] have tested Finer's relationship [equation (16)][286,287] for $^2J$(PNP) in triazatriphosphorines (16). The $\lambda$

$$^2J(\text{PNP}) = (\lambda_A + \lambda_B)(\lambda_C + \lambda_D) \qquad (16)$$

are empirical constants for the phosphorus substituents. Measurements on a wider range of compounds and with some hydrogen-substituted molecules show a rough relationship of $\lambda$ to electronegativity, $\chi$. The triazadiphosphorines (17) give $^2J$(PNP) values smaller than those calculated using equation (16), which reflects the smaller PNP angle in these compounds and shows

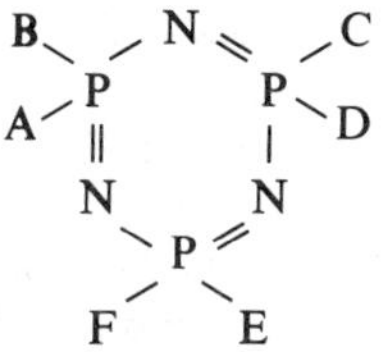

(16)

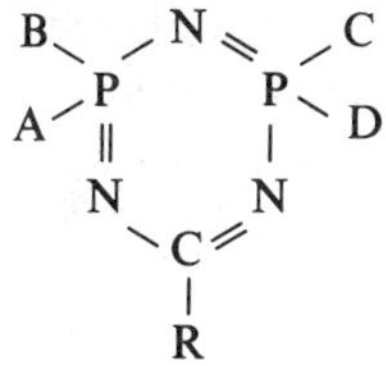

(17)

that the $\lambda$ values also depend upon the geometry of the ring. For these systems the authors propose the modification (17) of Finer's equation. With $f = 0.76$ good agreement with the experimental couplings is found. The values of

$$^2J(\text{PNP}) = (f\lambda_A + f\lambda_B)(f\lambda_C + f\lambda_D) = f^2(\lambda_A + \lambda_B)(\lambda_C + \lambda_D) \qquad (17)$$

$^2J(P^VNP^V)$ for cyclophosph(III)azones and their oxo- and thioxo-derivatives exceed those of $^2J(P^VNP^{III})$, in contrast to the situation in acyclic analogues.[288]

Several recent publications have been concerned with the detailed analysis of the spectra of phosphine complexes of the transition metals[123,125,184,185] and attempts to determine the sign and magnitude of $^2J$(PMP) by means of double resonance experiments.[181,185] For complexes of Pd, Pt, Rh, and Ir, with only alkyl or aryl groups attached to phosphorus in the ligands, the couplings are less than 30 Hz in magnitude in general, and can be negative[125,181] or positive.[185] Goodfellow, Hardy, and Taylor[184] have reinvestigated the $^1$H spectra of *cis*-dimethylbis(trimethylphosphine)platinum and some related compounds. They conclude that the range of $^2J$(PPtP) values should be 10—12 Hz, not 0—21 Hz as was previously suggested.

Grim *et al.*[186] have measured the (P,P) coupling in $Ph_2PCH_2CH_2OPPh_2$ and $Ph_2PCH_2OPPh_2$ and in transition-metal complexes of these ligands. In the first ligand no $^4J$(POCCP) is observed, but when the complexes are formed (P,P) couplings of 46.5 Hz (Cr), 35 Hz (Mo), 31 Hz (W), and 9 Hz (Pd) can be detected. In $Ph_2PCH_2OPPh_2$ $^3J$(POCP) = 16.5 Hz, but in the complexes the

[285] K. Schumann and A. Schmidpeter, *Phosphorus*, 1973, **3**, 51.
[286] E. G. Finer, *J. Mol. Spectroscopy*, 1967, **23**, 104.
[287] E. G. Finer and R. K. Harris, *Progr. N.M.R. Spectroscopy*, 1971, **6**, 61.
[288] R. Jefferson, J. F. Nixon, T. M. Painter, R. Keat, and L. Stobbs, *J.C.S. Dalton*, 1973, 1414.

(P,P) couplings are generally smaller than those cited above. The authors suggest opposing effects of $^3J$(POCP) and $^2J$(PMP) to explain these results.

(*ii*) *Elements of Group VIIB.* $^2J$(FPN) is negative.[113]

Other two-bond couplings to phosphorus may be found in the following papers: $^2J$(PPP) (ref. 171); $^2J$(POP) (ref. 202); $^2J$(PSP) (ref. 289); $^2J$(PMoP) (refs. 146 and 198); $^2J$(PCrP) (ref. 198); $^2J$(PRhP) (ref. 198); $^2J$(PNiP) (ref. 200); $^2J$(PPtP) (ref. 183); $^2J$(PPdP) (ref. 183); $^2J$(PBF) (ref. 86); and $^2J$(PCF) (refs. 197, 290, and 291).

**F. Two-bond ($^{19}$F,$^{19}$F) Coupling.**—Jolley, Sutcliffe, and Williamson[292] have noted that $^2J$(FCF) values in trifluoro-1,2-oxathietan 2,2-dioxides are much smaller than in the corresponding chlorotrifluorocyclobutanes, and in *o*-fluorophenyldifluorocarbenium ion a very large value, 246 Hz, has been measured.[293] Hinton and Jaques[94] have examined the temperature and solvent dependence of $^2J$(FCF) and $^3J$(FCCF) in 1-bromo-3,3,4-trifluorobut-4-ene. They discuss the solvent dependence in terms of the reaction field model.

Seppelt[187,189] has reported values of $^2J$(FSeF) for 28 compounds and of $^2J$(FTeF) for 18. The ranges of these couplings are surprisingly narrow,[187] being 216–231 Hz for the former and 116–192 Hz for the latter. Since the range of $^2J$(FSF) in molecules containing the $F_5SO$ group is 140—160 Hz, we have the sequence $|^2J(\text{FSeF})| > |^2J(\text{FTeF})| > |^2J(\text{FSF})|$.

$^2J$(FKrF) = 351 Hz in $[Kr_2F_3]^+$ and $^2J$(FXeF) = 308 Hz in $[Xe_2F_3]^+$.[190]

Further reports of two-bond (F,F) coupling may be found in the following papers: $^2J$(FCF) (ref. 294); $^2J$(FSiF) (ref. 194); $^2J$(FPF) (refs. 194 and 201); $^2J$(FSF) (ref. 194); $^2J$(FSeF) (refs. 188 and 203); $^2J$(FIF) (ref. 295); $^2J$(FXeF) (ref. 193); $^2J$(FTIF) (refs. 196 and 296); $^2J$(FWF) (refs. 206 and 208); and $^2J$(FTaF) (ref. 297).

**G. Miscellaneous Two-bond Couplings.**—Values of $^2J$(SnSnSn),[137] $^2J$(SnOSn),[137] $^2J$(SePF),[176] $^2J$(PtCF),[298] and $^2J$(RhPF)[198] have been reported.

## 5 Coupling between Nuclei Separated by Three Chemical Bonds, $^3J$

**A. Vicinal Proton–Proton Coupling *via* Carbon, $^3J$($^1$HCC$^1$H).**—Semi-empirical relationships connecting $^3J$(HCCH) to dihedral angle and substituent

289 E. R. Andrew, W. S. Hinshaw, and A. Jasinski, *Chem. Phys. Letters*, 1974, **24**, 399.
290 R. C. Dobbie and P. R. Mason, *J.C.S. Dalton*, 1973, 1124.
291 R. G. Cavell, D. D. Poulin, K. I. The, and A. J. Tomlinson, *J.C.S. Chem. Comm.*, 1974, 19.
292 K. W. Jolley, L. H. Sutcliffe, and K. L. Williamson, *Spectrochim. Acta*, 1974, **30A**, 511.
293 G. A. Olah and Y. K. Mo, *J. Org. Chem.*, 1973, **38**, 2686.
294 M. Green, S. Heathcock, and D. C. Wood, *J.C.S. Dalton*, 1973, 1564.
295 I. R. Beattie and G. J. Van Schalkwyk, *Inorg. Nuclear Chem. Letters*, 1974, **10**, 343.
296 P. A. W. Dean and B. J. Ferguson, *Canad. J. Chem.*, 1974, **52**, 667.
297 J.-Y. Calves and J.-.E. Guerchais, *J. Fluorine Chem.*, 1974, **4**, 47.
298 H. C. Clark and L. E. Manzer, *J.C.S. Chem. Comm.*, 1973, 870.

electronegativity have been proposed for dioxa-, diaza-, and dithia-phospholans[299] and for six-membered carbocyclic and heterocyclic systems.[300] Revel and Navech[299] have proposed equations (18) and (19) by means of which $^3J$(HCCH) can be corrected for the presence of substituents other than hydrogen on the two carbons. $\Delta\chi$ is the difference between the electronegativity of the substituent and that of hydrogen on the Pauling scale. Unfortunately, $\lambda_1$ and $\lambda_2$ are left undefined.

$$J_{cis}\ (\text{unperturbed}) = J_{cis}(\text{exp})/[1 - (\lambda_1 + \lambda_2)\Delta\chi] \qquad (18)$$

$$J_{trans}(\text{unperturbed}) = [J_{trans}(\text{exp})/2][1/(1 - 2\lambda_1\Delta\chi) + 1/(1 - 2\lambda_2\Delta\chi)] \qquad (19)$$

A relationship between $^3J$(HCCH) and dihedral angle, $\phi$, for the $HC_\alpha C_\beta H$ system of amino-acid residues in peptides has been derived experimentally from model compound measurements.[301] The result may be formulated as

$$^3J(\text{HCCH}) = 6.3 - 1.4\cos\phi + 4.7\cos2\phi \qquad (20)$$

equation (20). Manor *et al.*[302] have made a comparison of the *X*-ray diffraction and proton n.m.r. studies of 5′,2-*O*-cyclo-2′,3′-*O*-isopropylidine uridine. They find considerable discrepancies between the dihedral angles in the five-membered rings determined from $^3J$(HCCH) and those from the *X*-ray study and conclude that a modified Karplus equation is not appropriate for this task.

The connection between the vicinal coupling constants of substituted ethanes and their rotational potential functions has been discussed in considerable detail.[303,304] Studies of the variation of $^3J$(HCCH) with geometry in deuteriated t-butylcyclohexane,[222] and with temperature in [$^2H_2$]acrolein,[305] saturated five-membered rings,[306] and deuteriated butadienes[68] have been published. Viti *et al.*[307] have determined $^3J$(HCCH) in 2-methoxyethanol and 1,2-dimethoxyethane in solvents with dielectric constants ranging from 2 to 44, and over a wide range of temperatures. $^3J$(HCCH) for *trans*-difluoroethylene in a liquid crystal solvent has been measured.[308] Other workers have focused attention upon the dependence of $^3J$(HCCH) on angles other than the dihedral angle. The theoretical work of Solkan and Sergeyev[13] has been

299 M. Revel and J. Navech, *Bull. Soc. chim. France*, 1973, 1195.
300 N. M. Viktorova, M. A. Fedorovskaya and N. S. Zefirov, *Zhur. org. Khim.*, 1973, **9**, 1334.
301 K. D. Kopple, G. R. Wiley, and R. Tauke, *Biopolymers*, 1973, **12**, 627.
302 P. C. Manor, W. Saenger, D. B. Davies, K. Jankowski, and A. Rabczenko, *Biochim. Biophys. Acta*, 1974, **340**, 472.
303 W. C. Lin, *J. Chem. Phys.*, 1973, **58**, 4971.
304 W. F. Reynolds and D. J. Wood, *Canad. J. Chem.*, 1973, **51**, 2659.
305 R. L. Lipnick, *Tetrahedron Letters*, 1973, 931.
306 T. M. Filipova, B. D. Lavrukhin, and I. K. Shmyrev, *Org. Magn. Resonance*, 1974, **6**, 92.
307 V. Viti, P. L. Indovina, F. Podo, L. Radics, and G. Némethy, *Mol. Phys.*, 1974, **27**, 541.
308 G. J. den Otter and C. Maclean, *Chem. Phys.*, 1974, **3**, 119.

reported in Section 2B. Good correlations of $^3J$(HCCH) with CCH bond angles have been found for furan, thiophen, selenophen, and pyrrole.[309]

De Kowalewski *et al.* have found additivity for substituent effects on $^3J$(HCCH) in benzaldehydes and benzene derivatives[310] and in substituted pyridines.[311] Aksnes and Støgard[238] have found that values of vicinal (H,H) coupling in propyl and butyl halides do not support the thesis[312] that strongly electronegative substituents increase the coupling when they are separated from the HCCH fragment by an odd number of bonds but decrease it when an even number of bonds intervenes. Fay *et al.*[212] have noted that the effect of substituents on vicinal (H,H) coupling is not simply a question of electronegativity.

In molecules such as (18) Revel and his colleagues[313] find that, although the $^3J$(HCCH) values cannot be assigned to a particular $H^l$ or $H^k$, the couplings are very different in the two isomers when X = S, but for the minority isomer when X = lone pair the couplings are equal. No explanation of this observa-

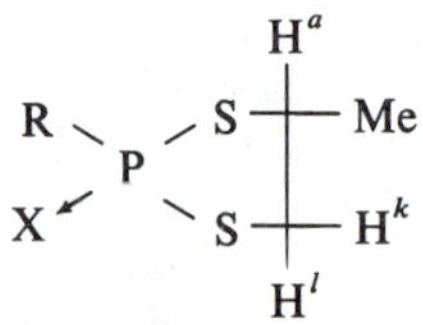


(18) R = $NMe_2$, OMe, or Cl; X = lone pair or S

tion is advanced, but it is pointed out that the situation in the dioxaphospholans is rather similar. The values of $^3J$(HC—CH) and $^3J$(HC=CH) in diene fragments have been discussed with reference to the delocalization of non-bonded electrons in unsaturated heterocycles.[314]

Measurements of $^3J$(HCCH) have been reported for disubstituted pentanes,[315] glutaric anhydrides,[316] tricarbonyliron adducts of cycloheptatriene and cyclo-octatetraene,[294] cyclononatrienyl anions,[317] and the 4-fluorobenzenium cation.[318]

Other reports of $^3J$(HCCH) values may be found in references 10, 26, 53, 92, 94—96, 107, 117, 216—219, 221, 223, 225—233, 235, 236, 239, 240, 242—245, 253, 263, and 319—333.

[309] F. Fringuelli and A. Taticchi, *Gazzetta*, 1973, **103**, 453.
[310] D. G. De Kowalewski, R. Buitrago, and R. Yommi, *J. Mol. Structure*, 1973, **16**, 395.
[311] D. G. De Kowalewski and R. H. Contreras, *J. Mol. Structure*, 1973, **16**, 451.
[312] G. Schrumpf, *J. Magn. Resonance*, 1972, **6**, 243.
[313] M. Revel, J. Roussel, H. Boudjebel, and J. Navech, *Compt. rend.*, 1973, **277**, *C*, 57.
[314] P. Crews, R. R. Kintner, and H. C. Padgett, *J. Org. Chem.*, 1973, **38**, 4391.
[315] T. Moritani and Y. Fujiwara, *J. Chem. Phys.*, 1973, **59**, 1175.
[316] F. J. Koer, T. M. W. van Asbeck, and C. Altona, *Rec. Trav. chim.*, 1973, **92**, 1003.
[317] S. W. Staley and N. J. Pearl, *J. Amer. Chem. Soc.*, 1973, **95**, 3437.
[318] G. A. Olah and Y. K. Mo, *J. Org. Chem.*, 1973, **38**, 3212.
[319] J. P. Dorie, M. L. Martin, S. Odiot, and F. Tonnard, *Org. Magn. Resonance*, 1973, **5**, 265.

**B. Three-bond Coupling of Protons *via* Other Atoms.**—Theoretical work on $^3J$(HNCH)[27,30] has been described in Section 2B. Seven $^3J$(HNCH) values have been measured in two tertiary amines of known, rigid conformation.[334] The couplings agree to within 0.5 Hz with a Karplus-type equation of the form (21). However, the authors suggest that the prediction of dihedral

$$J(\text{HNCH}) = 4.9 - 1.8\cos\phi + 4.9\cos2\phi \qquad (21)$$

angles from $^3J$(HNCH) values cannot be claimed to be better than $\pm10°$ even under optimum conditions. Shrestha and Henold[335] have measured $^3J$(HNCH) for a number of primary and secondary amines. They find a good correlation between the coupling and the carbon chemical shift [equation (22)]. This suggests that $J$ is related to the electron-withdrawing power of the

$$^3J(\text{HNCH}) = 12.02 - 0.04\delta_\text{C} \qquad (22)$$

substituent on nitrogen and that as this increases $J$ decreases. $^3J$(HNCH) values for a large number of peptides have been reported.[336]

$^3J$(HSiNH) is positive.[113]

Other values of three-bond (H,H) coupling may be found in the following publications: $^3J$(HCBH) (ref. 90); $^3J$(HSiCH) (ref. 114); $^3J$(HSnCH) (ref. 258); $^3J$(HNCH) (refs. 96, 117, 219, 261, 331, 337, and 338); and $^3J$(HSCH) (ref. 339).

**C. Three-bond Coupling of Other Nuclei to Protons.**—(*i*) *Elements of Group*


320 F. Borremans, M. Anteunis, and F. Anteunis-De Ketelaere, *Org. Magn. Resonance*, 1973, **5**, 299.
321 R. J. Abraham and J. R. Monasterios, *Org. Magn. Resonance*, 1973, **5**, 305.
322 J. Devillers, J. Roussel, J. Navech, and R. Burgada, *Org. Magn. Resonance*, 1973, **5**, 511.
323 C. H. M. Adams, D. J. Cawley, and K. Mackenzie, *J.C.S. Perkin II*, 1973, 909.
324 H. Paulsen and W. Greve, *Chem. Ber.*, 1973, **106**, 2124.
325 J. B. Lambert, J. J. Papay, E. S. Magyar, and M. K. Neuberg, *J. Amer. Chem. Soc.*, 1973, **95**, 4458.
326 J. A. Mosbo and J. G. Verkade, *J. Amer. Chem. Soc.*, 1973, **95**, 4659.
327 S. W. Staley and G. M. Cramer, *J. Amer. Chem. Soc.*, 1973, **95**, 5051.
328 D. W. Aksnes and O. Vikane, *Acta Chem. Scand.*, 1973, **27**, 2135.
329 L. Marchetti, L. Pentimalli, P. Lazzeretti, L. Schenetti, and F. Taddei, *J.C.S. Perkin II*, 1973, 1926.
330 P. Albriktsen and R. K. Harris, *Acta Chem. Scand.*, 1973, **27**, 3993.
331 J. Elguero, A. Fruchier, and M. del Carmen Pardo, *Org. Magn. Resonance*, 1974, **6**, 272.
332 J. Devillers, M. Cornus, J. Roussel, J. Navech, and R. Burgada, *Org. Magn. Resonance*, 1974, **6**, 205.
333 J. Peeling, L. Ernst, and T. Schaefer, *Canad. J. Chem.*, 1974, **52**, 849.
334 R. R. Fraser, R. N. Renaud, J. K. Saunders, and Y. Y. Wigfield, *Canad. J. Chem.*, 1973, **51**, 2433.
335 K. P. Shrestha and K. L. Henold, *J. Amer. Chem. Soc.*, 1973, **95**, 6699.
336 P. L. Wessels, J. Feeney, H. Gregory, and J. J. Gormely, *J.C.S. Perkin II*, 1973, 1691.
337 J. A. Zoltewicz, L. S. Helmick, T. M. Oestreich, R. W. King, and P. E. Kandetzki, *J. Org. Chem.*, 1973, **38**, 1947.
338 O. Dahl, J. Andersen, and O. Larsen, *Acta Chem. Scand.*, 1973, **27**, 2503.
339 M. Dagonneau, D. Paquer, and J. Vialle, *Bull. Soc. chim. France*, 1973, 1699.

*IVB*. Delbaere *et al.*[340] have determined the structures of two cyclonucleosides by *X*-ray diffraction, and since the structures are rigid the angles thus determined allow one to construct a plot of dihedral angle *versus* $^3J$(CCCH) which has ten points when some other similar data are included. A good relationship of the Karplus type is obtained, though two coupling constants are notably anomalous. Explanations of these exceptions are offered in terms of the electron-density distribution revealed by the crystallographic structure determination. Other discussions of the relationship between $^3J$(CCCH) and dihedral angle in nucleosides, glycosides, and peptides have been published.[341,342] Albriktsen[65] has noted a dependence of $^3J$(CCCH) on $\phi$ in cyclic sulphites and in thiet-1,1-dioxide; this coupling appears to be more sensitive to dihedral angle than to hybridization at C.[107]

The difference in $^3J$(CNCH) and $^3J$(CCCH) in nicotinamide (couplings *via* N are about twice those *via* C) has been noted.[105]

The dependence of $^3J(^{119}$SnCCH) in phenyltin halides upon added Lewis base has been investigated.[343] The coupling is much larger in the ethyltin trihalides. Values of $^3J(^{119/117}$SnNCH) have been reported for the first time. For compounds of the form $Me_{4-n}Sn(NR_2)_n$ (R = Me or Et, $n$ = 1—4) couplings in the range 42—53 Hz are found,[247] but $^3J(^{119/117}$SnN=CH) in dimethyltin dichloride–picolinaldimine adducts lies in the range 16—18 Hz.[259]

Kennedy and McFarlane[54] have measured the magnitudes and signs of many $^3J(^{119}$SnSeCH) couplings. The coupling follows $^1J$(SnSe) in its behaviour but a plot of $^3J$ against $^1J$ does not pass through the origin, indicating contributions to the coupling which are not FC.

(*ii*) *Elements of Group VB*. Some theoretical work on $^3J$(NCCH) and dihedral angle has been discussed in Section 2B, and Sogn, Gibbons, and Randall[116] have given a brief discussion of the problems of relating the coupling to this parameter. An accurate value of $^3J(^{15}$NCCH) for pyrrole has been measured.[96,117]

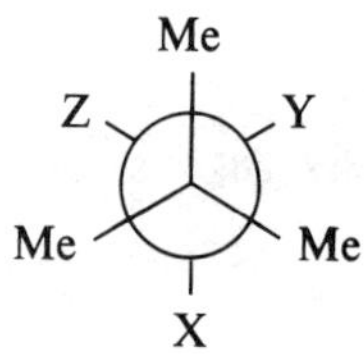


(19) X = S, $BH_3$, or lone pair;
Y, Z = Bu$^t$ or Cl

340 L. T. J. Delbaere, M. N. G. James, and R. U. Lemieux, *J. Amer. Chem. Soc.*, 1973, **95**, 7866.
341 R. U. Lemieux, *Ann. New York Acad. Sci.*, 1973, **222**, 915.
342 A. S. Perlin, N. Cyr, H. J. Koch, and B. Korsch, *Ann. New York Acad. Sci.*, 1973, **222**, 935.
343 C. W. Allen, A. E. Burroughs, and R. G. Anstey, *Inorg. Nuclear Chem., Letters*, 1973, **9**, 1211.

Bushweller and Brunelle[344,345] have discussed the stereochemical dependence of $^3J$(PCCH) in t-butylphosphine and certain of its derivatives (19). They report that the coupling of P to the methyl group *trans* to X is much smaller, in most cases, than the coupling to the *gauche* methyl group. For $X$ = S, however, the couplings are roughly equal. Similarly, in saturated cyclic phosphorus heterocycles, Bentrude and Tan[346] have found a dependence of $^3J$(POCH) upon both dihedral angle and the orientation of the phosphorus lone pair. Couplings in the presence of an axial lone pair (20a)

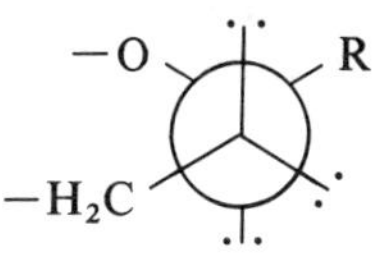

(20a)

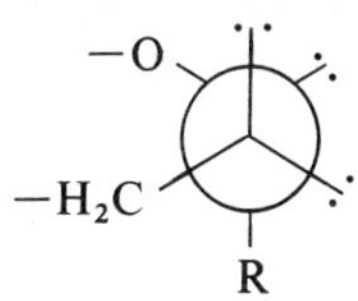

(20b)

are of the order of twice those for an equatorial lone pair (20b). This, as a comparison of structures (19) and (20) shows, represents exactly the same type of dependence of $^3J$(PCCH) and $^3J$(POCH) on the orientation of the phosphorus lone pair.

Paulsen and Greve[324] have discussed the value of $^3J$(POCH) in phosphorus-containing carbohydrates and its dependence upon dihedral angle. On the basis of their experimental results they propose equation (23) and note that the values of the coupling given by this equation closely follow those for $^3J$(PCCH)

$$^3J(\text{POCH}) = 11.2 - 8.2\cos\phi + 11.2\cos2\phi \qquad (23)$$

given by Benezra and his colleagues.[347] Benezra[348] has also found a Karplus-type relationship between $^3J$(PCCH) and dihedral angle in phosphonates. The curve has maxima at $\phi = 0°$ ($\sim$18 Hz) and $\phi = 180°$ ($\sim$41 Hz) and a minimum at $\phi = 90°$ ($\sim$0 Hz).

Radics *et al.*[349] have made accurate measurements of $^3J$(PCCH), $^4J$(PCCCH), and $^5J$(PCCCCH) in triphenylphosphine, together with all the (H,H) couplings. All the $^nJ$(P,H) values have the same sign, which the authors suggest is positive, and they are remarkably similar in magnitude to the corresponding $^nJ$(H,H). Hägele *et al.*[263] have also suggested that the signs of $^3J$(PCCH) and $^4J$(PCCCH) are positive, and the similarity of $|^3J(\text{PCCH})|$ for $P^{III}$ and $P^{V}$ has been noted.[253]

Jefferson *et al.*[288] have found that $^3J$(PNCH) and $^4J$(PNCCH) in cyclophosph(III)azones and their oxo- and thioxo-derivatives are larger when a

344 C. H. Bushweller and J. A. Brunelle, *J. Amer. Chem. Soc.*, 1973, **95**, 5949.
345 C. H. Bushweller and J. A. Brunelle, *Tetrahedron Letters*, 1974, 893.
346 W. G. Bentrude and H.-W. Tan, *J. Amer. Chem. Soc.*, 1973, **95**, 4666.
347 C. Benezra and G. Ourisson, *Bull. Soc. chim. France*, 1967, 624; C. Benezra, *Tetrahedron Letters*, 1969, 4471.
348 C. Benezra, *J. Amer. Chem. Soc.*, 1973, **95**, 6890.
349 L. Radics, E. Baitz-Găcs, and A. Neszmzélyi, *Org. Magn. Resonance*, 1974, **6**, 60.

phosphoryl phosphorus atom is coupled than when a thiophosphoryl atom is involved. This is the reverse of the order found in most acyclic phosphorus compounds. $^3J$(PNSiH) is positive.[113] A linear correlation of $^3J$(POCH) with $\delta_H$ has been reported.[270]

Kennedy, McFarlane, and Rycroft[350] have taken issue with the claim[351] that $^3J$(PWCH) in *trans*-($\pi$-$C_5H_5$)W$(CO)_2PR_3X$ is a 'through-space' coupling. The main evidence for this supposition, *i.e.* that $^2J$(PWC), which involves some of the same bonds, is effectively zero, can be explained by the fact that the sign of $^2J$(PWC) can change from positive to negative as the hybridization at P changes. They suggest that $^3J$(PWCH) is a normal 'through-bond' coupling.

(*iii*) *Elements of Group VIB.* Most values of $^3J$(SeCCH) are positive but a small negative value has been found in 2,4-dinitroselenophen.[254] $^3J$(TeCCH) usually lies in the range 22—23 Hz but small values of 1—3 Hz have been recorded.[255] Couplings to Te are about twice as large as those to Se.[255]

(*iv*) *Elements of Group VIIB.* Hall and Jones[352] have considered the effect of substituent electronegativity on the vicinal (F,H) coupling in 16 1,2-disubstituted derivatives of acenaphthene. They make the very interesting point that, although it has almost invariably been assumed that the relationship between electronegativity and coupling constant is a linear one, this assumption does not bear close scrutiny. If, for example, a substituent Q having electronegativity $\chi_Q$ reduces a coupling constant $J$ from $J_0$ to $(J_0 - q)$, then a second similar substituent would not be expected to reduce the coupling by the same amount, *i.e.* reduce it to $(J_0 - 2q)$. Rather, it should reduce the already diminished coupling by the same fraction, *i.e.* from $(J_0 - q)$ to $[(J_0 - q)/J_0](J_0 - q)$. This leads to an exponential dependence, equation (24), of coupling on electronegativity; $a$ and $k$ are constants. Hall and Jones then show

$$J = a\mathrm{e}^{-k\chi} \qquad (24)$$

that equation (24) fits the experimental data much better than a linear relationship, especially for $J_{trans}$, and suggest reasons for the apparent success of the linear correlations proposed hitherto. Important among these are (*a*) extensive use of $^3J$(H,H) which is not so sensitive to electronegativity; (*b*) investigation of a narrow range of electronegativities only; and (*c*) choice of dihedral angles, in the case of $^3J$(F,H), for which electronegativity dependence was diminished.

For phosphonodifluoridic acids, $R^1R^2C{=}CHP(O)F_2$, Zakharov *et al.*[199] generally found positive values of $^3J$(FPCH) in the range 0.4—6 Hz, though some small negative values (*e.g.* −0.9 Hz for $R^1$ = Cl, $R^2$ = Me or $Me_3C$) were also observed. They suggest that the negative values can be attributed to non-averaged dipolar interactions caused by magnetic anisotropy of the

350 J. D. Kennedy, W. McFarlane, and D. S. Rycroft, *Inorg. Chem.*, 1973, **12**, 2742.
351 T. A. George and C. D. Turnipseed, *Inorg. Chem.*, 1973, **12**, 394.
352 L. D. Hall and D. L. Jones, *Canad. J. Chem.*, 1973, **51**, 2925.

chemical bonds! An explanation based upon conformational preference and contributions to the coupling other than FC would seem to be simpler and more likely. $^3J$(FPNH) is positive in $PF_2NHSiH_3$.[113]

Other reports of three-bond coupling to protons may be found in the following papers: $^3J$(BSCH) (ref. 156); $^3J$(TlNCH) (ref. 353); $^3J$(TlSnCH) (ref. 258); $^3J$(CCCH) (refs. 26, 104—106, 110, 112, and 246); $^3J$(CCCD) (ref. 46); $^3J$(CNCH) (refs. 104, 110, and 354); $^3J$(COCH) (ref. 106); $^3J$(SiCSiH) (ref. 114); $^3J$(SiSeCH) (ref. 99); $^3J$(SnTlCH) (ref. 258); $^3J$(SnCCH) (ref. 283); $^3J$(SnSnCH) (refs. 93 and 258); $^3J$(SnOCH) (ref. 260); $^3J$(SnSCH) (ref. 55); $^3J$(SnSeCH) (ref. 99); $^3J$(NCCH) (refs. 96, 100, 118, and 139); $^3J$(NCSH) (ref. 118); $^3J$(PCCH) (refs. 93, 108, 115, 172, 186, 199, 229, 250, 262, 265, 266, 269, 271, 324, 355, and 356—358); $^3J$(PNCH) (refs. 201, 267, 271, 284, 322, 326, 332, 355, and 359—362); $^3J$(PPCH) (ref. 269); $^3J$(PAsCH) (ref. 269); $^3J$(POCH) (refs. 185, 186, 236, 239, 322, 326, 332, 356, 358—360, and 363—365); $^3J$(PSCH) (refs. 216, 239, 313, and 358); $^3J$(PPtCH) (refs. 184 and 277); $^3J$(PPtSiH) (ref. 122); $^3J$(SeMCH) (M = Si, Ge, or Sn) (ref. 99); $^3J$(FCCH) (refs. 92, 223, 227, 231, 243, 273, 308, 318, and 321); $^3J$(FPCH) (ref. 201); $^3J$(FCPH) (ref. 120); $^3J$(RhCCH) (ref. 279); $^3J$(PtCCH) (ref. 278); $^3J$(PtNCH) (refs. 275, 366, and 367); $^3J$(RhPCH) (ref. 181); $^3J$(PtPCH) (refs. 177, 181, 184, 266, and 268); $^3J$(PtAsCH) (ref. 153); and $^3J$(PtSCH) (ref. 126).

**D. Three-bond Coupling to Carbon.**—(*i*) *Elements of Group IIIB.* Values of $^3J(CCC^{10}B)$ and $^3J(CCC^{11}B)$ in sodium tetraphenylboron have been measured,[128] and in $Tl^{III}$ chelates of coproporphyrin derivatives $^3J$(TlNCC) values of 106, 109, and 147 Hz have been found.[280]

(*ii*) *Elements of Group IVB.* An INDO FPT analysis of the dependence of $^3J$(CCCC) on dihedral angle[32] has been described in Section 2B. In an experimental study of the same problem Marshall and Miller[132] investigated ten

353 M. G. Voronkov, V. A. Pestunovich, S. V. Mikhailova, J. Popelis, and E. Liepins, *Khim. geterotsikl. Soedinenii*, 1973, 312.
354 M. P. Schweizer, E. B. Banta, J. T. Witkowski, and R. K. Robins, *J. Amer. Chem. Soc.*, 1973, **95**, 3770.
355 J. Emsley and J. K. Williams, *J.C.S. Dalton*, 1973, 1576.
356 G. Hägele, *Z. Naturforsch.*, 1973, **28b**, 753.
357 C. W. Allen and A. J. White, *Inorg. Chem.*, 1974, **13**, 1220.
358 R. E. Ardey, J. Emsley, A. J. B. Robertson, and J. K. Williams, *J.C.S. Dalton*, 1973, 2641.
359 J. Devillers, M. Cornus, and J. Navech, *Org. Magn. Resonace*, 1974, **6**, 211.
360 J. Durrieu, R. Kraemer, and J. Navech, *Org. Magn. Resonance*, 1973, **5**, 407.
361 A. Wolff and J. G. Riess, *Bull. Soc. chim. France*, 1973, 1587.
362 G. J. Bullen, P. E. Dann, V. B. Desai, R. A. Shaw, B. C. Smith, and M. Woods, *Phosphorus*, 1973, **3**, 67.
363 T. Sato and K. Goto, *J.C.S. Chem. Comm.*, 1973, 494.
364 R. H. Sarma, R. J. Mynott, F. E. Hruska, and D. J. Wood, *Canad. J. Chem.*, 1973, **51**, 1843.
365 C. B. C. Boyce, S. B. Webb, L. Phillips, and P. A Worthington, *J.C.S. Perkin I*, 1973, 2881.
366 B. E. Reichert and B. O. West, *J. Organometallic Chem.*, 1973, **54**, 391.
367 B. Crociani and R. L. Richards, *J.C.S. Dalton*, 1974, 693.

labelled compounds. They assume the coupling to be positive, an assumption substantiated by the calculations[32] cited above, and find a curve showing maxima at $\simeq 0°$ (2 Hz) and $\simeq 180°$ (4—6 Hz) and a minimum at $\simeq 65°$ (0 Hz). This result is only in qualitative accord with the work of Doddrell *et al.*[32] and draws attention to the fact that the range of application of theoretical or experimental curves of this type may be very limited.

Doddrell and his co-workers[368] have also suggested an equation, (25), relating $^3J$(SnCCC) to dihedral angles; under the assumption that all the couplings have the same sign, positive. A strong dependence of $^3J$(SnCCC)

$$^3J(\text{SnCCC}) = 30.4 - 7.6\cos\phi + 25.2\cos2\phi \qquad (25)$$

on dihedral angle has been noted[136] in 14 organotin compounds. The large $^3J_{trans}$ of *ca.* 68 Hz seems to be little affected by a change of carbon hybridization from $sp^2$ to $sp^3$. Mitchell[133] has published a large quantity of data on $^3J$(SnCCC) and also some values of $^3J$(SnNCC).

(*iii*) *Elements of Group VB.* Theoretical work on $^3J$(NCCC)[28] has been discussed in Section 2B. In oximes $^3J$(NCCC) is greater for the *E*-isomer (8a) than for the *Z* (8b),[141] and the coupling is zero where the oximes are formed from saturated aldehydes or ketones. This may be related to the rather surprising absence of measurable $^3J$(NCCC) values in a series of saturated amides.[140]

Burdon *et al.*[281] have noted an interesting stereochemical dependence of of $^3J$(PNCC) in some aminophosphorus compounds [(13) and (14)]. For (13a) and (13b) four different $^3J$(PNCC) values are observed at −40 °C, one of 23 Hz and the others between 2 and 7 Hz. The averaged couplings at 55 °C indicate that both pairs of values have the same sign. For (14a) and (14b), on the other hand, all couplings are less than 8 Hz both at −40 °C and at 55 °C. The authors suggest that the large value of $^3J$(PNCC) results from a preferred molecular conformation which places a methyl group close to the phosphorus lone pair.

(*iv*) *Metals.* Good correlations between $^1J$(PtC), $^2J$(PtCC), and $^3J$(PtCCC) have been noted,[150] indicating a dependence of all these couplings upon similar factors. A surprisingly large value of $^3J$(PtCCC) (179 Hz) has been measured in a norbornenyl complex of $Pt^{II}$.[102] Measurements of $^3J$(PtCNC), $^3J$(PtPCC), and $^3J$(PtAsCC) in carbene complexes of 1,3-diorganoimidazolidin-2-ylidenes have been reported.[151] All three couplings are similar in magnitude, and always larger in the *cis*- than in the *trans*-isomers.

Other reports of three-bond coupling to carbon include $^3J$(SnCCC) (refs. 134, 135, and 283); $^3J$(PbCCC) (ref. 135); $^3J$(SnCOC) (ref. 135); $^3J$(PbCOC) (ref. 135); $^3J$(SnCSC) (ref. 135); $^3J$(PbCSC) (ref. 135); $^3J$(NCCC) (ref. 139); $^3J$(PCCC) (refs. 144, 146, and 151); $^3J$(NNCC) (ref. 141); $^3J$(POCC) (refs. 146,

[368] D. Doddrell, I. Burfitt, W. Kitching, M. Bullpitt, Che-Hung Lee, R. J. Mynott, J. L. Considine, H. G. Kuivila, and R. H. Sarma, *J. Amer. Chem. Soc.*, 1974, **96**, 1640.

282, and 369); $^3J$(PPtPC) (ref. 149); $^3J$(FCCC) (refs. 92, 101, 163, and 164); $^3J$(PtCNC) (ref. 149); $^3J$(PtNCC) (ref. 153); $^3J$(PtCOC) (ref. 149); and $J$(HgCXC) (X = C, O, or S) (ref. 135).

$^3J$(SnCCSn) has also been measured.[137]

**E. Three-bond Coupling to Phosphorus.**—(*i*) *Phosphorus.* $^3J$(POCP) has been observed to correlate with $^2J$(PCH) and $^3J$(POCH),[270] and is probably opposite in sign to $^2J$(PMP)[186] (see also Section 4E).

(*ii*) *Fluorine.* An attempt has been made to correlate $^3J$(PCCF) with dihedral angle[370] but it met with little success. This is perhaps not surprising in view of the theoretical evidence on the complexity of coupling to fluorine.[8]

Other recent reports on three-bond coupling to phosphorus concern $^3J$(POPF) (refs. 197 and 202) and $^3J$(PMPF) (M = Rh, Cr, or Mo) (ref. 198) and (M = Ni) (ref. 200).

**F. Three-bond Coupling to Fluorine.**—den Otter *et al.* have reported theoretical[9] and experimental[9,308,371] studies of the anisotropy of (F,F) coupling. The theoretical work has been described in Section 2B and concerns the anisotropy of $^3J$(FCCF) in *trans*-difluoroethylene. However, no anisotropy is found for the same coupling in *o*-difluorobenzene, though small anisotropies are found in $^4J$(F,F) and $^5J$(F,F) in the *meta*- and *para*-isomers.[371] $^3J$(FCCF) has been found to be extremely sensitive to substituent electronegativity, and no correlation of the coupling with dihedral angle was possible.[292] This cannot be considered surprising.[8] The temperature and solvent dependence of $^3J$(FCCF) in 1-bromo-3,3,4-trifluorobut-4-ene has been investigated.[94]

Abe and Shreeve[372] have reported $^3J$(FCSF) values for bis(perfluoroalkyl) sulphur tetrafluorides. The values range between 17 and 23 Hz.

Further reports of three-bond coupling to fluorine include $^3J$(FCCF) (ref. 163) and $^3J$(PtSCF) (ref. 123).

## 6 Coupling between Nuclei Separated by Four Chemical Bonds, $^4J$

**A. Proton–Proton Coupling *via* Carbon, $^4J(^1HCCC^1H)$.**—Theoretical work[13] on this coupling in cycloalkenes has been noted in Section 2B. Values of $^4J$(HCCCH) for cyclohexane have been compared[220] with values calculated by Barfield.[373] The comparison suggests that the theoretical values may show too low a maximum and too high a minimum.

De Kowalewski *et al.*[310,311] have investigated substituent effects on coupling in benzaldehydes, methylbenzenes, and pyridines. They find that the

369 M. Haemers, R. Ottinger, D. Zimmermann, and J. Reisse, *Tetrahedron*, 1973, **29**, 3539.
370 R. J. Burns, P. B. Bulkowski, S. C. V. Stevens, and M. C. Baird, *J.C.S. Dalton*, 1974, 415.
371 G. J. den Otter, J. Gerritsen, and C. Maclean, *J. Mol. Structure*, 1973, **16**, 379.
372 T. Abe and J. M. Shreeve, *Inorg. Nuclear Chem. Letters*, 1973, **9**, 465.
373 M. Barfield, *J. Chem. Phys.*, 1964, **41**, 3825; *J. Amer. Chem. Soc.*, 1971, **93**, 1066.

experimental couplings are in good agreement with values calculated using additivity relationships, provided that strong interactions between substituent groups are not present. $^4J$(HCCCH) values in norbornane-2,3-diones are found to be rather large (1.0—1.5 Hz) but an *X*-ray structure determination of 6-*endo*-bromo-1,5,5-trimethylbornane shows that there is only a slight distortion of the norbornane framework.[374] This result supports the idea that the large couplings are due to substituent effects rather than to stereochemical distortions. Fomichov *et al.*[375] have examined the dependence of $^4J$(HCCCH) on ring puckering angle in azetidines and oxetans. Both $^4J_{cis}$ and $^4J_{trans}$ are negative and decrease in magnitude with the puckering angle. The results appear to be at variance with the conclusions of Gamba and Mondelli.[376] Lambert and Magyar[377] have discussed the variation with geometry of $^4J$(HCCCH) in cyclopropyl systems. The coupling follows a non-W path by a $\sigma$–$\pi$ mechanism analogous to that in allylic couplings. The effect of geometry on $^4J$(HCCCH) in deuteriated t-butylcyclohexane has been investigated.[222]

Mark[378] has discussed four-bond (H,H) coupling in norbornenes (21), distinguishing between ‘W’, ‘sickle’ (˥⊔), and ‘fork’ (∪) paths. While fork

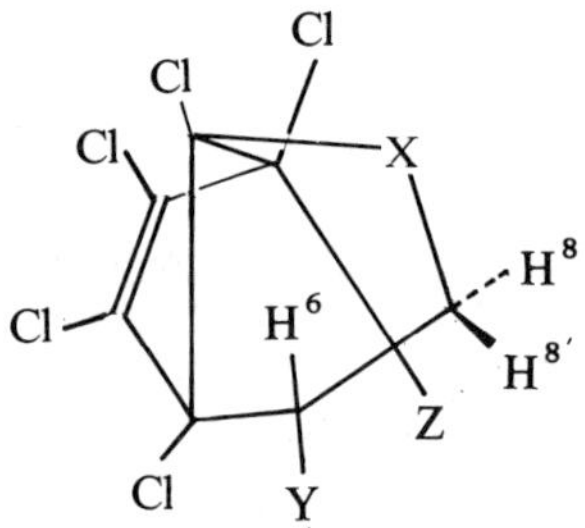

(21) X = O or S;
Y = Me or H;
Z = Cl, Me, Ph, or H

and some sickle paths show negligible coupling one type of sickle path consistently showed couplings whose magnitudes overlapped with those of the W path (1—4 Hz). Mark suggests the following explanation of this phenomenon. For the strong coupling ($H^6$,$H^8$) the H—C bond representing the ‘handle’ of the sickle points towards the hydrogen atom four bonds away, and comes close to it, whereas it misses considerably the corresponding proton in the non-coupled sickle arrangement (Y,$H^{8'}$). Thus a direct interaction between bonds to the coupled nuclei might be at work as in the case of W coupling. In

374 G. F. Hambly, J. Leitch, P. Yates, and S. C. Nyburg, *Canad. J. Chem.*, 1973, **51**, 4076.
375 A. A. Fomichov, I. A. Zon, I. M. Gella, R. G. Kostyanovsky, A. N. Saprin, and V. I. Markov, *Org. Magn. Resonance*, 1973, **5**, 263.
376 A. Gamba and R. Mondelli, *Tetrahedron Letters*, 1971, 2133.
377 J. B. Lambert and E. S. Magyar, *Org. Magn. Resonance*, 1973, **5**, 403.
378 V. Mark, *Tetrahedron Letters*, 1974, 299.

sickle coupling, however, it may be the rear lobe of one of the hydrogen-bearing $sp^3$-carbon orbitals and the *s*-orbital of the other hydrogen which are the interacting functions.

Cunliffe and Harris[237] have discussed the variation of $^4J$(H,H) in 1,3-dienes with the dihedral angle about the C-2—C-3 bond, and in *trans*-sorbaldehyde Albriktsen and Harris[330] have noted that coupling to the aldehydic proton is significantly smaller than that to the diene protons. The longest detectable coupling to this proton was $^4J$(H,H), which had a value of $-0.25$ Hz. By contrast, five- and six-bond couplings of the order of $\pm 0.66$ Hz were observed between other protons.

Green *et al.*[233] have reported an accurate analysis of the $^1$H spectrum of cyclobutanone. They find that $^4J$(HCCCH) is noticeably larger in magnitude than the corresponding cross-ring couplings in cyclobutanes. The allylic coupling in thiet-1,1-dioxide is not very different from that in cyclobutene.[107] $^4J$(HCCCH) and $^5J$(HCCCCH) for the benzyl proton to the *o*- and *m*-protons of the ring in 2-(2,4-dichlorophenyl)oxetan are of opposite sign and $|^4J| > |^5J|$.[242] This order of magnitudes is not invariably the case; it is not so in substituted benzaldehydes for instance. Curry and Sardella[213] have made a very detailed study of isopropylidenic and allylic $^4J$(H,H) in *β,β*-dimethylstyrenes. [See also Section 4A for an account of the relationship between $^4J$(H,H) and $^2J$(HCH).] Both types of $^4J$(H,H) are uniformly negative and the authors discuss the couplings and substituent effects on them. The effect of substituents on the allylic couplings is unexceptional, but the variation of the isopropylidenic $^4J$(H,H) cannot be simply explained in terms of $\sigma$- and $\pi$-electron contributions. The effect upon $^4J$(H,H), in systems of the form $Me_2C{=}X$, of changing X from carbon to oxygen is discussed in terms of $\sigma$ and $\pi$ contributions and the effect upon the latter of polarization of the C=X bond. The analysis gives a plausible explanation of the small positive $^4J$(H,H) value for acetone.

Other reports of $^4J$(HCCCH) may be found in references 10, 26, 72, 95, 96, 98, 117, 219, 227, 230, 235, 244, 245, 254, 263, 272, 294, 314, 319, 323, 328, 333, 360, and 379—381.

**B. Four-bond Coupling of Protons *via* Other Atoms.**—$^4J$(HCNCH) couplings in substituted pyridines are in good agreement with values calculated using additivity rules.[311] Brettle and Logan[382] have discussed $^4J$(HCOCH) in 2,5-disubstituted-1,3-dioxolan-4-ones, and Hobson *et al.*[380] have found that $^4J$(HOCCH) in ethyl acetoacetate is temperature-dependent. It is also concentration-dependent and its variation among *β*-diketones may be related to hydrogen-bond strength.[380] The theoretical analysis of $^4J$(HCOCH) and $^4J$(HCSCH) by Lozac'h and Braillon[26] has been described in Section 2B.

379 M. Brink, *Org. Magn. Resonance*, 1973, **5**, 427.
380 R. F. Hobson, L. W. Reeves, and R. C. Shaddick, *Org. Magn. Resonance*, 1974, **6**, 129.
381 D. Wendisch, W. Hartmann, and H.-G. Heine, *Tetrahedron*, 1974, **30**, 295.
382 R. Brettle and I. D. Logan, *J.C.S. Perkin II*, 1973, 687.

$^{4}J$(HCSCH) values for annelated thiophens have been discussed by Skramstad.[383]

Further values of four-bond (H,H) couplings may be found in the following papers: $^{4}J$(HCNCH) (refs. 96, 100, 219, 227, 331, and 384); $^{4}J$(HNCCH) (refs. 96, 117, 118, and 331); and $^{4}J$(HCOCH) (refs. 319 and 385).

**C. Four-bond Coupling of Other Nuclei to Protons.**—(*i*) *Elements of Group IIIB.* Abraham, Barnett, and Smith[386] have examined (Tl,H) coupling in thallium(III) porphyrins (22). Replacement of H by Me at the $\beta$-position

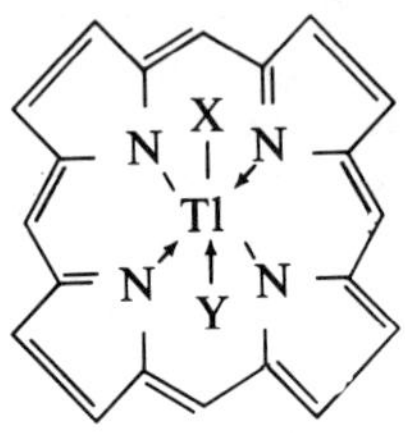


(22) X = $OCOCF_3$, OCOMe, OH, I, or CN; Y = $H_2O$ or absent

reduces the coupling from $^{4}J$(TlNCCH) $\simeq$ 70 Hz to $^{5}J$(TlNCCCH) $\simeq$ 9 Hz, which indicates a $\sigma$-mechanism. By contrast replacement of H by Me at the *meso*-position reduces the coupling from $^{4}J$(Tl,H) $\simeq$ 45 Hz to $^{5}J$(Tl,H) $\simeq$ 28 Hz, suggesting an important $\pi$-electron contribution to the coupling.

(*ii*) *Elements of Group IVB.* Matsubayashi, Koezuka, and Tanaka[387] have observed the long-range coupling between $^{119/117}$Sn and the methyl protons of *o*-, *m*-, and *p*-tolyltin trichlorides. The couplings are solvent-dependent, but in all solvents the order of magnitudes is $^{4}J$(Sn,H)(*ortho*) > $^{6}J$(Sn,H)(*para*) > $^{5}J$(Sn,H)(*meta*). These results follow the pattern of the (Tl,H) coupling in *o*-, *m*-, and *p*-tolylthallium cations and, indeed, the ratio $^{n}J(\mathrm{Tl,H})/^{n}J(\mathrm{Sn,H})$ = 8.5 $\pm$ 0.5 for the three positions of substitution, which strongly suggests a similar mechanism in the two types of coupling. The same group also reports[388] measurements of $^{4}J$($^{119/117}$SnCNCH) for diallylcarbamoyltin compounds (23). They find that $^{4}J$(SnCNCH) *cis* is always greater than the corresponding

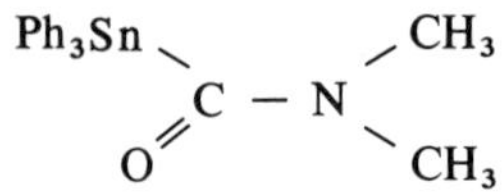


(23)

383 J. Skramstad, *Chem. Scripta*, 1973, **4**, 77.
384 D. B. Davies and M. Abu Khaled, *J.C.S. Perkin II*, 1973, 1651.
385 D. L. Davenport, C. H. Harrison, and D. W. Rathburn, *Org. Magn. Resonance*, 1973, **5**, 285.
386 R. J. Abraham, G. H. Barnett, and K. M. Smith, *J.C.S. Perkin I*, 1973, 2142.
387 G. Matsubayashi, H. Koezuka, and T. Tanaka, *Org. Magn. Resonance*, 1973, **5**, 529.
388 G. Matsubayashi, M. Hiroshima, and T. Tanaka, *J. Organometallic Chem.*, 1973, **59**, 207.

*trans* coupling by a factor of 1.2—2.0. This is also true of the one result for $^4J(^{207}PbCNCH)$.

(*iii*) *Elements of Group VB.* $^4J$(PCCCH) values in 1-aza-2-phosphacyclohepta-4,6-dienes range between 1.9 and 3.3 Hz while the $^5J$(P,H) couplings are of the order of 4.6—5.3 Hz[264] and $^4J$(PCCCH) appears to be positive in Me(*p*-$MeOC_6H_4)PS_2Na,2H_2O$.[263] A detailed study of $^nJ$(P,H) in triphenylphosphine[349] shows that all three-, four-, and five-bond (P,H) couplings have the same sign, presumably positive. The values of $^nJ$(P,H) are remarkably similar in magnitude to those of the corresponding $^nJ$(H,H).

Some observations on the magnitudes of $^3J$(PNCH) and $^4J$(PNCCH) in cyclophosph(III)azones[288] have been noted in Section 5C. Dahl, Anderson, and Larsen[338] have reported that in *PP*-disubstituted thiophosphinoyl-thioformhydrazides (24) $^4J$(PCNCH) is always larger in the *E*-isomer than in the *Z*; the reverse is true for the $^5J$(PCNNCH) coupling.

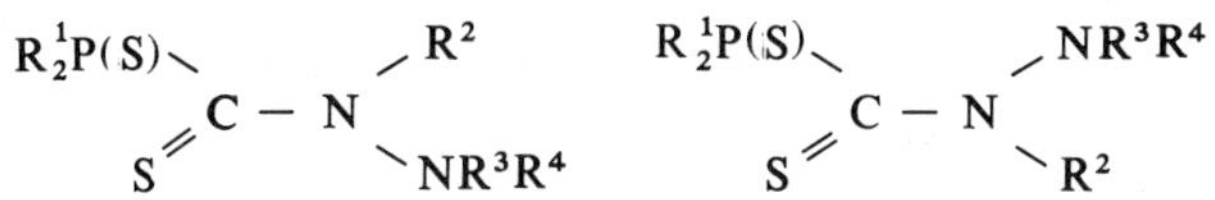

(a) *Z*-isomer (b) *E*-isomer

(24) $R^1$ = Et or Ph; $R^2$, $R^3$, $R^4$ = H, Me, Ph

Gaydou and Llinas[236] have observed that $|{}^4J(POCCH)_{cisoid}| < |{}^4J(POCCH)_{transoid}|$ in enolic phosphates (25) and that whereas the $^4J_{cisoid}$ increase in value with increasing temperature the $^4J_{transoid}$ decrease. They suggest that this is due to a $\pi$-electron contribution which is small at low temperatures and increases with increasing temperature. The $\pi$ contribution

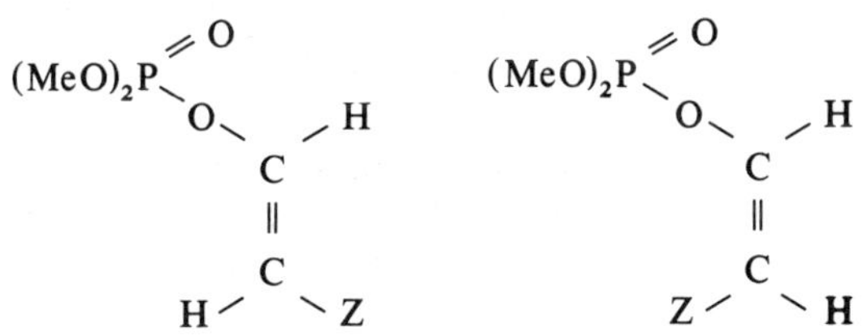

*cisoid* *transoid*

(25) Z = H, Cl, or Me

increases as the population of *gauche*-, as opposed to planar, isomers increases and it contributes a negative term to the coupling. This increases $J_{cisoid}$, which must therefore be negative, and decreases $J_{transoid}$. The negative $\pi$ contribution also accounts for the reversal in the order of the absolute values of $J_{cisoid}$ and $J_{transoid}$ which is brought about by bulky substituents at Z. $^4J$(POCCH) values in $\beta$-5′-nucleotides have been discussed by Sarma *et al.*[389]

389 R. H. Sarma, R. J. Mynott, D. J. Wood, and F. E. Hruska, *J. Amer. Chem. Soc.*, 1973, **95**, 6457.

(*iv*) *Elements of Group VIIB*. A good linear correlation has been noted[272] between $^4J$(FCCCH) for substituted acetones and $\Delta\chi$, the sum of the Huggins electronegativities of the atoms on the terminal carbon atoms relative to that for acetone itself. The coupling is positive and decreases as $\Delta\chi$ increases. $^4J$(FPNSiH) in silylaminodifluorophosphine is positive[113] and might be a 'through-space' coupling.

Other reports of four-bond coupling to protons may be found in the following papers. $^4J$(CCCCH) (refs. 105, 112, and 246); $^4J$(CCNCH) and $^4J$(CNCCH) (ref. 105); $^4J$(COCCH) and $^4J$(CSCCH) (ref. 246); $^4J$(SnCCCH) (refs. 343 and 283); $^4J$(PCCCH) (ref. 199); $^4J$(PNCCH) (ref. 332); $^4J$(PNSiCH) (ref. 288); $^4J$(PNPCH) (ref. 269); $^4J$(POCCH) (refs. 239, 326, 332, and 365); $^4J$(PSCCH) (refs. 216 and 358); $^4J$(PMCCH) (ref. 183); $^4J$(PMPCH) (M = transition metal) (ref. 181); $^4J$(FCCCH) (refs. 10, 92, 94, 227, and 231); $^4J$(FCCSiH) (ref. 390); $^4J$(FPCCH) (ref. 199); $^4J$(FPNCH) (refs. 201 and 288); $^4J$(FWNCH) (ref. 207); $^4J$(PtCCCH) (ref. 277); $^4J$(PtSiCCH) (refs. 115 and 122); $^4J$(PtCNC) (refs. 151 and 367); $^4J$(PtCOCH) (ref. 367); $^4J$(PtPOCH) (ref. 126); and $^4J$(HgCSiCH) (ref. 274).

**D. Four-bond Coupling to Carbon.**—(*i*) *Elements of Group IIIB*. Earlier[128] results[127] for $^4J$(CCCC$^{11}$B) in sodium tetraphenylboron have been corrected. Abraham, Hawkes, and Smith[280] have found that $^2J$(TlNC) $\simeq$ $^4J$(Tl-NCCC) in thallium(III) chelates of coproporphyrin. This is also the case in other ($^{13}$C,metal) couplings and in olefinic (H,H) coupling and supports the comparison of Tl and H coupling.

(*ii*) *Elements of Group IVB*. $^4J$(SnCCCC) values from several studies of organotin compounds have been reported.[133,134,136] In substituted aryltrimethyltins $^1J$(SnC) correlates well with the Hammett $\sigma$ values for the substituent, but the correlation falls off as the coupling constants get smaller and is very poor for $^4J$(SnCCCC).[134]

(*iii*) $^4J$(PtCCCC). The good linear relationships between $^1J$(PtC), $^2J$(PtCC), and $^3J$(PtCCC) noted by Clark and Ward do not extend to $^4J$(PtCCCC),[150] but this may be due to the large relative errors in the measurement of this small coupling.

Other reports of four-bond coupling to carbon include $^4J$(CCCCC) (ref. 132); $^4J$(NCCCC) (ref. 139); $^4J$(PCCCC) (ref. 144); $^4J$(POCCC) (ref. 369); $^4J$(PPdCNC) and $^4J$(PPtCNC) (ref. 151); and $^4J$(FCCCC) (ref. 92).

**E. Four-bond Coupling to Fluorine.**—(*i*) $^4J$(PCCCF). The value of this coupling in P (C≡C—CF$_3$)$_3$ has been found to be 6.4 Hz.[162]

(*ii*) (F,F) *Coupling*. The calculations of $^4J$(FCCCF) by Hirao and his co-workers[8] have been questioned with regard to both the signs and the magnitudes of the calculated couplings.[7] Shapiro *et al.*[272] have measured two

390 G. Hägele and M. Weidenbruch, *Org. Magn. Resonance*, 1974, **6**, 66.

values of $^4J$(FCCCF) in 1,1,3,3-tetrafluoroacetone and they show opposite and very sizeable solvent effects, although spectral analysis indicates that they must have the same sign. The anisotropy of $^4J$(FCCF) in *m*-difluorobenzene is small.[371]

Values of the *cis* (− 1.1 Hz) and *trans* (− 6.2 Hz) four-bond (F,F) coupling in perfluorocyclobutanone have been reported[163] and it is interesting to contrast these with the very large values (38 and 118 Hz) found in the *o*-fluorophenyldifluorocarbenium ion,[293] although the larger value is almost certainly a 'through-space' coupling.

Other reports of four-bond coupling to fluorine can be found in the following papers: $^4J$(POPCF) (ref. 197); $^4J$(PPtSCF) (ref. 123); $^4J$(FCCSiF) (ref. 390); $^4J$(FCPCF) (ref. 289); $^4J$(FPOPF) (ref. 202); and $^4J$(FPMPF) (M = Cr, Mo, or Rh) (ref. 198).

## 7 Coupling between Nuclei Separated by Five Chemical Bonds, $^5J$

**A. Proton–Proton Coupling *via* Carbon, $^5J(^1HCCCC^1H)$.**—Theoretical work[13,19] on this coupling has been reported in Section 2B. Additivity of substituent effects on the coupling in benzaldehydes,[310] methylbenzenes,[310] and pyridines[311] has been noted. Albriktsen and Harris have discussed (H,H) coupling in the linear conjugated systems *trans*-sorbaldehyde[330] and hexatriene and its 1,6-dimethyl derivative.[230] For the aldehyde it is interesting to note that no five-bond (H,H) coupling to the aldehydic proton was detected although other five- and six-bond couplings were readily measurable.[330] Values of $^5J$(HCCCCH) have been given for 1,3-alkoxybutadienes.[319] Wasylishen, Rowbotham, and Schaefer[391] have investigated both the signs and the magnitudes of (H,H) coupling in isothiazole (26), iso-oxazole (27), and ten of their alkyl derivatives. They find that five-bond (H,H) couplings between ring

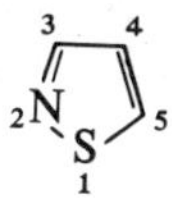

(26)

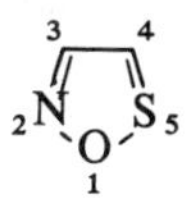

(27)

protons and methyl-group protons in these molecules are considerably larger than in furan and thiophen derivatives. Since the $\pi$-electron contribution would appear to be the same this indicates a $\sigma$-electron effect. The $^5J(CH_3, CH_3)$, on the other hand, seems to be primarily a $\pi$-electron coupling, but $\sigma$-electron effects cannot be ruled out.

Other reports of $^5J$(HCCCCH) can be found in references 10, 222, 242, 263, 314, 333, 377, and 392.

391 R. E. Wasylishen, J. B. Rowbotham, and T. Schaefer, *Canad. J. Chem.*, 1974, **52**, 833.
392 M. C. Fournié-Zaluski, C. Jaureguiberry, and B. Roques, *Tetrahedron Letters*, 1973, 4177.

**B. Five-bond Proton–Proton Coupling *via* Other Atoms.**—Davies and Khaled[384,393] have reported investigations of $^5J$(HCNCCH) in amides and peptides. They have used these measurements to assign the *cis* and *trans* conformations of peptide bonds in linear peptides containing N-methylated amino-acids.[169] This coupling has also been measured in the *E*- and *Z*-isomers of imidates.[394]

Vold and Vold[395] have demonstrated that spin-echo spectra may be used to obtain values of small ($< 0.2$ Hz) coupling constants and have used the technique to obtain some $^5J$(HCOCCH) values for simple esters.

Further five-bond (H,H) couplings may be found in the following references: $^5J$(HCNCCH) (refs. 314, 380, 396, 397) and $^5J$(HNCCCH) (ref. 398).

**C. Five-bond Coupling of Other Nuclei to Protons.**—(*i*) *Elements of Group IVB*. Some Japanese work[387,388] on $^5J$(Sn,H) has been described in Section 6C(*ii*).

(*ii*) *Elements of Group VB*. Paulsen and Thiem[399] have measured values of $^5J$(PCC=CCH) in phosphorus-containing carbohydrates. The coupling ranges between 2.5 and 5.3 Hz and shows a dependence upon angle similar to that of the $^5J$(H,H) homoallylic coupling. It is a minimum when the interacting nuclei lie in the plane of the C=C double bond, and a maximum when they lie out of this plane. Radics *et al.*[349] have noted a remarkable similarity in the magnitudes of $^nJ$(P,H) in triphenylphosphine and the corresponding $^nJ$(H,H) ($n$ = 3, 4, or 5). Values of $^5J$(PCCCCH) have also been reported for phosphoro-difluoridic acids[19] and 1-aza-2-phosphacyclohepta-4,6-dienes.[264]

In contrast to $^4J$(PCNCH), $^5J$(PCNNCH) is larger in the *Z*-isomers of *PP*-disubstituted thiophosphinoyl-thioformhydrazides (24a) than in the *E*-isomers (24b).[338]

(*iii*) *Elements of Group VIB*. Anteunis, van Cauwenberghe, and Becu[231] have observed some large (1.0—1.3 Hz) $^5J$(FCCOCH) values in substituted 1,3-dioxolans. The couplings do not appear to be favoured by a special path such as a zig-zag, nor are they understandable in terms of the converging vector rule.[400]

Other five-bond couplings may be found in the following papers: $^5J$-(TlNCCCH) (ref. 386); $^5J$(POCCCH) (refs. 185, 236, and 356); $^5J$(FCCCCH) (refs. 10, 94, 231, 318, and 401); and $^5J$(FPCCCH) (ref. 199).

**D. Five-bond Coupling to Carbon.**—Marshall and Miller[132] have noted that $^5J$(CCCCCC) is as large as $^4J$(CCCCC) in aliphatic carboxylic acids, and the

393 D. B. Davies and M. Abu Khaled, *Tetrahedron Letters*, 1973, 2829.
394 C. O. Meese, W. Walter, and M. Berger, *J. Amer. Chem. Soc.*, 1974, **96**, 2259.
395 R. L. Vold and R. R. Vold, *J. Magn. Resonance*, 1974, **13**, 38.
396 E. Dradi, G. Tosolini, and V. Vecchietti, *Gazzetta*, 1973, **103**, 609.
397 A. F. Bramwell and R. D. Wells, *Tetrahedron*, 1973, **29**, 3939.
398 T. Yokoyama, *Austral. J. Chem.*, 1974, **27**, 915.
399 H. Paulsen and J. Thiem, *Chem. Ber.*, 1973, **106**, 3850.
400 A. D. Cross and P. W. Landis, *J. Amer. Chem. Soc.*, 1964, **86**, 4005.
401 W. Adcock and S. Q. A. Rizvi, *Austral. J. Chem.*, 1973, **26**, 2659.

same is true of $^5J$(NCCCCC) and $^4J$(NCCCC) in 2,4,6-trimethylbenzonitrile $^{15}N$-oxide; in fact the former is larger.[139] $^5J$(NCCCCC) values have also been reported for oximes.[141]

**E. Five-bond ($^{19}$F,$^{19}$F) Coupling.**—A small anisotropy has been found in the (F,F) coupling of *p*-difluorobenzene.[317,402] Abraham and Loftus[92] have determined the chemical shifts and coupling constants of all the magnetic nuclei in 1,1,1,4,4,4-hexafluorobutane and a $^5J$(FPOPCF) value has been reported for $F_2POP(CF_3)_2$.[197]

**F. $^5J$($^{195}$PtSiCCC$^{19}$F).**—Signs (positive) and magnitudes for this coupling in bis(triethylphosphine) complexes of platinum have been reported by Anderson *et al.*[115,122]

**G. Other Five-bond Couplings.**—A value of $^5J$(PtOCOPtP) has been reported.[268]

## 8 Coupling between Nuclei Separated by Six Chemical Bonds, $^6J$

**A. Proton–Proton Coupling *via* Carbon, $^6J$($^1$HCCCCC$^1$H).**—It is interesting here to note that long-range (H,H) couplings of this type have attracted the attention of two groups who, by and large, have confined their attentions to linear conjugated systems in the case of Harris and his collaborators, and to cyclic systems in the case of Rowbotham and Schaefer and their colleagues. In 2,4-dimethyl-2,4-pentadiene, Cunliffe and Harris[237] were unable to measure a $^6J$(HCCCCCH) value directly but the appearance of the spectrum could only be rationalized by assuming appreciable (0.34 Hz) coupling to both distant methyl groups. In effect this is quite a large coupling since, in general, the spectra suggest that the molecule has an average dihedral angle of about 50° and this will markedly reduce the $\pi$-electron coupling. In hexatriene and 1,6-dimethylhexatriene[230] values of $^6J$(H,H) and $^7J$(H,H) were readily measured but were found to be much smaller than the couplings predicted theoretically, particularly so in the case of $^7J$. The authors comment that this discrepancy is consistent with some out-of-plane twisting, but the evidence is not definitive. $^6J$(H,H) was observed in sorbaldehyde,[330] but not to the aldehydic proton.

Rowbotham and Schaefer and their collaborators have analysed the (H,H) couplings in halogenotoluenes[10,333] and methyl-substituted aromatic systems[14] in terms of $\sigma$- and $\pi$-electron contributions to the coupling. In doing this they have made extensive use of CNDO and INDO/FPT calculations. In their study of $^6J(CH_3,CH_3)$ in 3,5-dimethyl-isothiazole and -iso-oxazole [see structures (26) and (27)] Wasylishen, Rowbotham, and Schaefer[391] make a rather interesting point. This coupling is of the order of 0.2 Hz and *negative*, which confirms that it passes through an *even* number of bonds. If it were

402 G. J. Den Otter, W. Heijser, and C. MacLean, *J. Magn. Resonance*, 1974, **13**, 11.

transmitted through the heteroatoms we would have a seven-bond coupling which would be expected to be positive as in *m*-xylene. However, in both the thiophen (28) and furan (29) derivatives $^{n}J(CH_3,CH_3) = +0.55$ Hz, suggesting that $n = 7$ rather than 6. It appears that the coupling 'prefers' the carbon-atom path.

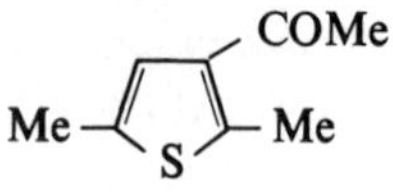

(28)

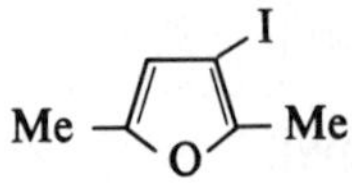

(29)

**B. Proton–Proton Coupling *via* Other Atoms.**—Bramwell and Wells[397] have reported a very marked steric effect on $^{6}J(H_bCNCCCH_a)$ in a series of 2-alkyl-3-methoxypyrazines (30). As the sizes of the alkyl groups $R^1$ and $R^2$ are increased $^{5}J(H_c,H_a)$ remains almost constant whereas $^{6}J(H_b,H_a)$ falls steadily.

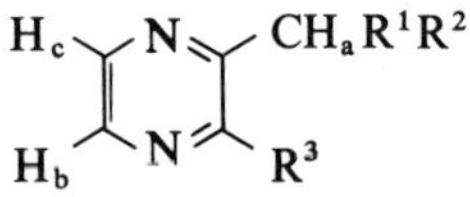

(30) $R^1$, $R^2$ = H, Me, or Et; $R^3$ = OMe

The $(H_b,H_a)$ coupling also shows a linear correlation with the difference between the chemical shift of $H_a$ in the molecule in question and in the parent 2-alkyl compound, (30; $R^3$ = H). The constancy of $^{5}J(H_c,H_a)$ lends support to the proposed[403] compensation of $\sigma$- and $\pi$-effects in couplings of this type. Values of $^{6}J$(HCNCCCH) have been reported and discussed by Rowbotham and Schaefer.[14]

Vold and Vold[395] have measured a very small value of $^{6}J$(HCOCCCH) ($0.068 \pm 0.001$ Hz) in methyl propiolate using a spin-echo technique. Values of this coupling have also been measured in pyrrole derivatives.[396] $^{6}J$(HCSC-NCH) in imidazo[2,1-*b*]thiazole (31) $\simeq$ 1.1 Hz[329] and this could also include a contribution from $^{6}J$(HCCNCCH), but the authors suggest that this would be much smaller, probably less than 0.5 Hz. This contrasts rather strangely with the $^{6}J$(H,H) coupling in 3,5-dimethyl-isothiazole and -iso-oxazole,[391] discussed in Section 8B above.

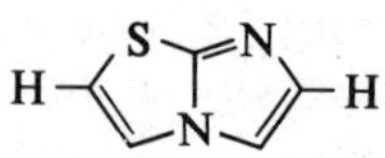

(31)

**C. Six-bond Coupling of Other Nuclei to Protons.**—The results of Matsubayashi *et al.*[387] on (Sn,H) coupling have been described in Section 6C(*ii*);

[403] T. Schaefer, C. M. Wong, and K. C. Tam, *Canad. J. Chem.*, 1969, **47**, 3688.

those of Schaefer, Rowbotham, *et al.*[10,333] on (F,H) couplings in halogeno-toluenes will be described in Section 10A. Values of 0.4—0.9 Hz for $^6J$-(FPCCCCH) have been measured in phosphonodifluoridic acids, $R^1R^2C{=}CHP(O)F_2$.[199]

## 9 Coupling between Nuclei Separated by Seven or More Chemical Bonds

**A. $^7J(^1H,^1H)$.**—The papers of Albriktsen and Harris[230] and of Rowbotham and Schaefer,[14] which include some discussion of $^7J$(H,H) values, have been described in Sections 7 and 8 above. They will also be included in Section 10, on $\pi$-electron systems.

**B. $^9J(^1HCCCCCCCC^1H)$.**—Curry and Sardella[213] have estimated a value of 0.05—0.15 Hz for this coupling in $\beta,\beta,p$-trimethylstyrene in agreement with the prediction of Barfield and Chakrabarti.[404]

**C. $^nJ(^{19}F,^{19}F)$; $n$ = 9—14.**—Mitchell, Phillips, Roberts, and Wray[405,406] have reported a remarkable series of (F,F) couplings in molecules of the form (32). Couplings of between 0.1 and 1.0 Hz are observed between the fluorine nuclei in molecules for which the group Z allows $\pi$-electron delocalization

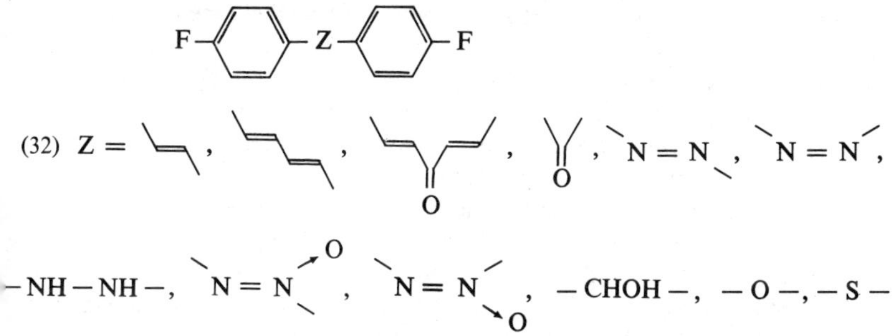


(32)

between the rings (*e.g.* where *Z* is —C=C—, —N=N—), but not in other cases (*e.g.* where Z is —CHOH—, —O—, —S—). Since the latter examples are also those for which the molecule is most flexible this is strong evidence against a 'through-space' interpretation of these coupling constants.

## 10 Coupling in $\pi$-Electron Systems

**A. Homo-aromatic Molecules.**—The theoretical work of Bildsøe and Schaumburg[16] has been described in Section 2B.

(H,H) couplings in 1,8-disubstituted naphthalenes[407] and (C,H) couplings

404 M. Barfield and B. Chakrabarti, *Chem. Rev.*, 1969, **69**, 757.
405 P. J. Mitchell, L. Phillips, S. J. Roberts, and V. Wray, *Org. Magn. Resonance*, 1974, **6**, 126.
406 P. J. Mitchell and L. Phillips, *J.C.S. Perkin II*, 1974, 109.
407 J.-B. Robert, J. S. Sherfinski, R. E. Marsh, and J. D. Roberts, *J. Org. Chem.*, 1974, **39**, 1152.

in 1,7- and 2,7-diacetoxynaphthalenes[112] have been measured. The latter paper reports that the couplings to the tertiary carbon atoms, C-9 and C-10, are similar to those involving the other carbons. There is also some evidence of inter-ring coupling. Rowbotham, Schaefer, and their co-workers[10,333] have given a detailed discussion of the long-range coupling between H and F in the side-chain and protons of the benzene ring in benzal fluoride and four of its dichloro-derivatives. In contrast to the (H,H) couplings, the (H,F) couplings are sensitive to substituent effects from the chlorine atoms in the ring. The Canadian workers have also examined methyl–methyl (H,H) coupling constants in some six-membered ring compounds[14] (see Section 2B).

A detailed study of coupling between alkyl and ring protons in 4-alkylnitrobenzenes has been reported by Beistel and Chen.[408] (H,H) coupling between the α-protons of the alkyl substituents and both the *ortho*- [$^4J$(HCCCH)] and *meta*- [$^5J$(HCCCCH)] ring protons were observed. Values are typically 0.3 and 0.7 Hz for coupling to the *meta*- and *ortho*-protons respectively. No coupling of ring protons with β-alkyl protons was observed. Couplings involving the methyl groups of β,β′-dimethylstyrenes have been discussed by Curry and Sardella.[213]

An analysis of the $^1$H spectrum of naphthalene in a nematic phase[409] has made it possible to determine the signs of all direct and indirect couplings, and Defay[410] has shown that the (H,H) couplings in heptahelicene have the same signs as similar couplings in naphthalene, acridine, and isoquinoline. The magnitudes are also similar, and there does not therefore appear to be any large dependence of the coupling upon the distortion of the aromatic system. Accurate values of $^nJ$(H,H) in *o*- and *m*-dicyanobenzene have been reported.[411]

Among the systems for which (H,H) couplings have been measured and discussed the following may be also noted: methylenecyclo-octatrienyl anion,[327] substituted ferrocenes,[412] and diphenyl sulphides and diphenyldisulphides.[413] (H,H) and (H,F) couplings in monofluoroanilines complexed with Pr(fod)$_3$ have been measured at 250 MHz,[414] and (F,F) couplings in pentafluorophenyl polysulphides have been reported.[415]

Studies of the effect of temperature on $^5J$(FCCCCH) in *ortho*- and *peri*-acetyl-substituted fluoronaphthalenes suggest that the coupling depends upon the populations of the *cis*- and *trans*-species and is therefore essentially a through-space coupling.[401]

The values of the coupling between $^{119/117}$Sn and the methyl group protons of *o*-, *m*-, and *p*-tolyltin trichlorides follow a pattern very similar to that of the corresponding (Tl,H) couplings in *o*-, *m*-, and *p*-tolylthallium cations. Indeed,

408 D. W. Beistel and H. E. Chen, *J. Chem. Phys.*, 1973, **59**, 4642.
409 J.-M. Dereppe, J. Degelaen, and M. van Meerssche, *J. Mol. Structure*, 1973, **17**, 225.
410 N. Defay, *Org. Magn. Resonance*, 1974, **6**, 221.
411 W. de Kieviet and C. A. de Lange, *Chem. Phys. Letters*, 1973, **22**, 378.
412 N. Kamezawa, *J. Magn. Resonance*, 1973, **11**, 88.
413 G. C. Pappalardo, *Spectrochim. Acta*, 1973, **29A**, 2055.
414 A. Forchioni and C. Chachaty, *Org. Magn. Resonance*, 1974, **6**, 267.
415 M. E. Peach, *Internat. J. Sulfur Chem.*, 1973, **8**, 27.

the ratio of the two types of coupling is equal to 8.5 ± 0.5 for the three positions of substitution, which strongly suggests similar mechanisms in the two cases.[387]

Values of $^1J$(CD) and $^3J$(CCCD) for some deuteriated aromatic hydrocarbons have been published.[46]

**B. Annulenes.**—Luz and Meiboom[416] have studied the proton spectrum of cyclo-octatetraene in nematic solvents, and several reports of (H,H) coupling in annulene derivatives have been published.[417,418]

**C. Heterocyclic Molecules.**—Proton–proton coupling in vinylpyridines[15] and methyl–methyl (H,H) coupling in substituted aromatic systems including borazine[14] have been examined both theoretically and experimentally by Rowbotham and Schaefer (see also Section 2B). Galasso's work on pyridine aldehydes,[19] pyridine,[20] and thiophens[17,18] has also been described in Section 2B.

Fringuelli and Taticchi[309] have found that there are good linear relationships between $^3J(H^3CCH^2)$ in furan, thiophen, selenophen, and pyrrole and the CCH and XCH bond angles (X = heteroatom), and a dependence of $^3J(H^3CCH^4)$ on ring planarity in heterocyclic molecules related to anthracene (33) and (34), has been noted.[419] $^3J(H^4,H^5)$ was independent of this factor, as

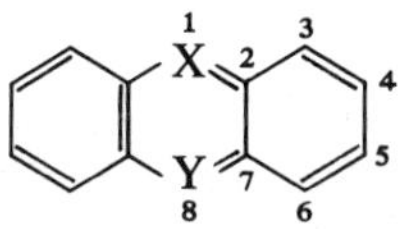


(33) X, Y = CH, CBr, CCl, N

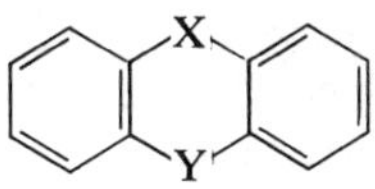


(34) X, Y = $CH_2$, O, S, CO, NH, CHOH

was the average of $^3J(H^3,H^5)$ and $^3J(H^4,H^6)$. Attempts to predict the couplings by use of additivity relationships gave disappointing results.[419]

Katayama *et al.*[420] have reported the effect of solvent upon the (H,H) coupling constants of 8-hydroxyquinoline, and the problems of spectral analy-

416 Z. Luz and S. Meiboom, *J. Chem. Phys.*, 1973, **59,** 1077.
417 G. Schröder, G. Kirsch, and J. F. M. Oth, *Chem. Ber.*, 1974, **107,** 460.
418 P. J. Beeby, R. T. Weavers, and F. Sondheimer, *Angew. Chem. Internat. Edn.*, 1974, **13,** 138; R. T. Weavers and F. Sondheimer, *ibid.*, p. 139, 141.
419 N. E. Sharpless, R. B. Bradley, and J. A. Ferretti, *Org. Magn. Resonance*, 1974, **6,** 115.
420 S. Katayama, Y. Akahori, and H. Mori, *Chem. and Pharm. Bull. (Japan)*, 1973, **21,** 2622.

sis in the $^{1}$H spectra of dibenzothiophen and benzo[*b*]fluoranthene have been pointed out.[421]

Elgeuro and his co-workers[331] have determined some coupling constants involving the protons bound to nitrogen from the spectra of benzimidazole, imidazole, 1-phenylimidazole, 1-phenylpyrazole, and some of their methyl derivatives in sulphuric acid. The work of Wasylishen, Rowbotham, and Schaefer[391] on the signs and magnitudes of (H,H) coupling over three to six bonds in isothiazoles and iso-oxazoles has been reported under the appropriate path in the preceding sections.

Other heteroaromatic molecules for which (H,H) couplings have been measured and discussed include thiophen,[17] tellurophen,[422] pyrrole,[96,117] pyrrole derivatives,[396] pyridine derivatives,[423] imidazo[2,1-*b*]thiazoles,[329] methoxypyrazines,[397] carbolines,[424] and thioheterohelicenes.[425]

Miyajima *et al.*[426] have reported values of (C,H) couplings in pyridine and have compared these results with those for benzene. They find that $^{2}J$(CCH) in pyridine is much greater than in benzene for ($C^{2}$,$H^{3}$) and ($C^{3}$,$H^{2}$) but less for ($C^{3}$,$H^{4}$) and ($C^{4}$,$H^{5}$). $^{3}J$(CCCH) is generally larger than $^{2}J$(CCH) as in benzene, but not always so because of the large $^{2}J$(CCH) values. The authors also compare $^{n}J$(F,C) values for 2-fluoropyridine and fluorobenzene. $|^{2}J(F,C)| > |^{3}J(F,C)|$ in both molecules and $^{2}J$($C^{3}$,$F^{2}$) and $^{3}J$($C^{6}$,$F^{2}$) are larger in the pyridine than in the benzene derivative. Hägele, Sartori, and Golloch[427] have determined the relative signs of the (F,F) couplings in 2,3,6-trifluoropyridine-4,5-dicarboxylic acid using the nuclear Overhauser effect. Comparison with the known negative sign for $^{3}J$(FCCF) in hexafluorobenzene then leads to the conclusion that $^{4}J$(FCNCF) and $^{3}J$(FCCF) are negative whereas $^{5}J$(FCNCCF) is positive. (F,F) Couplings have also been reported for polyfluoroheterocyclic compounds by Chambers *et al.*[428]

**D. Linear Molecules.**—Particular aspects of coupling in molecules of this type have been discussed in previous sections under the appropriate path. For ease of reference the linear conjugated systems for which particularly detailed studies have appeared in the past year are simply noted here again: hexatriene and dimethylhexatriene,[230] 2,4-dimethyl-2,4-pentadiene,[230] sorbaldehyde,[337] $\pi$-allylic palladium complexes,[72] and $\beta$,$\beta$-dimethylstyrenes.[213]

## 11 Through-space Coupling

The theoretical work of Mallory[37] has been described in Section 2B.

Hägele and Weidenbruch[390] have observed sizeable long-range (F,H) and

421 D. W. Jones, R. S. Matthews, and K. D. Bartle, *Spectrochim. Acta*, 1974, **30A**, 489.
422 A. D'annibale, L. Lunazzi, F. Fringuelli, and A. Taticchi, *Mol. Phys.*, 1974, **27**, 257.
423 U. Vögeli and W. von Philipsborn, *Org. Magn. Resonance*, 1973, **5**, 551.
424 F. Balkan and M. L. Heffernan, *Austral. J. Chem.*, 1973, **26**, 1501.
425 J. H. Dopper, D. Oudman, and H. Wynberg, *J. Amer. Chem. Soc.*, 1973, **95**, 3692.
426 G. Miyajima, K. Takahashi, and H. Sugiyama, *Org. Magn. Resonance*, 1974, **6**, 181.
427 G. Hägele, P. Sartori, and A. Golloch, *Z. Naturforsch.*, 1973, **28b**, 758.
428 R. D. Chambers, R. P. Corbally, T. F. Holmes, and W. K. R. Musgrave, *J.C.S. Perkin I*, 1974, 108.

(F,F) couplings in molecules of formulae $(C_6F_5)_nSiR_{4-n}$ ($n$ = 1 2, or 3; R = alkyl or aryl) and $C_6F_5[SiMe_2]_2R$ (R = Me or $C_6F_5$). They suggest that these long-range interactions [$^5J$(FCCSiCH) = 1.0—2.0 Hz; $^6J$(FCCSiSiCH) = 0.65 and 0.7 Hz] contain a significant contribution from through-space coupling. Similarly, $^6J$(FCCSiCCF), which is observable only as a line-broadening, is probably a through-space coupling between the *ortho*-fluorines of two $C_6F_5$ groups, and the broadening of the lines due to the *meta*- and *para*-fluorine atoms suggests interaction over even greater distances. However, as the authors point out, there are other phenomena which could also be responsible for this effect, relaxation processes or rotational motion for example. Chambers *et al.*[428] have observed (F,F) couplings in polyfluoroheterocyclic systems which probably contain a through-space element.

The extraordinary long-range (F,F) couplings observed by Phillips and his colleagues[405] (see Section 9C) appear to be through-bond rather than through-space interactions, and the suggestion[351] that some $^3J$(PWCH) values may contain a sizeable through-space contribution has been strongly contested[350] (Section 5C).

Adcock and Rizvi's study[401] of $^5J$(F,H) in *ortho*- and *peri*-acetyl-substituted fluoronaphthalenes has been noted above (Section 10A). The coupling of 0.30 Hz between terminal protons in a thioheterohelicene, for which the shortest path through bonds would comprise eight bonds, seems a clear case of a through-space effect.[425] The positive sign of $^4J$(FPNSiH) in silylaminodifluorophosphine, together with the similarity of the absolute values of $^4K$(F,H) and $^3K$(P,H) in this amine and related compounds, has led Anderson *et al.*[113] to propose that this coupling may be through-space. Clark and Manzer[298] have also suggested the presence of through-space contributions to coupling in fluoro-olefin complexes of platinum.

## 12 Recent Experimental Advances

The fundamental innovations in instrumentation which have overtaken n.m.r. spectroscopy in recent years have stimulated the development of a large number of ingenious experimental techniques, many of which are of potential relevance to those interested in the measurement and interpretation of coupling constants. The number of papers devoted to such techniques in the twelve months under review is almost twice that of the previous twelve-month period, and the Reporter must, with regret, admit that he is not qualified to assess this material in any meaningful way. It seems worthwhile, however, to give the references here under a set of brief headings:

(i) General experimental but essentially non-instrumental techniques: refs. 429—434.

[429] L. W. Reeves, A. S. Tracey, and M. M. Tracey, *J. Amer. Chem. Soc.*, 1973, **95**, 3799.
[430] T. Lindblom, *Chem. Scripta*, 1973, **4**, 103.
[431] C. A. Fyfe, M. Cocivera, and S. W. H. Damji, *J.C.S. Chem. Comm.*, 1973, 743.
[432] I. D. Campbell, C. M. Dobson, R. J. P. Williams, and A. V. Xavier, *J. Magn. Resonance*, 1973, **11**, 172.

(ii) A new instrumental magnetic resonance method: ref. 435.
(iii) Fourier-transform methods: refs. 436—441.
(iv) Double-resonance methods: refs. 69, 161, 437, 440, and 442—444.
(v) Spin-decoupling experiments: refs. 1, 92, 438, and 445—448.
(vi) Use of shift reagents: refs. 363 and 449—454.
(vii) CIDNP: refs. 38 and 39.
(viii) Spin-echo measurements: refs. 395 and 455.
(ix) Relaxation measurements: refs. 456 and 457.
(x) Spectra of solids: ref. 289.
(xi) Spectral analysis and deceptive simplicity: ref. 458.
(xii) A computer programme for the analysis of polymer spectra: ref. 459.

[433] H. Benoit and M. Rabii, *Chem. Phys. Letters*, 1973, **21**, 466.
[434] R. G. Lawler and M. Halfon, *Rev. Sci. Instr.*, 1974, **45**, 84.
[435] R. Freeman, *Chem. Abs.*, 1973, **79**, 110 226.
[436] G. E. Maciel and M. Borzo, *J.C.S. Chem. Comm.*, 1973, 394.
[437] J. Feeney and P. Partington, *J.C.S. Chem. Comm.*, 1973, 611.
[438] B. L. Tomlinson and H. D. W. Hill, *J. Chem. Phys.*, 1973, **59**, 1775.
[439] W.-K. Rhim, D. D. Elleman, and R. W. Vaughan, *J. Chem. Phys.*, 1973, **59**, 3740.
[440] K. G. R. Pachler and P. L. Wessels, *J. Magn. Resonance*, 1973, **12**, 337.
[441] C. Duret, P. Gonord, and S. K. Kan, *Compt. rend.*, 1974, **278**, *B*, 791.
[442] P. Bucci, P. A. Rolla, and C. A. Veracini, *Rev. Sci. Instr.*, 1973, **44**, 1423.
[443] P. Gonord, C. Duret, and S. K. Kan, *Compt. rend.*, 1974, **278**, *B*, 123.
[444] S. Sørensen, R. S. Hansen, and H. J. Jakobsen, *J. Magn. Resonance*, 1974, **14**, 243.
[445] A. Pines and J. D. Ellett, *J. Amer. Chem. Soc.*, 1973, **95**, 4437.
[446] S. Sørensen, R. S. Hansen, and H. J. Jakobsen, *J. Amer. Chem. Soc.*, 1973, **95**, 5080.
[447] R. C. Hewitt, S. Meiboom, and L. C. Snyder, *J. Chem. Phys.*, 1973, **58**, 5089.
[448] L. C. Snyder and S. Meiboom, *J. Chem. Phys.*, 1973, **58**, 5096.
[449] K.-T. Liu, *Tetrahedron Letters*, 1973, 2747.
[450] J. J. Uebel, C. Pacheco, and R. M. Wing, *Tetrahedron Letters*, 1973, 4383.
[451] G. P. Schiemenz, *J. Organometallic Chem.*, 1973, **52**, 349.
[452] G. P. Schiemenz, *Org. Magn. Resonance*, 1973, **5**, 257.
[453] K. Tori, I. Horibe, Y. Tamura, and H. Tada, *J.C.S. Chem. Comm.*, 1973, 620.
[454] G. P. Schiemenz and P. Klemm, *Org. Magn. Resonance*, 1974, **6**, 277.
[455] N. Boden and M. Mortimer, *Chem. Phys. Letters*, 1973, **21**, 538.
[456] F. W. Wehrli, *J.C.S. Chem. Comm.*, 1973, 379.
[557] L. G. Werbelow and A. G. Marshall, *Chem. Phys. Letters*, 1973, **22**, 568.
[458] D. E. Axelson and C. E. Holloway, *J.C.S. Chem. Comm.*, 1973, 455.
[459] R. C. Ferguson, *J. Magn. Resonance*, 1973, **12**, 296.

# 3
# Nuclear Spin Relaxation in Liquids

BY M. D. ZEIDLER

## 1 Introduction

During the review period, two areas of research produced a remarkable number of papers: studies of anisotropic rotational motions, and relaxation studies of heavy nuclei for which few results were known previously. The first type of problem was mentioned by Tomlinson in Volume 3 of this series as being one of the main areas of interest. Some papers go beyond reporting rotational correlation times and try to rationalize such results by application of specific models for the liquid state. In order for these models to be of value they must predict rotational correlation times over a wide temperature and pressure range, and therefore experimental work is now increasing in which not only the temperature but also the pressure range is extended.

Availability of Fourier transform and signal averaging techniques has made possible relaxation studies of nuclei like $^{29}Si$, $^{113}Cd$, $^{199}Hg$, $^{205}Tl$, and $^{207}Pb$. All of these nuclei have spin $\frac{1}{2}$ and thus relaxation mechanisms other than quadrupolar are effective. $^{29}Si$ relaxation studies were also discussed in Volume 3.

A new type of approach to obtain information on pair distribution functions in liquids by relaxation measurements has been presented in some papers. Evaluation of a set of atomic pair correlation functions for different atom positions in the molecule provides information on orientational correlations in the liquid.

## 2 Models for Molecular Rotation

Relaxation rates, *i.e.* the reciprocal values of the relaxation times $T_1$ and $T_2$, are usually a composite of contributions due to different relaxation mechanisms: dipole–dipole relaxation (DD), quadrupolar relaxation (QF), spin–rotational relaxation (SR), relaxations due to scalar coupling (SC) and to chemical shift (shielding) anisotropy (CSA).* In the case of dipole–dipole relaxation a further distinction between intramolecular and intermolecular contributions must be made. Before discussion of relaxation rates takes place separation of all contributions must be performed either by experimental techniques or by theoretical estimates. These techniques are well known (see

* These abbreviations will be used throughout this chapter.

the previous chapters on spin relaxation in this series, and basic literature cited therein). If one has knowledge of the specific interaction constants for these contributions corresponding correlation times can be obtained. Intramolecular DD, QF, and CSA relaxation all yield correlation times $\tau_2$ which describe the decay of spatial correlations due to molecular rotation. However, they depend on the position of the interaction tensor inside the molecule and thus are different in general if the rotational molecular motion is anisotropic. SR relaxation is related to a correlation time $\tau_J$ which describes the decay of angular momentum, but also to $\tau_2$. Therefore this contribution is difficult to evaluate. Models of rotational diffusion relating $\tau_2$ and $\tau_J$ are available in the literature (Vol. 3, p. 112).

$\tau_J$ may be considered to be the more fundamental or elementary quantity. Maryott *et al.*[1] compared $\tau_J$ with the mean time between molecular collisions in the liquid, $\tau_{coll}$, which was calculated either from a gas-like Enskog hard-sphere model or a solid-like cell model. Both times are related by an effective collision number describing the efficiency of randomizing angular momentum in a collision. $\tau_J$ values were calculated for liquids in which the SR relaxation mechanism strongly dominates and SR interaction constants are reliably known, as for the $^{19}F$ relaxation in $CF_4$, $SF_6$, $ClO_3F$, or $CCl_3F$, the $^{13}C$ relaxation in $CS_2$ or $CCl_4$, and the $^{31}P$ relaxation in $PH_3$, $PD_3$, or $PCl_3$. McClung's relation[2] between $\tau_J$ and $\tau_2$ for spherical and symmetric top molecules was used for evaluation of the SR contribution to the relaxation rate. Despite the fact that the hard-sphere and cell models yield different volume dependences for $\tau_{coll}$, namely $v_m$ and $v_m^{\frac{1}{3}}$, respectively, where $v_m$ is the average volume occupied by a molecule, the $\tau_J$ data can be fitted reasonably well to either model, with an effective collision number of *ca.* 2 for the hard-sphere model and *ca.* 1 for the cell model. These collision numbers indicate that collision efficiency is high.

The large-step random walk theory of rotational diffusion was applied by O'Reilly to anisotropic rotation (see Vol. 3, p. 111). Recently he extended his calculations to the case of a quadrupolar coupling tensor with asymmetry parameter, $\eta$, different form zero, whereas the previous expressions were confined[3] to $\eta = 0$. These results were applied to liquid water where proton, deuteron, and $^{17}O$ relaxation rates are available, so that calculated values of $\tau_2$ could be compared with 'experimental' ones. The mean time between collisions which occurs in the theoretical expression for $\tau_2$ was estimated from quasi-lattice models based on ice VII and Ic structures; the mean angle of a rotational jump was derived either using the assumption of inertial rotation or from molecular dynamics results of Rahman and Stillinger on liquid water. Both results are close to the 'experimental' $\tau_2$, for the evaluation of which quadrupole coupling constants and asymmetry parameters of ice I were employed. No anisotropy of $\tau_2$ is found, owing to large values of the

[1] A. A. Maryott, M. S. Malmberg, and K. T. Gillen, *Chem. Phys. Letters*, 1974, **25**, 169.
[2] R. E. D. McClung, *J. Chem. Phys.*, 1969, **51**, 3842; 1972, **57**, 5478.
[3] D. E. O'Reilly, *J. Chem. Phys.*, 1974, **60**, 1607.

rotation angle. From a diffusion model, on the other hand, larger anisotropies and values smaller by one order of magnitude for $\tau_2$ are calculated, in sharp contrast to the 'experimental' results.

Multiple internal rotations, *i.e.* superposition of rotational motions of several internal groups, *e.g.* of methyl and methylene groups along a carbon chain, have been treated on the basis of diffusion models. Wallach[4] considered isotropic rotation of the molecule as a whole and neglected asymmetry parameters of the quadrupolar coupling tensor, whereas Versmold[5] included anisotropic diffusion (symmetric top) and asymmetry parameters. Levine *et al.*[6] recently essentially reproduced Versmold's results. In a new paper, Levine[7] considers such internal rotations, either stochastic rotational diffusion about each bond or jumps between three conformations, in more detail by taking account of correlations of conformations about adjacent bonds. Now, two more parameters determine the internal motion: the population ratio of *gauche* to *trans* conformations and the average lifetime in the *trans* configuration. Since these additional parameters cannot be determined experimentally the simple previous approach must be retained; it seems that it gives a reasonable description of the properties of a real chain.

Versmold[8] has reconsidered time correlation functions for internal rotations by the jump mechanism. He showed that cross terms are retained in the correlation functions, the number of these terms decreasing with increasing number of possible equilibrium positions; *e.g.* there are four non-vanishing cross terms for $C_3$ symmetry. Their influence on the relaxation rate is rather involved; numerical examples show that in molecules with enough symmetry, as in toluene, the cross terms are of minor importance but may contribute in less symmetrical molecules as methanol.

## 3 Relaxation Studies in Pure Liquids

**A. Proton and Deuteron Relaxation.**—Kitchlew and Rao[9] measured the proton spin–lattice relaxation time of liquid furan in the temperature range 295—415 K. Contributing relaxation mechanisms are intramolecular and intermolecular DD and also SR. Their separation procedure is rather unsatisfactory: they estimated the intermolecular DD contribution from Torrey's formula[10] employing measured self-diffusion constants. After subtracting this contribution from the measured rate, intramolecular DD and SR contributions are separated in the usual manner using the minimum observed when plotting $(T_{1,\text{intra}})^{-1}$ as a function of temperature.[11] However, this procedure is based on a formula valid only for isotropic rotation. The analysis yields a

[4] D. Wallach, *J. Chem. Phys.*, 1967, **47**, 5258.
[5] H. Versmold, *Z. Naturforsch.*, 1970, **25a**, 367.
[6] Y. K. Levine, P. Partington, and G. C. K. Roberts, *Mol. Phys.*, 1973, **25**, 497.
[7] Y. K. Levine, *J. Magn. Resonance*, 1973, **11**, 421.
[8] H. Versmold, *J. Chem. Phys.*, 1973, **58**, 5649.
[9] A. Kitchlew and B. D. Nageswara Rao, *J. Chem. Phys.*, 1973, **58**, 4033.
[10] H. C. Torrey, *Phys. Rev.*, 1953, **92**, 962.
[11] J. G. Powles and R. Figgins, *Mol. Phys.*, 1966, **10**, 155.

spin–rotation interaction constant of (1/3) Tr $\boldsymbol{C} \leqslant 1.4$ kHz. Finally, the diffusion constant perpendicular to the ring is calculated from the intramolecular DD contribution showing no Arrhenius behaviour.

The pressure dependence of the proton spin–lattice relaxation time for eight different dicyclohexyl compounds (1) and for phenylcyclohexyl was studied by Brooks *et al.*;[12] the pressure range was 1—3000 bar and the

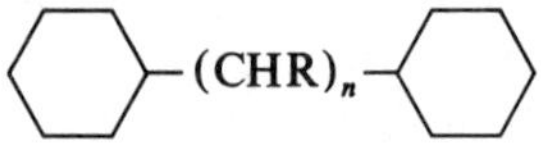

(1) R = H or Me

temperature was fixed at 28 °C. $T_1$ is an average over all protons, so sometimes non-exponential behaviour was found. Increasing pressure tends to increase the rotational correlation time $\tau_2$; since $\tau_2$ at normal pressure is already large for these big molecules (*ca.* 100 ps), the dispersion region $\omega_1\tau_2 \approx 1$ is approached and $T_1$-minima are indeed observed for three compounds. The dipolar coupling constant was evaluated from the $T_1$-minimum neglecting intermolecular DD contribution; the value is of comparable magnitude to the one calculated from a molecular model. Activation volumes extracted from the $\tau_2$ *vs.* pressure plot lie in the range 24—42 $cm^3\ mol^{-1}$.

Previously known proton $T_1$ data for water[13—16] have been checked by Hindman *et al.*[17] There is, in general, good agreement with the earlier measurements. Further considerations lead the authors to a separation of intramolecular and intermolecular DD and SR contributions. This was accomplished in the following way: (*a*) Using previously determined $^{17}O$ relaxation times[18] $\tau_2$ was calculated, with estimated values of the quadrupole coupling constant and asymmetry parameters (the latter were obtained by taking dielectric relaxation times over a smaller temperature range and assuming the diffusion limit relation $3\tau_2 = \tau_1$, where $\tau_1$ is the correlation time obtained from dielectric relaxation studies); (*b*) $\tau_J$ was calculated[2] from $\tau_2$; the relation between these two times in the diffusion limit ($\tau_J \ll \tau_2$) reduces to the so-called Hubbard relation;[19] (*c*) The SR interaction constant was found from the $T_1$-maximum observed by Hausser[13] and Smith *et al.*[14] in the usual way;[11] (*d*) The SR contribution was then calculated from $\tau_J$ and the interaction constant; (*e*) The intramolecular DD contribution was calculated from $\tau_2$ and the known proton–proton distance; (*f*) The intermolecular DD contribution was obtained as the difference between the measured $T_1^{-1}$ and the two calculated contributions. Typical values are $T_1 = 3.66$ s, $T_{1,\text{intra}} =$

12 A. A. Brooks, E. O. Stejskal, and V. W. Weiss, *J. Chem. Phys.*, 1973, **58,** 4045.
13 R. Hausser, *Z. Naturforsch.*, 1963, **18a,** 1143.
14 D. W. G. Smith and J. G. Powles, *Mol. Phys.*, 1966, **10,** 451.
15 K. Krynicki, *Physica*, 1966, **32,** 167 (see also older literature on water cited therein).
16 R. J. S. Brown, *J. Phys. Chem.*, 1969, **73,** 3157.
17 J. C. Hindman, A. Svirmickas, and M. Wood, *J. Chem. Phys.*, 1973, **59,** 1517.
18 J. C. Hindman, A. J. Zielen, A. Svirmickas, and M. Wood, *J. Chem. Phys.*, 1971, **54,** 621.
19 P. S. Hubbard, *Phys. Rev.*, 1963, **131,** 1155.

9.89 s and $T_{1,\text{inter}} = 5.83$ s at 26.1 °C. The reader should be aware of the conflict between the treatment given before by O'Reilly[3] and that presented here.

Hindman *et al.*[20] also extended the previously known relaxation times of deuterons in $D_2O$ to a lower temperature range. The measurements down to −37 °C were performed with emulsions of 25 and 50 volume % $D_2O$ in n-heptane. There was no detectable difference in $T_1$ between these two concentrations. Moreover, in the temperature region accessible by supercooling water no difference in $T_1$ was found between the emulsions and pure water. The low-temperature relaxation times show Arrhenius behaviour with an activation energy of *ca.* 58—63 kJ mol$^{-1}$, comparable to corresponding activation energies for $T_1$, dielectric relaxation, and diffusion in ice. This comparison indicates that the hydrogen-bonded structure is well defined in the liquid at these low temperatures.

Proton relaxation times, $T_1$, for some alkyl bromides ($C_nH_{2n+1}Br$ with $n = 2$ to 7) were measured by Sempère and Regnier[21] in the temperature range −60 to +100 °C. For three substances the intramolecular DD contribution was separated experimentally by dilution in the solvents $CCl_4$ and $C_2Cl_4$. The correlation times $\tau_2$ determined this way are compared with dielectric relaxation times and the relation $\tau_2 = \tau_1/3$ is verified. On the other hand, the Hill theory but not the (modified) Debye theory gives good agreement between $\tau_2$ values calculated from solution viscosities and measured ones.

Spin–lattice relaxation data for protons have been measured and compared with results from quasielastic incoherent neutron scattering of phenol by Grosescu.[22] The measurements were done between 200 and 400 K and the intramolecular DD contribution was separated by dilution of phenol with $CCl_4$. The intramolecular contribution in pure phenol was taken to be the same as in $CCl_4$, a rather suspicious assumption. Thus $\tau_2$ is obtained, 13 ps at 45 °C for example. In order to make comparisons between n.m.r. and neutron experiments, expressions derived for identical models must be used. Expressions exist for the jump–rotation model; for the n.m.r. correlation time $\tau_2$ by Hertz[23] as a simplified version of the Ivanov theory,[24] and for the neutron quasi-elastic linewidth by Larsson.[25] In the latter case, since no experimental separation of rotational and translational contributions to the linewidth, in contrast to the n.m.r. case, is possible, translation of phenol molecules has been assumed to be of the small-step diffusion type. All parameters entering both expressions for $\tau_2$ and the linewidth can be fitted to the experimental values, and one finds the result that rotation of phenol molecules occurs by finite jumps.

[20] J. C. Hindman and A. Svirmickas, *J. Phys. Chem.*, 1973, **77**, 2487.
[21] R. Sempère and J. F. Regnier, *Compt. rend.*, 1973, **276**, *C*, 1587.
[22] R. Grosescu, *Rev. Roumaine Phys.*, 1973, **18**, 133; V. Trepadus, S. Rapeanu, and R. Grosescu, *ibid.*, p. 314.
[23] H. G. Hertz, *Progr. N.M.R. Spectroscopy*, 1967, **3**, 159.
[24] E. N. Ivanov, *Soviet Phys. JETP*, 1964, **18**, 1041.
[25] K. E. Larsson, *Phys. Rev.*, 1968, **167**, 171.

Heatley[26] measured the relaxation times of single lines in a high-resolution proton spectrum by adiabatic fast passage and repetitive sweep progressive saturation methods for the liquids $CH_2Cl_2$, $CH_3I$, $CH_3CN$, $CH_3CCl_3$, $(CH_3)_3CCl$, TMS, and mesitylene. If the $T_1$ results of a certain (central) band and the corresponding $^{13}C$ satellite line (natural abundance) are compared, differences must be attributed mainly to ($^1H$,$^{13}C$) dipole–dipole coupling for the satellite spectrum, since all other DD and SR contributions are similar in both peaks. Evaluation of experiments is lengthy and must be computer-programmed since a coupled relaxation system must be solved. Finally one obtains correlation times $\tau_2$ for the H—C vector and also of course for the H—H vector from the central band after proper allowance for SR contributions, for the determination of which literature data of ($^{13}C$,$^1H$) NOE experiments were used.

Saitŏ *et al.*[27] tried to correlate deuterium and $^{13}C$ relaxation times; the former were measured by the inversion-recovery method and the latter were collected from the literature. Deuterium and $^{13}C$ relaxation are both intramolecular, deuterium relaxing by quadrupolar interaction and $^{13}C$ by dipolar interaction with directly attached protons, but in the latter case other mechanisms may also contribute. Sincc the z-axis of the field-gradient tensor should be directed along the C—D bond, both relaxation times measure the reorientation of the identical vector. Thus, by forming the ratio of relaxation times the correlation times $\tau_2$ cancel and one predicts numbers for this ratio which simply depend on the ratio of the deuteron quadrupole coupling constant and the inverse sixth power of the C—H distance. One indeed observes good coincidence between experimental and calculated ratios using $\chi e^2qQ/h$ = 170 kHz and $r_{CH}$ = 0.109 nm, except for phenylacetylene, where another relaxation mechanism, probably CSA, comes into play. The other investigated substances are *NN*-dimethylformamide, acetic acid, cyclohexane, and benzene.

In another investigation by the same authors,[28] deuteron relaxation times of different groups for a total of 23 compounds were measured, toluene and halogen-substituted ethanes being among them. Interpretation of the results is in terms of internal rotation and anisotropic rotational motion.

Pajak and co-workers[29] determined proton relaxation times for several methyl-substituted benzenes and pyridines over a large temperature range. Separate results for methyl and ring protons were obtained. A remarkable observation is that, for several compounds, breaks in the $T_1$ *vs.* temperature plot occur somewhat above the melting point. The methyl groups show internal rotation effects and a semi-empirical model is proposed[30] to rationalize the experimental data.

26 F. Heatley, *J.C.S. Faraday II*, 1974, **70**, 148.
27 H. Saitŏ, H. H. Mantsch, and I. P. C. Smith, *J. Amer. Chem. Soc.*, 1973, **95**, 8453.
28 H. H. Mantsch, H. Saito, L. C. Leitch, and I. C. P. Smith, *J. Amer. Chem. Soc.*, 1974, **96**, 256.
29 Z. Pajak, K. Jurga, J. Jurga, and W. Suchanski, *Inst. Nuclear Phys. Cracow*, Rep. 1973, No. 819 (PL)–(Pt. 1), 1.
30 W. Suchanski, Z. Pajak, K. Jurga, and J. Jurga, *Inst. Nuclear Phys. Cracow*, Rep. 1973, No. 819 (PL)–(Pt. 1) 11.

In another paper the same authors[31] have reported proton relaxation times in methyl halides, acetonitrile, and nitromethane between −100 and +150 °C. Dilution experiments in the corresponding deuteriated liquids were also done. SR contribution is evident in these liquids.

Angerer and Szczesniak[32] investigated the proton relaxation in pyrrole, thiophen, and furan between −120 and +150 °C. Relaxation is due mainly to DD interaction, but for furan SR becomes important above room temperature.

The pressure and temperature dependence of deuteron relaxation in chloroform was studied by van der Hart[33] in the range 30—165 °C and 1—5000 bar. $\tau_2$ was calculated from these data, using the quadrupole coupling constant, $\chi$, of 168 kHz of the solid. The assumed temperature independence of $\chi$ was proved by comparison with Shoup and Farrar's results[34] for $\tau_2$ derived from $^{13}C$ relaxation (note that again $\tau_2$ of the C—H vector is obtained) since in both cases comparable activation energies are derived. Variation of $\tau_2$ with temperature and pressure is complicated in detail: simple activation theory is unsatisfactory as, for example, the activation volume is pressure-dependent at lower pressures. Even if the Debye equation is modified to:

$$\tau_2 = 4\pi\,\eta\,a^3\,\kappa/3kT$$

(where $\eta$ is the viscosity and $a$ the molecular radius) by the addition of an empirical factor $\kappa$,[35] which gives the ratio of intermolecular torques to intermolecular forces,[36] a satisfactory description is still not obtained, since $\kappa$ is found to increase with higher temperature and decrease with pressure ($\kappa = 0.126$ at 30 °C and 1 bar). For the above approach to be applicable, rotation and translation of molecules must be strongly coupled, but coupling strength decreases for more symmetric molecules. Finally, cell theory for liquids has been used to estimate $\tau_J$ (see Section 2) and calculation of $\tau_2$ was done with McClung's relation[2] for the *J*-diffusion model. The cell model fails to predict the temperature dependence at constant volume correctly. Whether this is a shortcoming of the cell model itself or of the *J*-diffusion model cannot be decided.

Engelsmann *et al.*[37] studied the ethyl bromide system with the aim of getting information about relative molecular orientations. Proton relaxation times in the systems $CH_3CD_2Br + C_2D_5Br$, $CD_3CH_2Br + C_2D_5Br$, and $CH_3CD_2Br + CD_3CH_2Br$ were measured at room temperature by ARP techniques, so that $T_1$ data for the methyl and methylene groups were obtained separately for the last system. Dilution experiments allow separation

[31] W. Suchanski, K. Jurga, and Z. Pajak, *Inst. Nuclear Phys. Cracow*, Rep. 1973, No. 819 (PL)–(Pt. 1), 25.

[32] J. Angerer and E. Szczesniak, *Inst. Nuclear Phys. Cracow*, Rep. 1973, No. 819 (PL)–(Pt. 1), 58.

[33] D. L. van der Hart, *J. Chem. Phys.*, 1974, **60**, 1858.

[34] R. R. Shoup and T. C. Farrar, *J. Magn. Resonance*, 1973, **7**, 48.

[35] R. E. D. McClung and D. Kivelson, *J. Chem. Phys.*, 1968, **49**, 3380.

[36] D. Kievelson, M. G. Kivelson, and I. Oppenheim, *J. Chem. Phys.*, 1970, **52**, 1810.

[37] K. Engelsmann, H. G. Hertz, and M. D. Zeidler, *Z. Phys. Chem.*, 1974, **89**, 134.

of intramolecular and intermolecular DD contributions. Thus intermolecular proton rates on which attention is focused are available for methyl–methyl, methyl–methylene, and methylene–methylene protons. The intermolecular DD contribution is proportional to the space integral over the weighted nuclear pair distribution function $g(r)$[38,39] the weighting factor being $r^{-3/2} \times J_{5/2}(\rho r)$ ($J$ is here the Bessel function of specified order). Therefore only nearest neighbour effects in the distribution function are important. Assumption of certain forms for $g(r)$,[40] a step function in the simplest case, permits the determination of relevant parameters of $g(r)$, for example the distance of closest approach in the case of the step function, from the intermolecular relaxation rates. Having now the distances of closest approach for the different proton groups in ethyl bromide one can get some information on the relative orientation in a similar way as from diffraction data,[41] although the derivation of molecular angular pair distribution functions from a set of atomic pair distribution functions is not uniquely possible.[42] The results for ethyl bromide show that van der Waals (Br,Br) attraction and not dipole alignment is responsible for the angular correlations.

Some more proton and deuteron relaxation studies are discussed in Section 3E.

**B. $^{13}$C Relaxation.**—Noggle and co-workers[43] have added valuable information on one of the most extensively studied liquids: acetonitrile. So far proton, deuteron, and nitrogen resonances have been studied; now $^{13}$C relaxation data are also available. The authors covered the temperature range 239—341 K and separately studied both carbons in the molecule, including the determination of NOE factors. As expected, the methyl carbon relaxes strongly by intramolecular dipolar interaction with the directly attached protons but SR is also effective. The less shielded nitrile carbon, on the other hand, relaxes by intermolecular DD interaction besides some intramolecular DD contribution, but again SR is effective. In a very recent paper by von Goldammer *et al.*,[44] a CSA contribution is detected for the nitrile carbon in the low temperature region by measurement of $^{13}$C relaxation for both $CH_3CN$ and $CD_3CN$. The experimental results of both papers are in good agreement except for the nitrile carbon at lower temperatures. The papers differ, however, in the evaluation of the SR contributions: Noggle uses an inertial model for the methyl group[45] and an anisotropic diffusion model for the nitrile group,[46] whereas von Goldammer applies the latter model to both groups. Thus the

[38] I. Oppenheim and M. Bloom, *Canad. J. Phys.*, 1961, **39,** 845.
[39] J. F. Harmon and B. H. Muller, *Phys. Rev.*, 1969, **182,** 400.
[40] R. Göller, H. G. Hertz, and R. Tutsch, *Pure Appl. Chem.*, 1972, **32,** 149.
[41] A. H. Narten, *J. Chem. Phys.*, 1972, **56,** 5681.
[42] K. E. Gubbins, C. G. Gray, P. A. Egelstaff, and M. S. Ananth, *Mol. Phys.*, 1973, **25,** 1353.
[43] T. K. Leipert, J. H. Noggle, and K. T. Gillen, *J. Magn. Resonance*, 1974, **13,** 158.
[44] E. von Goldammer, H. D. Lüdemann, and A. Müller, *J. Chem. Phys.*, 1974, **60,** 4590.
[45] T. E. Burke and S. I. Chan, *J. Magn. Resonance*, 1970, **2,** 120.
[46] H. J. Bender and M. D. Zeidler, *Ber. Bunsengesellschaft phys. Chem.*, 1971, **75,** 236.

derived SR interaction constants are somewhat different. It is satisfying to learn that the intramolecular DD contribution in the methyl group essentially gives the same results as deuteron relaxation (except for the isotope effect between $CH_3CN$ and $CD_3CN$) since in the latter case some assumptions, now verified to be correct, about the position of the field-gradient tensor had to be introduced.

$^{13}C$ relaxation of all carbon atoms was studied separately in $C_6H_{14}$, $C_8H_{18}$, and $C_{10}H_{22}$ by Birdsall *et al.*[47] Evaluation of results was performed in terms of their multiple internal rotation model.[6] Overall rotation and internal rotations are described by two diffusion coefficients each; rotation about all internal bonds except the end bond carrying the methyl group is assumed to be determined by the same diffusion constant.

Grant *et al.*[48] examined anisotropic rotation of the molecules *trans*-decalin, cyclopentene, and norbornane. Application of the diffusion equation for an asymmetric rotor[49,50] requires $^{13}C$ relaxation data from three non-equivalent carbon nuclei which relax by DD interaction with directly bonded protons. Determination of NOE factors shows where this condition is fulfilled. The three principal diffusion constants for each molecule are given in the paper.

Dipolar interaction between heteronuclear spins, as in intramolecular ($^{13}C$,H) interaction, can lead to non-exponential relaxation. Therefore experimental errors in such measurements may be present, especially if the recovery is followed only over a time-span that is not much larger than the time-constant. Campbell and Freeman[51] have discussed possible errors in such cases.

**C. $^{29}$Si Relaxation.**—Levy *et al.*[52] measured $^{29}Si$ spin–lattice relaxation times and NOE factors in organosilicon compounds: $Si(CH_3)_4$, $Ph_2SiH_2$, $PhSi(CH_3)_2H$, and $Si(OC_2H_5)_4$. Since $^{29}Si$ relaxation times are much longer than the proton relaxation times in these compounds, recovering of magnetization is always exponential[51] for both decoupled and coupled conditions. From the NOE factors DD contributions are determined, which for example in $Ph_2SiH_2$ are the only interaction type. In other cases, as in TMS above $-50$ °C, SR contribution comes into play.

The long times in $^{29}Si$ relaxation are due to the small value of the magnetogyric ratio. Such effects are sometimes a problem in FT spectroscopy. An effective paramagnetic relaxation agent for $^{29}Si$ and $^{13}C$ spectroscopy was found in tris-acetylacetonatochromium.[53]

47 N. J. M. Birdsall, A. G. Lee, Y. K. Levine, J. C. Metcalfe, P. Partington, and G. C. K. Roberts, *J.C.S. Chem. Comm.*, 1973, 757.
48 D. M. Grant, R. J. Pugmire, E. P. Black, and K. A. Christensen, *J. Amer. Chem. Soc.*, 1973, **95,** 8465.
49 D. E. Woessner, *J. Chem. Phys.*, 1962, **37,** 647.
50 W. T. Huntress, jun, *J. Chem. Phys.*, 1968, **48,** 3524.
51 I. D. Campbell and R. Freeman, *J. Magn. Resonance*, 1973, **11,** 143.
52 G. C. Levy, J. D. Cargioli, P. C. Juliano, and T. D. Mitchell, *J. Amer. Chem. Soc.*, 1973, **95,** 3445.
53 G. C. Levy and J. D. Cargioli, *J. Magn. Resonance*, 1973, **10,** 231.

**D. $^{207}$Pb Relaxation.**—Relaxation times of $^{207}$Pb in liquid $PbCl_4$ have been studied by Hawk and Sharp.[54] $T_1$ has contributions from SR and SC whereas DD and CSA (the nuclear shielding tensor of lead is isotropic in tetrahedral $PbCl_4$) are unimportant. The scalar contribution arises from coupling to chlorine modulated by quadrupolar relaxation. By field-dependent measurements, where $1/T_{1SC}$ was found to vary as the inverse square of the field strength, the SC contribution can be extracted and corresponding parameters are obtained; for example at 25 °C the $^{35}$Cl relaxation time is 7.15 $\mu$s and the scalar coupling constant between $^{207}$Pb and $^{35}$Cl is 710 Hz. For $T_2$ SC is the only mechanism; $T_2$ is shorter than $T_1$ by three orders of magnitude. Finally, $\tau_2$ can be computed from the chlorine relaxation times and known values for the quadrupole coupling constant ($\tau_2 = 1.7 \times 10^{-11}$ s at 25 °C) and $\tau_J$ may be calculated,[2] the diffusion limit being approached.[19] The SR contribution to $T_1$ can now be used, employing calculated values of $\tau_J$ to derive the SR interaction constant. The results for C vary between 4.5 and 7.8 kHz depending on the diffusion model (*J* or *M* diffusion) used to estimate $\tau_J$. The surprisingly short times for $\tau_J$ ($10^{-14}$ to $10^{-15}$ s) are rationalized as being due to collision-induced distortions in the molecular structure. An absolute chemical shift scale for $^{207}$Pb was then deduced from the SR interaction constants (within the limits estimated), and this, together with the relative resonance frequencies of $^1H_2O$ and $^{207}Pb(ClO_4)_2$, allowed the absolute magnetic moment of $^{207}$Pb to be determined.

**E. Studies of Several Resonant Nuclei: $^1$H, $^2$H, $^{14}$N, $^{19}$F, $^{35}$Cl.**—O'Reilly and co-workers[55] investigated proton, deuteron, and $^{35}$Cl relaxation (the latter data are linewidth results) in t-butyl chloride over the total liquid range. The intramolecular DD contribution was separated by dilution in perdeuteriated t-butyl chloride. Correlation times for the different vectors were calculated; quadrupole coupling constants of 62.4 MHz (solid state value) and 174 kHz for $^{35}$Cl and $^2$H, respectively, have been used. The different values for $\tau_2$ are explained on the basis of an isotropically tumbling molecule with internal rotation of the methyl groups. Application of Woessner's formula[56,5] leads to $\tau_2^{int}$ values describing the internal rotation. They should be the same if calculated from proton and deuteron data; the observed small difference may be due to neglect of DD interaction between the methyl groups. The barrier height for methyl group rotation amounts to 0.72 kcal mole$^{-1}$ (3.0 kJ mol$^{-1}$).

Proton and $^{14}$N relaxation times in 3,5-dichloropyridine were studied by Zaucer *et al.*[57] Separate $T_1$ data for $\alpha$ and $\gamma$ protons were obtained by the ARP technique. Separation of intra- and inter-molecular proton contributions was not done in the pure liquid state but in solution with hexachlorobutadiene as a solvent; extrapolation to the pure liquid was then made by viscosity

[54] R. M. Hawk and R. R. Sharp, *J. Chem. Phys.*, 1974, **60**, 1009.
[55] D. E. O'Reilly, E. M. Peterson, C. E. Scheie, and E. Seyfarth, *J. Chem. Phys.*, 1973, **59**, 3576.
[56] D. E. Woessner, *J. Chem. Phys.*, 1962, **36**, 1.
[57] M. Zaucer, J. Perko, F. Jagodic, and A. Azman, *Z. Naturforsch.*, 1973, **28a**, 1731.

corrections. The total intramolecular proton contribution was found to be larger than the DD contribution estimated from the $^{14}N$ correlation time, and the difference was attributed to the SR mechanism. For the $\alpha$ protons SC to $^{14}N$ also seems to contribute.

As an extension of their previous work on pyridine, Kintzinger and Lehn[58] studied the planar symmetric-top molecule *sym*-triazine. Deuteron and $^{14}N$ relaxation times were measured between 90 and 150 °C. The deuteron quadrupole coupling constant was determined in the nematic phase (200 kHz) whereas the $^{14}N$ constant is a solid-state value. In these planar molecules diffusion constants $D_{\parallel}$ and $D_{\perp}$ can be calculated[50] only if the asymmetry parameters in the quadrupole coupling tensors for two relaxing nuclei differ, since the principal *z*-axes in both cases lie in the molecular plane. Fortunately, the asymmetry parameter of $^{14}N$ in these heterocyclic compounds is rather large ($\eta$ = 0.372), whereas for deuterons $\eta$ = 0. The results are discussed in relation to data obtained for two other planar six-membered cyclic molecules, pyridine[59] and benzene.[60] The motions are highly anisotropic, in-plane motions being faster (lower activation energy) than out-of-plane motions. One should be aware of the distinction from rotational behaviour in the gas phase: here the in-plane rotations, which are governed by the largest moments of inertia, are the slowest. Anisotropy seems to decrease in the direction benzene > triazine > pyridine. However, one should be careful with quantitative comparisons. Firstly, the results depend on the magnitude of the asymmetry parameters which may not be known accurately enough. Secondly, the data for benzene, where two different nuclei cannot be measured, were obtained from combined n.m.r. and light scattering results. Gillen and Griffiths obtained their results for the out-of-plane diffusion constant from Raman lineshapes ($D_{\perp} = 5 \times 10^{10}$ s$^{-1}$ at 20 °C, for example). In a recent paper by Litovitz and co-workers[61] a larger value for $D_{\perp}$ ($9 \times 10^{10}$ s$^{-1}$ at 20 °C) was found. Deuteron relaxation times yield $D_{\parallel}$ if $D_{\perp}$ is inserted from these other sources. A larger value of $D_{\perp}$ then gives a smaller $D_{\parallel}$; thus the anisotropy $D_{\parallel}/D_{\perp}$ comes out smaller than calculated by Gillen if Litovitz's results are used.

## 4 Relaxation Studies in Liquid Mixtures of Non-electrolytes

**A. Proton and Deuteron Relaxation.**—A proton and deuteron spin–lattice relaxation study of pyridine was performed by Tomchuk *et al.*[62] in the following solvents: acetone, benzene, chloroform, cyclohexane, carbon tetrachloride, and carbon disulphide. Perdeuterated compounds were used as solvents for the proton relaxation, whereas the normal solvents were used in the deuteron relaxation measurements. The total composition range was

58 J. P. Kintzinger and J. M. Lehn, *Mol. Phys.*, 1974, **27**, 491.
59 J. P. Kintzinger and J. M. Lehn, *Mol. Phys.*, 1971, **22**, 273.
60 K. T. Gillen and J. E. Griffiths, *Chem. Phys. Letters*, 1972, **17**, 359.
61 H. D. Dardy, V. Volterra, and T. A. Litovitz, *J. Chem. Phys.*, 1973, **59**, 4491.
62 E. Tomchuk, J. J. Czubryt, E. Bock, and N. Chatterjee, *J. Magn. Resonance*, 1973, **12**, 20.

covered. Extrapolation to infinite dilution of pyridine in the various solvents gives (in the case of the proton data) the intramolecular DD contribution, from which the proton correlation time $\tau_2^{H}$ is obtained. The corresponding deuteron correlation time $\tau_2^{D}$ was assumed to be equal to $\tau_2^{H}$, and consequently the deuteron quadrupole coupling constant of pyridine could be derived. The determined values range from 173 kHz in benzene to 185 kHz in acetone. The authors also performed the isotopic dilution experiment of $C_5H_5N$ in $C_5D_5N$, and found a quadrupole coupling constant of 177 kHz in pure pyridine which is in disagreement with earlier work.[63] An experimental determination of $\chi \approx 200$ kHz in the nematic phase,[59] however, seems to give support to the earlier work. If constancy of the coupling constant is assumed, which seems to be justified from comparison in the different solvents, then the deuteron relaxation rate is proportional to the correlation time. A drop of the rate with decreasing pyridine concentration, as is observed for all solvents except chloroform and carbon tetrachloride, then indicates faster rotational motion of pyridine molecules, *i.e.* weaker solute–solvent than solute–solute interaction. Between pyridine and chloroform rather strong intermolecular interactions must be present since the deuteron relaxation rate increases on dilution of pyridine. Finally, comparison with dielectric relaxation times shows good agreement with the small-step diffusion limit (where $\tau_2 = \tau_1/3$) at infinite dilution of pyridine in most of the investigated solvents.

A solution of 2,6-dimethyl-1,4-dimethoxybenzene in deuteriated chloroform was studied by Bovée and Smidt.[64] Separate proton relaxation rates for the ring protons and the three non-equivalent methyl groups were measured between $-20$ and 100 °C. By extrapolation to infinite dilution in chloroform the intramolecular contribution was separated and it was found that, contrary to the case for toluene, no SR is present and only DD interaction is effective. Dipolar interaction between protons in different methyl groups is of minor importance; the three protons in a methyl group relax themselves very efficiently because of their small separation in distance. The results were evaluated by application of Woessner's formula, treating the molecule as a symmetric-top rotor with superimposed internal rotations.[5,65] The numerical results of the correlation times for internal rotation of the methyl groups show the unexpected result that the methyl group belonging to the methoxy-group which has no directly neighbouring methyl groups rotates much slower than the other methoxy-group, which is sterically hindered by the direct neighbours. This was shown not to be a solvent effect by using carbon disulphide as a solvent.

Pajak and co-workers measured proton spin–lattice relaxation times of α-, β-, and γ-picolines[66] and of carboxylic acids (acetic, propionic, and

63 E. von Goldammer and M. D. Zeidler, *Ber. Bunsengesellschaft phys. Chem.*, 1969, **73**, 4.
64 W. M. M. J. Bovée and J. Smidt, *Mol. Phys.*, 1973, **26**, 1133.
65 D. E. Woessner, B. S. Snowden, and G. H. Meyer, *J. Chem. Phys.*, 1969, **50**, 719.
66 J. Jurga, Z. Pajak, K. Jurga, and S. Jurga, *Inst. Nuclear Phys. Cracow*, Rep. 1973, No. 819 (PL)–(Pt. 1), 46.

butyric)[67] dissolved in carbon tetrachloride. Individual rates of all different groups (methyl, methylene, hydroxyl, and ring protons) were obtained, and by extrapolation to infinite dilution of the solutes intramolecular contributions were separated. In the picolines SR contribution is evident. Information about the association of molecules could be derived; for the pure acids and for their concentrated solutions chain-type association is present.

Proton $T_1$ and $T_2$ measurements on the di-t-butyl nitroxide radical dissolved in carbon tetrachloride were made by Davis and Kreilick.[68] The dependence on the radical concentration was determined at three different frequencies. The results were interpreted in terms of the Solomon–Bloembergen theory taking account of electron–nuclear dipolar and scalar interactions; the latter only contributes to $T_2$. An approximate separation of intra- and inter-molecular contributions was achieved by estimating the intermolecular contribution from the dipolar relaxation of solvent protons ($C_6H_6$ or TMS) by the radical. The fit of the intramolecular term to the theoretical expression gives the rotational correlation time $\tau_2 = 15 \times 10^{-12}$ s and the average electron–proton distance $r = 0.357$ nm at 25 °C. The SC contribution is large for the transverse relaxation rate at concentrations below 1 mol $l^{-1}$.

Lee and Jonas[69] measured the deuteron relaxation rate in mixtures of [$^2H_8$]dioxan with normal water and of normal dioxan in heavy water. They covered the temperature range 30—50 °C and the pressure range up to 4.2 kbar. Normally in liquids increasing pressure produces an increase of the rotational correlation time and consequently an increase of the relaxation rate. Water behaves anomalously in this respect: the proton relaxation rate[70] in $H_2O$ decreases with increasing pressure up to 1.4 kbar at 0 °C, and only with further increase of pressure does the normal increase of the relaxation rate occur. At higher temperature, *e.g.* at 30 °C, this anomalous effect has disappeared, in agreement with the corresponding behaviour of the viscosity. On the other hand for the deuteron rate[71] in $D_2O$ this anomaly persists even at 90 °C, and no increase of the relaxation rate is observed at pressures up to 5 kbar, in contrast to the behaviour of the viscosity where the anomaly disappears above 30 °C (*ca.* 10 °C higher than for $H_2O$). This difference between the relaxation rates leads to the conclusion that the deuteron quadrupole coupling constant must decrease with increasing pressure. If dioxan is added to the water the anomaly in the deuteron relaxation rate of water disappears; already at 0.90 mole fraction of water the normal increase of the rate with increasing pressure is found at 30 °C. This is particularly interesting because, owing to the hydrophobic hydration[72] of dioxan molecules, the relaxation

67 E. Szczesniak, Z. Pajak, and S. Jurga, *Inst. Nuclear Phys. Cracow*, Rep. 1973, No. 819 (PL)–(Pt. 1), 36.
68 M. S. Davis and R. W. Kreillick, *J. Amer. Chem. Soc.*, 1973, **95**, 5514.
69 Y. Lee and J. Jonas, *J. Chem. Phys.*, 1973, **59**, 4845.
70 H. G. Hertz and C. Rädle, *Z. phys. Chem.*, 1969, **68**, 324.
71 Y. Lee and J. Jonas, *J. Chem. Phys.*, 1972, **57**, 4233.
72 F. Franks, in 'Water, A Comprehensive Treatise', ed. F. Franks, Plenum Press, New York, 1973, Vol. 2, p. 1.

rate of water is increased[73] at normal pressure in this composition region. Thus the effect of hydrophobic hydration yields the opposite pressure dependence to that obtained by lowering the temperature of water, a fact of interest in the discussion about the nature of the hydrophobic hydration.[74] The other component in the mixture, dioxan, behaves like a normal liquid; in the pure liquid as well as in the mixtures with water the relaxation rate of dioxan increases with increasing pressure. The deuteron quadrupole coupling constant of $[^2H_8]$dioxane is not expected to vary with pressure, so that correlation times $\tau_2$ could be calculated using a value of 173 kHz for the coupling constant. In contrast, no correlation times of water could be derived unambiguously, since the variation of the coupling constant is unknown. Correlation times were discussed in terms of the modified Debye equation (*cf.* p. 127) and the empirical factor $\kappa$[35] was calculated using measured values of the viscosity. At normal pressure, water obeys the Debye equation ($\kappa = 1$) at all compositions, but $\kappa \approx 0.1$ for dioxan, a typical value for normal liquids, and it depends on the composition. Furthermore, $\kappa$ for dioxan (and also for water after taking account of the decrease in the quadrupole coupling constant) depends on pressure, a decrease of $\kappa$ with increasing pressure being typical. At lower water content this decrease of $\kappa$ is most pronounced.

Oakes[75] performed relaxation time studies of hydroxylic protons in alcohol–water mixtures at 0 and 25 °C. The series methanol to butanol including i-propyl and t-butyl was investigated. If deuteriated alkyl groups are used only one relaxation rate (an average of alcohol hydroxyl and water proton rates) is observed at small alcohol concentrations whereas in undeuteriated samples two easily separable rates (the additional one arising from the alkyl protons) are found. At higher alcohol concentration, alcohol hydroxyl and water proton rates are separable, and extrapolation into the dilute region indicates that the water-proton relaxation rate, and consequently the rotational correlation time of water, is increased. To decide whether this increase is due to the alkyl group (hydrophobic hydration[72]) or to the alcohol hydroxy-group (hydrogen bonding to water) comparison was made between i-propyl alcohol $(CH_3)_2CHOH$ and acetone $(CH_3)_2CO$ (actually perdeuteriated acetone was used but this is irrelevant since no direct dipolar interaction between alkyl and water protons occurs). The result shows that hydrogen bonding between the alcohol hydroxy-group and water molecules is important. The effect of the alkyl groups on restricting the water mobility becomes more important only for longer alkyl chains.

In Section 3A a study of the intermolecular relaxation rate is reported which had the objective of determining relative molecular orientations in liquids. Another example is the determination of water orientation in the hydration sphere of neutral solute particles (or of ions, reported in Section

73 M. D. Zeidler, in ref. 72, p. 529.
74 H. G. Hertz, *Angew. Chem. Internat. Edn.*, 1970, **9**, 124.
75 J. Oakes, *J.C.S. Faraday II*, 1973, **69**, 1311.

5D). Hertz and Rädle[76] studied the hydrophobic hydration sphere of methanol $CHD_2OD$ and propionic acid $CHD_2CD_2CO_2D$. The solute species contain only one proton in order to make the intramolecular proton relaxation rate as small as possible. In one type of experiment $D_2{}^{17}O$ was used as a solvent and the proton rate of the solute caused by DD interaction with the $^{17}O$ nucleus was measured. In another experiment the proton DD interaction with a known amount of $HD^{16}O$ in $D_2{}^{16}O$ was determined. Comparison of both relaxation rates gives information regarding the (H,O) distance relative to the (H,H) distance between solute and water, and thus the water orientation relative to the solute is derived. The result of these experiments was that in the hydrophobic hydration sphere the water dipole points gently away from the solute, *i.e.* the negative end of the water molecule preferentially points towards the solute molecule; the angle between the methyl-water oxygen direction and the dipole vector was found to be *ca.* 110°. For comparison a molecule was studied which should not form a hydrophobic hydration sphere in aqueous solution: formic acid. It turned out that the corresponding angle here is indeed smaller, *ca.* 95°.

Further applications of the intermolecular proton relaxation rate to structural problems in aqueous solutions, specifically with respect to the hydrophobic association of carboxylic acids, are to be found in two papers by Hertz and co-workers.[77,78]

**B. $^{13}C$ Relaxation.**—Relative molecular orientation in ethanol–carbon tetrachloride mixtures was studied by Langer *et al.*[79] Again the method described in Section 3A of obtaining the orientation from intermolecular relaxation rates was used; this time $^{13}C$ DD relaxation was applied. Distances of closest approach were determined between the different protons in ethanol and the $^{13}C$ nucleus of $CCl_4$ from $^{13}C$ relaxation rates in mixtures of $^{13}C$-enriched carbon tetrachloride with $CH_3CD_2OD$, $CD_3CH_2OD$, and $CD_3CD_2OH$. These distances increase from the hydroxyl to the methyl protons. The results thus indicate the existence of a weak hydrogen bond EtOH . . . $ClCCl_3$, *i.e.* the hydroxy-group of ethanol points towards the carbon tetrachloride molecule.

**C. Studies of Several Resonant Nuclei: $^1H$, $^2H$, $^{14}N$, $^{19}F$.**—O'Reilly and Peterson[80] studied a solution of hexamethylenetetramine in chloroform. They wanted to find out whether the solute reorients by small diffusive steps or by larger jumps. By measuring $^1H$ and $^{14}N$ relaxation rates (the latter are line-width results) rotational correlation times $\tau_2$ of two different vectors in the solute molecule could be obtained: The proton–proton vector lies perpendicular to an edge; the principal $z$-axis of the electric field gradient tensor for $^{14}N$

76 H. G. Hertz and C. Rädle, *Ber. Bunsengesellschaft phys. Chem.*, 1973, **77**, 521.
77 H. G. Hertz, *J. Solution Chem.*, 1973, **2**, 239.
78 H. G. Hertz and R. Tutsch, *Ber. Bunsengesellschaft phys. Chem.*, 1973, **77**, 1013.
79 H. Langer, H. G. Hertz, and M. D. Zeidler, *Chem. Phys. Letters*, 1973, **23**, 417.
80 D. E. O'Reilly and E. M. Peterson, *J. Chem. Phys.*, 1973, **59**, 1551.

lies along a three-fold axis of the tetrahedron constituting the hexamethylenetetramine molecule. To obtain the $\tau_2$ data for the protons, extrapolation of the rates to infinite dilution of the solute in deuteriated chloroform was necessary to extract the intramolecular contribution. In an earlier paper, O'Reilly[81] had shown that for such a highly symmetric molecule the correlation times of both vectors are equal for small-step diffusion but unequal for large angle jumps. The measurements yielded equal correlation times; probably hydrogen bonding between chloroform and the nitrogen site in the solute molecule is decisive for the small-step diffusion.

Deuteron and fluorine relaxation rates of 1,3,5-trifluorobenzene dissolved in hexachlorobuta-1,3-diene were measured by Bull[82] in the temperature range 310—570 K. From the deuteron rates, assuming a quadrupole coupling constant of 181 kHz as in benzene, the orientational correlation time $\tau_2$ is obtained, whereas the angular momentum correlation $\tau_J$ is derived from the fluorine rates. These latter data are governed by SR interaction except below 390 K where intermolecular DD relaxation contributes. This was eliminated by extrapolation to infinite dilution of trifluorobenzene. The SR coupling constant required to calculate $\tau_J$ was obtained theoretically from chemical shift anisotropy data determined in the nematic phase. With these correlation time results McClung's theory[2] was tested and it was found that the extended *J*-diffusion model for a planar symmetric-top molecule explains the measurements satisfactorily.

## 5 Relaxation Studies in Electrolyte Solutions

**A. Proton Relaxation.**—Solvation of the paramagnetic cations $Co^{2+}$ and $Ni^{2+}$ in dimethylsulphoxide was studied by Boubel and Delpuech.[83] Relaxation rates of the solvent protons were measured in the range 8—60 MHz and 15—80 °C. New values of the relaxation rates in the solvation sphere and for the exchange rate between bound and bulk solvent molecules could be derived from the Swift—Connick equations, the contact shift necessary for evaluation being also measured. In the nickel solutions an anomalous increase of the proton relaxation rate with increasing frequency is observed. Relaxation is governed by dipolar interaction with the electronic spin; this implies a frequency dependence of the electron spin relaxation time, since this is the relevant correlation time for proton relaxation. It was assumed that electron spin relaxation is due to modulation of the quadratic crystalline zero-field splitting; thence a correlation time of 38 ps for electron spin relaxation was derived, which accounts for the impact of free solvent molecules against the solvated nickel cation.

Studies of the relaxation mechanism of water protons in aqueous solutions of neptunium and plutonium salts have been reported by Vdovenko *et al.*[84]

81 D. E. O'Reilly, *J. Chem. Phys.*, 1972, **57**, 885.
82 T. E. Bull, *J. Chem. Phys.*, 1973, **59**, 6173.
83 J. C. Boubel and J. J. Delpuech, *Mol. Phys.*, 1974, **27**, 113.
84 V. M. Vdovenko, V. B. Kolokol'tsev, and O. B. Stebunov, *Radiokhimiya*, 1973, **448**, 612.

**B. $^{13}$C Relaxation.**—Information on the segmental motion of unsymmetric tetra-alkylammonium bromides in aqueous solution was derived from $^{13}$C relaxation by Allerhand and co-workers.[85] By partially relaxed FT spectroscopy separate relaxation rates of all carbons in n-hexyl- and n-octyl-trimethylammonium salts were obtained. Relaxation occurs through intramolecular DD interaction with directly attached protons, and rotational correlation times could be calculated. There is a slight trend towards increased segmental mobility as one moves along the chain away from the polar head.

**C. Relaxation of Heavy Spin-½ Nuclei: $^{113}$Cd, $^{199}$Hg, $^{205}$Tl, $^{207}$Pb.**—Relaxation times of $^{113}$Cd [86] and $^{199}$Hg [87] were recently reported for the first time in a saturated aqueous cadmium chloride solution ($T_1 \approx 16$ s) and in an ethanolic solution of mercury chloride (1.4 molar, $T_1 \approx 1.4$ s).

Chan and Reeves[88] investigated $^{205}$Tl relaxation in aqueous solutions of thallium(I) nitrate and formate. $T_1$ and $T_2$ turned out to be independent of frequency, isotopic substitution of solvent ($D_2O$ for $H_2O$), concentration of thallium, and the nature of the anion. These quite remarkable facts are best accounted for by postulating that relaxation occurs through a transient spin–rotation mechanism similar to that observed for $^{129}$Xe.[89,90] Collisions take place between thallium and the solvent molecules or the anions so that the spherical charge distribution of the ion is distorted; rotation of the collision complex then produces the time-varying magnetic field necessary for relaxation. Comparison of data for larger and smaller anions shows that the former produce a larger distortion of the charge distribution, but the collision rate decreases so that relaxation is rather independent of anion size as observed. Substitution of the aqueous solvent by methanol produces a larger rate, probably owing to faster collisions. The influence of dissolved oxygen is extremely strong: two to three orders of magnitude larger than commonly observed. Equality of $T_1$ and $T_2$ indicates that not SC but DD interaction between the oxygen's unpaired electron and the thallium nucleus is governing relaxation. Thus the oxygen must penetrate deep into the hydration sphere but no complexes with oxygen are formed.

On the other hand, if instead of oxygen a nitroxide free radical is added to an aqueous solution of thallium(I) nitrate SC is effective, resulting in considerable shortening of $T_2$. Bangerter and Schwartz[91] observed an unusually strong SC effect between thallium and the unpaired electron in 4-hydroxy-2,2,6,6-tetramethylpiperidine-1-oxyl. $T_1$ is dominated by DD relaxation but is insensitive to temperature variation, a fact not yet understandable.

85 E. Williams, B. Sears, A. Allerhand, E. H. Cordes, *J. Amer. Chem. Soc.*, 1973, **95,** 4871.
86 G. E. Maciel and M. Borzo, *J.C.S. Chem. Comm.*, 1973, 394.
87 G. E. Maciel and M. Borzo, *J. Magn. Resonance*, 1973, **10,** 388.
88 S. O. Chan and L. W. Reeves, *J. Amer. Chem. Soc.*, 1974, **96,** 404.
89 H. C. Torrey, *Phys. Rev.*, 1963, **130,** 2306.
90 E. R. Hunt and H. Y. Carr, *Phys. Rev.*, 1963, **130,** 2302.
91 B. W. Bangerter and R. N. Schwartz, *J. Chem. Phys.*, 1974, **60,** 333.

Aqueous solutions of lead perchlorate were studied by Hawk and Sharp.[92] They measured the relaxation rates of $^{207}Pb$ in 1.15 and 3.50 molar solutions for the temperature range $-30$ to 100 °C at three magnetic field strengths between 0.6 and 1.7 *T*. At the lower concentration SR is the dominating relaxation mechanism over the whole temperature range and $T_1 = T_2$; DD contribution is not detectable. At 3.50 molar solution three mechanisms are important: SR at higher temperatures, CSA and DD at lower temperatures. Since the extreme narrowing condition is fulfilled only CSA can account for the observed inequality of $T_1$ and $T_2$. However, there is a small systematic difference between $T_1$ and $T_2$ even at higher temperature where CSA is small, and the authors therefore suggest that relaxation is due to scalar coupling to $^{17}O$ of water in natural abundance. The correlation time in this case is identified with the mean lifetime of water molecules in the first hydration sphere, and a value of *ca.* $4 \times 10^{-7}$ s is derived. Studies with $^{17}O$-enriched water which would unambiguously clarify the situation have not yet been carried out. The occurrence of CSA at the higher but not at the lower concentration is easily explained by anisotropy in the shielding tensor due to formation of cation–anion pairs which is strongly concentration dependent. Chemical shift measurements performed by the same authors confirm this view. The SR contribution, which is larger at the higher concentration, can be understood in two ways: a rigid (compared with the rotational correlation time) hydration sphere exists and reorients as a whole unit; here SR and DD contributions show opposite temperature dependence as implied by Hubbard's relation.[19] The second possibility is transient spin–rotation as discussed for $^{205}Tl$ relaxation (see above); this accounts for SR in non-rigid hydration spheres. The first alternative is more likely.

**D. Relaxation of Alkali-metal and Halogen Nuclei: $^{7}Li$, $^{23}Na$, $^{85}Rb$, $^{87}Rb$, $^{133}Cs$, $^{19}F$, $^{35}Cl$, $^{37}Cl$, $^{79}Br$, $^{81}Br$, $^{127}I$.**—A theoretical study of the enhancement of the $^{7}Li^{+}$ relaxation rate by manganous or nickelous ions in aqueous solutions was conducted by Friedman and co-workers.[93] Relaxation is governed in these systems by DD interaction between the electron spins of the paramagnetic ions and the lithium nucleus. As discussed in earlier sections, the intermolecular relaxation rate was used to test details of the pair correlation functions for the lithium–paramagnetic ion distribution. The authors point out the necessity of getting further knowledge of pair correlation functions from n.m.r. relaxation, since fitting to thermodynamic properties often does not give unique answers. Specifically they calculated the lithium relaxation rate as a function of field strength for two different models which are indistinguishable by thermodynamic comparisons: rigid hydration spheres or penetrable hydration spheres of the paramagnetic ions. A completely different field-dependence of the relaxation rates for solutions of manganous and

92 R. M. Hawk and R. R. Sharp, *J. Magn. Resonance*, 1973, **10**, 385; *J. Chem. Phys.*, 1974, **60**, 1522.
93 L. P. Hwang, C. V. Krishnan, and H. L. Friedman, *Chem. Phys. Letters*, 1973, **20**, 391.

nickelous ions is obtained due to differences in electronic properties of the paramagnetic species; for nickel the zero-field splitting is dominant, in contrast to manganese. In contrast to thermodynamic properties, where the ionic motion in solution does not matter, in n.m.r. relaxation it does. Usually to describe the relative motion between the interacting ions independent diffusion is assumed; this is a weak point in calculating the dispersion curves. Only very limited experimental data in these ionic solutions are available[40] and a decisive test of the presented theory is not yet possible.

There is an old controversy in the literature whether the orientation of the first-sphere hydration water of the fluoride ion is symmetric, such that the axis of the electric dipole moment connects the centres of the fluoride and the water oxygen, or whether asymmetric hydrogen-bonded orientation prevails, where one of the water protons lies on the fluoride–oxygen axis. A study of water orientation in the hydration sphere of neutral solute particles by Hertz and Rädle[76] is reviewed in Section 4A. In the same paper the above-mentioned controversy was resolved by investigating the $^{19}F^-$ relaxation rate in solutions of $D_2{}^{17}O$ and of $HD^{16}O$ in $D_2O$. Comparison of both results shows that the fluoride–proton interaction is much stronger than the (normalized) ($^{19}F$, $^{17}O$) interaction, and consequently the asymmetric form of water attachment is correct.

One of the problems of n.m.r. relaxation in liquids which was unsolved for a long time was the relaxation of quadrupolar nuclei in ions of noble gas structure in electrolyte solutions.[94] The nuclei concerned here are $^{35}Cl$, $^{37}Cl$, $^{79}Br$, $^{81}Br$, and $^{127}I$ in chloride, bromide, and iodide ions, respectively, and the cationic nuclei $^{7}Li$, $^{23}Na$, $^{85}Rb$, $^{87}Rb$, and $^{133}Cs$. The Group II and Group III metal ions should also be mentioned here; however, investigation of these nuclei has so far been less extensive.

The calculated relaxation rates for the limit of infinite dilution gave fair agreement with the experimental results for a considerable number of ions. The surprising fact was that the electrostatic theory yielded agreement even in those cases where the model underlying the theory was certainly incorrect. Thus a model of random distribution of water dipoles around the ion gave good agreement for $^{27}Al$ relaxation in $Al^{3+}$ in spite of the fact that it is a strongly hydrated ion. More severe difficulties appeared with concentrated electrolyte solutions. According to the electrostatic theory the ions as point charges should contribute to the relaxation rate, but in many cases they were found not to contribute at all and in others their contribution was too small.[95] In order to overcome both these difficulties Hertz[96,97] derived a new theory of quadrupolar relaxation in ionic solution. In the first part of his report the limiting case of high dilution was considered.[96] The main feature of the new theory is the inclusion of particle correlations higher than pair correlations. In

94 C. Deverell, *Progr. N.M.R. Spectroscopy*, 1969, **4**, 235 (and literature cited therein).
95 H. G. Hertz, G. Stalidis, and H. Versmold, *J. Chim. Phys.*, 1969, numeró spécial, 177.
96 H. G. Hertz, *Ber. Bunsengesellschaft phys. Chem.*, 1973, **77**, 531.
97 H. G. Hertz, *Ber. Bunsengesellschaft phys. Chem.*, 1973, **77**, 688.

this way more-or-less distorted symmetric water configurations may occur which quench the electric field gradient at the relaxing nucleus. Non-electrostatic contributions to the relaxation process are also discussed. The second part[97] treats solutions with high ion concentrations. Again higher ion–ion correlations are incorporated in the theory, which leads to a partial or total quenching of the electric field gradient caused by the other ions. The simplest example is the well-known ion cloud around an ion. Such an ion cloud causes the field gradient of a given ion to vanish if it is far enough away from the relaxing ion. In two further papers experimental data for the nuclei $^{35}Cl$, $^{81}Br$, and $^{127}I$ [98] in the corresponding anions and for the cation nuclei $^{7}Li$, $^{23}Na$ ,$^{85}Rb$, and $^{133}Cs$ [99] have been reported and interpreted in terms of the new theory. In the latter paper the interconnection between the anion and cation relaxation rates in the same solution was investigated. The detection of local higher ion–ion correlations was the main objective here. Actually evidence for such higher ion–ion correlations appeared for alkali halide solutions containing large ions and in sufficiently high concentration. From such observations it was predicted that in potassium, rubidium, and caesium fluoride solutions there should be rather close fluoride–fluoride approach. Indeed, in a study[100] of the ($^{19}F$,$^{19}F$) intermolecular relaxation rate in fluoride solutions in $D_2O$ it was found that the relaxation rate, which is caused by DD interaction, is much larger than it should be if the fluoride ions had a random uniform distribution.

Another puzzling problem in conjunction with the relaxation behaviour of quadrupolar nuclei in aqueous solution was the very strong increase in the halide ion relaxation caused by the presence of tetra-alkylammonium ions and other organic solutes containing methyl groups.[101] Hertz and Holz[102] performed further measurements on these systems, and from a comparison with the $^{87}Rb$ relaxation of $Rb^+$ added to the same solutions they were able to show that the strong relaxation effect is an effect of the 'dielectric holes' represented by the alkyl groups, which only attract the anions. Contact with the cations occurs to a much lesser degree. A corresponding study with the system $Rb^+$-$H_2O$–$CH_3OH$ has been undertaken by Neggia *et al.*[103]

[98] H. G. Hertz, M. Holz, R. Klute, G. Stalidis, and H. Versmold, *Ber. Bunsengesellschaft phys. Chem.*, 1974, **78**, 24.
[99] H. G. Hertz, C. Yoon, M. Holz, G. Keller, and H. Versmold, *Ber. Bunsengesellschaft phys. Chem.*, 1974, **78**, 493.
[100] H. G. Hertz and C. Rädle, *Ber. Bunsengesellschaft phys. Chem.*, 1974, **78**, 509.
[101] H. Wennerström, B. Lindman, and S. Forsén, *J. Phys. Chem.*, 1971, **75**, 2936 (and literature cited therein).
[102] H. G. Hertz and M. Holz, *J. Phys. Chem.*, 1974, **78**, 1002.
[103] P. Neggia, M. Holz, and H. G. Hertz, *J. Chim. Phys.*, 1974, **71**, 56.

# 4

# Bandshape Phenomena in Liquids

BY T. DRAKENBERG AND H. WENNERSTRÖM

## 1 Introduction

**A. Recent Developments.**—Total bandshape analyses of spectra showing chemical exchange effects are now performed more or less routinely in a large number of laboratories. This has made the use of approximate procedures still less popular. The tendency of putting the emphasis on exchange-mechanism studies rather than on barrier determinations can now also be seen in organic chemistry applications. A very clear trend during this review year is that the proton is losing its position as the outstanding nucleus in n.m.r. bandshape work. There have been numerous total bandshape studies using $^{13}C$ and a few have been performed for $^{2}H$, $^{19}F$, $^{23}Na$, and $^{31}P$. This trend is caused by both the increasing popularity of Fourier spectroscopy and the greater flexibility of the n.m.r. equipment available at most laboratories.

The use of simple linewidth determinations has decreased further in popularity except for the cases where it is not feasible to use pulse methods for measuring $T_2$. An area which has just started to develop is linewidth determinations in $^{13}C$ FT spectra. This technique may turn out to be a valuable complement to $T_1$ measurements, especially in biological applications, where often $T_2 \neq T_1$.

**B. Coverage of the Report.**—This Report is intended to cover most bandshape n.m.r. work for liquids published in periodical journals from June 1973 to May 1974, inclusive. The coverage is for obvious practical reasons not complete. For instance, only articles written in English, German, or French have been considered. There is a considerable time lag in the translation of Russian journals, and articles published in Russian during 1974 are not reviewed. The Reporters hope, in spite of the omissions made, that they have been able to cover the major part of the work done on n.m.r. bandshapes in liquids published during the year. There has been little significant bandscape work on gases, so discussion of this topic is held over until volume 5 of these Reports.

The subject coverage is, in general, the same as in volumes 1 and 2 of this series. Only those articles where non-trivial information has been obtained from some feature of the n.m.r. bandshape have been considered. This principle has, in some cases, resulted in an artificial division between related papers. We have regarded it as immaterial, from the point of the review, whether the bandshape has been obtained by a continuous-wave or by a pulsed Fourier-transform experiment. With two exceptions, only work done in isotropic systems has been considered. Special phenomena such as the CIDNP and Overhauser effects are not reviewed.

The largest part of the Report deals with the effect of chemical exchange on bandshapes. An attempt is made to reference all articles dealing with complete analysis of spectra showing magnetic-site exchange effects. Approximate single-parameter determinations of exchange times are mentioned only occasionally. In dealing with exchange-mechanism determinations we have put considerable emphasis on the chemical interest of the problem considered.

In the section dealing with relaxation and linewidths we include only the theory directly dealing with features observable from n.m.r. bandshapes. General relaxation effects are considered in Chapter 3. The discussion of articles using the Swift–Connick approach for interpreting linewidth data is intended to be complete. Simple linewidth determinations of other types are mentioned only when they, subjectively, are found interesting.

## 2 Experimental Aspects

**A. Fourier Transform *versus* Continuous Wave Spectra.**—The question of the equivalence between FT and CW spectra in systems showing intermolecular exchange, raised by Kaplan,[1] seems now to be resolved. In a note, Ernst[2] points out that the non-linear density-matrix equation of motion can be linearized for all practical purposes, and this ensures the equivalence between the FT and the CW spectra. In the answer to this note, Kaplan[3] claims that differences of up to 5—10% in spectral amplitudes can be obtained between the non-linear and the linearized equations. This conclusion seems, however, to have been in error.[4] Gupta *et al.*[5] were not able to detect any difference between the FT and CW spectra of the system $Me_3N$–$H_2O$. The authors also point out that the modified Bloch equations show that the FT and CW spectra should be identical in this simple case.

Ernst and co-workers have thoroughly analysed the conditions under which the CW and FT techniques give identical results and under which they do not.[6—8] Their conclusions, as summarized in ref. 7, are that CW and FT spectra are equivalent for single resonance at equilibrium without saturation, even in the presence of chemical exchange. The techniques are not equivalent, except for special cases, in Overhauser, double-resonance, and CIDNP

[1] J. I. Kaplan, *J. Chem. Phys.*, 1972, **57**, 5615.
[2] R. R. Ernst, *J. Chem. Phys.*, 1973, **59**, 989.
[3] J. I. Kaplan, *J. Chem. Phys.*, 1973, **59**, 990.
[4] J. I. Kaplan, P. P. Yang, and G. Fraenkel, *J. Chem. Phys.*, 1974, **60**, 4840.
[5] R. K. Gupta, T. P. Pitner, and R. Wasylishen, *J. Magn. Resonance*, 1974, **13**, 383.
[6] R. R. Ernst and R. E. Morgan, *Mol. Phys.*, 1973, **26**, 49.
[7] R. R. Ernst, W. P. Ane, E. Bartholdi, A. Höhener, and S. Schäublin, *Pure Appl. Chem.*, 1974, **37**, 47.
[8] S. Schäublin, A. Höhener, and R. R. Ernst, *J. Magn. Resonance*, 1974, **13, 196.**

experiments. Saturation effects are specifically treated in ref. 6. It is pointed out that there is a similarity between the effects of a high r.f. field and a short pulse spacing in CW and FT spectra, respectively. There is, however, no quantitative correspondence, and phenomena such as double-quantum transitions do not, of course, appear in FT spectra.

**B. Resolution Enhancement.**—Sardella[9] has suggested a method of measuring unresolved long-range couplings. The method relies on the possibility of making a double-resonance experiment to measure the linewidth and the peak height in the absence of couplings. The line spacing in the multiplet is then calculated from the linewidth and the peak height with long-range couplings, assuming Lorentzian shapes of the individual signals. The method requires that there is only one long-range coupling and that Overhauser effects can be neglected. It seems doubtful if the procedure suggested by Sardella[9] has any substantial advantages relative to a direct simulation of the bandshape, which is easily done when the widths of the individual peaks are known.

It is a common experience that proton resonance of proteins and other macromolecules gives spectra with low resolution. This is often due to the overlapping of different resonances with individually rather small natural linewidths. Campbell *et al.*[10] suggest two methods for getting resolution enhancement in this type of spectrum. The first procedure, which the authors call the convolution difference method, uses a mathematical trick especially designed for Fourier spectroscopy. The method consists of taking the free induction decay, storing it in duplicate, and multiplying the two identical decays with an exponential factor with different time constants for the two cases. Finally, one takes the difference between these modified decays and makes a Fourier transformation. This results in a better resolution at the cost of a decrease in signal to noise ratio. The second method that Campbell *et al.* suggest is chemical in nature. It consists of taking paramagnetic difference spectra,[10] *i.e.* recording a normal spectrum and then adding paramagnetic ions to the sample before taking a new spectrum. Then the difference spectrum between these two spectra is calculated; the peaks that have been substantially broadened by the paramagnetic ions now appear. This provides a possibility of sorting out a few of the peaks in the original spectrum. The method has, of course, the disadvantage that one cannot choose which resonances one wants to study, but in biological applications it often happens that the region around an ion binding site is the most interesting.

**C. Magnetic Field Inhomogeneities.**—A common experience in practical n.m.r. work is that the presence of small solid particles, such as from a precipitate, tends to decrease the resolution. Glasel and Lee[11] have made a thorough

[9] D. J. Sardella, *J. Magn. Resonance*, 1973, **10**, 188.
[10] I. D. Campbell, C. M. Dobson, R. J. P. Williams, and A. V. Xavier, *J. Magn. Resonance*, 1973, **11**, 172.
[11] J. A. Glasel and K. H. Lee, *J. Amer. Chem. Soc.*, 1974, **96**, 970.

theoretical and experimental study of this phenomenon. They find that the $^2H$ $T_2$, but not the $T_1$, of heavy water is strongly affected by the presence of glass beads in the solution. They conclude that this increase in $T_2$ is due to magnetic field inhomogeneities caused by the difference in magnetic susceptibility between the glass beads and the water. This susceptibility effect makes it difficult to study solid surface–water interactions by direct $T_2$ measurements since the magnetic field-inhomogeneity effect dominates. Glasel and Lee[11] do not take a definite stand on the question of whether the effects on $T_2$ arise from static inhomogeneities or whether it is a dynamic relaxation effect due to diffusion through field inhomogeneities. It appears from some passages in the rather involved paper that the authors regard the second alternative as the more probable. It was, however shown in a paper by Doskocilová and Schneider,[12] which Glasel and Lee apparently have overlooked, that the large proton linewidths in a sample containing solid ion-exchange resins are due to static inhomogeneities. This was done by performing a magic-angle rotation experiment,[13] in which a dramatic reduction of the linewidth was obtained. This result is in complete agreement with the reduction in linewidth by a factor of two found by Glasel and Lee[11] when the sample was rotated at right angles to the magnetic field.

It is often tacitly assumed that static magnetic field inhomogeneities lead to a symmetric, for example Gaussian, broadening of the n.m.r. signal. This is, however, not at all necessary, and Schoenberg *et al.*[14] have shown how a reasonable inhomogeneity distribution gives rise to a marked asymmetry in the n.m.r. signal. The paper serves mainly as a warning about incorrect interpretations of asymmetric signals.

## 3 Exchange of Magnetic Sites

**A. General Aspects.**—(i) *Theory*. It seems that a satisfactory theory[15] has been developed for the calculation of n.m.r. bandshapes in chemically exchanging systems in the case of single resonance and weak r.f. fields. Further advances are to be expected in the areas of double resonance, experiments in strong r.f. fields, and in computation techniques, so that, for example, it will be feasible to include non-trivial relaxation effects in the bandshape calculation. Herrmann *et al.*[16] solved numerically the equation of motion for the density matrix of a system with an $AB \rightleftharpoons BA$ exchange. They include chemical exchange effects by conventional methods, and the relaxation effects are treated approximately through a single relaxation time. The numerical solution makes it unnecessary to apply the steady-state approximation, and it is also a straightforward procedure to allow for strong r.f. fields. It is found in

[12] D. Doskocilová and B. Schneider, *Macromolecules*, 1972, **5**, 125.
[13] E. R. Andrew, *Progr. N.M.R. Spectroscopy*, 1971, **8**, 1.
[14] R. J. Schoenberger, R. B. Creel, B. K. Lunde, and R. G. Barnes, *J. Magn. Resonance*, 1973, **11**, 20.
[15] (*a*) J. Kaplan, *J. Chem. Phys.*, 1958, **29**, 462; (*b*) G. Binsch, *J. Amer. Chem. Soc.*, 1969, **91**, 1304.
[16] P. Herrmann, K. Arnold, and G. Klose, *Z. phys. Chem. (Leipzig)*, 1973, **253**, 257.

exploratory calculations that saturation phenomena occur in spectra broadened by chemical exchange, and also that the equations describe distortions of the bandshape caused by a finite sweep time. The value of these calculations seems somewhat limited owing to the inconsistent treatment of relaxation effects, since they should give more important contributions to the bandshape when saturation occurs. A further deficiency is that no double-quantum transitions seem to occur. This result is probably connected with the fact that the 1,4 element of the density matrix is neglected in the calculations by Herrmann *et al.*[16]

Even if the general problems of calculating bandshapes in chemically exchanging systems are solved, practical difficulties remain. Meakin and Jesson[17] were faced with the task of calculating the bandshape for an $A_3B_2X$ system showing intramolecular exchange to give an $A_5X$ system in an investigation of the complex $[Rh\{P(OMe)_3\}_5]^+$. After all the straightforward blockings of the matrices had been performed Meakin and Jesson found that the largest matrix to invert was of the order $100 \times 100$, and this was too large for their program to handle. The symmetry blocking usually possible in similar situations was not directly applicable since the complex possessed $D_{3h}$ symmetry. This is a non-Abelian point group, and the theorem normally used in symmetry blockings[18] requires Abelian groups. Meakin and Jesson[17] solved the problem by simply omitting from the calculations the lines with zero amplitude in the slow-exchange limit. The largest matrix to diagonalize was reduced to a $30 \times 30$ matrix in this way. The correctness of the procedure was not formally proven but rather shown empirically. In the Reporters' view a more satisfactory proof can be obtained if the problem is analysed relative to the two Abelian subgroups $C_{2v}$ and $C_3$ of the group $D_{3h}$.

Meakin and Jesson[17] also suggest two approximate procedures for making bandshape calculations on large spin systems feasible. The first, and more promising, approximation consists of omitting not only forbidden but also weak lines in the calculations. For this approximation to be possible one must choose a basis that makes the Liouville operator $\mathscr{L}_0$ diagonal. Meakin and Jesson demonstrate the validity of the approximation for a particular case. It is only in the limit of fast exchange that any differences can be detected between the exactly and the approximately calculated spectra. This deficiency occurs because of the violation of a detailed balance condition for transverse magnetization, and it can be corrected for. It is stated that the approximation is most reliable when each matrix block gives a single line in the fast-exchange limit.

When one has to make bandshape calculations for still larger spin systems, Meakin and Jesson[17] suggest that useful information can be obtained by omitting all off-diagonal elements of $[\mathscr{R} + \chi - i\mathscr{L}_0(\omega)]$ in a basis where $\mathscr{L}_0(\omega)$ is diagonal. ($\mathscr{R}$ is the relaxation and $\chi$ the exchange super-operator.)

[17] P. Meakin and J. P. Jesson, *J. Amer. Chem. Soc.*, 1973, **95,** 7272.
[18] P. Meakin, E. L. Muetterties, F. N. Tebbe, and J. P. Jesson, *J. Amer. Chem. Soc.*, 1971, **93,** 4701.

This approximation is equivalent to a first-order perturbation calculation in relaxation and exchange effects. The method can thus only be applied in the slow-exchange region where the signals start to broaden. The method has a limited value for determinations of rate constants and activation parameters, but it may be useful in investigating permutational mechanisms in exchange processes.

Brocas *et al.*[19] have derived some formal properties of magnetic-site exchange matrices. They considered in detail a special case of a trigonal-bipyramidal complex $ML^1{}_3L^2{}_2$ with two different types of ligand. One conclusion reached is that the magnetic-site exchange and isomer interconversion can be treated analogously.

(ii) *Methodological Aspects*. The problem of total bandshape analysis *versus* approximate treatments continues to arouse some interest. Dimitrov[20] suggested two new parameters for evaluating exchange times for a simple two-site exchange without spin coupling. The total bandshape equations are so simple in this case that it is questionable if there is any need for approximate formulae. Bickley and Serpone[21] made an extensive comparison between activation parameters obtained from approximate and total bandshape determinations of rate constants. They determined experimentally the bandshape for a number of six-co-ordinated inorganic complexes with acetylacetone as one of the ligands. The exchange time was determined for each spectrum by a number of different methods and then compared. Such studies have been performed previously,[22] but not for intermolecular exchange with a large negative entropy of activation. It is once again found that the approximate formulae are not sufficiently accurate to determine $\Delta H^{\ddagger}$ and $\Delta S^{\ddagger}$. Bickley and Serpone suggest that improved accuracy can be obtained if one takes the mean value of several approximate values. They find that this method gives deviations from the total bandshape values of 0—8 kJ $mol^{-1}$ in $E_a$ and 8—30 J $mol^{-1}$ $K^{-1}$ in $\Delta S^{\ddagger}$.

The use of lanthanide shift reagents (LSR) has not yet proved really popular for exchange studies. There is always the possibility that the shift reagent affects the exchange process, and this introduces an uncertainty in the investigations. Tanny *et al.*[23] have given a thorough discussion of the advantages and disadvantages of the use of shift reagents in the study of chemical exchange processes. They presented an experimental investigation of the internal rotation barrier in trimethyl carbamate. This barrier is difficult to study with conventional techniques owing to the small shift difference between the two methyl sites. Tanny *et al.*[23] determined the exchange rate at constant temperature but with varying concentration of shift reagent. It was found that

19 J. Brocas, P. Verdin, and R. Willem, *Bull. Soc. chim. belges*, 1974, **83**, 53.
20 V. S. Dimitrov, *Org. Magn. Resonance*, 1974, **6**, 16.
21 D. G. Bickley and N. Serpone, *Canad. J. Spectroscopy*, 1974, **19**, 40.
22 T. Drakenberg, K.-I. Dahlqvist, and S. Forsén, *Acta Chem. Scand.*, 1970, **24**, 694.
23 S. R. Tanny, M. Pickering, and C. S. Springer, jun., *J. Amer. Chem. Soc.*, 1973, **95**, 6227.

the exchange rate varied linearly with the LSR content, which showed that the exchange rate was different for free trimethyl carbamate and in a 1:1 complex with the LSR. The rate constant for free trimethyl carbamate could be accurately determined by extrapolating to zero LSR concentration. This method of determining the exchange rate for a free molecule is rather cumbersome if it is to be made at several temperatures. A practical problem in the bandshape calculation arises from the marked dependence of $T_2$ on the LSR concentration.

Kessler and Molter[24] report determinations of rotational barriers in methyl esters of some simple α-amino-acids substituted with t-butoxycarbonyl groups at the nitrogen. It was not feasible to investigate the rotational barrier around the C—N bond with conventional methods, but the use of an europium LSR made the barrier determination possible. Mannschreck *et al.*[25] used an optically active europium complex in the study of the interconversion between two stereoisomers of an optically active amine.

The use of liquid-crystalline solvents provides a potential possibility of extending the n.m.r. time-scale in chemical exchange studies. However, the low signal-to-noise ratios, extensive broadening, and other experimental problems make such studies difficult. Persson *et al.*[26] were able to detect broadening effects due to chemical exchange in a deuterium spectrum with quadrupole splittings. The exchange occurred between water and decanol deuterons in a lamellar lyotropic liquid crystal. The quadrupole coupling is markedly different in the two sites and this causes the broadening when exchange occurs. Persson *et al.*[26] performed a total bandshape calculation with the assumption that the two deuterium transitions $m_D = 1 \rightarrow m_D = 0$ and $m_D = 0 \rightarrow m_D = -1$ could be treated independently. The simple, two-site exchange equations were then valid for each transition separately and the bandshape calculation was easily performed.

Luz and Meiboom[27] have investigated the $^1H$ n.m.r. spectrum of cyclooctatetraene in nematic liquid-crystal solvents. They found that a bond-shift process gave rise to chemical exchange broadening in the spectrum. This exchange process can only be studied in isotropic solutions through the $^{13}C$ side-bands in the $^1H$ n.m.r. spectrum. An interesting feature of the spectrum in the liquid-crystalline solvent was that out of over one thousand transitions there were only three that were unaffected by the exchange process and remained sharp over the whole temperature region studied. The existence of these invariant lines gave a good check of the assumed mechanism of the exchange. It was not feasible to perform a total bandshape analysis but exchange rates were determined from approximate formulae in the fast- and slow-exchange limits.

[24] H. Kessler and M. Molter, *Angew Chem.*, 1973, **85**, 1059; *Angew. Chem. Internat. Edn.*, 1973, **12**, 1011.
[25] A. Mannschreck, V. Jonas, and B. Kolb, *Angew. Chem.*, 1973, **85**, 994; *Angew. Chem. Internat. Edn.*, 1973, **12**, 909.
[26] N.-O. Persson, H. Wennerström, and B. Lindman, *Acta Chem. Scand.*, 1973, **27**, 1667.
[27] Z. Luz and S. Meiboom, *J. Chem. Phys.*, 1973, **59**, 1077.

**B. Intramolecular Exchange.**—(i) *Hindered Rotation about* C—N *Bonds.* Reeves and co-workers are still active in this field and have now performed a careful study on *NN*-dimethylacetoacetamide (1),[28] which is shown to exist as a mixture of the keto and enol forms. Low-temperature $^1H$ n.m.r. spectra

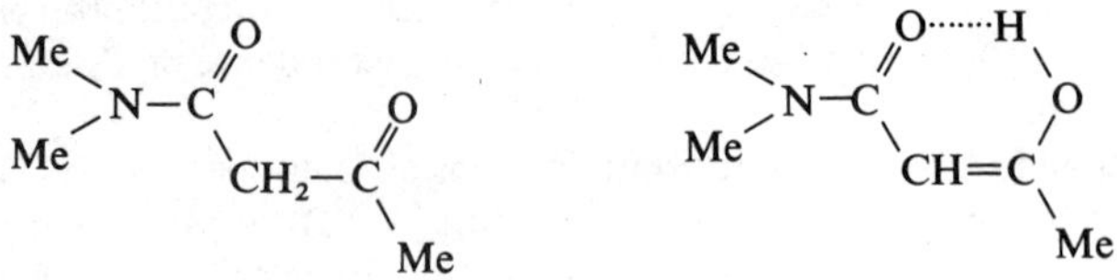

(1)

show that rotation about the C—N bond is slow, on the n.m.r. time-scale, in both keto and enol forms. The exchanges are of the type $A_3B_3X_2 \rightleftharpoons B_3A_3X_2$ for the keto-form and $A_3B_3X \rightleftharpoons B_3A_3X$ for the enol. Since there are no AB couplings and the AB parts are only weakly coupled to X, these systems can be and, judging from previous work, probably are treated as $ABX_2 \rightleftharpoons BAX_2$ and $ABX \rightleftharpoons BAX$, respectively, and only the AB parts arc used. The exchange rates are obtained from computerized fitting of calculated to experimental spectra, assuming a superposition of two independent two-site exchanges. The relative weights found for these two systems gave the keto–enol equilibrium constant.

Two more papers from Neuman and Jonas[29,30] dealing with internal rotation in amides have now appeared. These authors utilized total bandshape analysis with visual fitting of calculated to experimental spectra on [$^2H_1$]-*NN*-dimethylformamide. The deuteriated compound was used in order to circumvent the necessity of taking into account the long-range coupling. The authors claim that their activation parameters for the torsional barrier in *NN*-dimethylformamide may be the most accurate ones available. These Reporters, however, do not agree, since two other measurements using complete bandshape analysis have been performed,[31,32] which gave almost identical results, but not in agreement with those reported by Neuman and Jonas. Furthermore, the coalescence temperature reported by Neuman and Jonas is not in agreement with previous data.

Spassov *et al.*[33] have reported the values for the torsional barrier in the three phthalic tetramethylamides. The rate constants were obtained from a computerized fitting of calculated bandshapes to manually digitized experimental spectra. The exchanges were treated as uncoupled two-site exchanges,

[28] R. F. Hobson, L. W. Reeves, and R. C. Shaddick, *Org. Magn. Resonance*, 1974, **6**, 129.
[29] R. C. Neuman, jun. and V. Jonas, *J. Org. Chem.*, 1974, **39**, 925.
[30] R. C. Neuman, jun. and V. Jonas, *J. Org. Chem.*, 1974, **39**, 929.
[31] M. Rabinovitz and A. Pines, *J. Amer. Chem. Soc.*, 1969, **91**, 1585.
[32] T. Drakenberg, K.-I. Dahlqvist, and S. Forsén, *J. Phys. Chem.*, 1972, **76**, 2178.
[33] S. L. Spassov, V. S. Dimitrov, and I. Kantschovska, *Org. Magn. Resonance*, 1974, **6**, 20.

AX ⇌ XA. The authors found that the barrier in both the *meta-* and *para-*compounds agreed very well with that previously reported for dimethylbenzamide,[34,35] whereas the barrier in the *ortho*-compound is *ca.* 6 kJ $mol^{-1}$ higher owing to steric inhibition of the resonance between the benzene ring and the amide moieties.

Noe and Raben have studied the low-temperature spectra of some di- and tri-acylamines (2) and (3).[36,37] The distribution among the three possible

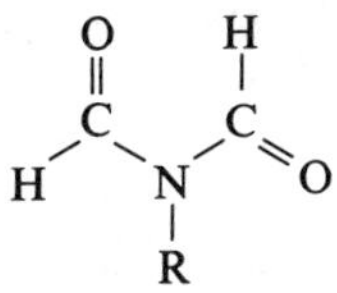

(2) a; R = H
b; R = Me

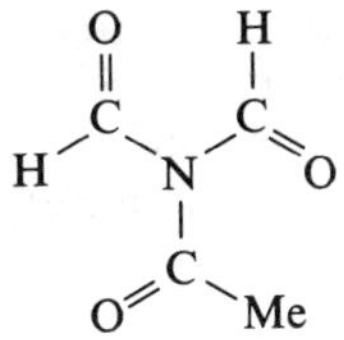

(3)

conformers in diformamide was found to be not in agreement with microwave spectroscopic data.[38] The low-temperature proton spectrum clearly shows that the major conformer has the (*E,E*) configuration.[36] The barrier to rotation in (2a) was obtained from total bandshape analysis, but the authors gave no details of the way in which this exchange was treated. For the other systems, (2b) and (3), only the rate constant at coalescence was calculated.

The barriers to internal rotation in several formyl- and acetyl-azepines (4)

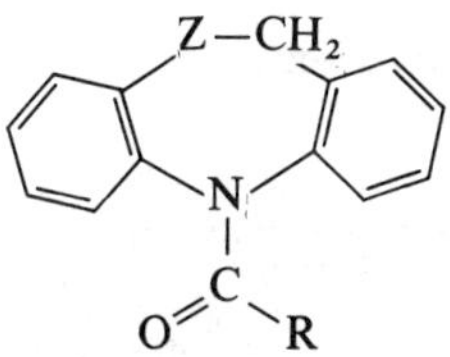

(4)

R = H or Me
Z = O or S

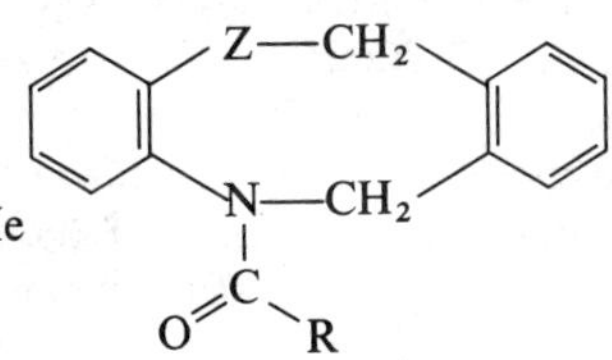

(5)

and -azocines (5) have been studied by complete bandshape analysis of the formyl or acetyl proton resonance, and the inversion barriers were estimated by coalescence-temperature measurements.[39] The reported rotational barriers ($\Delta G^{\ddagger}$) are similar to those measured on comparable aniline compounds.[40]

34 L. M. Jackman, T. E. Kavanaugh, and R. C. Haddon, *Org. Magn. Resonance*, 1969, **1**, 106.
35 K. Spargaaren, P. K. Korver, P. J. van der Haak, and Th. de Boer, *Org. Magn. Resonance*, 1971, **3**, 605.
36 E. Noe and M. Raban, *J. Amer. Chem. Soc.*, 1973, **95**, 6118.
37 E. Noe and M. Raban, *J. Amer. Chem. Soc.*, 1974, **96**, 1598.
38 W. E. Steinmertz, *J. Amer. Chem. Soc.*, 1973, **95**, 2777.
39 M. S. Puar, H. L. Yale, and A. I. Cohen, *Org. Magn. Resonance*, 1974, **6**, 106.
40 R. E. Carter, *Acta Chem. Scand.*, 1968, **22**, 2643.

However, the unusually large entropy values, both positive and negative, make the Reporters somewhat sceptical concerning the accuracy of these data.

Bushweller *et al.*[41] have observed that the barrier to rotation about the carbonyl carbon–nitrogen bond in two carbaborane cage molecules is considerably smaller than in normal amides. They interpret this as due to the aromaticity of the carbaborane cage. However, since this barrier is even lower than that in dimethylbenzamide,[34,35] this should indicate that the aromaticity of the carbaborane cage is stronger than in the benzene ring, unless the low barrier is due to steric interactions. Furthermore, a high barrier to rotation about the C—C bond was observed, but it was not possible for the authors to deduce its height.

Two papers dealing with the effect of complexation on the C—N rotation have appeared.[42,43] It has thus once more been shown that the amide rotational barrier in substituted ureas increases strongly with $SbCl_5$ adduct formation.[42] This paper is further discussed in Section 3C. In the second paper,[43] the torsional barrier about the N—C bond in *NNN′N′*-tetraethylthiuram disulphide (6) was studied in the free molecule and in complexes with

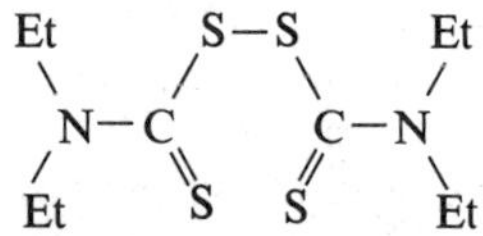

(6)

$ZnI_2$, $CdI_2$, and $HgI_2$. The data do not seem to be very accurate, but the activation energy ($\Delta G^{\ddagger}$) appears to increase with complex formation. (One data point in Table III of this paper is probably incorrect as compared with Figure 3.)

Once more, the great advantage of using $^{13}C$ instead of $^{1}H$ n.m.r. has been pointed out in a study of the barrier to internal rotation about the C—N bond in some substituted amidines (7).[44]

Liljefors[45] has performed a careful study of the barrier to internal rotation

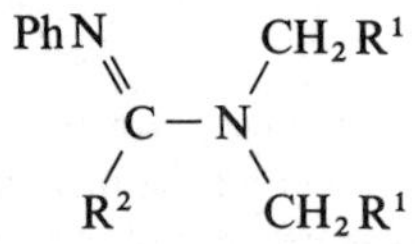

(7) $R^1$ = H or Me
$R^2$ = H, Et, Bu, Ph, or $C_3H_5$

[41] C. H. Bushweller, C. Y. Wang, W. J. Dewcett, G. Anderson, S. A. Daniels, and H. Beall, *J. Amer. Chem. Soc.*, 1974, **96**, 1589.
[42] P. Stilbs, *Tetrahedron*, 1973, **29**, 2269.
[43] A. M. Grotens and F. W. Pijpers, *Rec. Trav. chim.*, 1973, **92**, 619.
[44] M. L. Filleux, N. Naulet, J. P. Dorie, J. Pornet, and L. Miginiac, *Tetrahedron Letters*, 1974, 1435.
[45] T. Liljefors, *Org. Magn. Resonance*, 1974, **6**, 144.

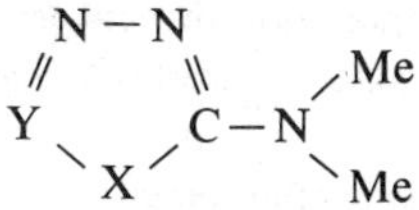

(8) X = O or S
Y = N or CR

about the C—N bond in some five-membered heterocyclic amines (8), utilizing automatic digitization of the experimental spectra and a computerized fitting procedure to obtain the rate constants. In these systems the author found an entropy of activation of *ca.* $-40$ J $mol^{-1}$ $K^{-1}$, which is unexpectedly low. The author thus made serious attempts to estimate the effect of systematic errors in $T_2$, $\Delta\nu$, and temperature on the activation entropy, but he finds it improbable that such errors could be responsible for these low activation entropies. These Reporters are, however, not convinced, since we believe that the uncertainties in $T_2$ and $\Delta\nu$ may be underestimated.

Another discussion of possible systematic errors in bandshape analysis has been presented by Colebrook *et al.*[46] in connection with their study of the relative effective sizes of methyl and chlorine as obtained from the barrier to internal rotation in 1-aryl- and 3-aryl-hydantoins (9) and (10). They found that the relative effective size can change depending on the type of interaction in which the methyl or chlorine takes part.

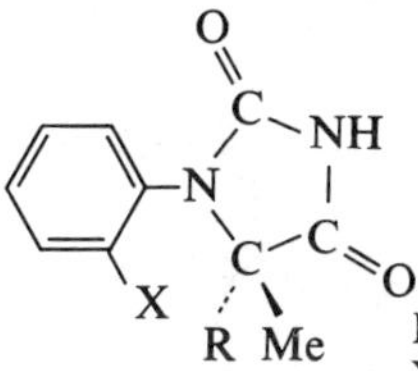

(9)

R = H or Me
X = Cl or Me

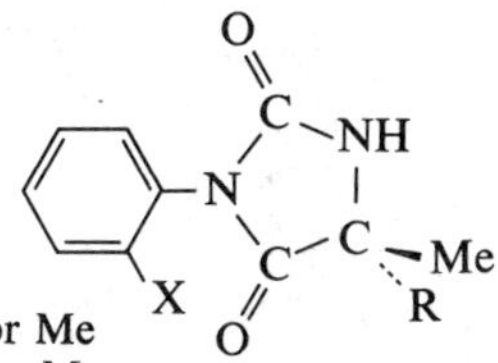

(10)

Buckley *et al.*[47] used a density-matrix program to calculate the bandshape for an exchange of the type ABCD $\rightleftharpoons$ BADC in *NN*-dimethyl- and *NN*-diethyl-*p*-nitrosoaniline (11). Instead of changing the rate constant in the

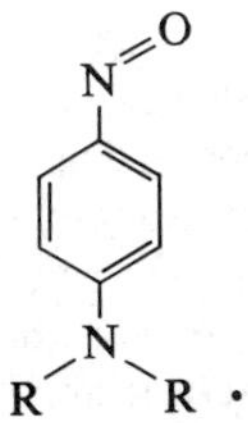

(11) R = Me or Et

[46] L. D. Colebrook, H. G. Giles, A. Granata, S. Idi, and J. R. Fehlner, *Canad. J. Chem.*, 1973, **51**, 3635.
[47] P. D. Buckley, A. R. Furness, K. W. Jolley, and D. N. Pinder, *Austral. J. Chem.*, 1974, **27**, 21.

calculated spectrum to obtain a good fit between experimental and calculated bandshapes, as is normally done, the authors changed the temperature to find the experimental spectra which gave the best fit to the calculated ones. This was done in order to save computer time.

(ii) *Hindered Rotation about* C—C *Bonds*. Some attention has been paid to the rotation of an aldehyde group bound to an aromatic system. Roques and co-workers[48] determined the torsional barrier in *N*-t-butyl-3-formylpyrrole from coalescence-temperature measurements, and they found that the barrier is slightly less than for 2-formylpyrrole.[49] Drakenberg *et al.*[50] have studied the barrier to internal rotation in protonated *p*-chlorobenzaldehyde, utilizing both $^{1}$H and $^{13}$C complete bandshape analysis. The exchange is treated as a four-site problem, where the rotations in both protonated and unprotonated aldehyde, as well as the proton exchange, are taken into account. It is thus shown that as soon as there is a fast proton exchange, the observed rotation rate is not that in the protonated species even though the population of the unprotonated species is extremely low. This shows clearly that in some cases even undetectable species have to be included in the bandshape analysis.

During the period covered in this Report, two papers have appeared dealing with the hindered rotation of a methyl group.[51,52] Very much the same strategy was applied to find suitable systems, and the molecules studied, (12) and (13), were also similar. The free energy of activation for (12) was reported to be

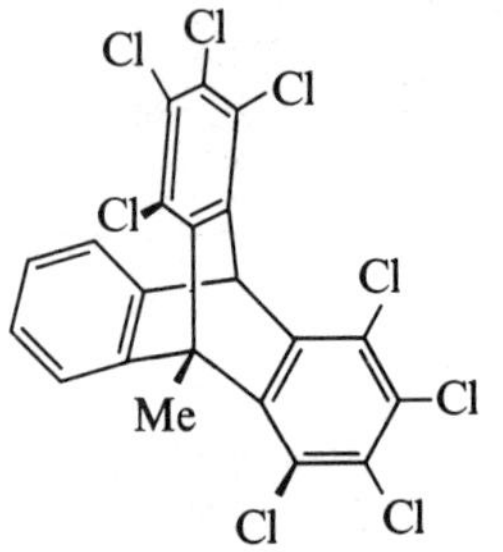


(12)

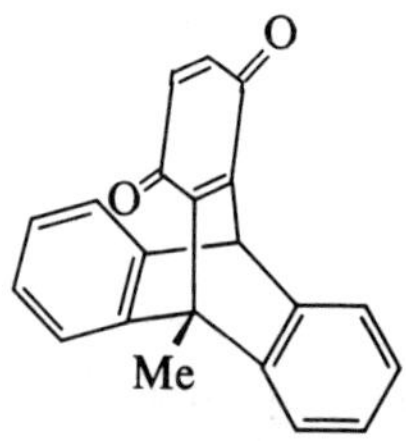


(13)

25.3 kJ mol$^{-1}$, and for (13) a value of 17.2 kJ mol$^{-1}$ is given. Both results were obtained from total bandshape analysis of the methyl proton signal. One of the groups of workers has also used the total bandshape technique to study t-butyl rotation in systems such as (14). They found that the barrier increases with increasing size of X except when X is changed from Cl to Br, as judged from the free energy of activation.[53] The authors do not attempt to

48 M. C. Fournié-Zaluski, C. Jaureguiberry, and B. Roques, *Tetrahedron Letters*, 1973, 4177.
49 L. Arlinger, K.-I. Dahlqvist, and S. Forsén, *Acta Chem. Scand.*, 1970, **24**, 672.
50 T. Drakenberg, S. Forsén, and J. M. Sommer, *J.C.S. Perkin II*, 1974, 520.
51 M. Nakamura, M. Oki, and H. Nakanishi, *J. Amer. Chem. Soc.*, 1973, **95**, 7169.
52 J. E. Anderson and D. I. Rawson, *J.C.S. Chem. Comm.*, 1973, 830.
53 M. Nakamura, M. Oki, and H. Nakanishi, *Tetrahedron*, 1974, **30**, 543.

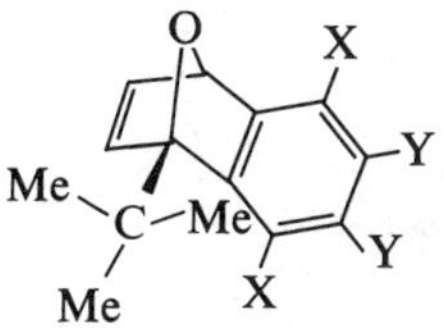

(14) X = H, OMe, F, or Cl
Y = H, F, or Cl

explain this anomaly. However, when the enthalpy of activation is used in this comparison, the barrier increases in the expected order. The steric hindrance to internal rotation of a t-butyl group has also been studied by Bushweller *et al.*[54] in compounds such as (15). They have found that the barrier for the

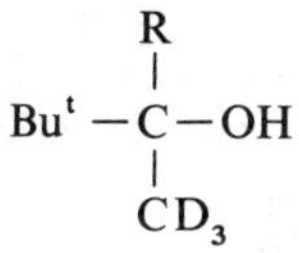

(15) R = $CD_3$, $C_2D_5$, or Bu

t-butyl rotation, as expected, increases with increasing size of the substituent R, and from comparison with other data it can be seen that the size of a fluorine is about the same as that of a hydroxy-group.

(iii) P—N, P—C, *and* P—P *Rotational Barriers*. DiStefano *et al.*[55] studied the barrier to internal rotation about the P—N bond in (16) as a function of both

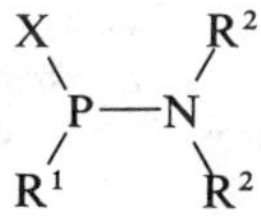

(16) X = Cl or F
$R^1$ = Me or Ph
$R^2$ = Me or $Pr^i$

P and N substituents. They report that a complete bandshape program was used to obtain the rate constants, but no further details are given.

Bushweller and Brunelle[56,57] used complete bandshape analysis to study the P—C rotation in t-butylphosphines and their complexes with borane. The exchange is treated as being between three sites. Further details are not given, but an exchange of the type shown in (17) was probably used.

[54] C. H. Bushweller, S. Hoogasian, and W. G. Anderson, *Tetrahedron Letters*, 1974, 547.
[55] S. DiStefano, H. Goldwhite, and E. Mazzola, *Org. Magn. Resonance*, 1974, **6**, 1.
[56] C. H. Bushweller and J. A. Brunelle, *J. Amer. Chem. Soc.*, 1973, **95**, 5949.
[57] C. H. Bushweller and J. A. Brunelle, *Tetrahedron Letters*, 1974, 893.

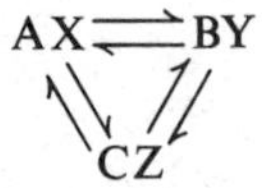

(17)

Aime *et al.*[58] used both $^1$H and $^{13}$C n.m.r. at low temperature in order to investigate the mechanism of exchange in tetra-t-butyldiphosphine. Their conclusion is that both P—C rotation and P-inversion are slow on the n.m.r. time-scale.

(iv) *Ring Inversion.* $^{13}$C dynamic n.m.r. has been increasingly used in studies of conformations and energy barriers in cyclic systems, especially by Anet's group.[59—62] They have thus, by means of $^{13}$C and/or $^1$H dynamic n.m.r., been able to detect barriers to conformational changes in six- and in eight- to sixteen-membered-ring ketones, and the lowest of these, 17.1 kJ mol$^{-1}$ for cyclohexanone, is probably the lowest observed by n.m.r. to date.[59] It is interesting to note that the cycloalkanones with odd numbers of carbon atoms in the ring always have a lower barrier than the nearest with an even number. In some of the systems it has been possible to observe two different conformational changes with $^1$H n.m.r., and the lowest of these is always the same as that observed with $^{13}$C. These observations have made it possible to deduce the ground-state conformation.

Dale and co-workers,[63—66] partly in collaboration with Anet, have used both $^{13}$C and $^1$H n.m.r. in their studies of different kinds of ring system. They have, for example, tried to evaluate the ground-state conformation adopted by 1,4,7,10-tetraoxacylododecane and its complexes with lithium salts.[63] Two different site-exchange processes were observed with $^1$H n.m.r.; the faster of these was also seen in $^{13}$C spectra. The lowest conformational barrier was found to be unaffected by complexation, whereas the higher barrier increased considerably on complex formation. These observations were interpreted in the following way: the low-energy form has a 'square' conformation with oxygen atoms at 'side' positions, all on the same side of the molecule, both for the free and complexed ring.

Bernard *et al.*[67,68] have studied the mechanism of conformational interconversion in some methyl-substituted six-membered rings with one or two

[58] S. Aime, R. K. Harris, and E. M. McVicker, *J.C.S. Chem. Comm.*, 1974, 426.
[59] F. A. L. Anet, G. N. Chmurny, and J. Krane, *J. Amer. Chem. Soc.*, 1973, **95**, 4423.
[60] F. A. L. Anet, A. K. Cheng, and J. Krane, *J. Amer. Chem. Soc.*, 1973, **95**, 7877.
[61] F. A. L. Anet and V. J. Basus, *J. Amer. Chem. Soc.*, 1973, **95**, 4424.
[62] F. A. L. Anet, J. Krane, J. Dale, K. Daastvatn, and P. O. Kristiansen, *Acta Chem. Scand.*, 1973, **27**, 3395.
[63] G. Borgen, J. Dale, F. A. L. Anet, and J. Krane, *J.C.S. Chem. Comm.*, 1974, 243.
[64] E. Augdahl, G. Borgen, J. Dale, and J. Krane, *Acta Chem. Scand.*, 1974, **B28**, 125.
[65] G. Borgen, *Acta Chem. Scand.*, 1974, **B28**, 13.
[66] G. Borgen, *Acta Chem. Scand.*, 1974, **B28**, 135.
[67] M. Bernard and M. St. Jacques, *Tetrahedron*, 1973, **29**, 2539.
[68] M. Bernard, L. Canuel, and M. St. Jacques, *J. Amer. Chem. Soc.*, 1974, **96**, 2929.

$sp^2$ carbons. They used total bandshape analysis to obtain the rate of interconversion. Normally the methyl proton spectrum was utilized, which gave simple exchanges of the type A⇌X. In one case, however, the methylene proton spectrum, of the type AB⇌BA, was used in the partly deuteriated 3,3-dimethylcyclohexene (18). Bernard *et al.*[68] used computerized fitting of the

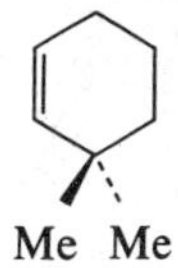


(18)

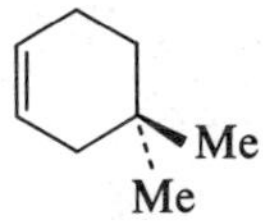


(19)

theoretical to the experimental spectra, with the digitization of the spectra performed manually. The effect of methyl substitution in various positions on the barrier to interconversion in the cyclohexene derivatives was interpreted as being in agreement with the assumption that the boat form is the transition state. This is not quite obvious, since compounds (18) and (19) have essentially the same barrier.

Larkin and Lord[69] have studied the low-temperature $^1$H n.m.r. spectrum of dioxen, with an exchanging spin system of the type $[AB]_2 \rightleftharpoons [BA]_2$. The values obtained for the activation parameters for the ring inversion were compared with those obtained from far-i.r. measurements.[70] The agreement is better than is normally found in this type of comparison, especially if it is considered that the i.r. data are from gas-phase studies, whereas the n.m.r. results are from studies in solution. The authors also state that 'the close correspondence of the two numbers is somewhat fortuitous'. It is, however, also pointed out that the same agreement is found for $\Delta^{2,3}$-dihydropyran.

Brickwood *et al.*[71] have studied the conformational behaviour of hexamethyl-5,6,11,12,17,18-hexahydrotribenzo[*a,e,i*]cyclododecaene (20). At low temperature the methyl protons give rise to three equally intense lines, showing that the ground state of this molecule cannot be the propeller-like con-

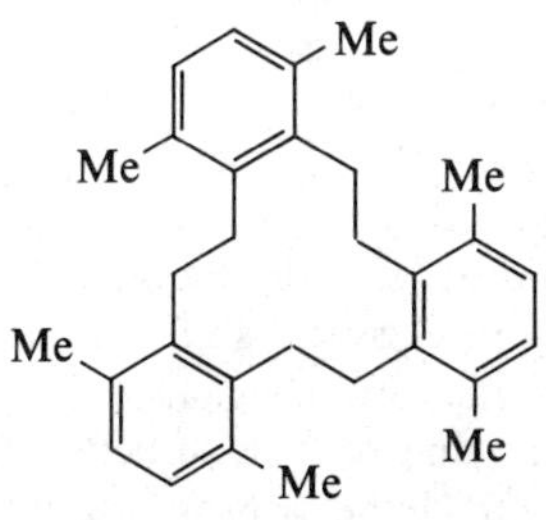


(20)

[69] R. H. Larkin and R. C. Lord, *J. Amer. Chem. Soc.*, 1973, **95**, 5129.
[70] R. C. Lord, T. C. Rounds, and T. Ueda, *J. Chem. Phys.*, 1972, **57**, 2572.
[71] D. J. Brickwood, W. D. Ollis, and J. F. Stoddard, *J.C.S. Chem. Comm.*, 1973, 638.

formation with $D_3$ symmetry. At high temperature a single line is observed, and the exchange is treated as being between three sites with a single rate constant.

Lindquist and Sandström[72] have studied the ring inversion in some lactams of dibenz[*b*,*e*]azepine (21) and dibenz[*b*,*f*]azocine (22). For one of these

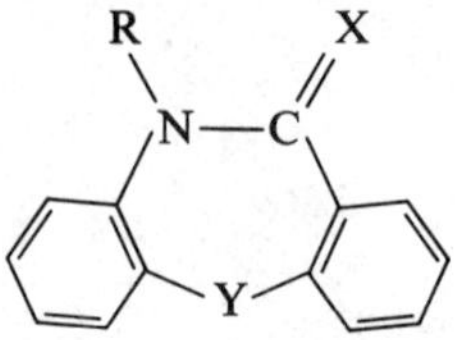


(21) R = H or hydrocarbon fragment;
X = O or S;
Y = $CH_2$ or CO

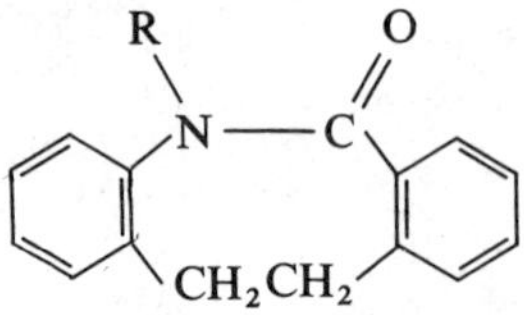


(22) R = H or hydrocarbon fragment

systems, complete bandshape analysis at several temperatures was performed in order to obtain the entropy and enthalpy of activation. In this case, as in most cases where a careful investigation of intramolecular exchange has been made, the entropy of activation is found to be indistinguishable from zero. For the other systems the barriers are only evaluated at the coalescence temperature.

(v) *Other Exchange Processes.* The Cope rearrangements of bullvalene (23) and semibullvalene (24) have been studied by means of both $^1$H and $^{13}$C dynamic n.m.r. Nakanishi and Yamamoto[73] used total bandshape analysis in their treatment of the $^{13}$C spectra of bullvalene (23). The assignment of the

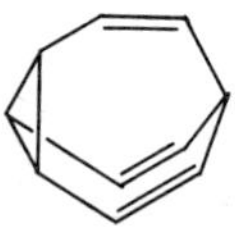

(23)

(24)

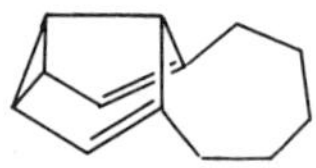

(25)

two types of vinylic carbon was based on the difference in broadening of their signals due to exchange in the slow-exchange region. Since only one of the type-2 carbons but all type-3 carbons change site in each exchange, the line due to type-3 carbons will broaden more than the line due to type-2 carbons. In systems of this kind, with chemical shift differences of the order of 100 p.p.m. between the exchanging carbon signals, it should be possible to obtain activation parameters, $\Delta H^{\ddagger}$ and $\Delta S^{\ddagger}$, with better than usual accuracy. For some reason this seems not to be the case, as judged from the large uncertainties quoted. For semibullvalene (24), a chemical shift difference between exchanging carbons of *ca.* 5000 Hz should make it possible to obtain rate

[72] Å. Lindquist and J. Sandström, *Chem. Scripta*, 1974, **5**, 52.
[73] H. Nakanishi and O. Yamamoto, *Tetrahedron Letters*, 1974, 1803.

constants as high as $10^7$ $s^{-1}$ to within better than 10%. Cheng *et al.*[74] have, however, only reported bandshape analysis of the proton spectra, which had to be treated in a somewhat approximate way since the exchange occurs within an eight-spin system.

An exchange between unequally populated sites in the substituted semi-bullvalene (25) has also been treated.[75] The isomer population at low temperature was determined, and the authors also claim that complete bandshape analysis at the coalescence temperature was utilized. No further details are given, but some approximations must have been used owing to the high complexity of the system.

Grunwell *et al.*[76] have studied the low-temperature $^1H$ n.m.r. spectra of two sulphur di-imides (26), for which three isomers are possible. It is found that (26a) exists mainly as the *cis–trans* conformer, with a small amount of a

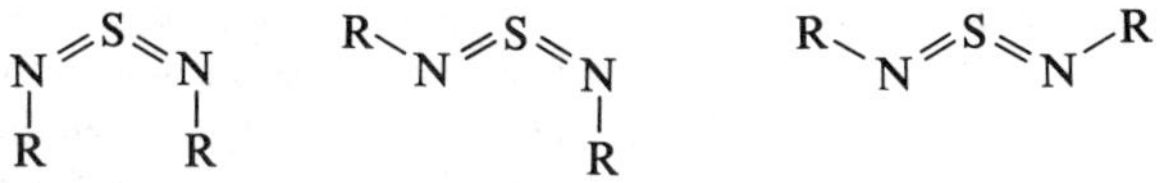

*cis-cis* *trans-cis* *trans-trans*

(26) a; R = Me
b; R = Bu$^t$

second isomer. For (26b) only two signals of equal intensity could be observed, and (26b) is assumed to be exclusively in the *cis–trans*-conformation.

Bushweller *et al.*[77] have been able to study both nitrogen inversion and ring inversion in *NN′N″*-trimethyl-1,3,5-triazane (27) with complete bandshape analysis. The data are in agreement with the assumption that only the monoaxial conformer is present in a detectable amount.

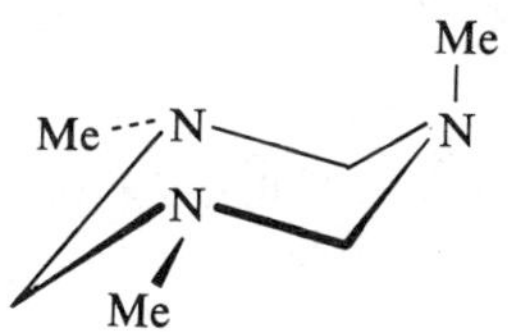

(27)

Gibson *et al.*[78] have used $^{19}F$ n.m.r. bandshapes to distinguish between inter- and intra-molecular exchanges in some sulphur, phosphorus, and silicon

[74] A. K. Cheng, F. A. L. Anet, J. Mioduski, and J. Meinwald, *J. Amer. Chem. Soc.*, 1974, **96,** 2887.
[75] R. K. Russel, L. A. Paquette, L. G. Greifenstein, and J. B. Lambert, *Tetrahedron Letters*, 1973, 2855.
[76] J. R. Grunwell, C. F. Hoyng, and J. A. Rieck, *Tetrahedron Letters*, 1973, 2421.
[77] C. H. Bushweller, M. Z. Lourandos, and J. A. Brunelle, *J. Amer. Chem. Soc.*, 1974, **96,** 1591.
[78] J. A. Gibson, D. G. Ibbott, and A. F. Janzen, *Canad. J. Chem.*, 1973, **51,** 3203.

fluorides. By observing the disappearance or existence of the splitting due to either (P,F) or (Si,F) couplings, the authors concluded that the intramolecular exchange was faster than the intermolecular exchange for both silicon and phosphorus fluorides.

Gilje *et al.*[79] have studied the fluxional rearrangement in trifluorophosphoranes by means of proton bandshape analysis, which shows that it is an intramolecular rearrangement in which the equatorial fluorine has an equal probability of exchanging with either of the axial fluorine ligands.

(vi) *Multisite Exchanges*. In a series of papers, Mislow and co-workers have discussed the dynamic stereochemistry of various propeller-like molecules.[80—88] Triarylmethanes with various degrees of symmetry have been studied. The shape of the $^1$H n.m.r. signals from methyl and methoxy-groups was used to obtain information about the mechanism of stereoisomerization. Even though no spin couplings could be observed in these signals, the spectra were in some cases quite complicated owing to the existence of various isomers. In the simplest case, (28), only three signals were observed in the slow-exchange region and normal two-site exchange analysis was used. For this example it was not possible to gain much information about the mechanism of isomerization from n.m.r. spectra. In the slightly more complex system (29), however, only

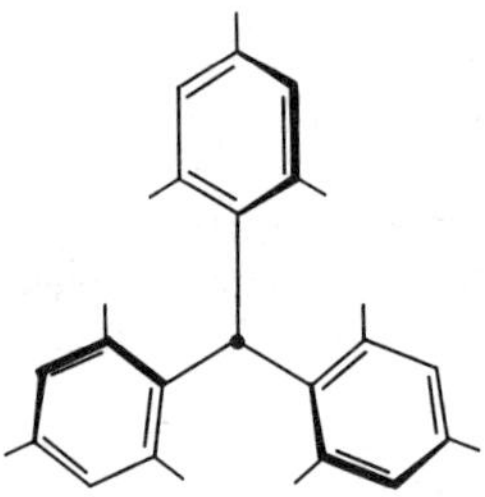

(28)

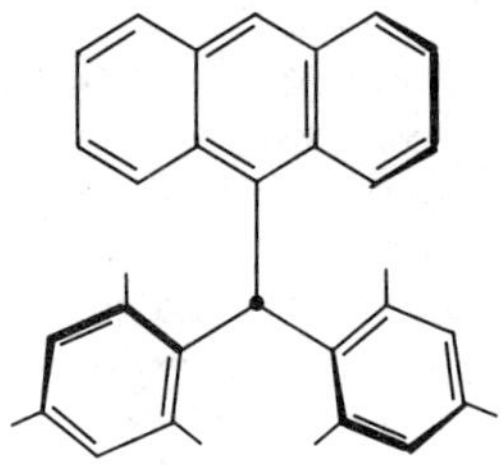

(29)

two mechanisms were in agreement with the observed spectra, provided there is no fortuitous equivalence of the rates for two processes. The four *o*-methyl groups gave rise to four signals at room temperature, which coalesced to a single line in a four-site exchange scheme with a single rate constant. In principle it should be possible to distinguish between the two remaining mechanisms from the shape of the aromatic proton spectrum from the anthryl

79 J. W. Gilje, R. W. Braun, and A. H. Cowley, *J.C.S. Chem. Comm.*, 1974, 15.
80 P. Finocchiaro, D. Gust, and K. Mislow, *J. Amer. Chem. Soc.*, 1973, **95,** 7029.
81 R. J. Boettcher, D. Gust, and K. Mislow, *J. Amer. Chem. Soc.*, 1973, **95,** 7157.
82 M. G. Hutchings, C. A. Maryanoff, and K. Mislow, *J. Amer. Chem. Soc.*, 1973, **95,** 7158.
83 P. Finocchiaro, D. Gust, and K. Mislow, *J. Amer. Chem. Soc.*, 1974, **96,** 2165.
84 P. Finocchiaro, D. Gust, and K. Mislow, *J. Amer. Chem. Soc.*, 1974, **96,** 2176.
85 P. Finocchiaro, D. Gust, and K. Mislow, *J. Amer. Chem. Soc.*, 1974, **96,** 3198.
86 P. Finocchiaro, D. Gust, and K. Mislow, *J. Amer. Chem. Soc.*, 1974, **96,** 3205.
87 J. P. Hummel, D. Gust, and K. Mislow, *J. Amer. Chem. Soc.*, 1974, **96,** 3679.
88 K. Mislow, D. Gust, P. Finocchiaro, and R. J. Boettcher, *Topics Current Chem.*, 1974, **47,** 1.

group. The complexity of the spin system has, however, precluded such a determination. More and more complex systems of this kind have been studied, up to a triarylmethane with no element of symmetry, (30), which is a 32-isomer system. The authors have shown that the isomers can be divided into

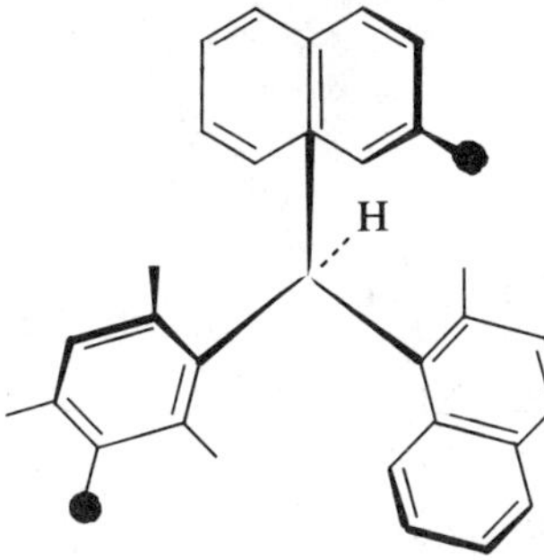


(30)

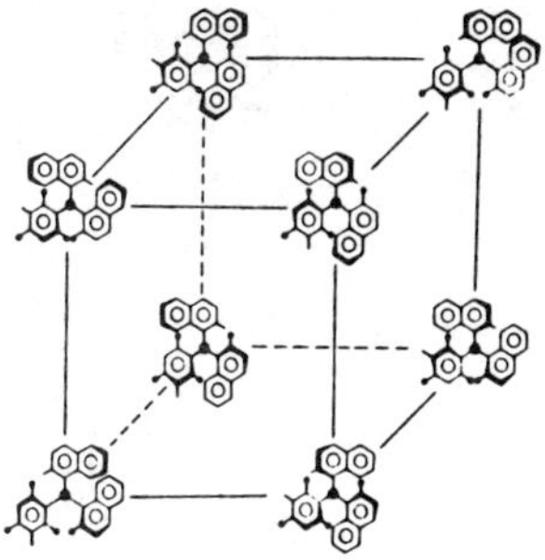

(31)

subgroups with eight isomers in each group. Within each group the exchange scheme is best represented by a cube with the isomers at the corners and the exchanges along the edges connecting them (31).* The exact treatment of this system would require an eight-site exchange for each of the chemically different methyl groups, but this has apparently not been done because the primary goal of the work was to study the mechanism of isomerization and not to determine the exact values for the isomerization barriers. This work provides very nice examples of the possibility of obtaining information about mechanisms from n.m.r. bandshapes without the need of sophisticated calculations.

Nilsson *et al.*[89—92] have continued their work on substituted 1,3,5,-trineopentylbenzenes (32). The torsional barrier for the neopentyl groups was studied with both $^{1}H$ and $^{13}C$ complete bandshape analysis. The authors have shown that the interchange of the 1- and 5-methylene protons in (32a) must be treated as an exchange among at least six sites. They have thus developed a computer program to calculate the bandshape for an AB system exchanging among six sites in a circular way with no cross-exchanges (33). The program was obtained by solving the two 12th-order equation systems necessary to describe the complex magnetizations involved in a six-site AB exchange. The only difference between the two systems of equations is in the sign before the coupling constant. The theoretical spectrum is found by a summation of the imaginary parts of the 24 complex magnetizations. Nilsson *et al.* have applied

[89] B. Nilsson, P. Martinsson, K. Olsson, and R. E. Carter, *J. Amer. Chem. Soc.*, 1973, **95**, 5615.
[90] B. Nilsson, *Mol. Phys.*, 1974, **27**, 625.
[91] B. Nilsson and T. Drakenberg, *Org. Magn. Resonance*, 1974, **6**, 155.
[92] B. Nilsson, P. Martinsson, K. Olsson, and R. E. Carter, *J. Amer. Chem. Soc.*, 1974, **96**, 3190.

* Reproduced by permission from *J. Amer. Chem. Soc.*, 1974, **96**, 3190.

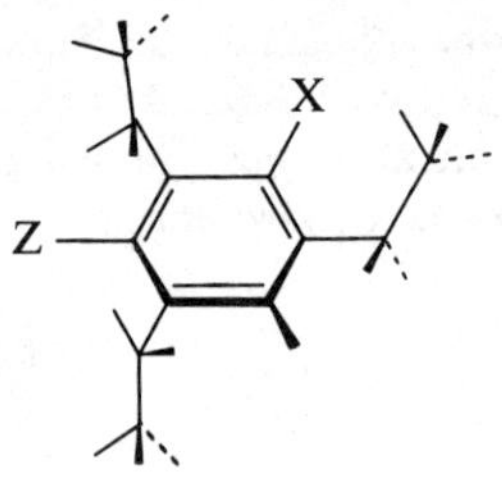


(32) a; X = Y = Br, Z = $NO_2$
b; X = Y = $NO_2$, Z = H

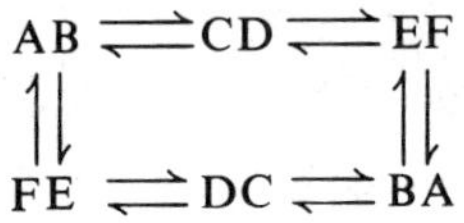


(33)

this program in the study of the barrier to internal rotation in (32a). In principle, the exchange in (32a) could be of the eight-site type, but since not more than three AB spectra were observed, the authors assumed that the conformer with all three t-butyl groups on the same side of the benzene ring had a higher energy than the other conformers, and consequently the former was not included in the bandshape analysis. It would have been more straightforward to use an eight-site exchange of type (31) to estimate the possible effect of an undetectable conformer. Complete bandshape analyses of the type (33) and of less complex systems have been performed and made it possible to evaluate the barrier to internal rotation for a neopentyl group past *ortho*-substituents ranging from H to I, with barriers from 22.6 to 78.6 kJ $mol^{-1}$.

Schaefer and co-workers[93,94] have carried out a complete bandshape analysis of the $^1H$ n.m.r. spectrum from (34), which was treated as a four-site

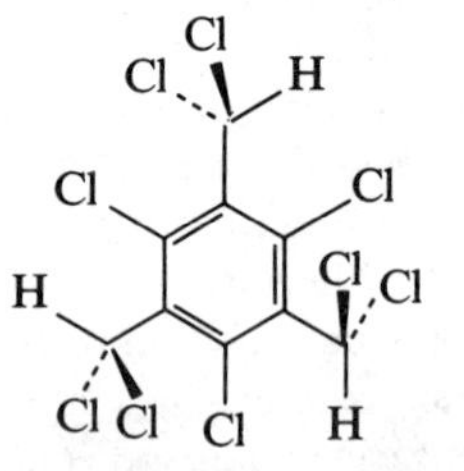


(34)

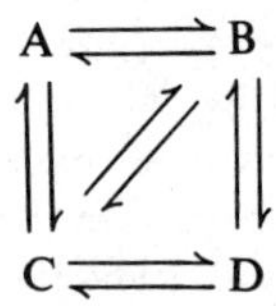


(35)

exchange (35). Agreement between calculated and experimental spectra could only be obtained when two different rate constants were used, one for the

[93] J. Peeling, B. W. Goodwin, T. Schaefer, and J. B. Rowbotham, *Canad. J. Chem.*, 1973, **51**, 2110.
[94] J. Peeling, L. Ernst, and T. Schaefer, *Canad. J. Chem.*, 1974, **52**, 849.

B→A, B→C, and B→D exchanges, where B is the conformer shown in (34), and another rate constant for the other two exchanges.

**C. Intermolecular Exchange.**—(i) *Ligand Exchange.* Matieu and Nixon[95] studied the ligand exchange in $Rh_2[P(OMe)_3]_8$ by means of variable-temperature $^{31}P$ n.m.r. It has thus been shown that only the equatorial ligands take part in an intermolecular exchange, whereas the axial ligands are not exchangeable at the temperatures studied.

Crea *et al.*[96,97] used a complete bandshape method, giving no details, to study the trimethyl phosphate exchange in magnesium(II) and beryllium(II) complexes. These authors attempt to rationalize the resulting positive activation entropies in terms of structural contributions to the activation energy, and show that there is a rough correlation between $\Delta H^{\ddagger}$ and $\Delta S^{\ddagger}$ for this process in different solvents. An alternative explanation, however, is that the observed effects may be due to systematic errors in the bandshape analysis, which are known to give rise to this type of correlation between $\Delta H^{\ddagger}$ and $\Delta S^{\ddagger}$.

Felten and Anderson[98] report a study of the exchange of 3,5-lutidine in complexes with tetrabenzylhafnium and tetrabenzylzirconium. They used total bandshape analysis for two samples, assuming two-site exchange. The resulting strongly positive entropy of activation and the fact that the mean lifetime of the complex is independent of concentration were taken as evidence for a dissociative mechanism in which the rate-determining step is

$$(PhCH_2)_4Hf,lut \rightleftharpoons (PhCH_2)_4Hf + lut$$

It was also shown that the benzyl groups are not exchangeable.

Benndorf *et al.*[99] used total bandshape analysis to study the ligand exchange in some phenanthroline–metal complexes. The authors also compared the results obtained from single-parameter formulae with those from total bandshape analysis. The difference in the results from the two methods is not very large, and the authors have thus used only the single-parameter formulae for other systems. Since it has been shown several times that the use of single-parameter formulae in this way often gives unreliable data for $\Delta H^{\ddagger}$ and $\Delta S^{\ddagger}$, the Reporters would prefer that only $\Delta G^{\ddagger}$ values were reported from this type of study. Benndorf *et al.* give no details about the bandshape analysis, but since they state that their computer program takes a maximum of three spins, the system is probably handled as a superposition of one ABC⇌DEF and one X⇌Y system.

The substrate exchange in some complexes with lanthanide shift reagents has been studied[100] utilizing total bandshape analysis in the intermediate-exchange region and one-parameter formulae in the slow-exchange region.

95 R. Matieu and J. F. Nixon, *J.C.S. Chem. Comm.*, 1974, 147.
96 J. Crea, S. F. Lincoln, and R. J. West, *Austral. J. Chem.*, 1973, **26**, 1227.
97 J. Crea and S. F. Lincoln, *J.C.S. Dalton*, 1973, 2075.
98 J. J. Felten and W. P. Anderson, *Inorg. Chem.*, 1973, **12**, 2334.
99 G. Benndorf, H. Rosenberger, M. Funk, and K. Madeja, *Org. Magn. Resonance*, 1973, **5**, 379.
100 D. E. Evans and M. Wyatt, *J.C.S. Dalton*, 1974, 765.

The authors discuss three different mean lifetimes, and they state that reliable mean lifetimes of the complex could not be obtained directly, but as far as the Reporters can see both the lifetime for the complex and that for the free substrate may be obtained with the same precision from the total bandshape analysis, as long as the population difference is not very large. The authors have found for all except two of the systems studied that the kinetics of the substrate exchange are either completely first-order or completely second-order, whereas for the remaining two systems, $Pr(fod)_3,2Me_2SO$ and $Nd(fod)_3,2Me_2SO$, a competition between a first- and a second-order mechanism had to be assumed.

(ii) *Proton Exchange*. Menger and Saito[101] studied the rate of proton exchange in some *o*-substituted *N*-methylbenzamides (36) as a function of pH. The rate

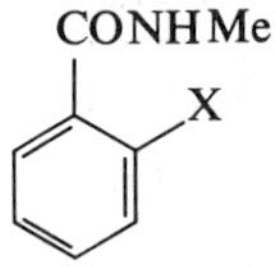


(36) a; X = H
b; X = OH
c; X = OMe
d; X = $CO_2^-$

constants were estimated by minimizing the deviation between some experimental and theoretical spectral parameters (bandwidth and peak-to-valley ratio). The authors claim that this gives rate constants within 15% of those derived from total bandshape analysis, but since the theoretical curve had to be calculated in any case, it is astonishing that the total bandshape has not been used in the evaluation of the rate constants. From the variation in the proton exchange rate with the substituents, the authors conclude that the larger rate constant for *N*-methylsalicylamide (36b) is due to intramolecular catalysis.

Whidby and Morgan[102] used $^1H$ n.m.r. to study the specific acid catalysis of acetic acid on the proton exchange of *NN'*-dimethylurea. The exchange rates were estimated by comparing experimental and theoretical spectra, but no details were given. The relative catalytic strengths of some substituted acetic acids were obtained.

From the variation of the $^1H$ n.m.r. spectrum of *meso*-tetraphenylporphyrin (37) with trifluoroacetic acid concentration and temperature, Abraham *et al.*[103] have concluded that the exchange between the uncharged and the diprotonated species is slow on the n.m.r. time-scale and that no monoprotonation occurs.

[101] F. M. Menger and G. Saito, *J. Amer. Chem. Soc.*, 1973, **95**, 6840.
[102] J. F. Whidby and W. R. Morgan, *J. Phys. Chem.*, 1973, **77**, 2999.
[103] R. J. Abraham, G. E. Hawkes, and K. M. Smith, *Tetrahedron Letters*, 1974, 71.

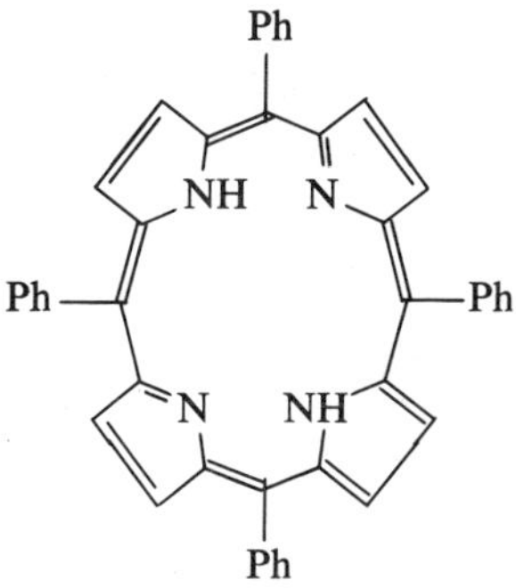

(37)

(iii) *Other Exchange Processes.* Alaluf *et al.*[104] studied the donor–acceptor exchange in the complexes between trimethylphosphine and trimethylborane, and between trimethylphosphine and trimethylalane. The $^1$H n.m.r. signal from free trimethylborane is a singlet, and that from the complex is a doublet, owing to spin coupling to phosphorus. The exchange was consequently treated as a three-site process, and the equation for this is given in some detail. A slightly negative entropy of activation for the trimethylphosphine–trimethylborane system and a strongly positive one for the trimethylphosphine–trimethylalane system were found. The authors suggest that this difference could be due to the fact that the free acceptor in the latter case is present as a dimer, hexamethyldialuminium. Donor–acceptor exchange has also been studied by Stilbs.[42] Some of the systems he studied had to be treated as exchanges among four sites. For amides there is a hindered rotation in the adduct as well as in the free molecules. There are thus three independent rate processes: rotation in the adduct, rotation in the free molecules, and donor exchange. Only for one system, $Me_2NCONHPr^i,SbCl_5$, was it possible to determine the rate constant for all three exchange processes. A good correlation was found between enthalpies of adduct formation and the enthalpies of activation for donor exchange.

Martin and Poignant[105] used both CW and FT $^1$H spectra to study the exchange reactions in chloro-iminium salts and between these salts and dimethylformamide. The authors state that data obtained from FT or CW spectra are consistent with each other, and one example is given. However, the uncertainties in the activation parameters, especially $E_a$ and $\Delta H^{\ddagger}$, seem to be unusually large. Computerized fitting of calculated to experimental bandshapes was utilized.

For the first time, as far as the Reporters are aware, a well-resolved doublet, due to slow exchange, has been observed in $^{23}$Na n.m.r.[106] Total bandshape analysis was used to obtain the rate constant for the sodium exchange between

[104] E. Alaluf, K. J. Alford, E. O. Bishop, and J. D. Smith, *J.C.S. Dalton*, 1974, 669.
[105] C. J. Martin and S. Poignant, *J.C.S. Perkin II*, 1974, 642.
[106] J. M. Ceraso and J. L. Dye, *J. Amer. Chem. Soc.*, 1973, **95**, 4432.

free ions and ions bound to a cryptate molecule. The authors conclude that the rate-determining step is the dissociation of the complex.

Variable-temperature $^1H$ n.m.r. studies on 5-phenyl-3,7-dimethyldibenziodole (38) have shown that an isomerization process is taking place at room

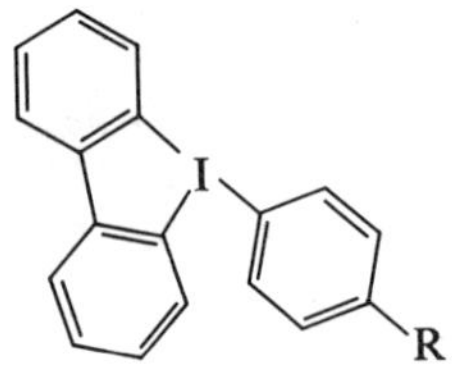

(38) R = H or $CF_3$

temperature.[107] From complete bandshape analysis at several temperatures, a strongly negative entropy of activation was obtained, indicating that the process is intermolecular. This conclusion has, however, not been confirmed since the rate of exchange is not concentration-dependent and no catalyst for this process was found.

(iv) *Biochemical Applications*. In this section a few papers dealing with biochemical processes will be briefly discussed. No real bandshape analysis has been performed, but as this method seems not to be often used among biochemists, the Reporters feel it useful to mention some of the papers in this area.

There is a current discussion in the literature as to whether $H_2CO_3$ or $HCO_3^-$ is the hydration product from the action of carbonic anhydrase on $CO_2$. Koenig *et al.*[108] have now used $^{13}C$ line-broadening to study this reaction, and their data are in agreement with the suggestion that $H_2CO_3$, and not $HCO_3^-$, is the hydration product.

The electron-transfer reaction in cytochrome *c* was studied by Stellwagen and Shulman,[109] who observed the line-broadening of several signals as a function of added $Fe^{III}$ and temperature.

Haar *et al.*[110] studied nucleotide–ribonuclease A complexes using $^{31}P$ n.m.r. For cytidine 2′-phosphate they observed a broad signal in the presence of ribonuclease A in a lower concentration than that of the nucleotide, whereas cytidine 3′-phosphate gave a relatively narrow signal even in the presence of the protein. For both nucleotides the shift changed with the addition of ribonuclease A. These observations show that the exchange between protein-bound and free nucleotide is fast on the n.m.r. time-scale for cytidine 3′-phosphate, whereas this rate is sufficiently slow to be determined for cytidine 2′-phosphate.

107 H. J. Reich and C. S. Cooperman, *J. Amer. Chem. Soc.*, 1973, **95,** 5077.
108 S. H. Koenig, R. D. Brown, T. E. Needham, and N. A. Matwiyoff, *Biochem. Biophys. Res. Comm.*, 1973, **53,** 624.
109 E. Stellwagen and R. G. Shulman, *J. Mol. Biol.*, 1973, **80,** 554.
110 W. Haar, J. C. Thompson, W. Maurer, and H. Rüterjans, *European J. Biochem.*, 1973, **40,** 259.

**D. Mechanistic Studies in Inorganic Chemistry.**—It has previously been observed that at least two exchange processes take place in the complex $(C_5H_5)_2Fe_2(CO)_4$:[111] (i) exchange between bridge and terminal CO groups in the *trans*-isomer, (ii) *cis–trans* equilibration, and maybe (iii) bridge–terminal CO group exchange in the *cis*-isomer. Adams and Cotton[112,113] have now proposed a mechanism for the exchange processes in this type of dinuclear metal carbonyl complex (39), which is consistent with all known experimental observations, mostly n.m.r. bandshape studies. The proposed mechanism explains very nicely the various dynamic n.m.r. spectra from three different complexes, $(C_5H_5)_2Fe_2(CO)_4$, $(C_5H_5)_2Fe_2(CO)_2(CNMe)_2$, and $(C_5H_5)_2$-$Mn_2(CO)_2(NO)_2$. In the last-mentioned complex all three possible exchange processes mentioned above could be observed.[114] It has also been shown that

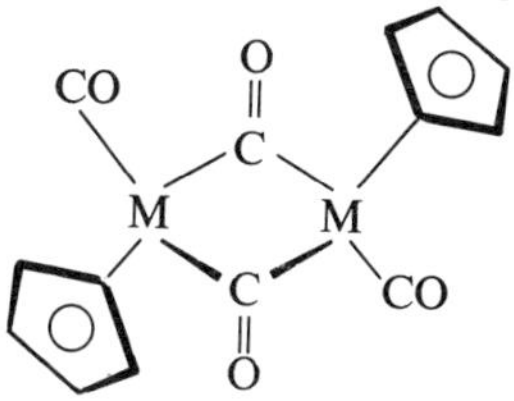


(39)

the *cis–trans* isomerization in $(C_5H_5)_2Fe_2(CO)_3(GeMe_2)$ is consistent with the same mechanism.[115] The $^1H$ dynamic n.m.r. spectra of $(C_5H_5)_2Cr_2(NO)_2$ have been reported, but these are too simple to give any mechanistic information.[114] It would be necessary to use $^{15}N$ n.m.r. in this case to gain further information. Cotton and co-workers[116] discussed the mechanism of intramolecular ligand scrambling in $(C_5H_5)_2Mo_2(CO)_3CNMe$. The solution of this complex is shown to exist as a mixture of non-bridged isomers, and the exchange among them is assumed to take place *via* a bridged intermediate or transition state.

Cotton and co-workers[117—119] studied the structure of some carbonyl complexes of the type $(C_5H_5)_2M_2(CO)_6$ and estimated the barrier to rotation about the metal–metal bond. The barrier increases with decreasing bond length, in the order $Cr < Mo < W$.

Some attempts have been made to study the mechanism of carbonyl exchange in tetranuclear metal–carbonyl complexes.[120—122] Evans *et al.*[120]

111 O. A. Gansow, A. R. Burke, and W. D. Vernon, *J. Amer. Chem. Soc.*, 1972, **94**, 2550.
112 R. D. Adams and F. A. Cotton, *J. Amer. Chem. Soc.*, 1973, **95**, 6589.
113 F. A. Cotton, *Bull. Soc. chim. France*, 1973, 2587.
114 R. M. Kirchner, T. J. Marks, J. S. Kristoff, and J. A. Ibers, *J. Amer. Chem. Soc.*, 1973, **95**, 6602.
115 R. D. Adams, M. D. Brice, and F. A. Cotton, *Inorg. Chem.*, 1974, **13**, 1080.
116 R. D. Adams, M. D. Brice, and F. A. Cotton, *J. Amer. Chem. Soc.*, 1973, **95**, 6594.
117 R. D. Adams and F. A. Cotton, *Inorg. Chim. Acta*, 1973, **7**, 153.
118 R. D. Adams, D. M. Collins, and F. A. Cotton, *J. Amer. Chem. Soc.*, 1974, **96**, 749.
119 R. D. Adams, D. M. Collins, and F. A. Cotton, *Inorg. Chem.*, 1974, **13**, 1086.
120 J. Evans, B. F. G. Johnson, J. Lewis, and J. R. Norton, *J.C.S. Chem. Comm.*, 1973, 807.
121 B. F. G. Johnson, J. Lewis, and T. W. Matheson, *J.C.S. Chem. Comm.*, 1974, 441.
122 P. E. Cattermole, K. G. Orrell, and A. G. Osborne, *J.C.S. Dalton*, 1974, 328.

showed that the $^{13}C$ bandshape of $Rh_4(CO)_{12}$ is consistent with a CO scrambling mechanism, which has a fully symmetric structure as an intermediate, and Johnson *et al.*[121] observed for $RhCo_3(CO)_{12}$ that two different processes are necessary to get complete scrambling of the carbonyl groups. One is a low-temperature process which exchanges all carbonyl groups bound to Co, and the other is a high-temperature process, in which all twelve CO groups participate.

Fluxional behaviour in various complexes between metal carbonyls and conjugated organic ligands has been observed. Carty *et al.*[123] used proton total bandshape analysis to study the low-temperature exchange process in (40). The authors found no isotope effect when the NH proton was exchanged for deuterium. The reported activation energy is similar to the values reported for (41) and other systems.[124—126]

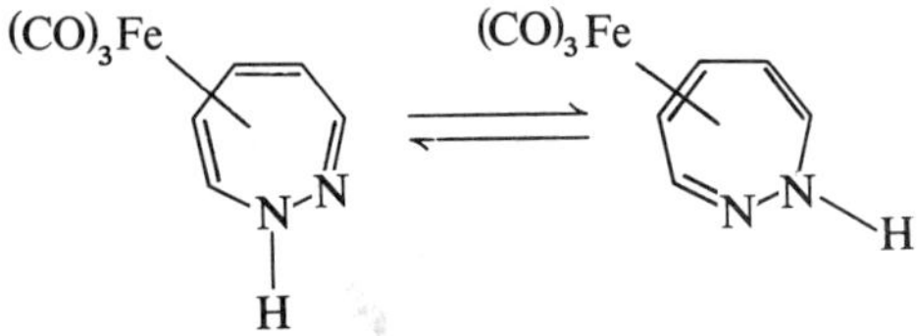


(40)

(41)

The intramolecular rearrangement process in some $ML_5$ complexes has been studied by Meakin and Jesson[17] and by Muetterties and Hirsekorn.[127] Meakin and Jesson performed a very careful analysis of the $^{31}P-\{^1H\}$ dynamic n.m.r. FT spectra for $[Rh\{P(OMe)_3\}_5]^+[BPh_4]^-$. The density-matrix program used by them to handle this $A_3B_2X$ system is discussed in Section 3A. The authors performed two calculations assuming different permutational sets: (i) processes involving simultaneous exchange of the axial ligands with two of the equatorial ligands and (ii) processes involving single axial–equatorial exchanges. The spectra calculated utilizing these two permutational sets showed only small differences at some exchange rates, with a slightly better fit between process (i) bandshapes and experimental spectra than for (ii). Jesson and Meakin[128] reported results for some other $ML_5$ complexes which are in agreement with the same permutational set.

The mechanistic bandshape work on complexes containing *NN*-disubstituted dithiocarbamato-ligands has continued. Pignolet and co-workers[129,130]

[123] A. J. Carty, R. F. Hobson, H. A. Patel, and V. Snieckus, *J. Amer. Chem. Soc.*, 1973, **95**, 6835.
[124] C. G. Kreiter and M. Lang, *J. Organometallic Chem.*, 1973, **55**, C27.
[125] C. G. Kreiter, S. Stüber, and L. Wackerle, *J. Organometallic Chem.*, 1974, **66**, C49.
[126] T. H. Whitesides and R. A. Budnik, *J.C.S. Chem. Comm.*, 1974, 302.
[127] E. L. Muetterties and F. J. Hirsekorn, *J. Amer. Chem. Soc.*, 1973, **95**, 5419.
[128] J. P. Jesson and P. Meakin, *Inorg. Nuclear Chem. Letters*, 1973, **9**, 1221.
[129] M. C. Palazzotto, D. J. Duffy, B. L. Edgar, L. Que, jun., and L. H. Pignolet, *J. Amer. Chem. Soc.*, 1973, **95**, 4537.
[130] L. Que, jun. and L. H. Pignolet, *Inorg. Chem.*, 1974, **13**, 351.

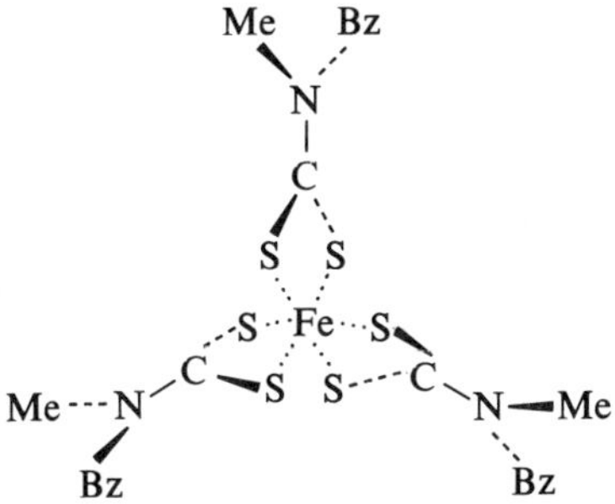

(42) a; $Fe^{III}$
b; $Fe^{IV}$,$BF_4$

have performed a careful examination of the spectra of several such complexes. They were able to show that the only reasonable mechanism for metal-centred rearrangement in (42) that is consistent with $^1H$ n.m.r. bandshapes at various temperatures is the trigonal twist pathway in which all diastereotopic pairs are averaged. The other $Fe^{III}$ complexes investigated have very similar $\Delta G^{\ddagger}$ values, so Pignolet assumes that the same mechanism is valid for them. The same mechanism was also assumed to be valid for the $Fe^{IV}$ complexes, since this had previously been shown to be the case for (42b). For the $Fe^{II}$ complexes, such detailed mechanistic information could not be obtained since this metal-centred rearrangement was found to be fast on the n.m.r. time-scale down to $-95$ °C.

Tolman *et al.*[131] utilized $^1H$ and $^{31}P$ n.m.r. to study the ligand exchange in RhCl[P($p$-tolyl)$_3$]$_3$ and $H_2$RhCl[P($p$-tolyl)$_3$]$_3$. Total bandshape analysis of the $^{31}P-\{^1H\}$ spectra is reported for two different mechanisms: either the degree of phosphine dissociation is negligible or a fraction of the complex has dissociated. The agreement between calculated and experimental spectra is obviously better using the second assumption. It was also shown that only one of the three phosphine groups was exchangeable.

Meakin *et al.*[132] studied the ligand-exchange processes in platinum, palladium, and nickel hydride phosphine complexes ($HM[PEt_3]_3^+$) by $^1H$ and $^{31}P$ n.m.r. Two processes could be observed: intermolecular exchange of two of the three phosphine groups and an intramolecular rearrangement. The relative rates of these processes change from one metal to another. The rate constant for the intramolecular rearrangement process increases in the order Pt < Pd < Ni.

Chan and Reeves[133] have continued their work on halide exchange in tin halides. They used computerized fitting of calculated to experimental spectra. The exchange is called a ten-site exchange, but it is in reality five overlapping

131 C. A. Tolman, P. Meakin, D. L. Lindner, and J. P. Jesson, *J. Amer. Chem. Soc.*, 1974, **96**, 2762.
132 P. Meakin, R. A. Schum, and J. P. Jesson, *J. Amer. Chem. Soc.*, 1974, **96**, 277.
133 S. O. Chan and L. W. Reeves, *Inorg. Chem.*, 1973, **12**, 1704.

two-site exchanges. The authors present a mechanism for the halogen exchange that is in agreement with the experimental results.

## 4 Relaxation Effects: Linewidths

**A. Theoretical Work.**—Goodwin and Wallace, in a series of papers,[134—136] have solved the equation of motion for the spin-density matrix using numerical methods. This approach enables them to avoid the steady-state approximation, and they can consequently account for the effect of finite sweep times on the bandshape. The first two articles in the series were reviewed in last year's Report. In the third article,[136] the authors include relaxation effects through the Bloch–Wangsness–Redfield formalism. Using this consistent description of relaxation effects, Goodwin and Wallace calculate how the bandshape for an AB spin system varies with the strength of the $B_1$ field for different relaxation models. The numerical method has in this case a definite advantage compared with, for example, the bandshape calculations by Hoffman,[137] since it is experimentally very difficult to achieve steady-state conditions at high r.f. fields. The question posed in last year's Report concerning the computer time required for the calculations remains unanswered, however.

In a paper dealing mainly with saturation effects in Fourier spectroscopy, Ernst and Morgan[6] also treat saturation effects in CW spectra. They present a theoretical analysis of the saturation behaviour of the bandshape for different relaxation models in an AB and an $AX_2$ system. It is found in an experimental investigation that the relaxation of the protons in 1,1,2-trichloroethane (in a sample doped with an organic radical relaxation catalyst) can be described using the random-field relaxation mechanism with complete correlation of the random fields at the protons.

The effect of internal rotations in organic radicals on the proton linewidths has been studied by de Boer *et al.*[138] It is well known from previous studies[139] that fast internal rotations must be included in a consistent analysis of relaxation times. The specific case of a paramagnetic molecule has, however, not been treated previously. It is found, as expected, that the linewidths depend on the internal rotation correlation time $\tau_m$ only in the region where $\tau_m$ is of the same order of magnitude as the total motional correlation time, which is effectively either the electron-spin relaxation time or the molecular tumbling correlation time. The authors[138] claim that the linewidths are always larger than or equal to the linewidth in the absence of internal rotations. The Reporters regard such a conclusion with scepticism, and can find no support for this claim in the calculations presented. In the experimental part of the paper, de Boer *et al.* discuss the linewidths of the proton signals in 3,3′-

134 B. W. Goodwin and R. Wallace, *J. Magn. Resonance*, 1972, **8**, 41.
135 B. W. Goodwin and R. Wallace, *J. Magn. Resonance*, 1973, **9**, 280.
136 B. W. Goodwin and R. Wallace, *J. Magn. Resonance*, 1973, **12**, 60.
137 R. A. Hoffman, *Adv. Magn. Resonance*, 1970, **4**, 87.
138 J. W. M. de Boer, M. R. Arick, and E. de Boer, *J. Chem. Phys.*, 1973, **59**, 638.
139 D. Wallach, *J. Chem. Phys.*, 1967, **47**, 5258.

dimethylbiphenyl. They determined the correlation times for the dipolar and scalar coupling relaxation mechanisms from the linewidths of the ring protons. They were then able to calculate the linewidths for the methyl protons assuming either slow or very fast methyl rotation. It was found that the experimental values were consistent with a fast ($\tau_m < 5 \times 10^{-12}$ s) methyl rotation.

The intricate question of the source of the $^1$H n.m.r. linewidths of sonicated lecithin dispersions has been discussed in three papers during the past year.[140—142] In thermodynamic equilibrium, lecithin–water systems are lamellar, and the n.m.r. spectrum is dominated by the effect of static dipolar couplings. When the samples are sonicated, vesicles are formed, and reasonably resolved n.m.r. spectra are observed. Chan and co-workers[140,141] conclude that the molecular motions in the lamellae and in the vesicles are markedly different. The main support for this conclusion is that a multiple-pulse experiment gives a linewidth of 100 Hz for the $CH_2$ resonance in the lamellae, whereas the corresponding linewidth is considerably smaller in the sonicated sample. The details of the multiple-pulse experiment have, however, not yet been published, and it seems doubtful that such an experiment gives the natural linewidth. Finer,[142] on the other hand, gives a theoretical analysis which shows that the linewidths in vesicle solutions are compatible with a model in which both the slow overall motion of the vesicles and the faster local motion contribute to the observed linewidth. Finer's discussion appears simple and convincing.

**B. Chemical Exchange Rates Determined by the Swift–Connick Approach.**— The Swift–Connick method[143] of determining chemical exchange rates continues to be a useful complement to the more accurate total bandshape analysis when the exchange is fast or when a bandshape analysis is not feasible for other reasons. The main application of the Swift–Connick method is to ligand exchange of paramagnetic transition-metal complexes.

Plotkin *et al.*[144] were able to observe the resonances from both solvent and co-ordinated methanol in the $[Co(MeOH)_5py]^{2+}$ complex. They were thus able to determine the rate constant for the ligand–solvent exchange by two independent methods, and good agreement was found. They also found that the ligand *trans* to the pyridine exchanges faster by a factor of three than the *cis* ligands. A notable feature in the investigation is the careful measurement of linewidths and shifts. The spectrum was recorded digitally by means of 1000 data points, which were analysed assuming Lorentzian shapes. A correction for incorrect phasing was also made. In this way a considerably improved accuracy of the measured shifts and linewidths was obtained. Five-co-ordinate transition-metal complexes contain two different ligand sites, and there is a potential possibility to observe two distinct ligand–solvent exchange

140 C. H. A. Seiter and S. I. Chan, *J. Amer. Chem. Soc.*, 1973, **95**, 7541.
141 G. W. Feigenson and S. I. Chan, *J. Amer. Chem. Soc.*, 1974, **96**, 1312.
142 E. G. Finer, *J. Magn. Resonance*, 1974, **13**, 76.
143 T. J. Swift and R. E. Connick, *J. Chem. Phys.*, 1962, **37**, 307; 1964, **41**, 2553.
144 K. Plotkin, J. Copes, and J. R. Vriesenga, *Inorg. Chem.*, 1973, **12**, 1494.

processes. Enriquez and Zumdahl[145] claim that they were able to observe two such processes for the complex $[Co(luno)_5]^{2+}$ (luno = 2,6-dimethylpyridine 1-oxide) in acetone solution with excess luno. The axial ligands are found to exchange rapidly with the free ligands over most of the temperature interval studied, while the exchange of the equatorial ligands goes from nearly fast exchange to slow exchange, and Enriquez and Zumdahl were able to measure the exchange rates for this process.

Lincoln and co-workers, in a series of papers,[146—148] have studied ligand exchange through measuring $^{14}N$ n.m.r. linewidths. The first paper[146] treats pyridine exchange in cobalt and nickel complexes with two acetylacetone and two pyridine ligands. The observed broadening of the $^{14}N$ signal is found to be caused by the shift difference between ligand and free pyridine. In the second study,[147] West and Lincoln investigated the acetonitrile exchange in copper(II) complexes. It is found that the exchange is too fast to give any detectable effects in the $[Cu(MeCN)_6]^{2+}$ complex. A substitution of four of the MeCN ligands for a multidentate ligand gives a marked decrease in the MeCN exchange rate, so that kinetic parameters could be determined. In this case shift differences between the bulk and ligand acetonitrile were neglected. The third paper[148] deals with the related subject of acetonitrile exchange in nickel(II) complexes of one terdentate and three acetonitrile ligands.

Kluiber *et al.*[149] studied ligand exchange of substituted pyridines in nickel(II) complexes analogous to those studied by Lincoln.[146] The introduction of a substituent in the 4-position in the pyridine ring results in a first-order proton resonance spectrum, and the exchange could be followed using n.m.r. The kinetic data were mainly derived from the measured shifts, but they were checked through the observed linewidths. Kluiber *et al.* were able to conclude that the exchange occurred *via* a dissociative mechanism by making the measurements with only a slight excess of ligand.

The activation parameters $\Delta H^{\ddagger}$ and $\Delta S^{\ddagger}$ were evaluated in all of the articles referred to above, and the value of $\Delta S^{\ddagger}$ was sometimes used in the discussion of possible reaction mechanisms. It is, however, notoriously difficult to decompose a measured temperature dependence of a rate constant into values for $\Delta H^{\ddagger}$ and $\Delta S^{\ddagger}$. This procedure requires accurately determined rate constants free from even small systematic errors. Approximations which are justified for the determination of the rate constants themselves can introduce serious errors when $\Delta H^{\ddagger}$ and $\Delta S^{\ddagger}$ are evaluated. A possible illustration of these dangers can be found in the work of Kluiber *et al.*[149] They determined exchange rates for a number of differently substituted ligands. The rate constant for the exchange at 25 °C was found to be in the interval 0.26—$1.33 \times 10^5\ s^{-1}$ for the eight cases studied, while $\Delta H^{\ddagger}$ and $\Delta S^{\ddagger}$ ranged from 36.5 to 50.0 kJ mol$^{-1}$ and from $-30.1$ to $+18.4$ J mol$^{-1}$ K$^{-1}$, respectively.

[145] P. M. Enriquez and S. S. Zumdahl, *Inorg. Chem.*, 1973, **12**, 2475.
[146] S. F. Lincoln, *J.C.S. Dalton*, 1973, 1896.
[147] R. J. West and S. F. Lincoln, *J.C.S. Dalton*, 1974, 281.
[148] S. F. Lincoln and R. J. West, *J. Amer. Chem. Soc.*, 1974, **96**, 400.
[149] R. W. Kluiber, R. Kukla, and F. Thaller, *J. Coordination Chem.*, 1973, **3**, 39.

The Reporters find it difficult to have confidence in these values of $\Delta H^{\ddagger}$ and $\Delta S^{\ddagger}$ before a careful error analysis has been performed. Such an analysis should preferably take into account not only the experimental errors in linewidths, shifts, and concentrations, but also errors introduced by inadequacies in the model used for analysing the data, *e.g.* the presence of additional complexes in small concentrations, neglect of outer-sphere co-ordination, and errors in the linewidths and shifts in the absence of exchange.

In a paper on relaxation effects in $^{13}C$ FT spectra of adenosine triphosphate, ATP, Lam *et al.*[150] give an example of the possibility of using the Swift–Connick method in investigations of biologically interesting systems. They give a detailed derivation of the Swift–Connick-type equations for a three-site exchange a $\rightleftharpoons$ b $\rightleftharpoons$ c with no direct a $\rightleftharpoons$ c exchange. The equations derived are used to analyse $^{13}C$ $T_1$ and $T_2$ data for a solution containing ATP and $Mn^{2+}$ ions. It was found that the manganese ion could bind either to the phosphate group or to the nitrogens in the adenine moiety. The intramolecular exchange of $Mn^{2+}$ between these two sites was found to be the rate-determining step in the exchange process observed through the relaxation times.

The Swift–Connick relations have also been used to study the exchange of quadrupolar nuclei. Dodgen *et al.*[151] reinvestigated the exchange between chloride ions and chlorine in aqueous solutions. This system had previously been studied by Hall *et al.*,[152] also using $^{35}Cl$ n.m.r. The interpretations of the two investigations are not in agreement, and it is not even clear which exchange process gives the observed enhancement in $^{35}Cl$ relaxation. In an earlier study of the corresponding iodide–iodine system, Genser and Connick[153] found that the exchange $I^{*-} + I_3^- \rightleftharpoons I_3^{*-} + I^-$ accounted for the $^{127}I$ relaxation data. The $^{35}Cl$ relaxation data do not, however, seem to be fully compatible with the analogous description of the exchange process.

**C. Linewidth Determinations in Paramagnetic Systems.**—The analysis of nuclear magnetic relaxation data from paramagnetic systems is complicated by the possibility of several different relaxation mechanisms. It is, however, often possible to extract useful information through a careful analysis of the experimental data. La Mar and Walker[154] measured $^1H$ magnetic resonance linewidths for a series of metalloporphyrins in high spin states. They performed the experiments with a constant porphyrin skeleton but varied the metal ion and the apical ligand. It was found that the linewidths depend strongly on the metal ion, and for the Mn and Fe systems also on the apical ligand. It was concluded that the proton relaxation is determined by the hyperfine coupling interaction for the Fe, and partly also for the Mn porphyrin, whereas the dipolar coupling gives the dominant contribution to the

150 Y.-F. Lam, G. P. P. Kuntz, and G. Kotowycz, *J. Amer. Chem. Soc.*, 1974, **96**, 1834.
151 H. W. Dodgen, A. D. Jordan, and R. B. Jordan, *J. Phys. Chem.*, 1973, **77**, 2149.
152 C. Hall, D. W. Crydon, R. E. Richards, and R. R. Sharp, *Proc. Roy. Soc.*, 1970, **A318**, 119.
153 E. E. Genser and R. E. Connick, *J. Chem. Phys.*, 1973, **58**, 990.
154 G. N. La Mar and F. A. Walker, *J. Amer. Chem. Soc.*, 1973, **95**, 6950.

relaxation in the Cr porphyrins. The apical ligand has a marked influence on the zero-field splitting parameter, which in turn determines the electron relaxation time. La Mar and Walker[154] point out that one can obtain a reasonably high-resolution n.m.r. spectrum of the porphyrins by choosing an apical ligand that gives a large zero-field splitting.

Canters and de Boer[155] present a theoretical and experimental investigation of the linewidths of the resonances of the counterions in organic radical-anion–alkali-metal-cation ion pairs. They consider contributions to the relaxation from the hyperfine coupling and the (electron spin, nuclear spin) dipole–dipole interaction, as well as from the quadrupole interaction. They are able to estimate the importance of the different relaxation mechanisms by making linewidth determinations for seven different alkali-metal isotopes. The linewidth data are also analysed with respect to two different models for the ion pairs, and it is found that a static model with a single ion-pair species is in best accordance with the experimental data.

**D. Effects of Scalar Coupling to Quadrupolar Nuclei.**—During the past year comparatively little work has been done on the n.m.r. spectra of spin-$\frac{1}{2}$ nuclei with scalar couplings to quadrupolar nuclei. The most important contribution in the field was made by Blackborow,[156] who studied adducts between tertiary amines and boron halides by both $^{1}H$ and $^{11}B$ magnetic resonance. It was possible to resolve a ($^{1}H$,$^{11}B$) scalar coupling at intermediate temperatures. Lowering the temperature resulted in a collapse of the spin multiplet due to $T_1$ relaxation of $^{11}B$, while an increase in temperature gave multiplet collapse due to intermolecular chemical exchange. Blackborow included both chemical exchange and $T_1$ effects in a total bandshape calculation and was thus able to determine exchange rates accurately. The matrix used in the bandshape calculation is given in the paper. The proton spectrum studied is a methyl signal with no scalar couplings except to the boron. This makes the inclusion of both chemical exchange and $T_1$ relaxation of the quadrupolar nucleus straightforward. More complex cases with proton–proton couplings are best treated using the Liouville representation.[15]

Bula and Hartman[157] investigated the chemically analogous problem of adducts between boron trihalides and dimethyl sulphide. The bandshape of the methyl signal varied with temperature in such a way that the quartet due to coupling to $^{11}B$ coalesced at both high and low temperatures in the same way as found by Blackborow. The boron relaxation rate was found to decrease in the sequence $BCl_3 > BBr_3 > BI_3$, and the same trend was found for the chemical exchange rates for the breaking of the donor–acceptor bond. No quantitative account for the bandshape was given. A similar study was performed by Bravo *et al.*[158] on the donor–acceptor adducts between $BF_3$ and

[155] C. W. Canters and E. de Boer, *Mol. Phys.*, 1974, **27**, 665.
[156] J. R. Blackborow, *J.C.S. Dalton*, 1973, 2193.
[157] M. J. Bula and J. S. Hartman, *J.C.S. Dalton*, 1973, 1047.
[158] R. Bravo, M. Duran, and J.-P. Laurent, *Org. Magn. Resonance*, 1973, **5**, 357.

sulphoxides, amine oxides, phosphine oxides, and arsine oxides. The observed bandshapes were only interpreted qualitatively.

Fronza *et al.*[159] investigated the ratio between the linewidths of the individual peaks in multiplets due to coupling to spins with $I \geqslant 1$. It is often found that the central peak(s) are broader than the outer peaks when a paramagnetic relaxation catalyst is added to the solution. The authors point out that this observation is not consistent with a quadrupole relaxation mechanism, but that it can be explained in terms of a relaxation of the spin with $I \geqslant 1$ through a scalar coupling to the unpaired electron. It is, for example, easily seen when $I = 1$ that the state with $m = 0$ has half the lifetime of the states $m = \pm 1$ for a general random-field relaxation mechanism. No selective broadening of the central peak was observed in the $^{13}C$ spectra of $(CD_3)_2CO$, $CD_3OD$, and $CD_3I$. This finding was stated to be in accordance with an electron dipole–nuclear dipole relaxation mechanism in these compounds, but the authors give no detailed arguments on this point.

[159] G. Fronza, R. Mondelli, and E. W. Randall, *J.C.S. Chem. Comm.*, 1974, 195.

# 5
# Multiple Resonance

BY W. McFARLANE AND D. S. RYCROFT

## 1 Introduction

The use made of multiple-resonance techniques has continued to increase although the rate of new developments has slowed down: this chapter therefore covers a two-year period—from June 1972 to May 1974. The coverage is confined to nuclear systems, and topics such as bandshape analysis and solid-state work which are more appropriately dealt with in other chapters are referred to only briefly. Similarly, routine applications (particularly those involving $^{1}H-^{1}\{H\}$, $^{13}C-\{^{1}H\}$, $^{31}P-\{^{1}H\}$, and $^{1}H-\{^{31}P\}$ for spectral simplification) have been generally ignored, although there is some mention of work involving less usual combinations of nuclei.

During the period under review several review articles on various aspects of magnetic multiple resonance have appeared.[1—6]

## 2 Theory of Multiple Resonance

The theory of multiple resonance was satisfactorily developed in the 1960's; recent developments have been largely confined to modifications needed for liquid-crystal work, to predictions of the results of pulsed experiments, and to more accurate treatment of relaxation effects using the density matrix.

Continuous-wave off-resonance proton decoupling is now routinely used for assignments in $^{13}C$ spectra: the NOE enhancement is retained and the number of protons attached to a particular carbon can be counted from the multiplicity of the residual splitting of its resonance. However, if this residual splitting is comparable with coupling between the decoupled protons and other protons in the molecule then a case of deceptive simplicity (strictly analogous to 'virtual coupling') may arise and the $^{13}C$ spectrum may be unexpectedly complex.[7] This has been demonstrated[7] for 1,2-dichloroethane

[1] (*a*) J. D. Memory, *Magn. Resonance Rev.*, 1972, **1,** 91; (*b*), *ibid.*, p. 185; (*c*), *ibid.* p. 333.
[2] R. L. Mieher, *Magn. Resonance Rev.*, 1972, **1,** 225.
[3] (*a*) L. R. Dalton, *Magn. Resonance Rev.*, 1972, **1,** 301; (*b*) L. R. Dalton and L. A. Dalton, *ibid.*, 1973, **2,** 361.
[4] W. McFarlane, in 'N.M.R. Spectroscopy of Nuclei other than Protons', ed. T. Axenrod and G. A. Webb, John Wiley, New York, 1974, Ch. 3, pp. 31–56.
[5] W. McFarlane, *Ann. Reports N.M.R. Spectroscopy*, 1972, **5A,** 353.
[6] D. Schmid, *Springer Tracts Modern Phys.*, 1973, **68,** 1.
[7] R. A. Newmark and J. R. Hill, *J. Amer. Chem. Soc.*, 1973, **95,** 4435.

and poly(*cis*-butadiene) and clearly is a possibility in many other molecules. (See also the section on Applications of $^{13}C$–{$^{1}H$} Decoupling, p. 183.)

The computer simulation of multiple-pulse experiments has been described by Meakin and Jesson in three papers.[8–10] In the first,[8] based on the Bloch equations, it was shown that the computer program can calculate the effect of any sequence of pulses and delays upon the magnetization vector using comparatively small amounts of computer time. However, tightly coupled spin systems cannot be dealt with, and the details of the relaxation processes were not unravelled. In the second,[9] the density matrix was used and specific relaxation processes were introduced. An important result from the simulation of random decoupling on an AX spin system suggests that when the magnetic field inhomogeneity dominates the linewidth (*i.e.* $1/T_2 \ll 1/T_2'$) an appropriate choice of decoupling power will permit observation of the NOE enhancement without any accompanying decoupling. This approach could be simpler than an alternately pulsed experiment, especially for $^{13}C$–{$^{1}H$} work. The effects of selective and non-selective pulses were also considered for AX and AB spin systems. In the third paper[10] the effects of a repetitive chain of pulses were considered and it was shown that gated irradiation experiments can give an NOE enhancement without any accompanying decoupling.

Triple-resonance experiments may be used to study spin relaxation in AX and AB spin systems. If two of the lines are simultaneously saturated the recovery rate of a third will depend upon the various transition probabilities. By selecting appropriate combinations of observed and saturated lines it is possible to get a set of four independent simultaneous equations which can be solved to give the individual transition probabilities and hence the details of the relaxation mechanism.[11] This technique has been applied to the proton resonance of dichloroacetaldehyde (AX), and similar experiments based on the NOE, rather than saturation recovery rates, have been used to study relaxation in 4-methyl-1,3-dioxan (AB).[12]

Goodwin and Wallace[13] have presented methods of solving the density matrix based upon a sophisticated description of the relaxation processes and have shown that the results are especially valuable when a finite scan rate produces distortions or when transient effects are important. Numerical solutions were obtained, and detailed predictions of the results of double-resonance experiments upon an AB spin system were presented. Good agreement with experiment was found for weak irradiating fields, but for strong fields there were some discrepancies. The possibility of distinguishing different types of relaxation mechanism by means of double-resonance experiments was shown.

[8] P. Meakin and J. P. Jesson, *J. Magn. Resonance*, 1973, **10**, 290.
[9] P. Meakin and J. P. Jesson, *J. Magn. Resonance*, 1973, **11**, 182.
[10] P. Meakin and J. P. Jesson, *J. Magn. Resonance*, 1974, **13**, 354.
[11] J.-C. Duplan, A. Briguet, and J. Delmau, *J. Magn. Resonance*, 1973, **12**, 270.
[12] J.-C. Duplan, C. Chapelet-Barbier, and J. Delmau, *Mol. Phys.*, 1972, **23**, 609.
[13] B. W. Goodwin and R. Wallace, *J. Magn. Resonance*, 1973, **12**, 60.

The same authors have also given a general method, using the density matrix, for the calculation of n.m.r. spectra obtained under any conditions (*i.e.* not merely steady state); multiple-resonance spectra were considered in the treatment.[14]

Difficulties can arise in the labelling of transitions and energy levels in tightly coupled spin systems in the limits of complete mixing of a pair of wavefunctions. These can also arise in double-resonance experiments[15] and may be avoided by an appropriate choice of wavefunctions.[16]

The applications of multiple-resonance experiments to partially oriented molecules should be straightforward, and the theoretical principles involved have been discussed in three papers. In an extension[17] of early work by Freeman and Anderson[18] on isotropic systems the double resonance spectra have been calculated for $A_n$–{X}, A–{$X_n$}, and $A_3$–{$X_2$} systems with $I_A = I_X = \frac{1}{2}$, and for A–{X} and $A_3$–{X} systems with $I_A = \frac{1}{2}$, $I_X = 1$. The experimental part of the paper[17] is confined to decoupling, which was achieved to varying extents in $CF_3CH_2OH$, $CHD_2OH$, $CH_3CD_2OH$, and $C_6D_5COCH_3$ in different nematic phases. The prediction has been made that improved decoupling would be achieved by phase modulation of the decoupling r.f. field;[19] this was achieved to some extent in partially oriented $CH_3OD$ in which the deuterium spectrum is in two bands separated by 24 000 Hz owing to the large $^2H$ nuclear quadrupole interaction. Snyder and Meiboom[20] have also considered the theory of decoupling deuterium in anisotropic systems and have shown that efficient decoupling can be achieved using a strong ($\gamma_D B_2/2\pi \sim 2700$ Hz) r.f. field centred on a double-quantum transition in the deuterium spectrum. The effects of lower r.f. amplitudes and frequency offsets have also been considered, and good agreement with experimental results has been found.[21] The value of modulation schemes was also examined, but these were found to be less effective.

Decoupling of one nucleus can also affect the apparent relaxation time of another. A theoretical treatment[22] of $^{13}C$ spin–lattice relaxation in the presence or absence of proton irradiation shows that the recovery of $^{13}C$ magnetization after a spin population inversion is not accurately exponential unless the protons are decoupled. Terms involving proton transition probabilities must also be considered, and an apparent $^{13}C$ relaxation time which is significantly different from that measured under proton-decoupling conditions can be derived. This apparent $T_1$ also depends upon the method of measurement (*e.g.* pulse or adiabatic rapid passage) and the theory is illustrated by experimental results for benzene containing different amounts of dissolved

[14] B. W. Goodwin and R. Wallace, *J. Magn. Resonance*, 1972, **8**, 41.
[15] R. A. Hoffman and S. Forsén, *Progr. N.M.R. Spectroscopy*, 1966, **1**, 15.
[16] T. J. Batterham and C. J. Bigum, *Org. Magn. Resonance*, 1972, **4**, 67.
[17] J. W. Emsley, J. C. Lindon, and J. M. Tabony, *J.C.S. Faraday II*, 1973, **69**, 10.
[18] R. A. Freeman and W. A. Anderson, *J. Chem. Phys.*, 1962, **37**, 85.
[19] W. A. Anderson and F. A. Nelson, *J. Chem. Phys.*, 1965, **42**, 1199.
[20] L. C. Snyder and S. Meiboom, *J. Chem. Phys.*, 1973, **58**, 5096.
[21] R. C. Hewitt, S. Meiboom, and L. C. Snyder, *J. Chem. Phys.*, 1973, **58**, 5089.
[22] T. D. Alger, R. Freeman, and D. M. Grant, *J. Chem. Phys.*, 1972, **57**, 2168.

oxygen. Similar treatments applied[23] to relaxation in the $^{13}C$–H system of formic acid and the homonuclear proton system of cytidine indicated that errors in $T_1$ of up to 10% may arise in the absence of proton decoupling. This problem has also been dealt with theoretically by Buchner,[24] who reaches similar conclusions. The effects of paramagnetic species upon NOE enhancements have been studied[25] theoretically: if the observed nuclei are embedded within the molecule while those irradiated are on the periphery then paramagnetic species can reduce or eliminate the NOE enhancement without significantly broadening the observed signal. Such a preferential retention of the NOE enhancement forms a delicate probe into ligand binding sites, and a plot of observed enhancement against the concentration of added paramagnetic species can also give spin–lattice relaxation times.

A detailed treatment of selective NOE experiments on coupled spin systems has been presented;[26] these experiments can be used to study relaxation processes. Experimental results were obtained for vinyl bromide (ABX) and 1,1,3-trichloroethane ($A_2X$), and it was concluded that the selective NOE may be useful when it is difficult to get conformational information from conventional NOE experiments.

Larger NOE's can be obtained in transient than in steady-state experiments, and a method of measuring the effect from signal recovery rates following 180° pulses has been described.[27]

In pulsed spin-decoupling experiments it is tempting to assume that as the pulse repetition period ($T_p$) tends to zero the problem can be analysed in terms of an average perturbation described as a continuous irradiation of intensity $\langle B_2 \rangle = B_2(\tau_p/T_p)$, where $\tau_p$ is the pulse duration.[28] It has been pointed out[29] that this is normally wrong (although it may succeed in a few cases) because it amounts to Fourier-analysing the r.f. pulse chain and discarding the sidebands; in fact the complete interaction between the r.f. and the spin system must be considered. Detailed examples have been described for both homo- and hetero-nuclear cases.[29]

Rotating-frame resonance effects have been observed[30] in the high-resolution $^{13}C$ spectrum of methyl iodide by the use of forced transitory precession techniques. For coherent levels of the proton irradiating field $B_2$ near to the rotating-frame resonance condition $\gamma_C B_1 = \gamma_H B_2$ the spin–spin relaxation time is anomalously fast and the signal loses intensity.

## 3 Experimental Techniques

Most advances in this area have been concerned with applications of various

[23] I. D. Campbell and R. Freeman, *J. Magn. Resonance*, 1973, **11**, 143.
[24] W. Buchner, *J. Magn. Resonance*, 1973, **12**, 82.
[25] H. S. Gutowsky and D. F. S. Natusch, *J. Chem. Phys.*, 1972, **57**, 1203.
[26] N. R. Krishna, P. P. Yang, and S. L. Gordon, *J. Chem. Phys.*, 1973, **58**, 2906.
[27] I. D. Campbell and R. Freeman, *J. Chem. Phys.*, 1973, **58**, 2666.
[28] J. P. Jesson, P. Meakin, and G. Kneissel, *J. Amer. Chem. Soc.*, 1973, **95**, 618.
[29] A. Pines and J. D. Ellett, jun., *J. Amer. Chem. Soc.*, 1973, **95**, 4437.
[30] R. Freeman, H. D. W. Hill, and J. Dadok, *J. Chem. Phys.*, 1973, **58**, 2107.

pulse and FT techniques, although some instrumentation for CW has also been described. Triple-resonance experiments involving three different nuclear species are comparatively rare (excluding here the experiments in which one of the nuclei merely supplies a locking signal, *e.g.* in $^{13}C$–{$^{1}H$} experiments with a deuterium-locked spectrometer) but a versatile system has been described by McFarlane and Rycroft.[31] A Schlumberger FS 30 generated one of the irradiating frequencies, and the other irradiating frequency was obtained from a crystal-controlled r.f. oscillator of which the output was mixed with an audio frequency in a balanced modulator circuit to generate two r.f. sidebands. Some difficulties can arise owing to unfortunate combinations of r.f. frequencies [*e.g.* $\nu(^{77}Se) + 2\nu(^{31}P) \simeq \nu(^{1}H)$] but these can often be mitigated by avoiding audio-modulation in the generation of the second irradiating frequency.

A schematic diagram has been given[32] of the equipment used for $^{1}H$–{$^{13}C$} and $^{1}H$–{$^{1}H$} tickling experiments on formamide under conditions of simultaneous $^{14}N$ decoupling

A modification to the probe of the Varian HA-100 spectrometer for $^{1}H$–{X} double-resonance experiments has been described.[33] Separate 'decoupler boxes' were used in an arrangement similar to that described earlier,[34] and r.f. fields higher than in an earlier adaptation[35] could be generated at the sample.

A versatile multinuclear CW or FT instrument recently described has full facilities for double-resonance experiments.[36] A detailed investigation has been undertaken into sources of instability in spectrometers for double resonance and some conclusions have been drawn, based on experiment.[37] Details have been given of a dual r.f. gating system with adjustable delays which can be used for double-resonance experiments on a FT spectrometer.[38]

It is usually easiest to interpret double-resonance spectra obtained at constant $B_0$, but this difficulty can be avoided on field-sweep instruments by synchronously sweeping $\nu_2$ as the field changes. This system has been adopted by JEOL in some of their instruments ('synchrotrack'), and details of an operational amplifier circuit which will provide this facility for homonuclear double-resonance experiments on the Varian HR-220 spectrometer have been published.[39]

The beat pattern which arises in homonuclear double-resonance experiments

[31] W. McFarlane and D. S. Rycroft, *J.C.S. Dalton*, 1973, 2162.
[32] V. Barboiu and V. Petrescu, *Org. Magn. Resonance*, 1973, **5**, 43.
[33] H. J. C. Yeh, R. G. Tschudin, D. N. Lincoln, and E. Lustig, *J. Magn. Resonance*, 1973, **10**, 237.
[34] R. Burton, L. D. Hall, and P. R. Stegner, *Canad. J. Chem.*, 1971, **49**, 588.
[35] A. Charles and W. McFarlane, *Mol. Phys.*, 1968, **14**, 299.
[36] A. G. Marshall, L. D. Hall, M. Hatton, and J. Sallos. *J. Magn. Resonance*, 1974, **13**, 392.
[37] A. Sügis and M. Alla, *Izvest Akad. Nauk Est. S.S.R., Ser. Tekh. i Fiz-mat. Nauk*, 1973, **22**, 65.
[38] H. C. Born, S. J. Uurtamo, and G. E. Maciel, *Rev. Sci. Instr.*, 1973, **44**, 128.
[39] R. C. Ferguson and H. M. Records, *J. Magn. Resonance*, 1973, **11**, 266.

when the observing and irradiating frequencies are close together can be eliminated by a scheme of time division or by phase-splitting circuitry.[40] Half of the modulation signal used to achieve decoupling is fed to the receiver phase-sensitive detector, where it cancels out the parasitic beat signal. The device permits the observation of weak signals close to the irradiated one in tickling experiments. A similar approach[41] has also been used to avoid beats in FT multiple-resonance experiments, except that here the phase of the applied r.f. field is inverted between successive pulses.

In conventional INDOR spectroscopy a particular spectral line is monitored continuously while the second irradiating field is swept through a region of interest, but it is not easy to do this when the observed resonance is obtained by pulsed FT spectroscopy. However, double-resonance Fourier difference spectroscopy can be used to avoid this difficulty, and to generate what closely resemble true INDOR spectra.[42] The single-resonance spectrum acquired from Fourier transformation of an appropriate number of traces is subtracted from the same spectrum obtained with simultaneous irradiation of a particular transition. In the resultant spectrum most of the signals are cancelled out, leaving only responses which correspond to the frequencies of transitions connected with the irradiated transitions. This technique should prove very useful for making assignments in FT spectra provided that the positions of the lines to be irradiated are already known.

Equivalent results can be obtained from selective population-inversion experiments in which a pulse is applied to a transition immediately prior to the measuring pulse so as to invert (rather than merely equalize) the populations of a pair of energy levels.[43] It should be noted that these various techniques are not fully equivalent to conventional INDOR, for if the irradiated spectrum has *additional* fine structure this will not in general be reflected in the observed response.

Two groups have given extensive details of FT spectrometers which incorporate facilities for double- and multiple-resonance experiments and for multiple-pulse sequences. In one of these[44] the decoupler can be operated in the time-shared mode by gating it at 4 kHz so that the decoupler is on only during the free precession period; there is also provision for destroying the signal induced by the double-resonance pulse before the observing pulse is applied. The other group[45] has described two spectrometers with which a wide variety of experiments may be performed; the authors have reported the design of components in considerable detail.

One of the advantages of FT experiments for double-resonance work is that direct interaction between the irradiating r.f. field and the receiver circuits

[40] P. Bucci, P. A. Rolla, and C. A. Veracini, *Rev. Sci. Instr.*, 1973, **44**, 1423.
[41] P. Gonord, C. Duret, and S. Kuoy Kan, *Compt. rend.*, 1974, **278**, *B*, 123.
[42] J. Feeney and P. Partington, *J.C.S. Chem. Comm.*, 1973, 611.
[43] K. G. R. Pachler and P. L. Wessels, *J. Magn. Resonance*, 1973, **12**, 337.
[44] A. G. Redfield and R. K. Gupta, *Adv. Magn. Resonance*, 1971, **5**, 81.
[45] J. D. Ellett, jun., M. G. Gibby, V. Haeberlen, A. Pines, and J. S. Waugh, *Adv. Magn. Resonance*, 1971, **5**, 117.

can be easily avoided by not having both switched on simultaneously. Details of a suitable timing scheme have been given[28] and applications to homonuclear decoupling and the elimination of solvent resonances have been described. It should be noted, however, that the background theory has been severely criticized[29] (see p. 117).

Tomlinson and Hill[46] have described in general terms a method for exciting an n.m.r. spectrum with a radiofrequency source having any desired frequency spectrum. The spectrum is obtained by Fourier transformation and the technique may be used for the suppression of solvent lines and for decoupling while observing the complete spectrum. For the latter application the r.f. excitation has a constant low-level power spectrum, for observation, with one or more components of much higher amplitude, which provide the decoupling field.

## 4 The Nuclear Overhauser Effect

**A. Intramolecular.**—The many routine uses of the $^{13}C-\{^{1}H\}$ NOE for assignments, and the $^{1}H-\{^{1}H\}$ NOE for establishing stereochemistry, are not dealt with here. Theoretical developments are summarized in an earlier section.[25—27]

As pointed out above, the presence of paramagnetic species can seriously reduce NOE enhancements,[25] and it has been found[47] in $^{1}H-\{^{1}H\}$ experiments that even when dissolved (paramagnetic) oxygen is removed from samples by a freeze–pump–thaw technique the values of $T_1$ (and hence of the NOE) may not be reproducible. However, distillation of the sample from $[Co^{II}(bipy)_3](ClO_4)_2$ + $NaBH_4$ before examination has been reported to remove *all* free oxygen and to lead to higher values of $T_1$ and reproducible NOE's.[47]

The use of paramagnetic reagents to reduce relaxation times and suppress the NOE is well established[48] in $^{13}C$ n.m.r., and successful applications have now been reported in $^{15}N$ work.[49] This is important because the negative magnetogyric ratio of this nucleus leads to a negative NOE, so that in certain circumstances inverted or extremely weak signals may be obtained. Similar use has also been made of a relaxation reagent for studying $^{29}Si$ spectra.[50]

In many complex molecules it is difficult to apply the NOE to stereochemical assignments because of overlapping resonances. However, the use of a paramagnetic shift reagent can alleviate these problems, although there is a reduction in NOE as the concentration of the shift reagent increases. This can be put to good use, as the decrease in effect is stated[51] to be proportional to the inverse sixth power of the distance between the observed proton and the lanthanide ion (however, compare ref. 25 cited above), and examples of this

46 B. L. Tomlinson and H. D. W. Hill, *J. Chem. Phys.*, 1973, **59**, 1775.
47 J. Homer, A. R. Dudley, and W. R. McWhinnie, *J.C.S. Chem. Comm.*, 1973, 893.
48 O. A. Gansow, A. R. Burke, and G. N. La Mar, *J.C.S. Chem. Comm.*, 1972, 456.
49 L. F. Farrell, E. W. Randall, and A. I. White, *J.C.S. Chem. Comm.*, 1972, 1159.
50 R. K. Harris, and B. J. Kimber, *J. Organometallic Chem.*, 1974, **70**, 43.
51 K. Tori, I. Horibe, Y. Tamura, and H. Tada, *J.C.S. Chem. Comm.*, 1973, 620.

approach have been given in the assignment of the 100 MHz $^{1}H$ spectra of some ten-membered-ring sesquiterpene lactones.[51]

The NOE can also be used to study relaxation behaviour, and $^{1}H-\{^{1}H\}$,[52] $^{1}H-\{^{13}C\}$,[53] and $^{13}C-\{^{1}H\}$[54,55] experiments have been reported. The intramolecular $^{1}H-\{^{13}C\}$ NOE has been used to estimate[56] the $^{13}C$ relaxation time in chloroform, but it has been pointed out that the intermolecular $^{1}H-\{^{1}H\}$ NOE should also be considered.[57]

In some work it may be necessary to suppress the NOE in order to get accurate intensity measurements, and this has been done by using a suitable pulsing scheme in studies of $^{13}C$ CIDNP during the photolysis of perfluorobenzoyl peroxide in the presence of chlorobenzene.[58]

The period under review has also included work on NOE enhancements of some of the less-common nuclei. $^{2}H-\{^{1}H\}$ experiments gave natural-abundance FT deuterium spectra in which the small size of the NOE was attributed to dominance of quadrupolar relaxation.[59] $^{29}Si-\{^{1}H\}$ experiments gave enhancements which varied from $-2.4$ to $-0.07$ in $Ph_3SiH$ and $PhSiH_3$ respectively ($^{29}Si$ has a negative magnetogyric ratio); the difference was attributed to different relaxation mechanisms being dominant.[60] In other work on silicon, paramagnetic reagents were added to shorten $T^{Si}_1$ and so suppress the NOE.[61,62] No inverted signals were obtained in proton-decoupled $^{113}Cd$ spectra in spite of the negative magnetogyric ratio of this nuclide.[63]

**B. Intermolecular.**—A density-matrix description of the intermolecular NOE has been developed[64] and applied to the proton resonance of 1,1,2-trichloroethane dissolved in $Me_4Si$ when the solvent resonance is irradiated. Analysis of the differential enhancements observed gave information about solute–solvent interactions.[64] The feasibility of using the intermolecular NOE for making spectral assignments has been explored.[65] The $[AB]_2$ proton spectrum of *o*-dichlorobenzene is symmetrical, but differential NOE's were obtained in TMS solution when the solvent resonance was irradiated, and these were used to probe the nature of the solute–solvent interactions and so assign the two

52 H. Ozawa, Y. Arata, and S. Fujiwara, *J. Chem. Phys.*, 1973, **58**, 4037.
53 J.-C. Duplan, A. Brignet, and J. Delmau, *J. Chem. Phys.*, 1973, **59**, 6269.
54 G. A. Gray and S. E. Cremer, *J. Magn. Resonance*, 1973, **12**, 5.
55 G. C. Levy, J. D. Cargioli, and F. A. L. Anet, *J. Amer. Chem. Soc.*, 1973, **95**, 1527.
56 H. Ozawa, Y. Arata, and S. Fujiwara, *J. Chem. Phys.*, 1972, **57**, 1613.
57 A. Briguet, J.-C. Duplan, and J. Delmau, *J. Chem. Phys.*, 1974, **60**, 719.
58 R. Kaptein, R. Freeman, H. D. W. Hill, and J. Bargon, *J.C.S. Chem. Comm.*, 1973, 953.
59 J. M. Briggs, L. F. Farnell, and E. W. Randall, *J.C.S. Chem. Comm.*, 1973, 70.
60 R. K. Harris and B. J. Kimber, *J.C.S. Chem. Comm.*, 1973, 255.
61 G. C. Levy, J. D. Cargioli, and T. D. Mitchell, *J. Amer. Chem. Soc.*, 1973, **95**, 3445.
62 C. R. Ernst, L. Spialter, G. R. Buell, and D. C. Wilhite, *J. Organometallic Chem.* 1973, **59**, C13.
63 G. E. Maciel and M. Borzo, *J.C.S. Chem. Comm.*, 1973, 394.
64 N. R. Krishna and S. L. Gordon, *J. Chem. Phys.*, 1973, **58**, 5687.
65 N. R. Krishna and S. L. Gordon, *J. Chem. Phys.*, 1973, **59**, 4569.

halves of the spectrum.[65] A negative NOE was observed for the $^{13}$C resonance of carbon disulphide when the proton resonance of the solvent, chloroform, was irradiated.[66] This was thought to be due to a time-dependent intermolecular scalar coupling resulting from the formation of a collision complex.[66] Pulsed intermolecular double-resonance experiments have been used to study relaxation and the NOE in DMF and dimethylnitrosamine.[67]

## 5 Chemical Applications of Multiple Resonance

**A. Decoupling: Simplification and Assignment.**—Although decoupling techniques are used extensively, just a selection of some of the more interesting examples is given here.

A–{$^1$H} *Experiments*. (*i*) A = $^1$H. Successful decoupling of C*H* protons has been achieved[68] with oligopeptides dissolved in the biologically important solvent, water, the N*H* protons being observed. The C*H* resonances occur near the water signal; the experiments are easier to perform if the C*H* peak to be irradiated either coincides with the water resonance frequency or is outside the water resonance peak, but there are problems of amplifier saturation if the peak to be observed is close to but not exactly coincident with the water peak. Values of *J*(HCNH) have been assigned by this technique and used in a discussion of the peptide conformations of angiotensin II (an octapeptide hormone) in aqueous solution;[69] the C*H* resonances had been assigned in earlier studies in $D_2O$. This paper[69] also used transfer of saturation from the water protons to show that the 2-Arg peptide N*H* exchanges with the solvent at a significantly more rapid rate than the other peptide N*H* protons.

The technique by which a saturated spin may be transferred to another site in a valence isomerization equilibrium has been applied[70] in the case shown in

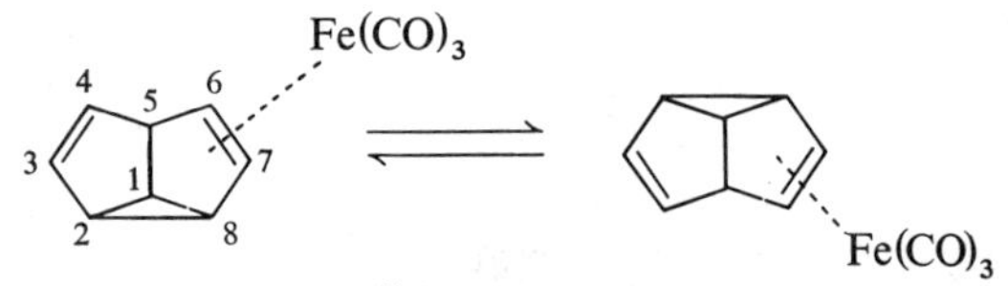


**Scheme 1**

Scheme 1. Saturation of the resonances of H$^4$, H$^5$, and H$^6$ diminishes the intensity of H$^1$, H$^2$, and H$^8$, showing that the isomerization is rapid compared with $1/T_1$ but slow compared with $\Delta\nu$, the chemical shift difference between the two kinds of proton.

(*ii*) A = $^2$H. Natural-abundance (0.016%) deuterium spectra have been observed[59] using a commercial FT spectrometer with proton noise decoupling.

[66] D. P. Miller, B. Ternai, and G. E. Maciel, *J. Amer. Chem. Soc.*, 1973, **95,** 1336.
[67] M. Alla, E. Lippmaa, and A. Sügis, in *Proceedings 16th Congress Ampère*, 1971, p. 950.
[68] J. Dadok, P. H. von Dreele, and H. A. Scheraga, *J.C.S. Chem. Comm.*, 1972, 1055.
[69] J. D. Glickson, J. Dadok, and G. R. Marshall, *Biochemistry*, 1974, **13,** 11.
[70] R. Aumann, *J. Organometallic Chem.*, 1974, **66,** C6.

These have the advantage over proton spectra of being first-order stick spectra (*cf.* natural-abundance $^{13}C-\{^{1}H\}$ spectra).

(*iii*) A = $^{11}B$. Some ($^{11}B$,$^{11}B$) coupling constants have been measured[71] from proton-decoupled $^{11}B$ spectra of some apically substituted pentaborane(9) derivatives. In another paper[72] simultaneous, selective homonuclear $^{11}B$ decoupling was used to confirm the existence of unresolved ($^{11}B$,$^{11}B$) coupling in $B_4H_{10}$ (this coupling had previously[73] been deduced to be zero). Proton-decoupled $^{11}B$ spectra were also used to obtain an upper limit to $J(^{11}B,^{11}B)$ in diborane, *via* comparisons of the linewidths of the $^{11}B-\{^{1}H\}$ resonances of $^{10}B^{11}BH_6$ and $^{11}B_2H_6$ (an earlier determination[74] of this coupling did not consider the natural linewidth of the $^{11}B$ spectrum).

(*iv*) A = $^{13}C$. Several types of $^{13}C-\{^{1}H\}$ experiment are used routinely as assignment aids by many workers in the $^{13}C$ n.m.r. field. Normally, off-resonance proton decoupling is used for assignments in $^{13}C$ spectra when the chemical shift differences in the proton spectrum are large. However, it has been shown that the residual splittings, given by $\nu_i - \nu_0 = \gamma B_2 x_i/2\pi\, {}^1J(CH)$, where $\nu_i - \nu_0$ is the proton frequency offset and $x_i$ is the residual splitting, can also be used in this way for small proton chemical shift differences provided that the proton spectrum is analysable.[75] The method has been applied to 1-nitronaphthalene, which has an ABCDEFG proton spectrum.

The need to achieve effectively complete proton decoupling before making detailed analyses based on linewidths has been emphasized.[76] In studies[77] of deuteriated long-chain aliphatic compounds the failure to observe signals from $^{13}C$ nuclei bearing two or three deuterons was attributed to increased relaxation times and the absence of any NOE. However, it must also be borne in mind that the signal would be split into many components by spin coupling to several deuterons. Simultaneous proton- and deuteron-decoupling of $^{13}C$ spectra has been described[78] for $^{2}H$-labelling studies of bile acids isolated from rats; deuterium content could only be estimated by comparing $^{13}C-\{^{1}H,\ ^{2}H\}$ spectra with $^{13}C-\{^{1}H\}$ spectra, because $^{13}C^{2}H$ lines were hidden under the $^{13}C^{1}H$ lines in the $^{13}C-\{^{1}H\}$ spectra and because NOE's were not suppressed. Ernst's noise-modulated decoupling experiment,[79] in which a component of the central line of an A triplet in an $AX_2$ spin system remains unaffected during perturbation of the system by the decoupling field,

[71] D. W. Lowman, P. D. Ellis, and J. D. Odom, *Inorg. Chem.*, 1973, **12**, 681.
[72] J. D. Odom, P. D. Ellis, D. W. Lowman, and M. H. Gross, *Inorg. Chem.*, 1973, **12**, 95.
[73] R. C. Hopkins, J. D. Baldeschwieler, R. Schaeffer, F. N. Tebbe, and A. Norman, *J. Chem. Phys.*, 1965, **43**, 975.
[74] T. C. Farrar, R. B. Johannesen, and T. D. Coyle, *J. Chem. Phys.*, 1968, **49**, 281.
[75] J. W. Emsley, J. C. Lindon, and D. Shaw, *J. Magn. Res.*, 1973, **10**, 100.
[76] D. E. Axelson, A. J. Oliver, and C. E. Holloway, *Org. Magn. Resonance*, 1973, **5**, 255; *ibid.*, 1974, **6**, 64.
[77] A. P. Tulloch and M. Mazurek, *J.C.S. Chem. Comm.*, 1973, 692.
[78] D. M. Wilson, A. L. Burlingame, T. Cronholm, and J. Sjövall, *Biochem. Biophys. Res. Comm.*, 1974, **56**, 828.
[79] R. R. Ernst, *J. Chem. Phys.*, 1966, **45**, 3845.

has been applied[80] to the identification of methylene carbons in polymers. In the noise-modulated, on-resonance partially proton-decoupled $^{13}C$ spectrum of poly(propylene oxide) the signal which remained sharp independent of the amplitude of $B_2$ was assigned to the methylene carbons; in another experiment, on propylene glycol, the amplitude of $B_2$ was kept constant ($\gamma B_2/2\pi$ = 1200 Hz) but the offset was gradually increased until (at 1000—1500 Hz) one signal only was left, that of the methylene carbons; the signals of the methyl and methine carbons had broadened. Single-frequency off-resonance decoupling gives complications when the methylene protons are non-equivalent, as in 2-methylnorbornan-2-ol, (1); in this case[81] second-order effects in the off-resonance proton-irradiated $^{13}C$ spectrum can be used to assign the methylene carbons.

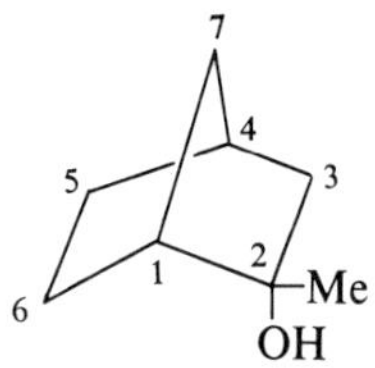


(1)

The X part of an ABX system (*e.g.* $^{13}C$ for a methylene group in an off-resonance proton-decoupled spectrum) is a well-resolved triplet if $J_{AX}$ and $J_{BX}$ are similar; broadening arises if $J_{AX}$ and $J_{BX}$ are considerably different. In the case of (1) methylene carbons C-3 and C-7 give sharp triplets but C-5 and C-6 occur as broad multiplets in the off-resonance proton-decoupled $^{13}C$ spectrum. In the $^{13}C$ spectra of alkylbenzenes[82] with decoupling of the aromatic protons, residual couplings to the alkyl protons (which are effectively off-resonance decoupled) may permit assignment of the aromatic $^{13}C$ resonances. A selective proton irradiation technique, gated spin tickling, has been used[83] to assign the C-2 and C-5 carbon signals of 3-methyl- and 3-t-butyl-furan; the conditions used are similar to those of homonuclear spin tickling, but the spin decoupler is on only during the pulse delay, and off during acquisition of the FID. The coupling remains but the spin population is perturbed: irradiation of one $^{13}C$ satellite in the $^1H$ spectrum gave an inverted peak for one half of the $^{13}C$ doublet and an enhanced peak (2.5 times) for the other half. It is possible to apply the technique to long-range ($^{13}C$,$^1H$) coupling (where the $^{13}C$ satellites are hidden under the $^1H$ peak) as well as to one-bond ($^{13}C$,$^1H$) coupling.

A method for increasing the *relative* intensity of $^{13}C$ signals from quaternary carbon atoms has been described.[84] Broad-band (2 kHz) decoupling applied

[80] J. Schaefer, *Macromolecules*, 1972, **5**, 590.
[81] J. B. Grutzner, *J.C.S. Chem. Comm.*, 1974, 64.
[82] S. Sørensen, M. Hansen, and H. J. Jakobsen, *J. Magn. Resonance*, 1973, **12**, 340.
[83] J. Runsink, J. de Wit, and W. D. Weringa, *Tetrahedron Letters*, 1974, 55.
[84] I. H. Sadler, *J.C.S. Chem. Comm.*, 1973, 809.

to the proton resonances at very low powers has the effect of decoupling the small proton couplings to quaternary carbons while producing incomplete decoupling of protons directly bound to $^{13}$C. Similar results have previously been obtained[85] by applying narrow-band (*ca.* 300 Hz) proton decoupling at higher power levels upfield of the proton signals, as in a normal off-resonance decoupling experiment. The 'fingerprints' observed[86] in the proton-coupled $^{13}$C spectra of compounds containing a HC═CH—CH═CH fragment with a plane of symmetry are retained, magnified, in the off-resonance proton-decoupled spectra.[87] Thus the use of off-resonance decoupling as an assignment technique has been extended, since it usually fails when proton chemical shift differences are small.[75] The fingerprints are observed even when the proton chemical shift difference is zero—this arises because the fingerprint is attributable to the different ($^{13}$C,$^{1}$H) couplings in the fragment, which induce effective chemical shift differences. Use of the technique[87] has shown[88] that the assignments of other workers[89] for $C_\alpha$ and $C_\beta$ of indan and benzocyclo-alkenes up to benzocyclo-octene must be reversed.

(*v*) A = $^{15}$N. More reports of the observation of natural-abundance (*ca.* 0.35%), proton-noise-decoupled $^{15}$N spectra have appeared.[49,90] In unsymmetrical dialkyl hydrazines, the use of proton decoupling enables the nitrogen resonances to be readily assigned since the negative NOE inverts the signals from nitrogens with directly bound protons while the signals from alkylated nitrogens remain non-inverted.[90]

(*vi*) A = $^{19}$F. Proton noise decoupling of fluorine spectra can be used to simplify accurate measurement of $^{19}$F chemical shifts, as in the case of some highly hindered aryl ethers with fluorine substituents.[91] An interesting example is the observation[92] of $^{13}$C satellites in proton-decoupled $^{19}$F spectra of compounds of the type $(p\text{-FC}_6H_4)_2Z$, where Z is, *e.g.*, *trans,trans*-disubstituted *trans*-buta-1,3-diene; the $^{13}$C satellites show the effects of ($^{19}$F,$^{19}$F) coupling, and the variation of *J*(F,F) with Z was studied.

(*vii*) A = $^{29}$Si. Reports of studies of $^{29}$Si–{$^{1}$H} NOE enhancements (which may be positive or negative) have appeared,[50,60—62,93—95] and relaxation times in simple molecules[93,94] and chemical shifts in polymers[95] have been reported.

85 E. Wenkert, A. O. Clouse, D. W. Cochran, and D. Doddrell, *J. Amer. Chem. Soc.*, 1969, **91**, 6879; A. Allerhand, R. F. Childers, and E. Oldfield, *Biochemistry*, 1973, **12**, 1335.
86 H. Günther, H. Schmickler, and G. Jikeli, *J. Magn. Resonance*, 1973, **11**, 344.
87 G. Jikeli, W. Herrig, and H. Günther, *J. Amer. Chem. Soc.*, 1974, **96**, 323.
88 H. Günther, G. Jikeli, and H. Schmickler, *Angew. Chem. Internat. Edn.*, 1973, **12**, 762.
89 E. L. Motell, D. Lauer, and G. E. Maciel, *J. Phys. Chem.*, 1973, **77**, 1865; G. W. Buchanan and R. H. Wightman, *Canad. J. Chem.*, 1973, **51**, 2357.
90 R. L. Lichter and J. D. Roberts, *J. Amer. Chem. Soc.*, 1972, **94**, 4904.
91 P. A. Lehmann F. and L. F. Johnson, *Org. Magn. Resonance*, 1973, **5**, 61.
92 P. J. Mitchell, L. Phillips, S. J. Roberts, and V. Wray, *Org. Magn. Resonance*, 1974, **6**, 126.
93 G. C. Levy, *J. Amer. Chem. Soc.*, 1972, **94**, 4793.
94 R. L. Scholl, G. E. Maciel, and W. K. Musker, *J. Amer. Chem. Soc.*, 1972, **94**, 6376.
95 H. Jancke, G. Engelhardt, M. Mägi, and E. Lippmaa, *Z. Chem.*, 1973, **13**, 392; *ibid.*, p. 435.

(*viii*) A = $^{31}P$. With the advent of modern instruments, phosphorus spectra have increasingly been measured with proton noise decoupling (comparison with the proton-coupled spectra being a useful assignment aid). Single-line spectra are useful for quantitative analysis of reaction product mixtures of compounds containing one phosphorus atom, *e.g.* as in studies[96] of the oxidative addition of bromine to compounds such as *trans*-$[Pt^{II}Cl_2(PEt_3)(py)]$, where five singlets (with $^{195}Pt$ satellites) corresponding to *trans*-$[Pt^{IV}Br_nCl_{4-n}(PEt_3)(py)]$ ($n$ = 0—4) were observed in the proton-decoupled $^{31}P$ spectrum, and the intensities showed that random scrambling of the halogens had occurred. 'White-noise' proton-decoupled $^{31}P$ spectra were used[97] to study the sensitivity of $^{31}P$ chemical shifts of enantiomeric pyrophosphoramides to differences in magnetic environment caused by association with chiral solvents; the small shift effects observed would have been difficult to quantify in the absence of proton decoupling. The values of $^1J(^{77}Se-^{31}P)$ in several triarylphosphine selenides were obtained[98] from $^{31}P$ spectra, and the advantage was noted of using proton noise decoupling to increase the signal-to-noise ratio (however, signal averaging was also necessary). Proton noise decoupling is generally appropriate when satellites from other rare nuclei (*e.g.* $^{183}W$ in $^{31}P$ spectra of tungsten complexes with phosphorus ligands[99]) are to be observed. The conformational equilibrium in 1-methylphosphorinan has been studied[100] using $^{31}P$ and $^1H$ n.m.r. The proton-decoupled $^{31}P$ spectrum was a sharp singlet at room temperature but separated into two sharp signals at low temperature; the advantage of proton decoupling was that the heights of the singlet signals could be used to assess the relative concentrations of the two conformers.

(*ix*) A = $^{77}Se$. Proton-decoupled FT $^{77}Se$ spectra have been reported.[101,102] The proton-decoupled spectrum of one of the products from the reaction of acetylene and selenium was a single line,[102] a result consistent with structure (2) but excluding structure (3), which had been proposed by earlier workers.

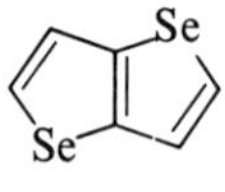


(2)

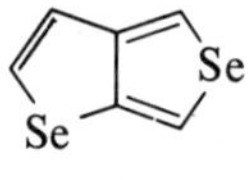


(3)

$^1H-\{X\}$ *Experiments.* (*i*) X = $^2H$. The technique of using deuterium-labelled compounds and deuterium decoupling is well known in proton n.m.r. spectroscopy. The deuterium-decoupled proton spectra of XCHD·CHDY

[96] B. T. Heaton and K. J. Timms, *J.C.S. Chem. Comm.*, 1973, 931.
[97] M. D. Joesten, H. E. Smith, and V. A. Vix, jun., *J.C.S. Chem. Comm.*, 1973, 18.
[98] R. P. Pinnell, C. A. Megerle, S. L. Manatt, and P. A. Kroon, *J. Amer. Chem. Soc.*, 1973, **95**, 977.
[99] R. L. Keiter and L. W. Cary, *J. Amer. Chem. Soc.*, 1972, **94**, 9232.
[100] S. I. Featherman and L. D. Quin, *J. Amer. Chem. Soc.*, 1973, **95**, 1700.
[101] S. Gronowitz, I. Johnson, and A.-B. Hörnfeldt, *Chem. Scripta*, 1973, **3**, 94.
[102] S. Gronowitz, T. Frejd, and A.-B. Hörnfeldt, *Chem. Scripta*, 1974, **5**, 236.

compounds are useful in stereochemical investigations and have been used in studies of the methoxymercuration of ethylene,[103] and the conversion of tyramine into tyrosol.[104] The assignment of the stereochemistry of the product, *cis*-3,5-[$^2H_2$]cyclopentene, resulting from the non-catalytic addition of molecular deuterium to cyclopentadiene, was based[105] on the analysis of the $A_2BC$ spectrum obtained when both deuterium and the olefinic protons were decoupled. Deuterium decoupling was used[106] to simplify the proton spectra of some polydeuteriated cyclohexane and cyclohexene derivatives and was also used in the complete analysis of the proton spectra of [3,3,4,4,5,5-$^2H_6$]- and [3,3,5,5-$^2H_4$]-t-butylcyclohexane.[107] A technique used in the analysis has been suggested as a general technique for measuring particular proton chemical shifts;[107] the protons of interest would be replaced *via* specific deuteriation and the spectrum of the deuteriated compound would then be subtracted from that of the normal compound. The technique is, however, limited by the accessibility of the deuteriated compounds. There have been more reports[17,20,108,109] of the use of deuterium substitution and decoupling to simplify the spectra of molecules partially oriented in the nematic phase.

(*ii*) X = $^{11}B$. One example[110] out of many is the $^{11}B$ decoupling from the proton spectra of some small *closo*-carbaboranes; decoupling revealed fine structure attributed to homonuclear, long-range proton coupling. The $^{11}B$-decoupled proton spectra of several *o*-, *m*-, and *p*-carbaborane derivatives have also been studied.[111] $^1H$–{$^{11}B$} and $^1H$–{$^{27}Al$} decoupling experiments were used[112] in studies of solutions of aluminium tetrahydroborate complexes.

(*iii*) X = $^{14}N$. Linewidth arising from incomplete relaxation induced by the $^{14}N$ quadrupole can be removed by irradiation at the $^{14}N$ resonant frequency, as in studies[113] of the rates of deprotonation of some *para*-substituted pyridinium ions, and studies[114] of C—N bond rotation in urea and its sulphur and selenium analogues and in primary amides and their thio-analogues. The proton spectrum of diethylmethylvinylammonium bromide has been analysed;[115] the parameters obtained from analysing the $^{14}N$-decoupled spectrum were used as trial parameters for an iterative analysis including $^{14}N$. Con-

[103] T. Ibusuki and Y. Saito, *J. Organometallic Chem.*, 1973, **56**, 103.
[104] C. Fuganti, D. Ghiringhelli, P. Grasselli, and A. Santopietro-Amisano, *J.C.S. Chem. Comm.*, 1973, 862.
[105] F. A. L. Anet and F. Leyendecker, *J. Amer. Chem. Soc.*, 1973, **95**, 156.
[106] J.-P. Aycard, R. Geiss, J. Berger, and H. Bodot, *Org. Magn. Resonance*, 1973, **5**, 473.
[107] V. R. Haddon and L. M. Jackman, *Org. Magn. Resonance*, 1973, **5**, 333.
[108] J. W. Emsley, J. C. Lindon, and J. Tabony, *Mol. Phys.*, 1973, **26**, 1485, 1499.
[109] R. Ader and A. Loewenstein, *Mol. Phys.*, 1974, **27**, 113.
[110] T. Onak and E. Wan, *J.C.S. Dalton*, 1974, 665.
[111] A. P. Tupciauskas, V. I. Stanko, Yu. A. Ustynyuk, and V. V. Khrapov, *Zhur. strukt. Khim.*, 1973, **13**, 823.
[112] G. N. Boiko, Yu. I. Malov, and K. N. Semenenko, *Izvest Akad. Nauk. S.S.S.R., Ser. khim.*, 1973, 1143.
[113] J.-J. Delpuech and G. Serratrice, *Org. Magn. Resonance*, 1972, **4**, 667.
[114] W. Walter, E. Schaumann, and H. Rose, *Tetrahedron*, 1972, **28**, 333; *Org. Magn. Resonance*, 1973, **5**, 191.
[115] M. Ueyama and K. Tori, *Org. Magn. Resonance*, 1972, **4**, 913.

clusive evidence that the origin of the width[116,117] of the hydride resonances of the *N*-bonded isomer of *trans*-[PtH(NCS)$L_2$] (L = $Et_3As$, $R_3P$) in solution is the $^{14}N$ quadrupolar effect has been given[118] using $^{14}N$ decoupling experiments; simultaneous $^{14}N$ and $^{31}P$ decoupling experiments have also shown[119] that phosphine exchange makes no additional contribution to the width of the peaks.

(*iv*) X = $^{19}F$. One of many examples is the use of fluorine decoupling to confirm the presence of long-range ($^{19}F$,$^{1}H$) coupling in pentafluorophenylsilane derivatives.[120] $^{1}H$–{$^{19}F$} and $^{1}H$–{$^{1}H$} experiments were used[121] to assign coupling constants and resonances in the spectra of some five-co-ordinate fluoro-olefin complexes of methylplatinum(II). The nitrogen methyl protons of $WF_4$(NMe)(NCMe) are coupled to both $^{14}N$ and $^{19}F$; this was confirmed[122] by $^{1}H$–{$^{19}F$} and $^{1}H$–{$^{14}N$} decoupling experiments, which also permitted accurate values of $^{2}J(^{14}N,^{1}H)$ and $^{4}J(^{19}F,^{1}H)$ to be obtained.

(*v*) X = $^{31}P$. A common use of $^{31}P$ decoupling is the simplification of spin systems prior to analysis and use of the parameters obtained in a full analysis, as in studies of some 1,3,2-diazaphosphorinans,[123] 1,3,2-oxazaphosphorinans,[124] 1,3,2-dithiaphospholans,[125] phosphonate and hydroxyphosphonate derivatives,[126] and triphenylphosphine.[127] (However, another analysis[128] of the proton spectrum of triphenylphosphine does not mention the use of phosphorus decoupling.) An apt use of $^{31}P$ decoupling is the establishment of unequivocal relationships between the $^{31}P$ and $^{1}H$ spectra of a compound, as in studies[129] of the two diastereomers of *sym*-diphenyldiphosphine which were present in an equilibrium mixture with phenylphosphine and pentaphenylcyclopentaphosphine, and in studies[100] of the conformational equilibrium in 1-methylphosphorinan mentioned earlier. An example of chemical interest where only the proton spectrum was observed directly occurred during studies[130] of intermolecular association in diphosphines. It was found that the methyl resonance (a broad singlet) of a mixture of $Me_4P_2$ and $Ph_4P_2$

[116] A. Pidcock, *J.C.S. Chem. Comm.*, 1973, 249; M. W. Adlard and G. Socrates, *J.C.S. Chem. Comm.*, 1972, 17; *J.C.S. Dalton*, 1972, 797; *J. Inorg. Nuclear Chem.*, 1972, **34,** 2339.
[117] J. Powell and B. L. Shaw, *J. Chem. Soc.*, 1965, 3879.
[118] B. E. Mann, B. L. Shaw, and A. J. Stringer, *J. Organometallic Chem.*, 1974, **73,** 129.
[119] W. McFarlane and D. S. Rycroft, unpublished work.
[120] G. Hägele and M. Weidenbruch, *Org. Magn. Resonance*, 1974, **6,** 66.
[121] H. C. Clark and L. E. Manzer, *J.C.S. Chem. Comm.*, 1973, 870.
[122] M. Harman, D. W. Sharp, and J. M. Winfield, *Inorg. Nuclear Chem. Letters*, 1974, **10,** 183.
[123] R. O. Hutchins, B. E. Maryanoff, J.-P. Albrand, A. Cogne, D. Gagnaire, and J.-B. Robert, *J. Amer. Chem. Soc.*, 1972, **94,** 9151.
[124] J. Durrieu, R. Kraemer, and J. Navech, *Org. Magn. Resonance*, 1972, **4,** 709.
[125] J.-P. Albrand, D. Gagnaire, J. Martin, and J.-B. Robert, *Org. Magn. Resonance*, 1973, **5,** 33.
[126] L. Evelyn, L. D. Hall, P. R. Steiner, and D. H. Stokes, *Org. Magn. Resonance*, 1973, **5,** 141.
[127] L. Radics, E. Baitz-Gács, and A. Neszmélyi, *Org. Magn. Resonance*, 1974, **6,** 60.
[128] S. Sørensen and H. J. Jakobsen, *Acta Chem. Scand.*, 1974, **A28,** 249.
[129] J. P. Albrand and D. Gagnaire, *J. Amer. Chem. Soc.*, 1972, **94,** 8630.
[130] H. C. E. McFarlane and W. McFarlane, *J.C.S. Chem. Comm.*, 1972, 1189.

in $CH_2Cl_2$ was associated with two distinct phosphorus frequencies, corresponding closely to the chemical shifts of the separate biphosphines; after several days the spectrum changed to that of $Me_2PPPh_2$ (a double doublet). These and other observations showed that intermolecular association of $Me_4P_2$ with itself and with $Ph_4P_2$ was occurring in the polar medium dichloromethane.

*Homonuclear Experiments* (*Other than* $^{1}H$–{$^{1}H$}). (*i*) $^{11}B$–{$^{11}B$}. The energy-level diagram for a pair of coupled $^{11}B$ nuclei ($I = \frac{3}{2}$) has been derived, and it has been shown that when spin population transfer effects predominate then net intensity changes will arise only in experiments involving the outermost lines of the observed and irradiated multiplets. Thus interpretation of INDOR experiments is straightforward and has been used to assign the $^{11}B$ spectrum of 6-methyldecaborane, in which there is extensive coupling among the boron nuclei.[131,132]

(*ii*) $^{13}C$–{$^{13}C$}. A homonuclear double-resonance experiment which can be useful for making $^{13}C$ n.m.r. peak assignments has been described.[133] It requires the observation of the $^{13}C$ satellites of the proton-decoupled $^{13}C$ signals: $^{13}C$ irradiation of one satellite, brought about by adiabatic rapid passage (40 Hz $s^{-1}$) prior to the pulse used to obtain the FT spectrum, perturbs the populations of the energy levels and alters the intensities of related satellites, enhancing one and diminishing the other. Hence if the assignments of one of the $^{13}C$ resonances and the ($^{13}C$,$^{13}C$) coupling are known, an unknown $^{13}C$ peak can be picked out and assigned (the enhancement of one of its satellites makes this possible even when the $^{13}C$ satellites are not very clear initially). [$^{13}C_2$]Ethyl acetate was used as an artificial illustration of the technique, which was then applied to the real problem of the assignment of the bridge $^{13}C$ resonances of $MeOC(CH_2CH_2)_3CMe$.

Details of the circuitry used to produce the ARP irradiation have been published.[38]

(*iii*) $^{19}F$–{$^{19}F$}. Homonuclear fluorine-decoupling experiments have been used[134] to confirm assignments in pentafluorophenyl derivatives of metal carbonyls, *e.g.* in (4) all five ring fluorines are non-equivalent, and one *ortho*-

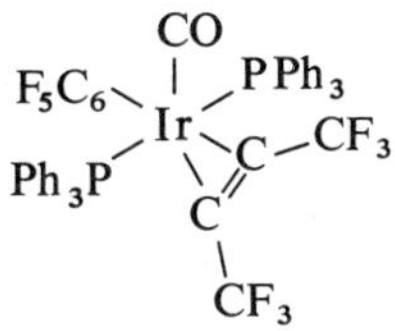


(4)

[131] R. F. Sprecher and J. C. Carter, *J. Amer. Chem. Soc.*, 1973, **95**, 2369.
[132] R. F. Sprecher, B. E. Aufderheide, G. W. Luther tert., and J. C. Carter, *J. Amer. Chem. Soc.*, 1974, **96**, 4404.
[133] H. C. Dorn and G. E. Maciel, *J. Phys. Chem.*, 1972, **76**, 2972.
[134] R. L. Bennett, M. I. Bruce, and R. C. F. Gardner, *J.C.S. Dalton*, 1973, 2653.

fluorine is coupled to one $CF_3$ group. On $^{31}P$-decoupling, the $CF_3$ multiplets appear as a quartet and a doublet of quartets.

**B. Selective Irradiation Experiments: Measurements of Magnitudes and Relative Signs of Coupling Constants.**—Numerous applications of selective irradiation techniques, both homonuclear and heteronuclear, have been reported during the review period, although few new selective irradiation experiments have been published. The off-resonance decoupling technique reported[135] in 1972 has been applied[136] to the determination of the relative signs of the phosphorus–carbon couplings in triphenylphosphine; it has been demonstrated[137] that the method can be extended to include cases where the $^{13}C$ nuclei involved do not have directly-bound protons. The technique involves using a weak decoupling r.f. field (typically $\gamma B_2/2\pi$ = 100—150 Hz), which is set only slightly off-resonance from the proton to be decoupled. In the example chosen the signs of $^2J(^{31}P,^{13}C\text{-}3)$ and $^4J(^{31}P,H_5)$ were compared for (5), in which C-3 does not carry a proton. Previous results obtained in this way have been reported for a set of related phosphorus compounds.[138]

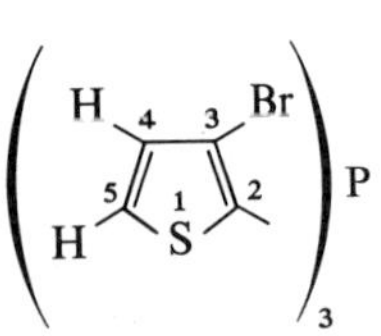

(5)

Similar information has also been obtained from (5) and other molecules by use of selective population-transfer $^{13}C$–{$^1H$} FT experiments. Results obtained for compound (5) are illustrated in Figure 1, in which a weak field applied to a transition in the proton spectrum has been gated so that line-splitting effects are avoided, but effects due to changes in energy-level populations are retained.[139]

A novel method of measuring the sign and magnitude of $J_{XX}$ in deceptively simple $[A_nX]_2$ spin systems [$J_{AA} = 0$, and $|J_{XX}| \gg |J_{AX} - J'_{AX}|$] has been described.[140] Earlier methods of determining $J_{XX}$ in these systems suffer from several disadvantages: (i) bandshape analysis[141,142] is difficult to perform and has limited accuracy;[143] (ii) observation of weak A lines with low transition probabilities is limited to neat liquids or very concentrated

135 H. J. Jakobsen, T. Bundgaard, and R. S. Hansen, *Mol. Phys.*, 1972, **23**, 197.
136 T. Bundgaard and H. J. Jakobsen, *Acta Chem. Scand.*, 1972, **26**, 2548.
137 S. Sørensen, R. S. Hansen, and H. J. Jakobsen, *J. Amer. Chem. Soc.*, 1973, **95**, 5080.
138 S. Sørensen, R. S. Hansen, and H. J. Jakobsen, *J. Amer. Chem. Soc.*, 1972, **94**, 5900.
139 S. Sørensen, R. S. Hansen, and H. J. Jakobsen, *J. Magn. Resonance*, 1974, **14**, 243.
140 W. McFarlane and D. S. Rycroft, *J.C.S. Faraday II*, 1974, **70**, 377.
141 D. G. Hendricker, R. E. McCarley, R. W. King, and J. G. Verkade, *Inorg. Chem.*, 1966, **5**, 639.
142 E. G. Finer and R. K. Harris, *Mol. Phys.*, 1967, **12**, 457.
143 F. B. Ogilvie, J. M. Jenkins, and J. G. Verkade, *J. Amer. Chem. Soc.*, 1970, **92**, 1916.

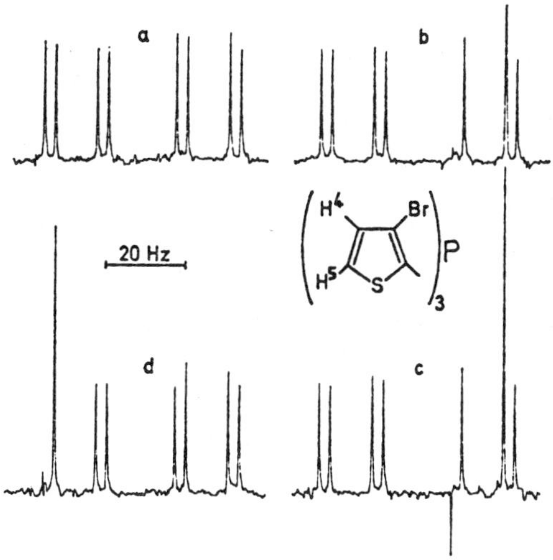


**Figure 1** *25.16-*MHz *double-resonance FT n.m.r. spectra of the* C-3 *carbon in tris-(3-bromo-2-thienyl)phosphine obtained using* $\gamma B_2/2\pi = 0.2$ Hz *under various conditions. (a) Spin decoupler offset by 2000* Hz *from the* $^1$H *region; (b) and (c) irradiation of the highest frequency line in the* H-5 *satellite spectrum; (b) continuous irradiation (spin-tickling); (c) gated irradiation,* $\tau = 2$ s; *(d) gated irradiation of the highest frequency line in the* H-4 *satellite spectrum,* $\tau = 2$ s. *All spectra were obtained using an acquisition time of 8* s *and 300 transients. In (b) a pulse delay of 2* s *was applied.*

(Reproduced by permission from *J. Magn. Resonance*, 1974, **14,** 243)

solutions,[142] even in favourable cases where $J_{XX}$ is not very large;[144] and (iii) double-resonance experiments[144—146] involving $^{13}$C satellites observed in the A spectrum enable the sign as well as the magnitude of $J_{XX}$ to be obtained, but these are limited by the need for suitably situated carbon atoms and the need to observe the $^{13}$C satellites at a reasonably good signal-to-noise ratio, which again restricts measurements to concentrated samples. The new method[140] depends upon the determination by double resonance of the positions of weak lines at about $\pm J_{XX}$ from the centre of the X spectrum; it suffers from none of the previous disadvantages and has indeed been applied routinely to a wide variety of systems. Observation is confined to strong A lines and the relative signs of $J_{XX}$ and $[J_{AX} + J'_{AX}]$ are obtained; dilute solutions can be studied and the presence of $^{13}$C or other rare magnetic isotopes is unnecessary. Although the irradiated lines have low transition

[144] R. D. Bertrand, F. B. Ogilvie, and J. G. Verkade, *J. Amer. Chem. Soc.*, 1970, **92,** 1908.
[145] R. D. Bertrand, F. B. Ogilvie, and J. G. Verkade, *Chem. Comm.*, 1969, 756.
[146] E. G. Finer and R. K. Harris, *Mol. Phys.*, 1967, **13,** 65.

moments, high precision is possible because large perturbations of the observed lines may be obtained *via* suitably large amplitudes of the r.f. field used to irradiate X. The method has been used to study the solvent- and concentration-dependence of $^2J$(PHgP) in $[(MeO)_2P(O)]_2Hg$; other examples given were the measurement of the sign and magnitude of $^2J$(PMP) in molybdenum, platinum, and rhodium complexes containing two identical *trans* phosphorus ligands, and the determination of $^1J$(PP) in diastereomeric symmetrical dimethyldiphenyldiphosphines.[147] Effects essentially similar to those observed and used in this work[140] have been reported without details or discussion by other workers,[148,149] who observed asymmetric $^{31}$P INDOR spectra in platinum complexes having two *cis* phosphorus ligands. The $[A_nX]_2$ spin system is not deceptively simple in these cases because *cis*-$J$(PPtP) is relatively small and the central lines of the X spectrum overlap with the 'weak' outer lines, of which only one set is observed when one line of the N doublet in the A spectrum is monitored. These systems provided scope for the application of a known[150] double-resonance technique, *viz.* homonuclear proton tickling experiments, to confirm the presence and to measure the position of the weak outer lines in the proton spectra of some of the *cis*-complexes and so enable accurate values of $^2J$(PP) to be obtained.[148,149]

Nuclear spin coupling constants between directly bound elements are especially interesting from a theoretical point of view, and signs and magnitudes are necessary to provide an experimental basis. Multiple-resonance techniques have been used in several recent sign determinations, and work still remains to be done in this area. The signs of $^1J(^{11}B–^1H)$ and $^1J(^{31}P–^{11}B)$ have been obtained[151] in the first reported use of selective, heteronuclear triple resonance to remove the degeneracy arising from a negligibly small $J_{AX}$ in an AMX type of spin system;[152] the compound studied was $(MeO)_3$-$PBH_3$. In this compound $^4J(^{11}B,^1H)$ is zero, and irradiation at the $^{11}$B frequency has no effect on the methoxy proton resonance. The degeneracy in the $^{11}$B spectrum can be lifted by selective irradiation in the $^{31}$P spectrum so that only those parts of the boron spectrum that are associated with the same total proton spin orientation as the irradiated phosphorus transitions are split; the methoxy proton lines are also perturbed by the phosphorus irradiation. Selective irradiation in the $^{11}$B spectrum then gives differential effects according to the phosphorus spin orientation associated with the boron lines irradiated, and the sign of $^1J(^{31}P–^{11}B)/^3J(^{31}P,^1H)$ may be obtained. $^1H–\{^{31}P\}$ and $^1H–\{^{11}B\}$ experiments gave the signs of $^1J(^{11}B–^1H)/^1J(^{31}P–^{11}B)$ and

147 H. C. E. McFarlane and W. McFarlane, *Chem. Comm.*, 1971, 1589.
148 R. J. Goodfellow, M. J. Hardy, and B. F. Taylor, *J.C.S. Dalton*, 1973, 2450.
149 P. L. Goggin, R. J. Goodfellow, S. R. Haddock, J. R. Knight, F. J. S. Reed, and B. F. Taylor, *J.C.S. Dalton*, 1974, 523.
150 D. A. Duddell, J. G. Evans, P. L. Goggin, R. J. Goodfellow, A. J. Rest, and J. G. Smith, *J. Chem. Soc. (A)*, 1969, 2134.
151 H. C. E. McFarlane, W. McFarlane, and D. S. Rycroft, *J.C.S. Faraday II*, 1972, **68**, 1300.
152 A. D. Cohen, R. Freeman, K. A. McLauchlan, and D. H. Whiffen, *Mol. Phys.*, 1963, **7**, 45.

${}^{1}J({}^{31}\mathrm{P}-{}^{11}\mathrm{B})/{}^{2}J({}^{31}\mathrm{P},{}^{1}\mathrm{H})$ respectively. The sign of ${}^{3}J({}^{31}\mathrm{P},{}^{1}\mathrm{H})$ is well established[153] as positive in this type of phosphorus compound, and the absolute signs of ${}^{1}J({}^{31}\mathrm{P}-{}^{11}\mathrm{B})$ and ${}^{1}J({}^{11}\mathrm{B}-{}^{1}\mathrm{H})$ were shown to be positive. Earlier determinations[154] of the signs of these couplings were relative to signs which were not firmly established. This work therefore shows the sign of ${}^{1}J({}^{19}\mathrm{F}-{}^{11}\mathrm{B})$ to be definitely negative. Other one-bond couplings whose signs have been determined for the first time are: ${}^{1}J({}^{13}\mathrm{C}-{}^{7}\mathrm{Li})$(positive)[155] in ${}^{13}$C-enriched methyl-lithium tetramer dissolved in THF at $-50$ °C [fine structure due to ${}^{2}J({}^{7}\mathrm{Li},{}^{1}\mathrm{H})$ was not resolved, but ${}^{7}$Li decoupling almost halved the width of the proton resonance]; ${}^{1}J({}^{199}\mathrm{Hg}-{}^{31}\mathrm{P})$(positive)[140] in $[(\mathrm{MeO})_2\mathrm{P(O)}]_2\mathrm{Hg}$ (this sign probably applies to direct ${}^{199}$Hg–${}^{31}$P coupling in general); ${}^{1}J({}^{183}\mathrm{W}-{}^{31}\mathrm{P})$ (positive)[156] in complexes of hexacarbonyltungsten(0) with phosphorus ligands; ${}^{1}J({}^{77}\mathrm{Se}-{}^{31}\mathrm{P})$ (negative)[31,157,158] in phosphine selenides (this was already known[159] to be negative in other bonding situations); ${}^{1}J({}^{119}\mathrm{Sn}-{}^{77}\mathrm{Se})$ (positive)[160] in a series of methyltin selenides; ${}^{1}J({}^{29}\mathrm{Si}-{}^{15}\mathrm{N})$(positive)[161] in ${}^{15}$N-enriched trisilylamine; and ${}^{1}J({}^{195}\mathrm{Pt}-{}^{29}\mathrm{Si})$(negative)[162] in a silylplatinum complex.

${}^{1}J({}^{13}\mathrm{C}-{}^{15}\mathrm{N})$ in the (*Z*)-oxaziridine (6) was found[163] to be $+4.9$ Hz; this is the first example of a negative, one-bond, reduced coupling constant between carbon and nitrogen; ${}^{2}J({}^{15}\mathrm{N},{}^{1}\mathrm{H})$ in (6) is $-5.0$ Hz. The magnitude of ${}^{1}J({}^{15}\mathrm{N}-{}^{13}\mathrm{C})$ in the (*E*)-isomer is 3.1 Hz, but the sign could not be determined as ${}^{2}J({}^{15}\mathrm{N},{}^{1}\mathrm{H})$ is unresolvably small.

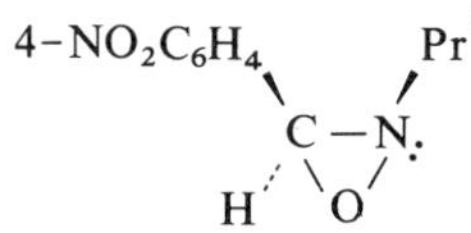


(6)

The magnitude of ${}^{1}J({}^{195}\mathrm{Pt}-{}^{14}\mathrm{N})$ in three platinum amine complexes was obtained[149] from the fine structure exhibited by the ${}^{195}$Pt INDOR spectra. The sensitivity of the double-resonance technique was used during isotope-effect studies[164] in the measurement of ${}^{1}J({}^{31}\mathrm{P}-{}^{1}\mathrm{H})$ and ${}^{1}J({}^{31}\mathrm{P}-{}^{2}\mathrm{H})$ in *P*-

153 W. McFarlane, *Proc. Roy. Soc.*, 1968, **A306**, 185.
154 R. W. Rudolph and C. W. Schultz, *J. Amer. Chem. Soc.*, 1971, **93**, 6821.
155 W. McFarlane and D. S. Rycroft, *J. Organometallic Chem.*, 1974, **64**, 303.
156 W. McFarlane and D. S. Rycroft, *J.C.S. Chem. Comm.*, 1973, 336.
157 W. McFarlane and D. S. Rycroft, *J.C.S. Chem. Comm.*, 1972, 902.
158 D. W. W. Anderson, E. A. V. Ebsworth, G. D. Meikle, and D. W. H. Rankin, *Mol. Phys.*, 1973, **25**, 381.
159 W. McFarlane and J. A. Nash, *J.C.S. Chem. Comm.*, 1969, 913.
160 J. D. Kennedy and W. McFarlane, *J.C.S. Dalton*, 1973, 2134.
161 D. W. W. Anderson, J. E. Bentham, and D. W. H. Rankin, *J.C.S. Dalton*, 1973, 1215.
162 D. W. W. Anderson, E. A. V. Ebsworth, and D. W. H. Rankin, *J.C.S. Dalton*, 1973, 2370.
163 W. B. Jennings, D. R. Boyd, C. G. Watson, E. D. Becker, R. B. Bradley, and D. M. Jerina, *J. Amer. Chem. Soc.*, 1972, **94**, 8501.
164 W. McFarlane and D. S. Rycroft, *Mol. Phys.*, 1972, **24**, 893.

protio- and *P*-deuterio-dimethyl phosphite in dilute solutions of various solvents.

A few cases are now known where the signs of coupling constants can have diagnostic value. Selective irradiation experiments have been used[165] to determine the stereochemical dependence of the absolute signs [relative to $^{1}J(^{195}Pt-^{31}P)$ positive] of $^{3}J(^{31}P,^{1}H)$ in some methylplatinum(II) complexes containing tertiary phosphines [(7)—(9)]. It was found that $^{3}J(^{31}P,Me_{cis})$

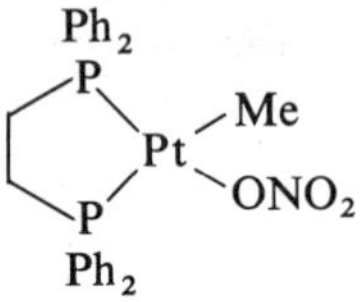

(7)

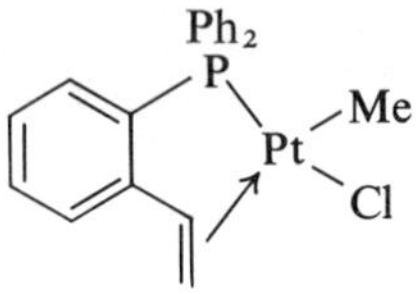

(8)

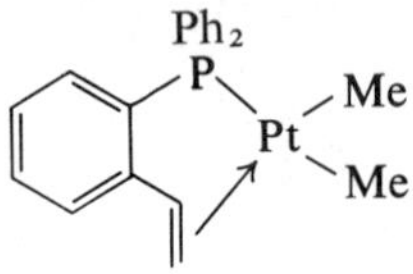

(9)

is positive and $^{3}J(^{31}P,Me_{trans})$ is negative, and this fact was used for the purpose of assignment.[148] The signs of the coupling constants in the $^{13}C$- and $^{15}N$-enriched isomers of an *N*-6-butyl-aldimine, (10; R = 9-anthryl), have been determined[166] relative to $^{1}J(^{13}C-^{1}H)$: $^{2}J(^{1}H-C{=}^{15}N)$ is positive in the *E*-isomer, (10a), and negative in the *Z*-isomer, (10b). Simultaneous homonuclear decoupling of the t-butyl protons was carried out

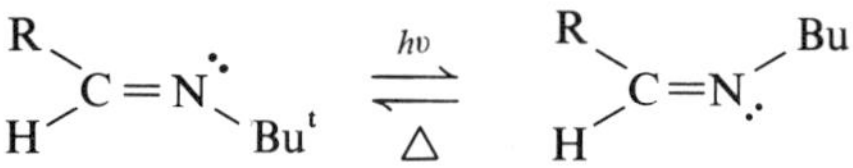

(10a) (10b)

during the $^{1}H-\{^{15}N\}$ irradiation in order to simplify the experiments. $^{1}J(^{13}C{=}^{15}N)$ is negative in both isomers. During studies[167] of coupling constants involving $^{77}Se$, it was found that the effect of electron lone pairs was very similar to the case of coupling involving $^{31}P$. Two-bond X–C–$^{1}H$ coupling is usually found to have the opposite sign to the corresponding one-bond $^{13}C$–X coupling; however, in double-resonance studies[155] of methyllithium tetramer, selective $^{1}H-\{^{13}C\}$ irradiation experiments gave bandshape changes which were interpreted as showing that $^{2}J(^{7}Li,^{1}H)/^{1}J(^{13}C-^{7}Li) > 0$, even though the actual magnitude of $^{2}J(^{7}Li,^{1}H)$ is still unknown.

In a study[168] of spin–spin coupling in the $^{15}NH_2$ fragment, it was found that $^{2}J(^{1}HN^{1}H)$ was positive relative to $^{1}J(^{15}N-^{1}H)$ negative in *o*-acetylaniline.

[165] M. A. Bennett, R. Bramley, and I. B. Tomkins, *J.C.S. Dalton*, 1973, 166.
[166] H. J. C. Yeh, H. Ziffer, D. M. Jerina, and D. R. Boyd, *J. Amer. Chem. Soc.*, 1973, **95,** 2741.
[167] W. McFarlane and D. S. Rycroft, *J.C.S. Chem. Comm.*, 1973, 10.
[168] R. E. Wasylishen and T. Schaefer, *Canad. J. Chem.*, 1973, **51,** 3087.

Other examples of relative sign determinations using multiple-resonance techniques are studies of some 2,4-disubstituted selenophens,[169] some cyclic phosphorus esters,[124,125] some phosphonate derivatives,[126] some phosphonic acid difluorides,[170] trimethylmethylenephosphorane and its nickeltricarbonyl complex,[171] some organometallic compounds,[172] 1,8-difluoronaphthalene,[173] and some products of the reactions of $PF_2X$ (X = Cl or Br) with $(MH_3)_2Y$ (M = Si or Ge; Y = O, S, Se or Te).[174]

It has been reiterated[161] that in sign determinations using double-resonance techniques it is the relative signs of the *reduced* couplings $K$ $[K_{AB} = J_{AB}.4\pi^2/h\gamma_A\gamma_B]$ that are obtained, and it has been suggested[161] that work[175] which reported negative signs for $^2J(^{31}P,^1H)$ and $^2J(^{31}P,^{19}F)$ in $H_2NP(CF_3)_2$ did not consider the negative magnetogyric ratio of $^{15}N$ in the interpretation of the double-resonance experiments.

$^1H-\{^2H\}$ selective irradiation and INDOR experiments have been used[108] to obtain the magnitudes and relative signs of dipolar and quadrupolar couplings in $CH_3CD_2OH$ and $CD_3COCH_3$ oriented in the nematic phase.

A partial analysis[176] of the planar $[AX_9]_4$ spin system present in the $[Pt(PMe_3)_4]^{2+}$ cation was helped by information obtained from $^1H-\{^{195}Pt\}$ INDOR experiments: plotting the height of the $^{195}Pt$ INDOR lines observed by monitoring various points across the proton spectrum (Figure 2) made it possible to resolve the broad proton resonance into the contributions from molecules with different phosphorus total spin quantum numbers

**C. Chemical Shift Determination.**—In principle, any multiple-resonance experiment can give a frequency which may be converted into a chemical shift of the nucleus concerned if the operating frequency of the spectrometer used to record the directly observed spectrum is known. The widely quoted formula (1)[177] which can be used to effect this conversion was designed for simple manual calculations; now that the use of cheap electronic calculators is widespread, it is probably just as quick to use the exact relationship (2)

$$\Xi = X_{obs}[1 - (f - 100\delta)/10^8] \quad (1)$$

$$\Xi = X_{obs}10^8/(10^8 + f - 100\delta) \quad (2)$$

where $\Xi$ is the required resonant frequency in a polarizing field in which TMS

169 M. P. Simonnin, M. J. Pouet, and C. Paulmier, *Tetrahedron Letters*, 1973, 4487.
170 V. I. Zakharov, Yu. V. Belov, B. I. Ionin, and A. A. Petrov, *Doklady Akad. Nauk. S.S.S.R.*, 1973, **209**, 1343.
171 K. Hildenbrand and H. Dreeskamp, *Z. Naturforsch.*, 1973, **28b**, 226.
172 F. J. Weigert and J. D. Roberts, *Inorg. Chem.*, 1973, **12**, 313.
173 S. L. Mannatt, M. A. Cooper, C. W. Mallory, and F. B. Mallory, *J. Amer. Chem. Soc.*, 1973, **95**, 975.
174 D. E. J. Arnold, J. S. Dryburgh, E. A. V. Ebsworth, and D. W. H. Rankin, *J.C.S. Dalton*, 1972, 2518.
175 A. H. Cowley, J. R. Schweiger, and S. L. Manatt, *J.C.S. Chem. Comm.*, 1970, 1491.
176 P. L. Goggin, R. J. Goodfellow, J. R. Knight, M. G. Norton, and B. F. Taylor, *J.C.S. Dalton*, 1973, 2220.
177 W. McFarlane, *Ann. Rev. N.M.R. Spectroscopy*, 1968, **1**, 135.

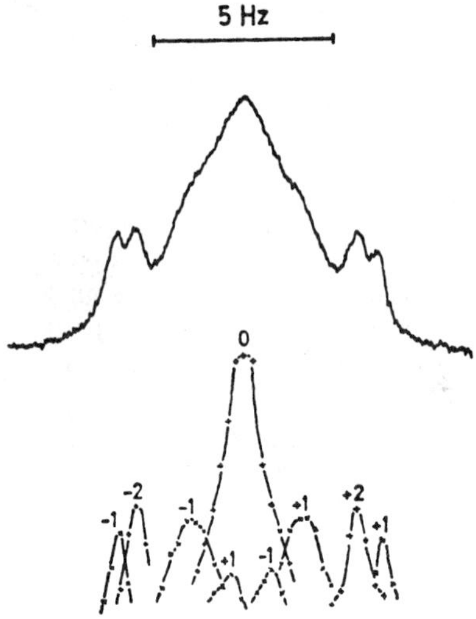


**Figure 2** $^1H$ *N.m.r. spectrum of* $[Pt(PMe_3)_4][BF_4]_2$ ($D_2O$ *solution, main resonance only*) *with* (*below*) *plot of height of each line in the* $^1H$–{$^{195}Pt$} *INDOR spectrum against proton frequency. The numbers are the sums of the phosphorus spin eigenvalues for the various lines.*

(Reproduced from *J.C.S. Dalton*, 1973, 2220)

gives a proton resonance at exactly 100 MHz, $X_{obs}$ is the observed frequency, $\delta$ is the chemical shift in p.p.m. of the proton locking reference, and $f$ is the sideband modulation frequency (spectrometer operating at 100 MHz).

*New Methods.* Chemical shifts in incomplete spin systems have been measured[31,157] by use of a selective heteronuclear triple-resonance technique, similar in principle to the method mentioned earlier which was used[151] to study $(MeO)_3PBH_3$. The compounds examined were a series of organophosphorus selenides $R^1_nR^2_{3-n}PSe$. Unless $R^1$ = H, there is no spin coupling between $^{77}Se$ and protons in $R^1$, so that $^1H$–{$^{77}Se$} double-irradiation experiments cannot be used to determine $^{77}Se$ chemical shifts. However, there is spin coupling between $^{31}P$ and $^{77}Se$, and irradiation of $^{77}Se$ perturbs the $^{31}P$ transitions. Therefore if the proton spectrum is already perturbed by irradiation of $^{31}P$, additional irradiation of $^{77}Se$ will affect the $^{31}P$ transitions and cause the perturbation in the proton spectrum to change. All the compounds examined in this way had methyl groups which were coupled to phosphorus. Using a partial decoupling technique previously described in the literature, part of the methyl doublet was removed to give a peak at the centre; this peak was monitored while the $^{77}Se$ frequency was swept. This procedure has the advantage that only peaks arising from molecules containing $^{77}Se$ are monitored. The alternative would be to monitor the peaks arising from molecules containing non-magnetic selenium and look for perturbations of the superimposed $^{77}$Se peaks (natural abundance of $^{77}$Se = *ca.* 7.6%); this was demonstrated[119] to be feasible by performing the selective triple-resonance experiment[152] necessary to compare the signs of $^{1}J(^{77}\text{Se}-^{31}\text{P})$ and $^{3}J(^{31}\text{P},^{1}\text{H})$.

*General Examples.* Reports of the use of multiple-resonance methods to determine the chemical shifts of a wide variety of nuclei have appeared during the review period and are mentioned here in order of increasing atomic mass (this cover is not exhaustive, but is more complete in the case of the more unusual nuclei).

(*i*) $^{1}$H. INDOR techniques have been used[178] to measure the positions of some hidden and unresolved lines in the proton spectrum of the pyrylium cation, $C_5H_5O^+$; this was done to increase the number of assigned lines which could be used in an iterative analysis of the spectrum.

(*ii*) $^{2}$H. Deuterium chemical shifts have been measured[179] from the $^{1}$H–{$^{2}$H} INDOR spectra obtained by monitoring the peaks in the proton spectra of monoprotio-containing 'impurities' in commercially available perdeuteriated solvents; the INDOR spectra had linewidths as small as 0.5 Hz. A $^{1}$H–{$^{31}$P,$^{2}$H} triple-resonance experiment was used in studies[164] of *P*-deuterio-dimethyl phosphite to measure the $^{2}$H resonance frequency, using the technique described above,[31,157] and so obtain $\gamma(^{1}\text{H})/\gamma(^{2}\text{H})$ for this compound and *P*-protio-dimethyl phosphite.

(*iii*) $^{7}$Li. The resonance frequency (with respect to that of protons in TMS) in methyl-lithium tetramer has been reported.[155]

(*iv*) $^{10,11}$B. The precision attainable in double-resonance experiments has been illustrated by studies[180] of the $^{10}$B–$^{11}$B isotope effect on boron nuclear shielding. Primary isotope effects on nuclear shielding are readily measured by double resonance as long as the spectra to be studied are not complicated by the presence of large numbers of lines and several different couplings. $^{1}$H–{$^{10}$B}, $^{1}$H–{$^{11}$B}, $^{19}$F–{$^{10}$B}, and $^{19}$F–{$^{11}$B} tickling experiments were used to show that the boron chemical shift between $BH_4^-$ and $BF_4^-$ is 0.11 ($\pm$ 0.03) p.p.m. greater for $^{10}$B than $^{11}$B. The $^{11}$B resonance frequency (with respect to the $^{1}$H signal of TMS) in trimethyl phosphite borane has also been reported.[151]

(*v*) $^{13}$C. Miscellaneous measurements have been reported.[151,155,156,160,181]

(*vi*) $^{14}$N. The $^{14}$N chemical shifts of several five- and six-membered rings (*e.g.* pyrroles, pyridines, diazines) have been measured[182] to an accuracy of *ca.* 1 p.p.m. (protons observed at 100 MHz), sufficient for studies of solvent effects and hydrogen bonding. Most of the compounds did not have direct N–H bonds, so the spin interactions used were 2- and 3-bond ($^{14}$N,$^{1}$H) couplings:

178 L. Radics and J. Kardos, *Org. Magn. Resonance*, 1973, **5**, 251.
179 J. R. Campbell, L. D. Hall, and P. R. Steiner, *Canad. J. Chem.*, 1972, **50**, 504.
180 W. McFarlane, *J. Magn. Resonance*, 1973, **10**, 98.
181 J. D. Kennedy and W. McFarlane, *J.C.S. Perkin II*, 1974, 146.
182 H. Saitô, Y. Tanaka, and S. Nagata, *J. Amer. Chem. Soc.*, 1973, **95**, 324.

when the C–H proton peaks were monitored the peak height was increased by 10—70% on $^{14}N$ decoupling.

(*vii*) $^{15}N$. The $^{15}N$ chemical shifts in $^{15}N$-enriched *trans*-$Co^{III}$(glycinato)$_3$ have been measured[183] using $^{1}H$–{$^{15}N$} techniques. In the *trans*-isomer all three chelate rings are different, and in $D_2SO_4$ solution three NH proton signals were observed. Three $^{15}N$ chemical shifts were obtained (the corresponding $^{14}N$ compound had been studied previously but the $^{14}N$ peaks were unresolved). $^{15}N$ chemical shifts of some $^{15}N$-enriched silylamine derivatives have been reported.[161]

(*viii*) $^{29}Si$. $^{1}H$–{$^{29}Si$} experiments have been used[184] to measure $^{29}Si$ chemical shifts in series of methylsilicon carboxylates $Me_nSi(O_2CR)_{4-n}$ ($n$ = 1—3). The range of shifts observed was relatively small and consequently it was important to locate the centre of the silicon spectrum accurately by identifying the individual components correctly, as an error of $^2J(^{29}Si,^1H)$ corresponded to an error of nearly 1 p.p.m. at the observing field strength used. $^{1}H$–{$^{29}Si$} INDOR spectra have been measured[185] for methylsilicon halides; the multiplicities observed enabled $Me_3Si$, $Me_2Si$, and MeSi moieties to be readily distinguished. The $^{29}Si$ chemical shifts of trisilylamine[161] and a silylplatinum complex[162] have also been measured using double resonance.

(*ix*) $^{31}P$. Phosphorus chemical shifts measured by double-resonance techniques have been reported in refs. 130, 140, 149, 151, 156—158, 161—165, 174, 176, and 186—188.

(*x*) $^{77}Se$. Selenium chemical shifts in phosphine selenides,[157,158] organotin selenides,[160] and $R_2Se$ (where R = $SiH_2$, $GeH_3$, and $PF_2$)[174] have been reported.

(*xi*) $^{119}Sn$. Measurements of tin compounds are now quite common, and a compilation of chemical shifts, many of which were determined by double resonance, has been published.[189] In addition to studies designed to elucidate the factors affecting nuclear shielding,[160,181,189—192] the tin chemical shift has been used as a probe to study chemical systems.[193—196] The dependence

[183] B. M. Fung, S. C. Wei, T. H. Martin, and I. Wei, *Inorg. Chem.*, 1973, **12**, 1203.
[184] W. McFarlane and J. M. Seaby, *J.C.S. Perkin II*, 1972, 1561.
[185] E. V. van den Berghe and G. P. van der Kelen, *J. Organometallic Chem.*, 1973, **59**, 175.
[186] H. Steinberger and W. Kuchen, *Z. Naturforsch.*, 1973, **28b**, 44.
[187] J. A. Mosbo and J. G. Verkade, *J. Amer. Chem. Soc.*, 1973, **95**, 204.
[188] P. Junkes, M. Baudler, J. Dobbers, and D. Rackwitz, *Z. Naturforsch.*, 1972, **27b**, 1451.
[189] P. J. Smith and L. Smith, *Inorg. Chim. Acta, Rev.*, 1973, **7**, 11.
[190] (*a*) W. McFarlane, J. C. Maire, and M. Delmas, *J.C.S. Dalton*, 1972, 1862; (*b*), E. V. van den Berghe and G. P. van der Kelen, *J. Organometallic Chem.*, 1973, **61**, 197.
[191] W. McFarlane and R. J. Wood, *J. Organometallic Chem.*, 1972, **40**, C17.
[192] L. Verdonck and G. P. van der Kelen, *J. Organometallic Chem.*, 1972, **40**, 139.
[193] J. D. Kennedy, W. McFarlane, P. J. Smith, R. F. M. White, and L. Smith, *J.C.S. Perkin II*, 1973, 1785.
[194] E. V. van den Berghe and G. P. van der Kelen, *J. Organometallic Chem.*, 1974, **72**, 65.
[195] A. G. Davies, L. Smith, and P. J. Smith, *J. Organometallic Chem.*, 1972, **39**, 279.
[196] P. J. Smith, R. F. M. White, and L. Smith, *J. Organometallic Chem.*, 1972, **40**, 341.

of tin chemical shifts on co-ordination number has been used[193] in studies of auto-association in butyltin trialkoxides; the sensitivity of the heteronuclear double resonance method is such that it was possible to make measurements by observing the complex $^{119}$Sn satellites of butyl proton signals at several dilutions and temperatures, and so to obtain thermodynamic parameters for dimer–monomer equilibria. The $^{1}$H–{$^{119}$Sn} INDOR spectra of the binary systems $Me_{4-n}SnX_n$–$Me_{4-n}Sn(SMe)_n$ (X = halogen, $n$ = 1—4) show[194] single broad resonances with chemical shifts intermediate in value between those of the isolated components, indicating the presence of exchange equilibria rapid on the n.m.r. timescale.

(*xii*) $^{125}$Te. $^{1}$H–{$^{125}$Te} experiments have been used to measure $^{125}$Te chemical shifts in a representative series of organotellurium compounds.[197] Comparison with corresponding selenium compounds showed that a plot of $^{125}$Te *vs.* $^{77}$Se chemical shifts is linear.

(*xiii*) $^{183}$W. The $^{183}$W resonance frequency (with respect to the $^{1}$H signal of TMS) in dimethylphenylphosphinepentacarbonyltungsten(0) has been reported.[156]

(*xiv*) $^{195}$Pt. Studies of $^{195}$Pt nuclear shielding have been popular, and papers[162,198] listing 44 and 47 chemical shifts have been published. Measurements of $^{195}$Pt spectra and chemical shifts have been used, firstly, to identify species present in mixtures of complexes {*e.g.* six species were identified[198] in a mixture of $(PhMeS)_2PtCl_2$ and $(PhMeS)_2PtBr_2$, *viz. cis-* and *trans*-[$(PhMeS)_2$-PtXY] where XY = $Cl_2$,$Br_2$, and BrCl}, and secondly, to identify species present in reaction product mixtures {*e.g.* it was possible to deduce[199] the number of hydride ligands bound to platinum in the products resulting from addition of HX and $X_2$ to *trans*-[$PtHY(PEt_3)_2$] and of HX to *trans*-[$PtY_2$-$(PEt_3)_2$] (X, Y = Cl, Br, I)}. Products were also identified[200] *in situ* in the reaction of $Pt_2Cl_4(SMe_2)_2$ and $(PPh_4)Cl$. $^{195}$Pt chemical shifts of some methylplatinum(II) complexes[165] and of some trimethyleneplatinum(IV) complexes[201] have been reported although there are believed[202] to be some errors in the reported[201] shifts.

$^{1}$H–{$^{195}$Pt} INDOR spectra have been reported in two papers[149,176] but the $^{195}$Pt chemical shifts were not reported.

(*xv*) $^{199}$Hg. The variation of the $^{199}$Hg chemical shift with solvent and concentration has been studied[149] in $[(MeO)_2P(O)]_2Hg$ using $^{1}$H–{$^{199}$Hg} experiments.

(*xvi*) $^{207}$Pb. Studies of $^{207}$Pb chemical shifts, including solvent and concentration effects, in $Pb^{IV}$ compounds having methyl, propynyl, vinyl, 4-fluoro-


197 H. C. E. McFarlane and W. McFarlane, *J.C.S. Dalton*, 1973, 2416.
198 W. McFarlane, *J.C.S. Dalton*, 1974, 324.
199 D. W. W. Anderson, E. A. V. Ebsworth, and D. W. H. Rankin, *J.C.S. Dalton*, 1973, 854.
200 P. L. Goggin, R. J. Goodfellow, and F. J. S. Reed, *J.C.S. Dalton*, 1974, 576.
201 P. W. Hall, R. J. Puddephatt, and C. F. H. Tipper, *J. Organometallic Chem.*, 1974, **71**, 145.
202 R. J. Puddephatt, personal communication.

phenyl, chloride, and acetate substituents have been reported, $^{1}H-\{^{207}Pb\}$ techniques being employed.[203]

## 6 Solid-state Studies

This area is discussed in detail in another chapter, but the recent use of multiple-resonance pulse techniques to give high-resolution spectra from solids is so important as to warrant brief consideration here. Normally the information which can be obtained from solids is severely limited by large linewidths arising from dipolar interactions, but recently-developed techniques make it possible to obtain information about weak effects such as chemical shifts by eliminating the dipolar interactions.[204—207] One typical scheme suggested[207] for the detection of the resonance of low-abundance spins $S$ in the presence of abundant spins $I$ is as follows:

(i) Polarize the $I$ spins in a high field.
(ii) Cool the $I$ spins to a low spin temperature in the rotating frame.
(iii) Establish cross-relaxation between the $I$ and $S$ spin systems.
(iv) Record the $S$ spin free induction decay signal with simultaneous decoupling of the $I$ spins by a suitable pulse sequence.

Fourier transformation then yields a high-resolution spectrum with linewidths (in the case of $I = {}^{1}H$, $S = {}^{13}C$) which are typically some tens of Hz and may be as small as 5 Hz. Full details of the method have been published[208] and the technique has also been patented.[209] Most applications so far have been in the determination of complete $^{13}C$ shielding tensors in solids[210—213] but other nuclei have also been studied.[214,215] It has been pointed out that the detected nuclear species need not be naturally rare and that artificial dilution can be used.[216] Many variations on the pulsing scheme used are possible,[217—222] and an important one which yields the same signal-to-noise


[203] M. J. Cooper, A. K. Holliday, P. H. Makin, R. J. Puddephatt, and P. J. Smith, *J. Organometallic Chem.*, 1974, **65**, 377.
[204] P. Mansfield and P. K. Grannell, *J. Phys.* (*C*), 1971, **4**, L197.
[205] H. E. Bleich and A. G. Redfield, *J. Chem. Phys.*, 1971, **55**, 5405.
[206] C. S. Yannoni and H. E. Bleich, *J. Chem. Phys.*, 1971, **55**, 5406.
[207] A. Pines, M. G. Gibby, and J. S. Waugh, *J. Chem. Phys.*, 1972, **56**, 1776.
[208] A. Pines, M. G. Gibby, and J. S. Waugh, *J. Chem. Phys.*, 1973, **59**, 569.
[209] M. G. Gibby, A. Pines, and J. S. Waugh, U.S.P. 3 792 346 (1974).
[210] S. Kaplan, A. Pines, R. E. Griffin, and J. S. Waugh, *Chem. Phys. Letters*, 1974, **25**, 78.
[211] J. S. Waugh, M. G. Gibby, S. Kaplan, and A. Pines, in *Proceedings 17th Congress Ampère*, 1972, p. 11.
[212] A. Pines, M. G. Gibby, and J. S. Waugh, *Chem. Phys. Letters*, 1972, **15**, 373.
[213] J. J. Chang, R. G. Griffin, and A. Pines, *J. Chem. Phys.*, 1974, **60**, 2561.
[214] P. Van Hecke, J. C. Weaver, B. L. Neff, and J. S. Waugh, *J. Chem. Phys.*, 1974, **60**, 1668.
[215] M. G. Gibby, R. G. Griffin, A. Pines, and J. S. Waugh, *Chem. Phys. Letters*, 1972, **17**, 80.
[216] C. S. Yannoni, *J. Chem. Phys.*, 1973, **58**, 1773.
[217] N. Boden and M. Gibb, *Mol. Phys.*, 1974, **27**, 1359.
[218] H. Ernst, D. Fenzke, W. Heink, H. Pfeifer, and H. Schmiedel, in *Proceedings 17th Congress Ampère*, 1972, p. 251.

ratio (*ca.* $10^3$ times that otherwise attained for $S = {}^{13}C$, $I = {}^{1}H$) has been reported; it utilizes the effect of the dilute spins upon the abundant ones whose resonance is detected.[223]

[219] J. D. Ellett, jun., R. G. Griffin, and J. S. Waugh, *J. Amer. Chem. Soc.*, 1974, **96**, 345.
[220] P. Mansfield and V. Haeberlen, *Z. Naturforsch.*, 1973, **28a**, 1081.
[221] V. Haeberlen and V. Kohlschütter, *Chem. Phys.*, 1973, **2**, 76.
[222] A. Pines and T. W. Shattuck, *Chem. Phys. Letters*, 1973, **23**, 614.
[223] P. K. Grannell, P. Mansfield, and M. A. B. Whittaker, *Phys. Rev.* (*B*), 1973, **8**, 4149.

# 6
# Macromolecules

BY I. D. ROBB AND G. J. T. TIDDY

## 1 Introduction

This chapter reviews reports dealing with the n.m.r. of macromolecules, published between June 1973 and May 1974. It is divided according to the nature of the material under investigation, and then by type of nucleus and technique.

The studies under review have in general been concerned with the determination of the structures and conformations of macromolecules or with the interactions between small molecules and macromolecules.

## 2 Reviews

Most reviews of the application of n.m.r. to macromolecules have been concerned with the determination of structure or conformation of biopolymers. Dwek[1] has surveyed the use of small molecules, particularly paramagnetic ions and spin labels, in determining the structure of proteins and enzymes. The conformation of proteins, their pH dependence and the influence of inhibitors has also been reviewed by Ruterjans[2] and a short survey[3] of the applications of $^{1}H$ and $^{13}C$ resonance to protein conformation has been presented. Kearns and Shulman[4] have given an account of the recent results of high-resolution techniques in the study of transfer ribonucleic acid molecules, while a review of the helix-coil transition in polynucleic acids has been presented[5] which covers both pulsed and CW techniques. The application of $^{1}H$ and $^{13}C$ resonance to a wide range of water- and oil-soluble polypeptides[6] and to polysaccharides[7] has been reviewed. A very useful collection[8] of short reviews concerning biopolymers has been published, and a small section on the analysis of synthetic polymers has been included within a general review of analysis by n.m.r.[9]

[1] R. A. Dwek, 'Nuclear Magnetic Resonance in Biochemistry', Clarendon Press, Oxford, 1973.
[2] H. Ruterjans, *Dechema Monograph.*, 1973, **71**, 219.
[3] K. Wüthrich, *Pure. Appl. Chem.*, 1974, **37**, 235.
[4] D. R. Kearns and R. G. Shulman, *Accounts Chem. Res.*, 1974, **7**, 33.
[5] T. Wilczok and B. Lubas, *Studia Biophys.*, 1973, **38**, 85.
[6] E. M. Bradbury, P. D. Cary, C. Crane-Robinson, and P. G. Hartman, *Pure Appl. Chem.*, 1973, **36**, 53.
[7] M. Vincendon, *Bull. Soc. chim. France*, 1973, **12**, 3501.
[8] *Ann. New York Acad. Sci.*, 1973, **222.**
[9] J. R. Wasson and D. K. Johnson, *Analyt. Chem.*, 1974, **46**, 314R.

## 3 Synthetic Macromolecules

**A. [1]H High-resolution Work.**—Papers in this section are grouped into studies on the analysis of polymers, and other reports.

*Analysis.* One of the most important and useful applications of high-resolution techniques is in the determination of the structure of polymers and the distribution of various units within the polymer. In Table 1 are summarized

**Table 1** *Polymer Analysis by* $^1$H *High-resolution N.M.R.*

| *Polymer* | *Comments* | *Ref.* |
|---|---|---|
| Poly(methacrylic ester) | Effect of steric requirement of various ester groups on tacticity determined from $\alpha$-$CH_3$ resonance. | 10 |
| Poly(methyl methacrylate) | Time and temperature dependence of chemical shifts induced by $Eu(thmd)_3$ measured; pentad configurations obtained. | 11 |
| Polymethacrylonitrile | $\alpha$-$CH_3$ resonance used to determine tacticities of polymers prepared from different initiators. | 12 |
| Poly(vinyl chloride) | Influence of preparation method on tacticity investigated. | 13 |
| Poly(2,3-dimethyl-1,3-butadiene) | Dyad and triad structures calculated; *cis* and *trans* contents estimated. | 14 |
| Polybutadiene | 300 MHz spectra and computer simulation of olefinic resonances indicated Bernoullian distribution of *cis*-1,4 and *trans*-1,4 units. | 15 |
| Polybutadiene | 300 MHz spectra used to determine *cis–trans* isomeric content. | 16 |
| Isotactic poly(but-1-ene) and deuteriated analogues | Chemical shifts and coupling constants used to obtain possible configurations of polymers. | 17 |
| Oligomers of oxy-2,2-dimethylethylene carbonyl | Spectral assignments made; end-group and solvent effects noted. | 18 |

[10] K. Matsuzake, T. Kanai, K. Yamawaki, and K. P. Samre-Rung, *Makromol. Chem.* 1973, **174**, 215.
[11] S. Amiya, I. Ando, and R. Chûjô, *Polymer J.* 1973, **4**, 385.
[12] T. Suzuki, S. Koshiro, and Y. Takegami, *Polymer*, 1973, **14**, 549.
[13] A. M. Hassan, *J. Polymer Sci., Part A-2, Polymer Phys.*, 1974, **12**, 655.
[14] D. Blondin, J. Regis, and J. Prud'homme, *Macromolecules*, 1974, **7**, 187.
[15] E. R. Santee, V. D. Mochel, and M. Morton, *J. Polymer Sci., Part B, Polymer Letters*, 1973, **11**, 453.
[16] E. R. Santee, R. Chang, and M. Morton, *J. Polymer Sci., Part B, Polymer Letters*, 1973, **11**, 449.
[17] J. G. Murray, J. Zymonas, E. R. Santee, and H. J. Harwood, *Polymer Preprints*, 1973, **14**, 1157.
[18] E. Ogawa, S. Kondo, and K. Tanaka, *Makromol. Chem.*, 1973, **169**, 261.

**Table 1** *continued*

| *Polymer* | *Comments* | *Ref.* |
|---|---|---|
| Epoxy resins and polycaprolactone polyols | Number of repeat units calculated from ratios of terminal: central protons. | 19 |
| Novolac resins | Solvent shifts of pyridine bases utilized to improve spectral resolution. | 20 |
| $Ti(NEt_2)_4$–$AlMe_3$ Catalyst for styrene | Spectrum of the catalyst alone allowed the species responsible for the polymerization of styrene to be identified. | 21 |
| Cellulose triacetate | Coalescence of two acetate resonances at high temperature incorrectly assigned to higher population of boat conformation. Spectra show chemical shift variations rather than collapse of resonances. | 22 |
| Copolymer of 1,1,4,4-[$^2H_4$]butadiene and acrylonitrile | 300 MHz spectra showed two olefinic AB patterns which were attributed to random configurations of neighbouring acrylonitrile methine carbons. | 23 |
| Copolymer of butadiene and acrylonitrile | Triad and pentad sequences determined with the aid of partially deuteriated samples for polymers from different initiating systems. | 24 |
| Copolymer of vinyl chloride and propene | Reactivity ratios calculated, and methine resonance of vinyl chloride interpreted in terms of triad structures. | 25 |
| Copolymer of vinyl chloride and propene | Cotacticity parameters measured and activation energies for propagation steps calculated. | 26 |
| Copolymer of THF and diketene | Segment distribution shown to be random, and fractions of two possible addition modes were determined. | 27 |

[19] A. D. Hammerich and F. G. Willeboordse, *Analyt. Chem.*, 1973, **45,** 1696.
[20] T. Yoshikawa and K. Kimura, *Makromol. Chem.*, 1974, **175,** 1001.
[21] N. Koide, K. Iimura, and M. Takeda, *J. Polymer Sci., Part A-1, Polymer Chem.*, 1973, **11,** 3161.
[22] K. Ogura, H. Sobue, M. Kasuga, and S. Nakamura, *J. Polymer Sci., Part B, Polymer Letters*, 1973, **11,** 421.
[23] G. A. Lindsay, E. R. Santee, and H. J. Harwood, *Polymer Preprints*, 1973, **14,** 646.
[24] T. Suzuki, Y. Takegami, J. Furukawa, E. Kobayashii, and Y. Arai, *Polymer J.*, 1973, **4,** 657.
[25] G. Sielaff and D. O. Hummel, *Makromol. Chem.*, 1974, **175,** 1561.
[26] G. Sielaff and D. O. Hummel, *Makromol. Chem.*, 1974, **175,** 1579.
[27] M. Okada, Y. Yokoyama, and H. Sumitomo, *Makromol. Chem.*, 1973, **170,** 131.

**Table 1** *continued*

| *Polymer* | *Comments* | *Ref.* |
|---|---|---|
| Copolymer of vinyl chloride and ethylene | Spectral analysis showed polymerization followed Bernoullian statistics. | 28 |
| Copolymer of $\beta$-[$^2H_2$]styrene and [$^2H_5$]methacrylonitrile | Spectra indicated highly alternating structure. | 29 |
| Copolymer of $\alpha$-[$^2H_1$]styrene and acrylic esters | Brief spectral analysis of backbone $CH_2$ resonances in terms of dyad and triad sequences. | 30 |
| Copolymer of styrene and acrylates | Sequence distribution from 300 MHz spectra; possible reaction processes proposed. | 31 |
| Copolymer of styrene and methacrylic acid | Methine and $\alpha$-$CH_3$ resonances of copolymers of styrene–methyl methacrylate copolymers compared with title compounds. | 32 |
| Copolymer of $\alpha$-phenyl acrylate and methyl methacrylate | Composition of copolymer determined as a function of reaction temperature. | 33 |
| Copolymer of methacrylic acid and benzyl methacrylate<br>Copolymer of methacrylic acid and diphenyl methacrylate | Copolymer preparation by reaction of PMAA with diazotoluene or diazophenylmethane. Spectral analysis showed the aromatic groups were substituted according to Bernoullian statistics. | 34 |
| Copolymer of methacrylic acid and methyl methacrylate | Copolymer resulting from reaction of PMAA with diazomethane analysed in terms of triad distribution; Bernoullian statistics found. | 35 |

references to the analyses of various polymers, together with appropriate remarks.

*Other Studies.* In this section, mainly theoretical papers dealing with chemical shifts or molecular motion are considered. The n.m.r. spectrum[36] of a linear chain containing two types of spin-$\frac{1}{2}$ nuclei, with slightly different chemical shifts and coupled by a scalar interaction between nearest neighbours, was

[28] T. K. Wu, *Macromolecules*, 1973, **6**, 737.
[29] R. C. Chang and H. J. Harwood, *Polymer Preprints*, 1973, **14**, 31.
[30] K. Yokota, T. Hirabayashi, and K. Takahashi, *J. Polymer Sci., Part B, Polymer Letters*, 1974, **12**, 49.
[31] F. A. Blouin, R. C. Chang, M. H. Quinn, and H. J. Harwood, *Polymer Preprints*, 1973, **14**, 25.
[32] R. E. Bockrath and H. J. Harwood, *Polymer Preprints*, 1973, **14**, 1163.
[33] K. Hatada, J. Ohshima, T. Komatsu, S. Kokan, and H. Yuki, *Polymer*, 1973, **14**, 565.
[34] D. Strasilla and E. Klesper, *Makromol. Chem.*, 1974, **175**, 535.
[35] E. Klesper, D. Strasilla, and W. Regel, *Makromol. Chem.*, 1974, **175**, 523.
[36] Y. Ayant, E. Belorizky, R. Casalengo, D. Gagnaire, and M. Taieb, *J. Chem. Phys.*, 1973, **59**, 4735.

calculated and an expression for the first 10 moments given. A detailed theoretical description[37] of the effect of anisotropic motion in molten polymers on n.m.r. signals has been given, using a model for chain fluctuations, analogous to that of Rouse. Predictions from the theory were in agreement with observations on molten poly(*cis*-buta-1,4-diene). Theoretical calculations[38] of methylene chemical shifts in long hydrocarbons were made, using Pople's approximation and the Monte Carlo method. In the absence of averaging processes, doublets were obtained which were in reasonable agreement with experimental results for linear hydrocarbons in α-chloronaphthalene. Zachmann[39] obtained moderately narrow resonances (*ca.* 50 Hz) from partially crystalline polyethylene in solution and in the molten state. The spectra were approximately resolved into two Lorenzian lines, arising from the amorphous regions, and information was obtained on the structure of the solid polymer.

**B. Broadline N.M.R.**—Broadline studies have mainly been used to determine either the nature of the molecular motions taking place in solid polymers, or the relative amounts of various phases present in solids.

There are several controversies concerning the structure of partially crystalline polymers, one of them being whether or not the polymer chains exist in bundles. A phase-composition study[40] of partially crystalline polymers such as nylon and polyethyleneterephthalate indicated that the polymer chains did exist in bundles but that the lifetime of each bundle is $<10^{-4}$ s. In a similar paper,[41] broadline spectra of several partially crystalline polymers were decomposed into two or more components, and the crystallinity estimated. Both broadline[42] and high-resolution n.m.r. were used to investigate the structure of partially crystalline polymers such as polyethyleneterephthalate. Fractions of crystalline and amorphous component were determined, and the nature of the mobile fraction discussed. Phase transitions[43] in poly(*trans*-buta-1,4-diene) crystals, induced by temperature changes or addition of solvent, were studied by broadline n.m.r. and differential scanning calorimetry techniques, and information was obtained about the fraction and location of the amorphous material present. Polyethylene oxide[44] produced both a broad and a sharp line below 330 K and these were identified with the bulk and surface phases, respectively, of the polymer. Complex changes between two solid polymer phases were invoked to explain annealing behaviour below 330 K. The broadline[45] spectra of four n-paraffins, having

37 J. P. Cohen-Addad, *J. Chem. Phys.*, 1974, **60**, 2440.
38 I. Ando and A. Nishioka, *Makromol. Chem.*, 1973, **171**, 195.
39 H. G. Zachmann, *J. Polymer Sci., Part C, Polymer Symposia*, No. 42, 1973, 693.
40 H. G. Zachmann, *Kolloid-Z.*, 1973, **251**, 951.
41 K. Bergmann, *Kolloid-Z.*, 1973, **251**, 962.
42 H. G. Zachmann, *J. Polymer Sci., Polymer Symposia*, No. 43, 1973, 111.
43 S-B. Ng, J. M. Stellman, and A. E. Woodward, *J. Macromol. Sci.*, 1973, **B7**, 539.
44 Ch. Fritzsche and E. W. Fischer, *Kolloid-Z.*, 1973, **251**, 721.
45 O. Phaovibul, H. Čačković, J. Loboda-Čačković, and R. Hosemann, *J. Polymer Sci., Part A-2, Polymer Physics*, 1973, **11**, 2377.

chain lengths about $C_{30}$, gave 3 components which were attributed to rigid chains (*ca.* 95%), chains of intermediate mobility (*ca.* 4%), and mobile chains (*ca.* 1%). These data were used, together with *X*-ray results, in one determination of various crystal parameters; unfortunately, however, criteria for the purity of the samples are not given. Rotation at the magic angle[46] of polymers such as butadiene–styrene copolymers, above their glass transition temperatures, was used to eliminate the dipolar contribution to the linewidth, in a similar manner to earlier reports by Schneider and co-workers.

**C. Spin-echo N.M.R.**—Spin-echo techniques have been used in two main areas of study: the structure of solid polymers and the kinetics of polymerization processes.

Proton $T_1$ and $T_2$ measurements were used[47] to investigate the effects of carbon black filler on the molecular mobility of natural rubber. The effects of swelling with $CCl_4$ were also examined. The bulk of the rubber acted similarly to normal rubber but there was a strongly immobilized layer adjacent to the carbon particles. Broadline[48] and pulsed techniques were used to monitor the curing of unsaturated polyester resin. Three components were detected, which were unreacted monomer, free polyester, and cured, glass-like resin, and their behaviour was followed as a function of various curing processes. For mixtures[49] of epoxy resins and styrene–methylmethacrylate copolymers, the room temperature $T_1$ minimum occurred at higher temperatures than for the pure resin. An increase in the mobility of the copolymer in boundary regions was noticed. The curing[50] of the diglycidyl ether of bisphenol-A with 4,4′-methylenedianiline was monitored by $T_1$, $T_{1\rho}$, and $T_2$ measurements. The observed time-dependent correlation frequencies were expressed in terms of reduced time functions, and chemical rate constants and activation energies were estimated for the curing process. Both $T_1$ and $T_{1\rho}$ were measured[51] on bisphenol-A polycarbonate homopolymers and copolymers as a function of temperature. Several transitions were observed and correlated with other physical properties such as the impact strength of the material. In a related paper[52] $T_1$ and $T_{1\rho}$ measurements were made on bisphenol-A polycarbonate blends, copolymers, and composites as a function of temperature. The various transitions noted were related to the different phases present and the data were correlated with transmission and scanning electron micrographs. The effect of mixing[53] and subsequent irradiation of a 2:1 mixture of PVC and tetraethylene glycol dimethacrylate has been investigated by pulsed techniques.

[46] H. Benoit and M. Rabii, *Chem. Phys. Letters*, 1973, **21,** 466.
[47] T. Nishi, *J. Polymer Sci., Part A-2 Polymer Phys.*, 1974, **12,** 685.
[48] K. Bergmann and K. Demmler, *Colloid and Polymer Sci.*, 1974, **252,** 193.
[49] Yu. S. Lipatov and F. G. Fabulak, *Vysokomol. Soedineniya (A)*, 1973, **15,** 1272.
[50] D. W. Larsen and J. H. Strange, *J. Polymer Sci., Part A-2, Polymer Phys.*, 1973, **11,** 1453.
[51] D. Stefan and H. L. Williams, *J. Appl. Polymer Sci.*, 1974, **18,** 1279.
[52] D. Stefan and H. L. Williams, *J. Appl. Polymer Sci.*, 1974, **18,** 1451.
[53] D. D. Davies and W. P. Slichter, *Macromolecules*, 1973, **6,** 728.

The complex behaviour of both $T_1$ and $T_2$ as a function of temperature was interpreted in terms of ordering of polymer chains or segmental motion of the various groups of the polymer. Werbelow and Marshall[54] have studied theoretically the time dependence of the longitudinal and transverse magnetization for methyl groups attached to a symmetric top molecule and applied the results to macromolecules in solution. For small molecules, spin-rotation tends to mask an important non-exponential behaviour of the magnetization. In another paper,[55] the authors qualify the previously published derivation of the effect of methyl rotation. Pitfalls in obtaining $\tau_c$ from $T_1$ and $T_2$ data are discussed and it is shown how the angle between methyl rotation and macromolecular rotation can be determined. The restricted anisotropic motion of liquid polymers can give rise to non-averaged dipolar interactions. A note[56] records a theoretical expression and results describing the signal observed after a $\frac{\pi}{2}$–$\tau$–$(\frac{\pi}{4})^{90}$–$\tau'$–$(\frac{\pi}{4})^{90}$ pulse sequence. As an example results for liquid poly(*cis*-buta-1,4-diene) are given, although no details about chain motion are discussed.

**D. $^{13}$C N.M.R.**—Papers describing $^{13}$C n.m.r. studies have been grouped into sections on structural analysis and other studies.

*Analysis and Structure Determination.* One of the most useful applications of $^{13}$C n.m.r. is for the simple analysis of polymers particularly where the proton spectra are poorly resolved. Compared to previous years proportionally greater emphasis has been given to $^{13}$C studies than $^{1}$H studies.

**Table 2** *Synthetic Polymer Studies by N.M.R.*

| *Polymer* | *Comments* | *Ref.* |
|---|---|---|
| Poly-buta-1,4-diene | 66.9 MHz spectra, giving improved detailed assignments. | 57 |
| Polybutadiene | Influence of *cis–trans*-isomerization on methylene resonance reassigned. | 58 |
| Polybutadiene | Spectra from model olefinic compounds indicated that earlier assignments for 1,4 polymers were incorrect. | 59 |
| Hydroxy-terminated polybutadiene | $^{13}$C and $^{1}$H spectra indicated terminal groups are primary alcohols; $^{19}$F spectrum of reaction with $CF_3CO_2H$ supported this conclusion. | 60 |

54 L. G. Werbelow and A. G. Marshall, *J. Magn. Resonance*, 1973, **11**, 299.
55 L. G. Werbelow and A. G. Marshall, *J. Amer. Chem. Soc.*, 1973, **95**, 5132.
56 J. P. Cohen-Addad and J. Duhterian, *Compt. rend.*, 1973, **277**, *C*, 1315.
57 K. F. Elgert, B. Stützel, P. Frenzel, H. J. Cantow, and R. Streck., *Makromol. Chem.*, 1973, **170**, 257.
58 Y. Tanaka and K. Hatada, *J. Polymer Sci., Part B, Polymer Letters*, 1973, **11**, 569.
59 A. D. M. Clague, J. A. M. van Broekhoven, and J. W. de Haan, *J. Polymer Sci., Part B, Polymer Letters*, 1973, **11**, 299.
60 K. C. Ramey, M. W. Hayes, and A. G. Altenau, *Macromolecules*, 1973, **6**, 795.

**Table 2** *continued*

| *Polymer* | *Comments* | *Ref.* |
|---|---|---|
| Polybuta-1,4-diene prepared with $\pi$-alkyl nickel trifluoroacetate | Spectral analysis gave distribution of *cis* and *trans* units. | 61 |
| Equibinary (*cis*-1,4–1,2)polybutadiene | Spectral analysis consistent with random distribution of monomers. | 62 |
| Polybuta-1,4-diene prepared by bis-($h^3$-alkyl nickel) trifluoroacetate | Short note indicating influence of polymerizing conditions on the distribution of *cis* and *trans* units along chain. | 63 |
| Polypropylene | Short note giving additional peak assignments and information on stereochemical statistics. | 64 |
| Polypropylene | Spectra used to identify pentad structure of amorphous polymer. | 65 |
| Polyisoprene initiated by t-butyl-lithium | $^1H$ and $^{13}C$ spectra, with aid of model compounds, indicated that the chain end has predominantly *trans* structure. | 66 |
| Poly(vinyl chloride) | Spectra interpreted in terms of triads and tetrads; indicated Bernoullian statistics at 273 K for radical polymerization in bulk. | 67 |
| Poly(vinyl chloride) | Resonances of methylene carbons interpreted in terms of tetrads, and of methyne carbons in terms of pentads. | 68 |
| Polystyrene | Structure of metallated polymer confirmed by using a $^{13}C$-enriched methyl group. | 69 |
| Poly(vinyl methyl ether) | Shifts of methylene carbons calculated by CNDO/2 method and found to be in reasonable agreement with experiment. | 70 |
| Polymethacrylonitrile | Spectra interpreted in terms of pentads, allowing propagation mechanisms to be determined for different initiating systems. | 71 |

[61] F. Conti, A. Segre, P. Pini, and L. Porri, *Polymer*, 1974, **15**, 5.
[62] J. Furukawa, E. Kobayashi, N. Katsuki, and T. Kawagoe, *Makromol. Chem.*, 1974, **175**, 237.
[63] M. Julémont, E. Walckiers, R. Warin, and P. Teyssié, *Makromol. Chem.*, 1974, **175**, 1673.
[64] A. Zambelli, D. A. Dorman, A. I. R. Brewster, and F. A. Bovey, *Macromolecules*, 1973, **6**, 925.
[65] J. C. Randall, *J. Polymer Sci., Part A-2, Polymer Physics*, 1974, **12**, 703.
[66] S. Bronstein, S. Bywater, and D. J. Worsfold, *Macromolecules*, 1973, **6**, 715.
[67] Q-T. Pham, J. L. Millan, and E. L. Madruga. *Makromol. Chem.*, 1974, **175**, 945.
[68] C. J. Carman, *Macromolecules*, 1973, **6**, 725.
[69] D. C. Evans, L. Phillips, J. A. Barrie, and M. H. George, *J. Polymer Sci., Part B, Polymer Letters*, 1974, **12**, 199.
[70] I. Ando, M. Kondo, T. Ikuta, and A. Nishioka, *Makromol. Chem.*, 1973, **169**, 285.
[71] Y. Indue, K. Koyama, R. Chûjô, and A. Nishioka, *Makromol. Chem.*, 1974, **175**, 277.

**Table 2** *continued*

| *Polymer* | *Comments* | *Ref.* |
|---|---|---|
| Polypentenamer | *Cis* and *trans* content determined from intensities of methylene carbon resonances. | 72 |
| Polypentenamer | *Cis–trans* sequence distribution determined and compared with data from polybutadienes. | 73 |
| Poly(vinyl alcohol) | Tacticities determined by $^{13}C$ and $^{1}H$ resonance and found to be in good agreement; sequences used to derive information on the statistics of polymerization mechanism. | 74 |
| Polyethyleneimine | Spectral assignments made; deshielding effect of the different nitrogen atoms on the backbone and side-chain resonances compared. | 75 |
| Polymers of mono-substituted thiiranes | Both $^{13}C$ and $^{1}H$ spectra used to deduce tacticities and details of the propagating process. | 76 |
| Phenol–formaldehyde resins | Structure determined with aid of model compounds; chemical shifts correlated with the reactivity of the formaldehyde with the model compounds. | 77 |
| Polyalkenylenes and some diolefin polymers and copolymers | Simple spectral assignment made, distinguishing between 1,2- and 1,4-*cis* and -*trans* structures. | 78 |
| Copolymers of ethylene and propylene | Assignments made and dyad sequence lengths and distributions calculated. | 79 |
| Terpolymers of ethylene and propylene | Spectral analysis yielded improved sequence distributions and reactivity ratios | 80 |
| Long-chain branched alkanes | Spectra useful as basis for interpretation of ethylene–propylene co-polymers. | 81 |

[72] C. J. Carman and C. E. Wilkes, *Macromolecules*, 1974, **7,** 40.
[73] H. Y. Chen, *J. Polymer Sci., Part B, Polymer Letters*, 1974, **12,** 85.
[74] T. K. Wu, and D. W. Ovenall, *Macromolecules*, 1973, **6,** 582.
[75] G. M. Lukovkin, V. S. Pshezhetsky, and G. A. Murtazaeva, *European Polymer J.*, 1973, **9,** 559.
[76] K. J. Ivin, E. D. Lillie, and I. H. Petersen, *Makromol. Chem.*, 1973, **168,** 217.
[77] Y. Mukoyama, T. Tanno, H. Yokokawa, and J. Fleming, *J. Polymer Sci., Part A-1, Polymer Chem.*, 1973, **11,** 3193.
[78] G. Gatti and A. Carbonaro, *Makromol. Chem.*, 1974, **175,** 1627.
[79] Y. Tanaka and K. Hatada, *J. Polymer Sci., Part A-1, Polymer Chem.*, 1973, **11,** 2057.
[80] C. E. Wilkes, C. J. Carman, and R. A. Harrington, *J. Polymer Sci., Polymer Symposia*, No. 43, 1973, 237.
[81] C. J. Carman, A. R. Tarpley, jun., and J. H. Goldstein, *Macromolecules*, 1973, **6,** 719.

**Table 2** *continued*

| *Polymer* | *Comments* | *Ref.* |
|---|---|---|
| Copolymers of ethylene and vinyl acetate | Backbone and side-chain resonances interpreted in terms of triad or pentad sequences with aid of model compounds. | 82 |
| Copolymers of styrene and butadiene | Dyad populations and average block lengths calculated. | 83 |
| Copolymer of styrene and acrylonitrile | Average acrylonitrile sequence lengths and distribution calculated. | 84 |

*Other Studies.* A theory[85] describing $T_1$, $T_2$, and NOE values in terms of a distribution of correlation times has been developed. A chi-square distribution of correlation times was used which showed good agreement with results from isotactic polystyrene (in solution), solid polyisoprene, and solid *cis*-polybutadiene. Four parameters were used for the equations deriving the values for $T_1$ and $T_2$. Relaxation times[86] have been measured for polyisobutylene, polyacrylonitrile, poly(vinyl chloride), and poly(vinyl alcohol). They were shown to be independent of stereochemical configuration or mol. wt. (above about 3000), and were determined by ($^{13}C$,$^{1}H$) dipole–dipole interactions.

The relaxation rates[87] for polypropylene and polyethylene in solution and the molten state were measured, allowing calculation of the $\tau_c$ and $E_a$ for the polymer segmental motion and rotation of the methyl group. Correlation times for the backbone nuclei were also calculated and were compared with the results of other workers.

For poly(2-vinyl pyridine)[88] in $CDCl_3$ and $CD_3OD$, $T_1$ was found to be almost independent of molecular weight or tacticity, and correlation times for the backbone nuclei were slightly longer than for the ring carbons. A short note[89] presented $T_1$ values of poly(n-butylmethacrylate) and discussed qualitatively the distribution of correlation times from the backbone along the butyl chain.

**E. $^{19}$F N.M.R.**—In a short note,[90] the effect of pressures of up to 6000 kg $cm^{-2}$ on the molecular motion of PTFE was described using broadline techniques. Motional narrowing of the broad component was observed,

82 T. K. Wu, D. W. Ovenall, and G. S. Reddy, *J. Polymer Sci., Part A-2. Polymer Phys.*, 1974, **12**, 901.
83 A. R. Katritzky and D. E. Weiss, *J.C.S. Chem. Comm.*, 1974, **10**, 401.
84 E. O. Stejskal and J. Schaefer, *Macromolecules*, 1974, **7**, 14.
85 J. Schaefer, *Macromolecules*, 1973, **6**, 882.
86 Y. Inoue, A. Nishioka, and R. Chûjô, *J. Polymer Sci., Part A-2, Polymer Phys.*, 1973, **11**, 2237.
87 Y. Inoue, A. Nishioka, and R. Chûjô, *Makromol. Chem.*, 1973, **168**, 163.
88 C. Chachaty, A. Forchioni, and J. C. Ronfard-Haret, *Makromol. Chem.*, 1973, **173**, 213.
89 G. C. Levy, *J. Amer. Chem. Soc.*, 1973, **95**, 6117.
90 C. Nakafuku, S. Taki, and T. Takemura, *Polymer*, 1973, **14**, 558.

indicating that the interchain potential is fairly anharmonic and that the vibration of the chain along the interchain direction became more incoherent with increasing pressure.

The structure, sequence distribution, and configuration of tetrafluoroethylene–propylene copolymers were deduced from the variation of the $^{19}F$ chemical shift differences between nonequivalent nuclei.[91] Deuteriated polymers assisted in the simplification of the spectra.

## 4 Biological and Related Macromolecules

**A. $^1H$ N.M.R. of Synthetic Polypeptides.**—The controversy over the assignment of the double α-CH resonance has attracted more attention in the past year. Tam and Klotz[92] have presented further data to support the assignment of the resonances to protonated and nonprotonated amide residues; they dispute the claim of Goodman and Ueyama[93] that one of the peaks is due to residual water and impurities in TMS. Ferretti and collaborators[94—96] favour the assignment of the resonances to different conformational states involving the helix and coil forms. Theoretical calculations[94] indicate that the peaks are due to the all-random-coil species and conformations containing at least one helix unit, Using poly(α-benzyl-L-glutamate) (PBLG) samples with different molecular weights, the double peaks were shown to have chain-length-dependent behaviour.[95] Bandshape analysis using a two-site chemical exchange reproduced the observed resonances and enabled lifetimes in the two sites to be determined. Evidence against polydispersity in the PBLG sample being the cause of the double peak was obtained[96] when the α-CH region was observed not to change after the sample had been fractionated using gel permeation chromatography. However, Nagayama and Wada have reported[97] that the α-CH bandshape is dependent on sample polydispersity, and that single broad resonances can be observed for monodisperse samples of PBLG. Bradbury and co-workers[98,99] also favour polydispersity as the explanation of the double peak. From spectra of fractionated and unfractionated PBLG they conclude[98] that the double peak is the result of conformational heterogeneity arising from molecular weight polydispersity. They also demonstrate[99] that double peaks are observed in non-protonating solvent mixtures, showing that the 'protonation' explanation is unlikely. Only a very small solvent dependence of the α-CH helix-coil chemical shift difference was

[91] K. Ishiguve, Y. Tebata, and K. Oshima, *Macromolecules*, 1973, **6**, 584.
[92] J. W. O. Tam and I. M. Klotz, *Biopolymers*, 1973, **12**, 2435.
[93] M. Goodman and N. Ueyama, *Biopolymers*, 1973, **12**, 2639.
[94] J. A. Ferretti, R. L. Jernigan, and G. H. Weiss, *J. Polymer Sci., Part C, Polymer Symposia*, No. 42, 1973, 1051.
[95] J. A. Ferretti and R. L. Jernigan, *Macromolecules*, 1973, **6**, 687.
[96] J. B. Milstien and J. A. Ferretti, *Biopolymers*, 1973, **12**, 2335.
[97] K. Nagayama and A. Wada, *Biopolymers*, 1973, **12**, 2443.
[98] E. M. Bradbury, C. Crane-Robinson, and P. G. Hartman, *Polymer*, 1973, **14**, 543.
[99] E. M. Bradbury, C. Crane-Robinson, L. Paolillo, and P. A. Temussi, *Polymer*, 1973, **14**, 303.

observed with halogenoacetic acids and various cosolvents.[100] The existence of helix and coil forms of random D,L-copolypeptides was indicated by the α-CH resonance.[101] For poly(α-benzyl-D,L-glutamate) the conformation was fully helical in chloroform and DMF, but only the coil form existed in DMSO. Sample polydispersity was also discussed in a further paper; proton relaxation measurements on the benzyl ring of PBLG in different $CDCl_3$–TFA solutions were interpreted to show disruption of the side-chain secondary structure followed by the helix-coil conformational change.[102]

In a study of a different polypeptide, poly(α-phenacyl-L-glutamate), using other techniques in addition to n.m.r., the polymer was shown[103] to undergo a helix-coil transition, and also to form gels where no high-resolution spectrum was observed. It has been suggested that the triple helix formation for a polytripeptide (Pro-Pro-Gly)$_n$ at low temperatures involves the conversion of *cis*-proline peptide bonds to the all-*trans* form, following observation of the the α-CH resonance and the appearance of a new glycine NH peak.[104] *cis–trans*-Isomerization about amide CN bonds was reported[105] for model compounds of (*N*-alkylglycine) polymers. The equilibrium populations of the different conformers were almost independent of the size of the alkyl group, unlike the polymers, but the $\Delta G^{\ddagger}$ values for the isomerization were virtually the same as that of polysarcosine.

Chemical shift changes were used to investigate the interaction between $Co^{2+}$ ions and polyglutamic acid.[106] The shifts were attributed to the contact or pseudo-contact mechanisms and it was concluded that $Co^{2+}$ complexed with the carbonyl group and the peptide NH.

**B. $^1$H N.M.R. of Natural and Related Polypeptides and Polysaccharides.**—Broadline[107] n.m.r. was used in conjunction with dynamic mechanical testing to investigate relaxation in different collagen samples. Above 250 K, a narrow n.m.r. resonance was observed in addition to the broad line, and assignments of this resonance, together with motions over the range 120—360 K, were discussed. The structure of a cross-link,[108] isolated from borohydride-reduced cow skin collagen, was determined from simple $^1$H and $^{13}$C n.m.r. spectra and mass spectrometry data.

The dependence of the chemical shifts of the peptide protons of a synthetic polypeptide [resembling elastin, and containing (L-Val-L-Pro-Gly-L-Val-Gly)$_n$ in [$^2H_6$]DMSO upon the temperature and upon the methanol:tri-

100 L. Ferrara and P. A. Temussi, *Biopolymers*, 1973, **12**, 1451.
101 L. Paolillo, P. A. Temussi, E. Trivellone, E. M. Bradbury, and C. Crane-Robinson, *Macromolecules*, 1973, **6**, 831.
102 Mingjien Chien, E. T. Samulski, and C. G. Wade, *Macromolecules*, 1973, **6**, 638.
103 F. Heitz, G. Spach, C. Hélène, and J. L. Dimicoli, *Macromolecules*, 1973, **6**, 845.
104 Y. Kobayashi and Y. Kyogoku, *J. Mol. Biol.*, 1973, **81**, 337.
105 M. Sisido, Y. Imanishi, and T. Higashimura, *Biopolymers*, 1973, **12**, 2375.
106 O. Iwaki, K. Hikichi, M. Kaneko, S. Shimizu, and T. Maruyame, *Polymer J.*, 1973, **4**, 623.
107 H. Stefanou, A. E. Woodward, and D. Morrow, *Biophys. J.*, 1973, **13**, 772.
108 E. Hunt and H. R. Morns, *Biochem. J.*, 1973, **135**, 833.

fluoroethanol ratio has been analysed[109] and found to be consistent with a $\beta$-turn in the chain, with the sequence L-Pro-Gly at the corners. This conclusion was supported by $^3J$(HNCH) for the valine residues.

Peak assignments of the 300 MHz[110] spectra of oxytocin and 4-leucine oxytocin have been made, and from the values of $^3J(HNC_\alpha H)$ it was concluded that oxytocin was a flexible macromolecule, having several different conformations. In a similar paper[111] the conformations of oxytocin and some analogues in DMSO were studied.

The rates of amide[112] proton–$^2H_2O$ exchange, together with spin decoupling and pH studies, enabled the assignments of the amide resonances of angiotensin II polypeptide to be made. Comparative experiments were made with derivatives of the polypeptide. In another paper[113] dealing with angiotensin II, a complete assignment of the NH resonances was made, by double irradiation of the $\alpha$-CH resonances. Transfer of saturation from the $H_2O$ protons indicated that the 2-Arg peptide NH exchanged more rapidly with the solvent than the other peptide NH protons.

The 220 MHz spectra[114] of the potassium–valinomycin complex showed $^3J(HC_\alpha C_\beta H)$ to be 11.0 $\pm$ 0.3 Hz rather than about 3.5—4.0 Hz as previously determined, and a model of the conformation taking this difference into account was proposed. The products[115] of the $\beta$-lactamase-catalysed hydrolysis of penicillins have methyl resonances chemically shifted from the original penicillins. The area of the different methyl resonances during the hydrolysis of penicillins was used to follow the kinetics of the reaction under various conditions. The slow exchange between low-spin dicyanide species and high-spin hydroxyaquo species of haemin complexes in $^2H_2O$ allowed an estimation[116] to be made of the relative amounts of these species from the appropriate chemically shifted haemin-methyl resonances.

The high-resolution spectra of malformin A, a cyclic pentapeptide, were measured[117] in DMSO. From the temperature dependence of the chemical shift of the peptide protons, their exchange rate and H—N—$C_\alpha$—H angles, it was concluded that malformin A is rigid, with the disulphide bridge exposed to interact with biological receptors. Using similar techniques, the solution conformation[118] of an 11-unit peptide, telomycin, was studied in different solvent mixtures.

[109] D. W. Urry, W. D. Cunningham, and T. Ohnishi, *Biochemistry*, 1974, **13,** 609.
[110] A. I. R. Brewster and V. J. Hruby, *Proc. Nat. Acad. Sci. U.S.A.*, 1973, **70,** 3806.
[111] A. I. R. Brewster, V. J. Hruby, J. A. Glasel, and A. E. Tonelli, *Biochemistry*, 1973, **12,** 5294.
[112] H. E. Bleich, R. E. Galardy, M. P. Printz, and L. C. Craig, *Biochemistry*, 1973, **12,** 4950.
[113] J. D. Glickson, J. Dadok, and G. R. Marshall, *Biochemistry*, 1974, **13,** 11.
[114] D. W. Urry and N. G. Kumar, *Biochemistry*, 1974, **13,** 1829.
[115] F. K. Schweighardt, J-M. Guo, S. M. Yang, and N. L. Li, *J. Magn. Resonance*, 1974, **14,** 52.
[116] W. Schossler, R. Hintsche, K. Po-merening, and P. Mohr, *Studia Biophys.*, 1973, **39,** 25.
[117] M. Ptak, *Biopolymers*, 1973, **12,** 1575.
[118] N. G. Kumar and D. W. Urry, *Biochemistry*, 1973, **12,** 4392.

Košiková *et al.*[119] have used high-resolution spectra to investigate the structure of lignin–carbohydrate complexes. From comparison with model compounds, spectral assignments of the complexes were made and information on their structure in solution obtained.

**C. $^1$H N.M.R. of Proteins.**—Mayer *et al.* have published an important paper[120] dealing with the assignment of high- and low-frequency resonances from haemoglobin M Iwate, which remains in the deoxy quaternary structure when the $\beta$ subunits are ligated to $O_2$, CO, or $CN^-$. Three spectral features (the paramagnetically shifted resonances from the $\beta^{III}$CN subunits, the low frequency resonance from Val E11 methyl proton in the $O_2$ and CO derivatives, and a resonance from an exchangeable proton at 14 p.p.m. to high frequency of DSS) confirmed previous assignments of the deoxy quaternary structure. A similar paper[121] has been published, concerning a comparison with haemoglobin Kansas.

Chemical shifts[122] of both high- and low-frequency resonances from deoxy haemoglobin Kempsey showed that it was significantly different from normal haemoglobin. The structure of both the $\alpha$ and $\beta$ chains and their capacity for binding ligands were altered. The authors have presented a similar paper[123] dealing with the structural nonequivalence of the $\alpha$ and $\beta$ chains in several other haemoglobins and have discussed the mechanism for transfer of information between subunits during ligand binding. The pH-induced $R \rightarrow T$ conformational changes[124] of high ligand affinity deoxyhaemoglobins have been studied by high-resolution techniques. Spectral changes as a function of pH, particularly in the high frequency region, were observed, and information on the kinetics of the process was obtained from line broadening.

Some interesting preliminary ENDOR experiments[125] have been reported for haemoglobin A and haemoglobin $M_{Hyde\ Park}$. The hyperfine interactions from $^{14}$N, $^1$H, and $^{57}$Fe were obtained from frozen solutions, and changes in the histidine $^{14}$N in the met subunit upon oxidation of the normal $\alpha$ subunits were observed.

The magnetic susceptibility[126] of weak paramagnetic haemoprotein solutions was measured by using an internal and external marker, ethane diol. Results were in good agreement with the Gouy method, although the n.m.r.


[119] B. Košiková, J. Polcin, and D. Joniak, *Holzforschung*, 1973, **27**, 59.
[120] A. Mayer, S. Ogawa, R. G. Shulman, and K. Gersonde, *J. Mol. Biol.*, 1973, **81**, 187.
[121] R. G. Shulman, S. Ogawa, A. Mayer, and C. L. Castillo, *Ann. New York Acad. Sci.*, 1973, **222**, 9.
[122] T. R. Lindstrom, J. J. Baldassare, H. F. Bunn, and C. Ho, *Biochemistry*, 1973, **12**, 4212.
[123] C. Ho, T. R. Lindstrom, J. J. Baldassare, and J. J. Breen, *Ann. New York Acad. Sci.*, 1973, **222**, 21.
[124] S. Ogawa, D. J. Patel, and S. R. Simon, *Biochemistry*, 1974, **13**, 2001.
[125] G. Feher, R. A. Isaacson, C. P. Scholes, and R. Nagel, *Ann. New York Acad. Sci.*, 1973, **222**, 86.
[126] K. D. Bartle, B. J. Dale, D. W. Jones, and S. Maricic, *J. Magn. Resonance*, 1973, **12**, 286.

technique was more simple to perform. A general method[127] has been described for the evaluation of the principle magnetic susceptibilities and dipolar shifts in $d^5$ low-spin haemin systems. The method was applied to ferricytochrome *c* and metmyoglobin cyanide, and the results showed that the second order Zeeman effect and the in-plane anisotropy were important in evaluating dipolar shifts. The high-resolution spectrum[128] of cytochrome *c*-557 from *Crithidia encopelti* showed that the co-ordination of the haeme was essentially the same as in other vertebrate cytochromes *c*, and information on the molecular arrangements of the haeme iron was obtained. A detailed study[129] of horse cytochrome *c*, giving many resonance assignments of the ferri and ferro forms, showed that the interior peptide NH protons of native ferrocytochrome exchanged slowly at room temperature. This was interpreted as indicating little unfolding of the protein. Oxidation to the native ferric form caused contact shifts of several resonances, and the effects of temperature and pH on the protein were discussed. The high-resolution spectra[130] of two cytochromes $c_3$ were measured as a function of temperature, pD and oxidation state, and molar magnetic susceptibilities were calculated. Chemical shifts, particularly of contact shifted resonances, were used to identify various ligands around the haeme group. Comparison[131] of the spectra from cytochrome $b_2$ core and cytochrome $b_5$ indicated that the two proteins had similar structures, allowing the *X*-ray data available on the latter to be tentatively extended to the former. Flavodoxins[132] exist in oxidized, semiquinone (paramagnetic) and reduced forms. The similarity of the oxidized and reduced forms suggests that there is little difference in their conformation, and that the resonance broadening in the semiquinone form is due to its paramagnetism. This, together with *X*-ray data, was used to assign some resonances from the proximity of flavon mononucleotide.

The temperature dependence[133] of the eight highest-frequency contact-shifted resonances of oxidized *Clostridium pasteurianum* ferredoxin showed a monotonic increase in the magnitudes of the contact shifts, a result compatible with antiferromagnetic spin-exchange coupling of the component iron atoms. Conformational changes that occurred during denaturation in DMSO solutions were also discussed.

The oxidized and reduced form of the iron–sulphur protein, *Bacillus polymyxa* ferredoxin I, were studied[134] by n.m.r. under high-resolution, and the resonances assigned, partly by comparison with other similar proteins. The

[127] W. D. Horrocks, jun., and E. S. Greenberg, *Biochim. Biophys. Acta*, 1973, **322,** 38.
[128] R. M. Keller, G. W. Pettigrew, and K. Wüthrich, *F.E.B.S. Letters*, 1973, **36,** 151.
[129] C. C. McDonald and W. D. Phillips, *Biochemistry*, 1973, **12,** 3170.
[130] C. C. McDonald, W. D. Phillips, and J. Le Gall, *Biochemistry*, 1974, **13,** 1952.
[131] R. Keller, O. Groudinsky, and K. Wüthrich, *Biochim. Biophys. Acta*, 1973, **328,** 233.
[132] T. L. James, M. L. Ludwig, and M. Cohn, *Proc. Nat. Acad. Sci., U.S.A.*, 1973, **70,** 3292.
[133] C. C. McDonald, W. D. Phillips, W. Lovenberg, and R. H. Holm, *Ann. New York Acad. Sci.*, 1973, **222,** 789.
[134] W. D. Phillips, C. C. McDonald, N. A. Stombaugh, and W. H. Orme-Johnson, *Proc. Nat. Acad. Sci., U.S.A.*, 1974, **71,** 140.

variation[135] of the 270 MHz spectrum of metallothionine with counter–ion concentration and pH enabled information on the binding sites of the metal ions to be obtained, and gave results in agreement with previous c.d. and *X*-ray data. Some p$K_a$ values[136] were measured from the high frequency resonances of two histidine residues in virgin and modified soybean trypsin inhibitor (Kunitz). Linewidth changes with pH were interpreted as indicating a change in the conformational equilibrium.

The high-resolution[137] spectra of glutamine binding protein from *E. coli* in the presence and absence of glutamine were interpreted as showing that binding of the substrate induced conformational changes in the protein, though the linewidths showed only minimal changes.

Linewidths and chemical shifts obtained from the wool fractions[138] α-keratose, urea soluble (US)3, and γ-keratose indicated various helix contents and were consistent with the view that the low sulphur proteins (high helical content) were extracted from helical wool fibrils whereas high sulphur materials came from the amorphous matrix region.

**D. $^1$H N.M.R. of Enzymes.**—One of the most significant papers[139] dealing with the structure of biopolymers has been concerned with human and hen egg white lysozyme. Paramagnetic ions were added to lysozyme, and the resulting partially broadened spectrum subtracted from the spectrum of pure lysozyme. The difference spectrum showed resonances from groups in the vicinity of the paramagnetic ion, and allowed the identification of resonances previously inaccessible. Methylation of the lysine residues and *N*-terminal amino groups of lysozyme allowed these groups to be easily identified, and their p$K_a$ values to be measured.[140] From comparison with model compounds, and because of methylation, a correction factor of 0.6 pH units had to be applied, and assignment of resonances to specific residues was accomplished using the selective broadening effect of $Gd^{3+}$.

The distance between various protons[141] in phosphorylase b and either paramagnetic ions or spin labels has been estimated from the change in relaxation rates with concentration of paramagnetic species. The correlation times were estimated from $T_1:T_2$ ratios.

The p$K_a$ values[142] for des-(121—124, Asp-Ala-Ser-Val-)RNAase A have been measured from the pH dependence of the chemical shift, and these

135 H. Rupp, W. Voelter, and U. Weser, *F.E.B.S. Letters*, 1974, **40,** 176.
136 J. L. Markley, *Biochemistry*, 1973, **12,** 2245.
137 G. P. Kreishman, D. E. Robertson, and C. Ho, *Biochem. Biophys. Res. Comm.*, 1973, **53,** 18.
138 B. J. Dale and D. W. Jones, *Polymer*, 1973, **14,** 523.
139 I. D. Campbell, C. M. Dobson, R. J. P. Williams, and A. V. Xavier, *J. Magn. Resonance*, 1973, **11,** 172; *Ann. New York Acad. Sci.*, 1973, **222,** 163.
140 J. H. Bradbury and L. R. Brown, *European J. Biochem.*, 1973, **40,** 565.
141 A. Bennick, R. A. Dwek, J. R. Griffiths, and G. K. Radda, *Ann. New York Acad. Sci.*, 1973, **222,** 175.
142 V. G. Sacharovsky, I. I. Chervin, G. I. Yakovlev, S. M. Dudkin, M. Ya. Karpeisky, and S. V. Shliapnikov, *F.E.B.S. Letters*, 1973, **33,** 323.

values correlated with differences in biological activity. The interaction[143] between pancreatic RNAase A and various nucleotides was studied by measuring the pH dependence of the chemical shifts of some of the enzyme protons. Information was obtained about the spatial arrangements of the nucleotides at the active site of the enzyme. In a similar paper,[144] the binding of cytidine-3′-monophosphate and cytidine-2′-monophosphate to bovine and rat ribonuclease was investigated. The unfolding of ribonuclease under the influence of various denaturants has been followed.[145] Simple spectra were used to identify the different stages of unfolding, and the results were correlated with existing knowledge of the process. The high-frequency resonances of chymotrypsin have been used[146] to study the hydrogen bonding involved in the binding of substrates to the molecule's active site. It was concluded that proton transfer from the histidine to the leaving group of the substrate was facilitated by hydrogen bonding between this group and the histidine. Preliminary information on the residence time of α-keto glutarate bound to glutamate dehydrogenase was obtained from line broadening.[147]

**E. $^1$H N.M.R. of Polynucleotides.**—Most papers in this section deal with the structure and interactions of tRNA molecules. A brief summary[148] has been made of the work reported in the last few years on tRNA particularly as inferred from the high-frequency resonances. Previous assignments of the high-frequency NH peaks to protons of the Watson–Crick base pairs were used[149] and it was concluded that no change in secondary structure occurs when yeast $tRNA^{Phe}$ is aminoacylated; evidence was also presented indicating that most of the tertiary structure is also unaltered. The chemical shift of the Y base from yeast $tRNA^{Phe}$ has been attributed[150] to ring current effects. From this, a change in the interbase separation in the anticodon stem of 0.02 nm was estimated. The high-frequency[151] region of the 220 MHz and 300 MHz spectra of fragments of yeast $tRNA^{Phe}$, of chemically modified $tRNA^{Phe}$, and intact $tRNA^{Phe}$ were examined and it was shown that first-order ring current shift theory, in conjunction with the clover leaf model for the molecule could explain the observed resonances to within 0.2 p.p.m. The resonance positions of the intact molecule were, to a close approximation, the sum of those from the appropriate fragments, indicating that some of the structural features existing in the intact molecule were also present in the fragments. In

143 W. Haar, W. Manrer, and H. Ruterjans, *European J. Biochem.*, 1974, **44,** 201.
144 C. Migchelsen and J. J. Beintema, *J. Mol. Biol.*, 1973, **79,** 25.
145 G. C. K. Roberts and F. W. Benz, *Ann. New York Acad. Sci.*, 1973, **222,** 130.
146 G. Robbillard and R. G. Shulman, *Ann. New York Acad. Sci.*, 1973, **222,** 220.
147 P. J. Andree and A. Zantema, *F.E.B.S. Letters*, 1974, **40,** 49.
148 D. R. Kearns, D. R. Lightfoot, K. L. Wong, Y. P. Wong, B. R. Reid, L. Cary, and R. G. Shulman, *Ann. New York Acad. Sci.*, 1973, **222,** 324.
149 Y. P. Wong, B. R. Reid, and D. R. Kearns, *Proc. Nat. Acad. Sci. U.S.A.*, 1973, **70,** 2193.
150 D. R. Kearns, K. L. Wong, and Y. P. Wong, *Proc. Nat. Acad. Sci., U.S.A.*, 1973, **70,** 3843.
151 D. R. Lightfoot, K. L. Wong, D. R. Kearns, and B. R. Reid, *J. Mol. Biol.*, 1973, **78,** 71.

a related paper[152] the chemical shifts of the ring NH protons of six different tRNA's were analysed and compared with predictions based on appropriate clover leaf models. The shifts were determined mainly by ring current effects from nearest neighbours, and good agreement was found between calculated and observed spectra.

Jones and Kearns[153] have shown that the lanthanide group ions can cause chemical shifts of the high-frequency resonances of $tRNA^{Phe}$, thus assisting the structure determination of the molecule.

The spectrum[154] from another fraction, and anticodon hairpin from $tRNA^{fMet}$, was recorded at various temperatures, and the high-frequency NH resonances from the Watson–Crick base pairs were accounted for by a semi-empirical ring current shift theory. The order of disappearance of the resonances at high temperature gave information on the melting behaviour of the molecule. The 300 MHz spectrum[155] of the hydrogen-bonded NH ring protons of $tRNA^{Phe}$, measured as a function of temperature, indicated melting of base pairs between 329 K and 331 K. The different melting behaviour of $tRNA^{Phe}_{Yeast}$ and $tRNA^{fMet}_{Coli}$, which have the same base-pairing scheme, was discussed in terms of tertiary structure.

**F. $^{13}$C Studies.**—Carbon-13 n.m.r. is a technique that should enable the controversy concerning the assignment of α-CH proton resonance peaks of polypeptides to be resolved. Allerhand and Oldfield[156] have reported a detailed study of PBLG, where relaxation rates and the NOE were measured for three samples of different mol. wt. in TFA– $CDCl_3$ mixtures. Peaks were assigned to helix and coil forms, and $\tau_c$ values for different groups were deduced. Over the helix-coil transition region, two sets of α-C peaks have been observed for poly-*N*5-(3-hydroxypropyl)-L-glutamine[157] and poly-L-methionine.[158] In both cases, the peaks were assigned to the helix and coil forms. For poly-L-lysine the $^{13}$C $T_1$ values were shorter in the helix form than in the coil form.[159] However, chemical shift changes observed with pD alteration required to induce the helix-coil transition were attributed to a pD effect rather than to conformational changes. The use of $^{13}$C n.m.r. to study conformations of polyprolines has been discussed.[160,161] Correlation times for polymer backbone motions were estimated from $T_1$ measurements, and intensity measurements gave information on the conformations.[160]

152 R. G. Shulman, C. W. Hilbers, D. R. Kearns, B. R. Reid, and Y. P. Wong, *J. Mol. Biol.*, 1973, **78**, 57.
153 C. R. Jones and D. R. Kearns, *J. Amer. Chem. Soc.*, 1974, **96**, 3651.
154 K. L. Wong and D. R. Kearns, *Biopolymers*, 1974, **13**, 371.
155 C. W. Hilbers, R. G., Shulman, and S. H. Kim, *Biochem. Biophys. Res. Comm.*, 1973, **55**, 953.
156 A. Allerhand and E. Oldfield, *Biochemistry*, 1973, **72**, 3428.
157 R. Di Blasi and A. S. Verdini, *Biopolymers*, 1974, **13**, 765.
158 S. Tadokoro, S. Fujiwara, and Y. Ichihara, *Chem. Letters*, 1973, 849.
159 H. Saito and I. C. P. Smith, *Arch. Biochem. Biophys.*, 1973, **158**, 154.
160 D. A. Torchia and J. R. Lyerla, jun., *Biopolymers*, 1974, **13**, 97.
161 F. A. Bovey, *Polymer Preprints*, **14**, (1), 163.

Helix-coil transitions are also observed for polynucleotides, and the transition to the ordered structure for polyuridylic acid caused resonance broadening owing to restricted motion.[162] Chemical shift changes showed that the bases interact below the transition temperature.

In a series of papers on small polypeptides, Smith and co-workers[163—167] have reported chemical shift assignments and relaxation rate ($T_1$) measurements. The results were used to determine the configurations and conformations in different solvents.

From a comparison of the spectrum of gelatin with that of poly(hydroxy-L-proline) and published data on other polyamino-acids, tentative peak assignments have been made.[168] Changes as a function of pH were observed, and $T_1$ values showed that the mobility of the polymer groups was that expected for a random coil conformation. The formation of a helix structure by a semi-synthetic ribonuclease-S′ polypeptide was studied by measuring the $^{13}C$ spectrum of glycine residues with specific $^{13}C$ enrichment.[169] Chemical shift and linewidth changes of the two specifically enriched carbon sites indicated the helix form. Enrichment of a single histidine C-2 carbon in an α-lytic protease was carried out,[170] and from the pH dependence of chemical shifts and coupling constants, the imidazole ring protonation and aspartic acid ionization were measured; a new model for the role of histidine in the catalytic mechanism was proposed. Single carbon resonances from non-protonated aromatic carbons of amino-acids and haeme residues have been reported for natural abundance spectra of cytochrome *c*[171,172] and other proteins.[172] Peak assignments were made, and it was observed that only a few of the resonances could be distinguished in spectra of the oxidized form of the protein.[171] Peak assignments and resonances from single types of carbon atoms have been reported[173] for several denatured proteins. Assignments were made by comparing spectra with those of small peptides.

Carbon-13 resonance has been used to elucidate the structures of polysaccharides. Bundle *et al.*[174] have deduced the structure of several bacterial

162 G. Govil and I. C. P. Smith, *Biopolymers*, 1973, **12**, 2589.
163 R. Walker, K. U. M. Prasad, R. Deslauriers, and I. C. P. Smith, *Proc. Nat. Acad. Sci. U.S.A.*, 1973, **70**, 2086.
164 R. Deslauriers, R. Walker, and I. C. P. Smith, *Proc. Nat. Acad. Sci. U.S.A.*, 1974, **71**, 265.
165 R. Deslauriers, R. Walker, and I. C. P. Smith, *Biochem. Biophys. Res. Comm.*, 1973, **53**, 244.
166 R. Walker, I. C. P. Smith, and R. Deslauriers, *Biochem. Biophys. Res. Comm.*, 1974, **58**, 216.
167 I. C. P. Smith, R. Deslauriers, H. Saito, R. Walter, C. Garrigou-Lagrange, H. McGregor, and D. Sarantakis, *Ann. New York Acad. Sci.*, 1973, **222**, 597.
168 J. C. W. Chien and W. B. Wise, *Biochemistry*, 1973, **12**, 3418.
169 I. M. Chaiken, *J. Biol. Chem.*, 1974, **249**, 1247.
170 M. W. Hunkapiller, S. H. Smallcombe, D. R. Whitaker, and J. H. Richards, *Biochemistry*, 1973, **12**, 4732.
171 E. Oldfield and A. Allerhand, *Proc. Nat. Acad. Sci. U.S.A.*, 1973, **70**, 3531.
172 A. Allerhand, R. F. Childers, and E. Oldfield, *Ann. New York Acad. Sci.*, 1973, **222**, 764.
173 J. H. Bradbury and R. S. Norton, *Biochim. Biophys. Acta.*, 1973, **328**, 10.
174 D. R. Bundle, I. C. P. Smith, and H. J. Jennings, *J. Biol. Chem.*, 1974, **249**, 2275.

polysaccharides from the complete assignments of all the spectral resonances, and also reported some conformational information. Peak assignments have also been made for some sulphated polysaccharides.[175]

**G. Other Nuclei.**—This section includes studies made using $^{31}P$, $^{2}H$, and $^{19}F$ resonances. Raftery and Huestis have summarized[176] their studies of ligand binding in haemoglobin using a derivative with a trifluoroacetonyl label at cysteine $\beta$ 93. Conformational changes with the binding of various ligands were detected by chemical shift changes and the appearance of additional resonances. Sykes *et al.*[177] have biosynthesized the enzyme alkaline phosphatase containing *m*-fluorotyrosine. The enzyme does have some differences in activity from a normal enzyme, especially at low pH. The $^{19}F$ spectrum shows eleven different resonances and is dependent on the presence or absence of metal ions.

The $^{31}P$ and $^{1}H$ chemical shifts of 1:2 actinomycin D when complexed with different oligonucleotides have been reported by Patel.[178—180] Although the $^{31}P$ data are more difficult to interpret than the $^{1}H$ data, the results were consistent with the phenoxazone ring of actinomycin D intercollating between GC and CG base pairs. The broad resonances observed for phosphorus in tRNA were attributed[181] to a distribution of chemical shifts, and line-narrowing was observed on denaturation. Since the authors do not eliminate proton–phosphorus dipolar relaxation by proton double resonance as they claim, the broad line may be caused by rapid relaxation.

Deuteron n.m.r. has been used to study the interaction of neurophysin with specifically deuteriated oxytocin and related molecules.[182,183] Linewidth measurements were used to obtain information about local segmental motion of the oxytocin and its exchange with neurophysin.

## 5 Small Molecules

The papers in this section are concerned with effects of macromolecules on the spectra of small molecules and, as for the previous review, are arranged according to the nucleus studied, with a separate section on hydration.

**A. $^{1}H$ Studies of Hydration.**—*General Studies.* N.m.r. is frequently used to study disperse systems such as water in polymer membranes. Shporer *et al.*[184]

175 S. E. Lasker and M. L. Chien, *Ann. New York Acad. Sci.*, 1973, **222**, 971.
176 M. Raftery and W. H. Huestis, *Ann. New York Acad. Sci.*, 1973, **222**, 40.
177 B. D. Sykes, H. I. Weingarten, and M. J. Schlesinger, *Proc. Nat. Acad. Sci. U.S.A.*, 1974, **71**, 469.
178 D. J. Patel, *Biochemistry*, 1974, **13**, 1476.
179 D. J. Patel, *Biochemistry*, 1974, **13**, 2388.
180 D. J. Patel, *Biochemistry*, 1974, **13**, 2396.
181 L. M. Weiner, J. M. Backer, and A. I. Rezvukhin, *F.E.B.S. Letters*, 1974, **41**, 40.
182 J. A. Glasel, V. J. Bruby, J. F. McKelvy, and A. F. Spatola, *J. Mol. Biol.*, 1973, **79**, 555.
183 J. A. Glasel, J. F. McKelvy, V. J. Hruby, and A. F. Spatola, *Ann. New York Acad. Sci.*, 1973, **222**, 778.
184 M. Shporer, A. J. Vega, and M. A. Frommer, *J. Polymer Sci., Part A-2, Polymer Phys.*, 1974, **12**, 645.

have drawn attention to the fact that variations in bulk diamagnetic susceptibility can cause splitting of water resonances, and have demonstrated this for several cellulose membranes. Separate peaks for water inside and outside various cation exchange resins have been observed,[185] and similar results have also been reported for alcohols and alcohol–water mixtures. The shifts were related to the solvation of sulphonic acid groups on the resin, while resonance broadening was a function of particle size and cross links. High-resolution linewidths of water in cellulose as a function of water content were interpreted as indicating the presence of bound and free water.[186] Difficulties in determining $T_2$ for the bound water were a limiting factor in estimating the fraction of water that is bound.

A correlation between rheological properties and water linewidths in polysaccharide gels has been reported,[187] with the more rigid gels giving broader resonances. The authors did not consider the effects of proton exchange between polymer and water hydroxyl groups, and concluded that the state of water is closely related to the rheological properties. Linewidths have also been reported for agar solutions as a function of temperature.[188] A hysteresis effect on heating and cooling is observed as a consequence of the gel–sol transition, and the authors prefer to explain their results in terms of one collective water structure in the gel, rather than the more familiar model of bound and free water.

A fraction of water that does not freeze at 273 K is often observed when aqueous solutions are cooled. In an important paper, Ramirez and co-workers have demonstrated that this phenomenon also occurs for inorganic salt solutions and small organic molecules in solution, and interpretation of this narrow resonance as water of hydration of the macromolecule may be misleading.[189] In a detailed study using both proton and deuteron resonances of water in collagen fibres, rotational correlation times, proton exchange rates, and activation energies for these processes were determined.[190] A model with two specific water binding sites was consistent with the experimental results. The central peak observed in collagen with 25% added water at 295 K was shown to be due to collagen protons, these probably being at the mobile protein side chains. This contradicts the report of Gabuda and Ratnikova,[191] who assigned the peak to capillary water from a proton broadline study. Water relaxation rates and self-diffusion coefficients of DNA solutions have been reported, and were found to be almost invariant over the concentration range studied.[192]

185 W. J. Casey and D. J. Pietrzyk, *Analyt. Chem.*, 1973, **45,** 1404.
186 J. E. Carles and A. M. Scallan, *J. Appl. Polymer. Sci.*, 1973, **17,** 1855.
187 M. Aizawa, S. Suzuki, and T. Toyama, *Bull. Chem. Soc. Japan*, 1973, **46,** 1638.
188 M. S. Micciancio-Giammarinaro, S. Micciancio, M. B. Palma-Vittorelli, M. U. Palma, and M. L. Marino, *Ann. New York Acad. Sci.*, 1973, **204,** 134.
189 J. E. Ramirez, J. R. Cavanaugh, and J. M. Purcell, *J. Phys. Chem.*, 1974, **78,** 807.
190 C. Migchelsen and H. J. C. Berendsen, *J. Chem. Phys.*, 1973, **59,** 296.
191 S. P. Gabuda and N. V. Ratnikova, *Biofizika*, 1973, **18,** 981.
192 G. Kh. Bozhko, B. I. Verkin, and B. G. Yemets, *Dopovidi Akad. Nauk Ukrain. R.S.R., Ser. B,* 1973, **35,** 446.

Counter-ion binding to various polyelectrolytes has been studied by following the chemical shift changes in the water resonance on addition of paramagnetic ions to the polymer solution.[193] Addition of $CoCl_2$ caused shifts of ~0.3 p.p.m. in the $H_2O$ resonance, and from the smaller shifts observed in the presence of polymers the number of water molecules released on binding of $Co^{2+}$ to the polymer and the fraction of bound $Co^{2+}$ were determined.[193]

*Proton Relaxation Enhancement.* In Table 3 papers describing the use of proton relaxation enhancement (PRE) by paramagnetic ions to obtain details of macromolecule–small molecule interactions are listed. The use of $H_2O$–$^2H_2O$ mixtures as proposed by Sherry *et al.*[194] should be of value in increasing the accuracy of binding constant measurements by this technique.

**Table 3** *Relaxation Enhancement Studies*

| *Macromolecule* | *Comments* | *Ref.* |
|---|---|---|
| Bovine serum albumin (BSA) | Binding constant for $Mn^{2+}$ binding to BSA determined using $H_2O$–$^2H_2O$ mixtures. Greater accuracy obtained than with $H_2O$ alone. | 194 |
| Pyruvate kinase | Binding constants for enzyme $Mn^{2+}$ small molecule complexes in the presence of monovalent cations determined. Limits on $\tau_c$ and number of bound water molecules determined from temperature and frequency dependence. | 195 |
| Pyruvate kinase | Investigation of $Gd^{3+}$ binding in binary and ternary complexes, plus effects of pH and other ions. | 196 |
| Isocitrate dehydrogenase | Number of bound water molecules in binary and ternary complexes determined. Information on structure of ternary complex deduced from relaxation rates of bound α-ketoglutarate | 197 |
| Alcohol dehydrogenase | Water PRE and substrate relaxation rates in presence of spin-labelled enzyme enabled a model of the conformation of bound NADH[a] to be derived. | 198 |
| Ceruloplasmin, azurin superoxide dismutase, human carbonic anhydrase B and transferrin | PRE as a function of frequency used to obtain information on different types of bound $Cu^{2+}$ ions. | 199 |

[193] P. Spegt, C. Tondre, G. Weill, and R. Zana, *Biophys. Chem.*, 1972, **1**, 55.
[194] A. D. Sherry, E. R. Birnbaum, and D. W. Barnall, *Analyt. Biochem.*, 1973, **52**, 415.
[195] T. L. James, J. Reuben, and M. Cohn, *J. Biol. Chem.*, 1973, **248**, 6443.
[196] K. M. Valentine and G. L. Cottam, *Arch. Biochem. Biophys.*, 1973, **158**, 346.
[197] J. J. Villafranca and R. F. Colman, *Biochemistry*, 1974, **13**, 1152.
[198] D. L. Sloan and A. S. Mildvan, *Biochemistry*, 1974, **13**, 1711.
[199] S. H. Koenig and R. D. Brown, *Ann. New York Acad. Sci.*, 1973, **222**, 752.

**Table 3** *continued*

| *Macromolecule* | *Comments* | *Ref.* |
|---|---|---|
| Azurin, umeycanin, benzylamine oxidase and superoxide dismutase | The presence or absence of a PRE enabled the type of $Cu^{2+}$ binding site (type I or II) to be distinguished. | 200 |
| Superoxide dismutase | The lack of a PRE due to bound $Co^{2+}$ ions indicated no direct interaction between these ions and water. | 201 |
| Transcarboxylase | Number of $Co^{2+}$ bound water molecules determined for binary and ternary complexes. Distances of bound pyruvate from $Co^{2+}$ deduced from pyruvate $^{13}C$ relaxation rates. | 202 |
| Alkaline phosphatase | From $T_1:T_2$ ratio and PRE the $Cu^{2+}$–water proton distance was estimated. | 203 |
| Hexose-1,6-diphosphatase | Binding of $Mn^{2+}$ in binary and ternary complexes investigated. | 204 |
| Metmyoglobin fluoride | Mechanism of proton relaxation proposed; pH dependence indicated that proton-iron distance altered twice with increasing pH. | 205 |
| Ferrimyoglobin | From PRE in the presence of in-organic salts it was concluded that the exchanging proton at the haeme pocket was not water bound at the sixth ligand site of the haeme iron. | 206 |
| Adenylate kinase | PRE used to derive binding constants for enzyme complexes. | 207 |

[a] NADH = nicotinamide adenine dinucleotide.

*Water in Tissues.* In the interpretation of $T_2$ values for water in tissues the contribution from diffusion through locally inhomogeneous magnetic fields has been a controversial issue. Packer has proposed a method to calculate the contribution, and concluded that in the case of striated muscle the contribution is negligible.[208]

Most relaxation time results for water in tissues are interpreted using a two-site model (bound and free water) with fast exchange. Thompson and co-

[200] N. Boden, M. C. Holmes, and P. F. Knowles, *Biochem. Biophys. Res. Comm.*, 1974, **57**, 845.
[201] A. Rigo, M. Terenzi, C. Franconi, B. Mondovi, L. Calebrese, and G. Rotilio, *F.E.B.S. Letters*, 1974, **39**, 154.
[202] Chien-Hung Fung, A. S. Mildvan, and J. S. Leigh, jun., *Biochemistry*, 1974, **13**, 1160.
[203] R. S. Zukin and D. P. Hollis, *J. Biol. Chem.*, 1974, **249**, 656.
[204] H. J. Kolb and H. Kolb, *Z. physiol. Chem.*, 1973, **354**, 331.
[205] M. E. Fabry and M. Eisenstadt, *J. Biol. Chem.*, 1974, **249**, 2915.
[206] G. Pifat and S. Maricic, *Biophys. Chem.*, 1973, **1**, 112.
[207] N. C. Price, G. H. Reed, and M. Cohn, *Biochemistry*, 1973, **12**, 3322.
[208] K. J. Packer, *J. Magn. Resonance*, 1973, **9**, 438.

workers[209,210] have made $T_1$ and $T_{1\rho}$ measurements of water in mouse tissues as a function of frequency. They interpreted their results using an exchange time between free and bound water ($7 \times 10^{-6}$ s) and rotational correlation times for the free and bound water ($<10^{-10}$ s and $2 \times 10^{-8}$ s respectively). The authors point out that this is not a unique fit to the data, and that the exchange process will result in an underestimation of the fraction of water bound. A number of studies have been reported[211—214] where $T_1$ measurements were made on various tissues, all giving results similar to previous reports ($T_1 > T_2$, indicating bound and free water with rapid exchange). Hansen and Yellin,[211] however, using i.r. and n.m.r., have distinguished two types of bound water for stratum corneum. The higher $T_1$ values for cancerous tissue than for those of healthy tissue have also been widely studied.[215—222] The greater water content of the cancerous tissue is now thought to be responsible for the smaller relaxation rates,[215—217] and measurements on living mice demonstrate that the difference is present *in vivo*.[221] Fully cancerous and healthy tissue gave single exponential decays, but non-exponential relaxation was observed for partially cancerous tissue. Several authors suggest that relaxation time measurements on excised cancerous tissue could be used during operations.[216,219,220] However, Parrish *et al.* add a cautionary note, by reporting overlap in the ranges of $T_1$ values observed for healthy and cancerous brain tissue,[222] and it should be noted that $T_{1\rho}$ was longer for healthy muscle than for cancerous muscle.[210]

A difference in water n.m.r. linewidths has been reported for normal and sickle red blood cells.[223] The broad resonance from sickle erythrocytes was shown to be due to the presence of the haemoglobin-S. In a second study of erythrocytes, the dependence of the water lineshape and width on sugar con-

[209] R. T. Thompson, R. R. Knispel, and M. M. Pintar, *Chem. Phys. Letters*, 1973, **22,** 335.
[210] R. R. Knispel, R. T. Thompson, and M. M. Pintar, *J. Magn. Resonance*, 1974, **14,** 44.
[211] J. R. Hansen and W. Yellin, Water Structure, Water–Polymer Interface Symposium Proceedings, ed. H. G. Jellinek, Plenum, New York, 1972, p. 19.
[212] R. L. Vick, D. C. Cheng, B. L. Nichols, C. F. Hazlewood, and M. C. Harvey. *Ann. New York Acad. Sci.*, 1973, **204,** 575.
[213] R. Cooke and R. Wien, *Ann. New York Acad. Sci.*, 1973, **204,** 197.
[214] G. Bystrov, N. I. Nikolaev, and G. I. Romanenko, *Biofizika*, 1973, **18,** 484.
[215] R. E. Block, *F.E.B.S. Letters*, 1973, **34,** 109.
[216] W. Bovee, P. Huisman, and J. Smidt, *J. Nat. Cancer Inst.*, 1974, **52,** 595.
[217] L. A. Sargen, D. P. Hollis, J. S. Economou, and J. C. Eggleston, *J. Nat. Cancer Inst.*, 1974, **52,** 599.
[218] N. Iijima, S. Saitoo, Y. Yoshida, N. Fujii, T. Koike, K. Osanai, and K. Hirose, *Physiol. Chem. and Phys.*, 1973, **5,** 431.
[219] R. Damadian, K. Zaner, D. Hor and T. DiMaio, *Proc. Nat. Acad. Sci. U.S.A.*, 1974, **71,** 1471; *Physiol. Chem. and Phys.*, 1973, **5,** 381.
[220] R. Damadian, K. Zaner, D. Hor, T. DiMaio, L. Minkoff, and M. Goldsmith, *Ann. New York Acad. Sci.*, 1973, **222,** 1048.
[221] I. D. Weisman, L. H. Bennett, and L. R. Maxwell, *IEEE Trans. Magnetics*, 1973, (9), 454.
[222] R. G. Parrish, R. J. Kurland, W. W. Janese, and L. Bakay, *Science*, 1974, **183,** 438.
[223] A. Zipp, I. D. Kuntz, S. J. Rehfeld, and S. B. Shohet, *F.E.B.S. Letters*, 1974, **43,** 9.

centration was interpreted to indicate the formation of a ternary glucose–water–protein complex.[224]

N.m.r. self-diffusion measurements have been used to detect barriers to diffusion in biological tissues.[225] Cooper *et al.* have reported restricted diffusion in model and biological systems,[225] but no restricted diffusion was detected for muscle.[226]

**B. $^1$H N.M.R.—General Studies.**—The majority of papers in this section are concerned with the binding of small molecules to enzymes and other biological macromolecules. In a study of trifluoroacetic acid (TFA) and poly-α-benzyl-L-glutamate (PBLG), Brosio *et al.*[227] have reported TFA carboxyl proton shifts for TFA–$CDCl_3$–PBLG mixtures. The results are discussed in terms of the formation of TFA monomers and dimers, and their interactions with PBLG. Interactions between PBLG and dichloromethane caused the occurrence of a frequency-dependent spin-lattice relaxation process in the lyotropic l.c. phase,[228] and possible mechanisms were discussed. In a study of a synthetic polynucleotide,[229] linewidths and chemical shifts were interpreted to show that *N*(6)-methyl adenine forms a complex with polyuridylic acid of 1:1 adenine:uracil base pairs. The binding of peptides and oligopeptides to synthetic and natural polynucleic acids has been investigated in detail by Dimicoli and Hélène,[230—234] again using chemical shifts and linewidths. The importance of electrostatic influences was demonstrated by ionic strength and pH effects, and possible structures of the complexes formed were discussed. Electrostatic effects were also shown to be important in the binding of surfactants to proteins.[232] Oakes used linewidths and chemical shifts of various surfactants in aqueous bovine-serum albumin (BSA) solutions, together with e.s.r. studies of spin-labelled surfactants, to demonstrate that both electrostatic and hydrophobic interactions were important in the binding of surfactants to BSA.

The difference between $T_2$ values measured using pulse sequences in the presence and absence of $^{14}$N decoupling, has been used to study the binding of methyl isonitrile to haemoglobin β chains.[233] The presence of two types of binding was demonstrated, and $T_2$ values were used to determine the dissociation rate constant for the specific binding to the haeme iron. Linewidth


224 R. J. Naftalin, P. Seeman, N. L. Simmons, and M. C. R. Symons, *Biochim. Biophys. Acta*, 1974, **352**, 146.
225 R. L. Cooper, D. B. Cheng, A. C. Yound, C. J. Markin, and B. Ancker-Johnson, *Biophys. J.*, 1974, **14**, 161.
226 D. C. Chang, H. E. Rorschach, B. L. Nichols, and C. F. Hazlewood, *Ann. New York Acad. Sci.*, 1973, **204**, 434.
227 E. Brosio, M. Delfini, A. de Paolis, M. Paci, and F. Conti, *Biopolymers*, 1974, **13**, 745.
228 W. Hines and E. T. Samulski, *Macromolecules*, 1973, **6**, 793.
229 B. W. Bangerter and Z. Modestas, *Biopolymers*, 1974, **13**, 567.
230 J. L. Dimicoli and C. Hélène, *Biochemistry*, 1974, **13**, 714.
231 J. L. Dimicoli and C. Hélène, *Biochemistry*, 1974, **13**, 724.
232 J. Oakes, *Eur. J. Biochim.*, 1973, **36**, 553.
233 B. A. Manuek, J. B. Maloney, jun., and B. D. Sykes, *J. Mol. Biol.*, 1973, **81**, 199.
234 P. J. Andree and A. Zantema, *F.E.B.S. Letters*, 1974, **40**, 49.

changes have also been used in a similar way to determine changes in the residence time of α-ketoglutarate bound to spin-labelled glutamate dehydrogenase.[234] No line broadening was observed for lysine vasopressin in the presence of bovine neurophysin, and this was taken to indicate slow exchange between the bound and free forms.[235] Fast exchange was observed for di- and tri-peptide analogues, and details of the binding mechanism were deduced. Chemical shifts and linewidths were used to study binding of nucleotides to cytochrome-*c*, and binding constants were evaluated.[236,237] No binding of AMP was observed,[236] but broadening of AMP resonances has been used to study[238] the effects of SH groups on its binding to glycogen phosphorylase b. Addition of the blocking agent decreased the AMP–enzyme interaction. From linewidth changes, Mathews *et al.*[239] obtained binding constants for an inducer and an anti-inducer bound to the lac repressor protein, and $T_2$ values for the bound fractions were estimated.

From proton relaxation rates of succinate ions in the presence of aspartate transcarbamylase catalytic sub-unit, a complex model involving equilibria between four different enzyme species was proposed.[240] Proton $T_1$ values have also been used to obtain $\tau_c$ values for several pyrimidine nucleotides bound to aspartate transcarbamylase, and competitive binding studies with ATP were carried out.[241] The NOE and INDOR effects observed in spectra of formate ions in the presence of creatine kinase and a modified enzyme were used to identify the residue responsible for the double resonance effects.[242] Bothner-By and Glassend[243] have discussed the use of the NOE to determine the exchange rate between free and bound substrates. They give data for the case of d-tryptophan bound to α-chymotrypsin, and concluded that the exchange rate is too rapid to be determined by this technique.

Lysozyme rapidly catalyses the degradation of a tetrasaccharide, and the fraction of substrate bound increases as the reaction proceeds.[244] From changes in the resonance intensity the proportion of unreacted substrate was measured and the time dependence of the coupling constants enabled the conformation of the bound substrate to be determined.

Bennick and co-workers[245] have summarized their studies of enzyme–small molecule binding in the phosphorylase system, including the use of line broadening effects by spin labels or paramagnetic ions. The PRE caused by

[235] P. Balaram, A. A. Bothner-By, and E. Breslow, *Biochemistry*, 1973, **12**, 4695.
[236] L. A. Sibeldina, L. P. Kayashin, M. S. Okon, and A. U. Stepanyants, *Studia. Biophys.*, 1973, **38**, 177.
[237] L. P. Kayushin and Y. I. Ajipa, *Ann. New York Acad. Sci.*, 1973, **222**, 255.
[238] R. Gaspar and S. Damjanovish, *Biochim. Biophys. Acta*, 1973, **315**, 191.
[239] H. R. Mathews, H. W. Thielmann, K. S. Mathews, and O. Jardetzky, *Ann. New York Acad. Sci.*, 1973, **222**, 226.
[240] C. B. Beard and P. G. Schmidt, *Biochemistry*, 1973, **12**, 2255.
[241] R. E. London and P. G. Schmidt, *Biochemistry*, 1973, **13**, 1170.
[242] T. L. James and M. Cohn, *J. Biol. Chem.*, 1974, **249**, 2599.
[243] A. A. Bothner-By and R. Glassend, *Ann. New York Acad. Sci.*, 1973, **222**, 668.
[244] S. L. Patt, D. Dolphin, and B. D. Sykes, *Ann. New York Acad. Sci.*, 1973, **222**, 211.
[245] A. Bennick, R. A. Dwek, J. R. Griffiths, and G. K. Radda, *Ann. New York Acad. Sci.*, 1973, **222**, 175.

paramagnetic $Mn^{2+}$ ions has also been used by Nowak to study the binding of methylammonium ions to pyruvate kinase.[246] $T_1$ measurements were used to determine the electron–$CH_3$ distance, and a model for the active site of the enzyme was proposed.

Translational diffusion measurements have been reported for benzene and acetone as solvents of poly(methyl methacrylate).[247] A series of four gradient pulses before and after the $\pi$ pulse in a normal $\frac{\pi}{2}$–$\pi$ sequence was used, and the concentration dependence of $D$ was discussed in terms of the free volume of the solutions.

**C. $^{13}C$ N.M.R. Studies**—One obvious application of $^{13}C$ n.m.r. is to the study of small molecule or anion binding to macromolecules using enriched materials. Antonini and co-workers[248] have measured the $^{13}C$ chemical shifts of enriched $^{13}CO$ in the presence of separate $\alpha$ and $\beta$ haemoglobin chains, and were able to assign the peaks observed in mixtures of the two. Similar results were reported independently,[249] but in this case the third peak observed by Antonini *et al.* in mixtures of $\alpha$ and $\beta$ chains was not apparent. From changes in the $^{13}C$ linewidths of $^{13}CO_2$ and $H^{13}CO_3^-$ in the presence of different concentrations of carbonic anhydrase B, rate constants were obtained[250] for the enzyme catalysed reaction:

$$CO_2 + H_2O \rightleftharpoons HCO_3^- + H^+$$

Carbon-13 relaxation rates of bicarbonate in the presence of ribulose 1,5-diphosphate carboxylase were used to calculate the $Mn^{2+}$–bicarbonate distance in the enzyme complex.[251] The binding of $Mn^{2+}$ to the enzyme in the presence of bicarbonate was determined using PRE. Spin–lattice relaxation rates were also used[252,253] to determine the $Mn^{2+}$–C distances for $\alpha$-methyl-D-glucopyranoside in a ternary complex with concanavalin A. Both natural abundance[252] and enriched[253] samples were used, and the distances calculated were in general agreement, although they differed in the values determined for specific carbons.

In the second study the binding of the $\beta$-methyl derivative was also investigated using an enriched sample.[253] Measurements of values for $J(^{13}CNC_{\alpha}H)$ in $N$-acetyl(carbonyl-$^{13}C$)L-tryptophan bound to chymotrypsin have been reported by Rogers and Roberts.[254] The dependence of $J$ on the

246 T. Nowak, *J. Biol. Chem.*, 1973, **248,** 7191.
247 R. Kosfeld and J. Schlegel, *Angew. makromol. Chem.*, 1973, **29/30,** 105.
248 E. Antonini, M. Brunon, F. Conti, and G. Geraci, *F.E.B.S. Letters*, 1973, **34,** 69.
249 P. J. Vergamini, N. A. Matwiyoff, R. C. Wohl, and T. Bradley, *Biochem. Biophys. Res. Comm.*, 1973, **55,** 453.
250 S. Koenig, R. D. Brown, T. E. Needham, and N. A. Matwiyoff, *Biochem. Biophys. Res. Comm.*, 1973, **53,** 624.
251 H. M. Mitiorko and A. S. Mildvan, *J. Biol. Chem.*, 1974, **249,** 2743.
252 J. J. Villafranca and R. E. Viola, *Arch. Biochem. Biophys.*, 1974, **160,** 465.
253 C. F. Brewer, D. Marcus, A. P. Grollman, and H. Sternlicht, *Ann. New York Acad. Sci.*, 1973, **222,** 979.
254 P. Rodgers and G. C. K. Roberts, *F.E.B.S. Letters*, 1973, **36,** 330.

fraction of substrate that is bound was used to determine the coupling constant for the bound fraction, and from this the conformation in the bound site was determined.

The natural abundance $^{13}C$ spectra of retinal rod outer segment membranes show only lipid resonances, and $T_1$ values were unaltered by bleaching.[255] Bleaching of the membrane protein rhodopsin is the initial visual event, and the absence of any change in $T_1$ indicates that this does not cause detectable changes in the membrane structure. For canine sciatic nerve the natural abundance $^{13}C$ spectrum contained resonances from triglycerides and choline groups.[256] No cholesterol or other lipid resonances were detected, indicating that these are present in an immobilized state, and that observed signals are due to fat cells in the nerve interstices.

**D. Other Nuclei.**—The papers in this section are concerned with studies utilizing fluorine, phosphorus, oxygen, and deuteron n.m.r. Mildvan *et al.* have summarized[257] their previous studies on the role of divalent cations in enzyme mechanisms. Results discussed include data on $^{31}P$, $^{19}F$, $^{13}C$, and $^{1}H$ relaxation times, which have been used to obtain information on ternary enzyme complexes. Griffin *et al.* have reviewed[258] their determinations of the structures of ribonuclease–dinucleoside phosphonate complexes using $^{31}P$ and $^{1}H$ resonances.

Phosphorus-31 resonance has been used to study the binding of nucleotides to ribonuclease-A. Two groups[259,260] have independently demonstrated that the chemical shift variation of 3′-cytidine monophosphate as a function of pH shows that the nucleotide binds as the dianion. Structures of other bound nucleotides have also been inferred from their $^{31}P$ chemical shifts.[260] An interesting application of phosphate $^{31}P$ resonance is in the determination of the intracellular pH. Although the chemical shifts are modified by interactions with macromolecules, Moon and Richards have determined the intracellular pH of erythrocytes from the shifts of 2,3-diphosphoglycerate.[261] Phosphate relaxation enhancement of phosphate ions in the presence of a ternary phosphate–$Mn^{2+}$–alkaline phosphatase complex was interpreted as showing the formation of an outer-sphere complex.[262] In a detailed study of methyl phosphate binding to phosphoglucomutase, $^{31}P$ and $^{1}H$ relaxation rates of the

[255] F. Millett, P. A. Hargrave, and M. A. Raftery, *Biochemistry*, 1973, **12,** 3591.
[256] E. Williams, J. A. Hamilton, M. K. Jain, A. Allerhard, E. H. Cordes, and S. Ochs, *Science*, 1973, **181,** 869.
[257] T. A. S. Mildvan, T. Nowak, and Chien-Hung Fung, *Ann. New York Acad. Sci.*, 1973, **222,** 192.
[258] T. J. H. Griffin, A. N. Schechter, and J. S. Cohen, *Ann. New York Acad. Sci.*, 1973, **222,** 693.
[259] D. Gorenstein and A. Wyrwica, *Biochem. Biophys. Res. Comm.*, 1973, **54,** 976.
[260] W. Haar, J. C. Thompson, W. Maurer, and H. Ruterjans, *European J. Biochem.*, 1973, **40,** 259.
[261] R. B. Moon and J. H. Richards, *J. Biol. Chem.*, 1973, **248,** 7276.
[262] R. S. Zukin, D. P. Hollis, and G. A. Gray, *Biochem. Biophys. Res. Comm.*, 1973, **53,** 238.

substrate, in addition to PRE measurements, have been reported.[263] The results were used to derive $Mn^{2+}$–methyl phosphonate distances in the ternary complex, and a model of enzyme action has been presented. $^{31}P$ spectra have been reported[264] for various human serum lipoprotein fractions, and the resonances of lecithins and sphingomyelins have been identified. The effect of added paramagnetic ions on the spectrum of high density lipoprotein indicated that all the phospholipid phosphorus is located at the surface.

The single report involving $^{19}F$ resonance concerns investigation of the binding of $\omega$-trifluoromethyl-alkyl sulphates to BSA.[265] Chemical shift changes of the $C_{12}$- and $C_{13}$-surfactants as a function of surfactant–protein molar ratio were interpreted to show that binding of the $C_{13}$-surfactant causes a conformational change in the polymer, whereas binding of the $C_{12}$ derivative does not.

The reports of solvent $^{2}H$ and $^{17}O$ resonances are concerned with aqueous systems. From measurements of $^{2}H$ relaxation rates and self-diffusion coefficients in poly(acrylic acid) solutions, it was demonstrated[266] that hydroxyl proton exchange is responsible for the observed relaxation rate changes. No information can be deduced about associated water molecules, contrary to a previous publication. A theoretical description of $^{2}H_2O$ relaxation in polysaccharide solutions has been reported, involving diffusion of partially oriented water molecules through oriented polymer domains.[267] The experimental results were best fitted by a tetrahedral model of water structure. Deuteron splittings of water in collagen in the presence of various salts have been reported, and the results did not fit a simple model involving only exchange between bound and free water molecules.[268]

Measurements of $^{17}O$ for $H_2{}^{17}O$ in muscle showed a shorter $T_2$ than for normal water, indicating the existence of a fraction of water bound to the protein.[269] Non-exponential $T_1$ recovery curves have been reported for fresh muscle samples indicating a heterogeneity of populations of water.[270] The non-exponential character vanished over a period of *ca.* 2 days, with deterioration of the sample.

**E. Counter-ion N.M.R.**—The fact that quadrupole splitting can lead to an apparent intensity loss of resonance signals from nuclei with $I > \frac{1}{2}$ is now generally accepted. Berendsen and Edzes[271] have outlined a theoretical treat-

[263] W. J. Ray, jun., and A. S. Mildvan, *Biochemistry*, 1973, **12**, 3733.
[264] G. Assmann, E. A. Sokoloski, and H. B. Brewer, jun., *Proc. Nat. Acad. Sci. U.S.A.*, 1974, **71**, 549.
[265] N. Muller and R. J. Mead, jun., *Biochemistry*, 1973, **12**, 3831.
[266] J. J. van der Klink, J. Schreiver, and J. Leyte, *Ber. Bunsengesellschaft phys. Chem.*, 1974, **78**, 369.
[267] D. E. Woessner and B. S. Snowden, jun., *Ann. New York Acad. Sci.*, 1973, **204**, 113.
[268] R. E. Dehl, *Biopolymers*, 1973, **12**, 2329.
[269] T. E. Swift and E. M. Barr, *Ann. New York Acad. Sci.*, 1973, **204**, 191.
[270] M. M. Civan and M. Shporer, *Biochim. Biophys. Acta*, 1974, **343**, 399.
[271] H. J. C. Berendsen and H. T. Edzes, *Ann. New York Acad. Sci.*, 1973, **204**, 459.

ment applicable to situations where partially averaged quadrupole splittings are observed, and this theory also includes the fact that the central resonance peak may be observed[272] to have different saturation properties from the outer peaks. The application of the theory is illustrated by the sample of sodium in tissues,[271] and Cope[273] has shown that the existing data can still include a fast exchange between free and bound sodium ions. The $^{23}Na$ signal for NaCl in the presence of gram-negative bacteria was less intense than expected, showing that sodium binding to the cell walls occurs.[274] Only broadening of the resonance was observed with gram-positive bacteria. Both $^{23}Na$ and $^{39}K$ resonances have been reported for a number of tissues[275,276] with emphasis on etiolated pea stems.[276] All the sodium and $>80\%$ of the potassium contributed to the unsplit resonances, and experiments demonstrated that the two ions were taken up by the cells independently.

Cope and Damadian[277,278] have reported potassium relaxation rates for various tissues and simple electrolytes. The results for the tissues are similar to those of ion exchange resins, supporting the concept that $K^+$ is associated with charged macromolecular sites in the tissues. Measurement of $^{23}Na$ relaxation times ($T_1$) for sodium ions in the presence of an enzyme (sodium potassium transport adenosine triphosphatase) showed that little or no sodium binding to the pure enzymes occurred unless ATP was present.[279]

The $T_1$ values of $^{23}Na$ ions in solutions of polyphosphates were independent of the degree of polymerization above a certain value, although the $T_1$ values did become smaller on increasing the polymer concentration.[280] Sodium $T_1$ values were observed to be dependent on the degree of neutralization of poly(acrylic acid), and reasonable agreement between theoretical and observed values was obtained.[281] The effect of added salt may be described by an 'additivity rule'. The broadline $^{23}Na$ resonances of a dehydrated zeolite powder showed two quadrupole splittings,[282] assigned to $^{23}Na$ ions in front of six and eight oxygen atoms.

Chlorine-35 linewidths have been reported[283] for $Cl^-$ in the presence of alkaline phosphatase, and, while Cl ions do bind to the enzyme, it was shown that the site is not at the protein-bound $Zn^{2+}$. Linewidths of $^{35}Cl$ have also been reported for a mercuriated copolymer of cystine and glutamic acid.[284]

[272] J. L. Czeisler and T. J. Swift, *Ann. New York Acad. Sci.*, 1973, **204**, 261.
[273] F. W. Cope, *Ann. New York Acad. Sci.*, 1973, **204**, 416.
[274] N. S. Magnuson and J. A. Magnuson, *Biophys. J.*, 1973, **13**, 1117.
[275] J. A. Magnuson and N. S. Magnuson, *Ann. New York Acad. Sci.*, 1973, **204**, 297.
[276] J. A. Magnuson, N. S. Magnuson, D. L. Hendrix, and N. Higinbotham, *Biophys. J.*, 1973, **13**, 763.
[277] F. W. Cope and R. Damadian, *Physiol. Chem. and Phys.*, 1974, **6**, 17.
[278] R. Damadian and F. W. Cope, *Physiol. Chem. and Phys.*, 1974, **6**, 511.
[279] F. Ostray, T. L. James, J. H. Noggle, A. Sarrif, and L. E. Hokin, *Arch. Biochem. Biophys.*, 1974, **162**, 421.
[280] H. S. Kielman and J. C. Leyte, *J. Phys. Chem.*, 1973, **77**, 1593.
[281] J. J. van der Klink, L. H. Zuiderweg, and J. C. Leyte, *J. Chem. Phys.*, 1974, **60**, 2396.
[282] H. J. Hennig and H. Lechert, *Z. phys. Chem. (Frankfurt)*, 1973, **87**, 104.
[283] J. E. Norne, H. Csopak, and B. Lindman, *Arch. Biochem. Biophys.*, 1974, **162**, 552.
[284] D. E. Carlstrom, W. G. Miller, and R. G. Bryant, *J. Phys. Chem.*, 1973, **77**, 2759.

They were independent of molecular weight ($>20\,000$), and a model involving motion of a monomeric unit plus side-chain rotation sufficed to account for the observed values. Chlorine binding to glutamate decarboxylase was studied[285] by following changes in the $^{81}$Br $T_2$ values in the presence of chloride and bromide ions. Bromine-81 relaxation rates were correlated with enzyme activity as a function of pH.

[285] J. T. Gerig and L. Kwock, *Biochem. Biophys. Res. Comm.*, 1973, **54**, 937.

# 7

# Liquid Crystals and Micellar Solutions

BY G. J. T. TIDDY

## 1 Introduction

N.m.r. has been applied to the study of normal solids and liquids for many years. Recently, however, liquid crystals have received attention from a significant and growing number of n.m.r. spectroscopists. In many of the early investigations liquid crystals were used as anisotropic solvents, the main interest being focused on the spectra of solute molecules. Today this situation has altered, and now many reports are published concerning the n.m.r. spectra of bulk liquid-crystalline (l.c.) phases. These articles form the subject material of this chapter, together with articles on micellar solutions and emulsions. Although this chapter reviews publications appearing between June 1973 and May 1974 it is intended that in future the subject will be reviewed biennially. Papers concerned with solutes in l.c. phases will be reviewed in a separate article to appear in alternate volumes of these Reports.

In this Report articles are grouped under main headings according to the phase structure of the material studied (lyotropic, thermotropic, micellar solutions, reversed micellar solutions, and emulsions). Further sub-division has been made where necessary according to the type of l.c. phase and component (lipid or water) investigated.

Two major developments have occurred in the past year in the area of lyotropic l.c. phases. The use of deuteriated lipids has enabled an accurate estimation of the order parameter pertaining to $CH_2$ (or $CD_2$) groups at different positions along the lipid chain to be made.[1—4] With the exception of the terminal $CH_3$ and $CH_2$ groups only a slight variation of the order parameter was observed, in contrast to the large variation measured using spin probes. Another major development concerns the theoretical description of the linewidth-narrowing caused by sonication. Two independent derivations of equations describing these effects have been published,[5—7] but the authors

[1] J. Charvolin, P. Manneville, and B. Deloche, *Chem. Phys. Letters*, 1973, **23**, 345.
[2] J. Seelig and W. Niederberger, *J. Amer. Chem. Soc.*, 1974, **96**, 2069.
[3] J. Seelig and W. Niederberger, *Biochemistry*, 1974, **13**, 1585.
[4] J. Seelig and A. Seelig, *Biochem. Biophys. Res. Comm.*, 1974, **57**, 406.
[5] C. H. A. Seiter and S. I. Chan, *J. Amer. Chem. Soc.*, 1973, **95**, 7541.
[6] S. I. Chan, M. P. Sheetz, C. H. A. Seiter, G. W. Fergenson, Ming-chu Hsu, A. Lau, and A. Yau, *Ann. New York Acad. Sci.*, 1973, **222**, 499.
[7] E. G. Finer, *J. Magn. Resonance*, 1974, **13**, 76.

draw opposite conclusions concerning the similarity of alkyl-chain mobility in a vesicle or a continuous lamellar phase.

## 2 Reviews

Two major reviews covering magnetic resonance in thermotropic phases (including e.s.r.) and n.m.r. of lyotropic phases have been published.[8,9] In the first of these[8] Luckhurst gives a detailed account of the general theory describing the effects of the liquid-crystal anisotropy on magnetic resonance spectra, followed by a discussion of published data on nematic, smectic, and cholesteric mesophases. The second review[9] describes n.m.r. studies of amphiphile l.c. phases, concentrating on non-biological lipid systems. This review is a detailed description of published articles, but some sections are out of date because of the rapid development of the field. Luckhurst has published a shorter review on magnetic resonance of thermotropic l.c. phases,[10] this being the text of a plenary lecture to the 4th International Liquid Crystals Conference. Short introductory articles on n.m.r. of lyotropic[11] and thermotropic[12] phases have also appeared.

## 3 Lyotropic Liquid Crystal Phases

In lyotropic l.c. phases the alkyl chains occur in regions physically distinct from the water and counter-ions. The n.m.r. properties of the lipid regions are very different from those of the aqueous regions, and for this reason papers on lyotropic l.c. phases have been sub-divided into separate sections dealing with alkyl chains, water, and counter-ions. Sonication of lamellar phase + water emulsions causes dramatic narrowing of alkyl-chain resonances, and this is a popular technique for the study of phospholipid bilayers. Since these are two-phase, non-equilibrium systems with qualitatively different n.m.r. properties, they are also reviewed in a separate section.

**A. Alkyl-chain Resonances.**—The most important development in the past year is the use of deuteron n.m.r. to derive information on the order parameter of $CD_2$ groups in lamellar phases. The order parameter ($S$) is a measure of the degree of orientation of particular groups and is related to the measured quadrupole splitting [$\Delta\nu(\theta)$], as shown in equation (1) for an aligned sample ($\theta$ is the angle between the l.c. director and the magnetic field). The symbol $\chi$ in (1) refers to the nuclear quadrupole coupling constant, $e^2qQ/h$.

$$\Delta\nu(\theta) = \tfrac{3}{4}\chi S(3\cos^2\theta - 1) \qquad (1)$$

[8] G. R. Luckhurst in 'Liquid Crystals and Plastic Crystals', ed. G. W. Grey and P. A. Winsor, Ellis Harwood Ltd., Chichester, 1974, Vol. 2, p. 144.
[9] Å. Johansson and B. Lindman, ref. 8, p. 192.
[10] G. R. Luckhurst, *Mol. Crystals Liquid Crystals*, 1973, **21**, 125.
[11] L. W. Reeves, *Cienc. Cult.* (*São Paulo*), 1973, **25**, 516.
[12] A. Loesche, *Comments Solid State Phys.*, 1973, **5**, 119.

In a powder sample the term ($3\cos^2\theta - 1$) is omitted. The proton CW line-shape for alkyl chains in a lamellar phase is 'super lorentzian' with little structure observable. It is possible to resolve some structure with samples where the bilayers are aligned (for example by glass slides), but even in this case only two separate $CH_2$ peaks can be resolved for $C_8$–$C_{18}$ alkyl chains. Charvolin and co-workers[1] have reported that the deuteron spectrum of an aligned potassium [$^2H_{23}$]dodecanoate lamellar-phase sample shows seven different doublets. The peaks are resolved because the inter-alkyl-group dipolar coupling is much reduced compared with the corresponding proton spectra. By assuming that the order parameter decreases as the distance from the head group increases, the orientation profile shown in the Figure was derived. The order parameter is similar for the first seven alkyl groups and only becomes much smaller for the terminal three alkyl groups. This is very different from the continuous decrease in order with increasing distance from the head group obtained in similar systems using e.s.r. and nitroxide-labelled lipids.

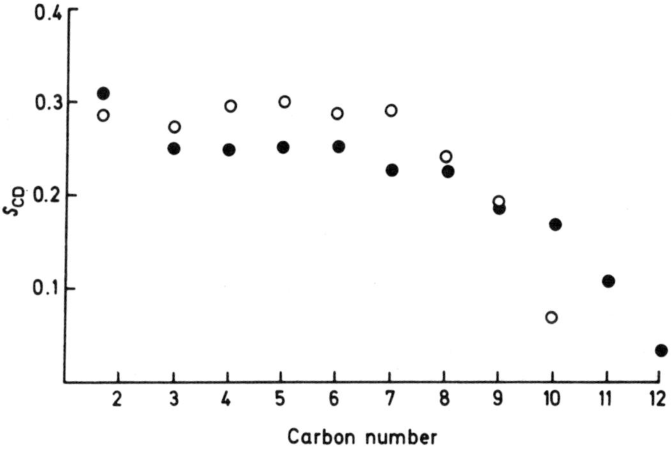


**Figure 1** *Variation of the order parameter* ($S_{CD}$) *with the position of the* $CD_2$ *group along the alkyl chain in lamellar-phase samples* (●, *potassium* [$^2H_{23}$] *dodecanoate, from Charvolin* et al., *ref.* 1; ○, *sodium* [$^2H_2$] *decanoate, from Seelig and Niederberger, ref.* 3).

Seelig and Niederberger have avoided using the assumption of a decrease in order with increasing distance from the head group by using a series of specifically labelled sodium decanoates.[3] The position of the deuteriated alkyl group was varied from C-2 to C-10, and by incorporating the decanoates into lamellar phases prepared using 1,1-dideuteriodecanol and water, the order parameters shown in the Figure were obtained.[3] These show essentially the same variation as the data of Charvolin *et al.*, but if the small variations in *S* are real, Charvolin's assumption that $S$ decreases continuously as the distance from the head group increases may not be valid. A less extensive study[2] on sodium octanoate–decanol–water lamellar phase demonstrated a similar dependence of $S$ on alkyl-group position. In a further communication,[4] Seelig's group have reported preliminary results on two di-$C_{16}$-lecithin (1) lamellar phases with specific dideuteriation at C-2 and C-5. For the C-5 derivative the order parameter was similar to the values for carboxylate lamellar phases. However, three separate splittings were observed for the C-2-deuteriated derivative, the largest being similar to that of the C-5-labelled lecithin. This may indicate that the distribution of alkyl-group order parameters is similar in lecithin and simple amphiphile bilayers, but the additional resonance requires explanation before definite conclusions can be drawn.

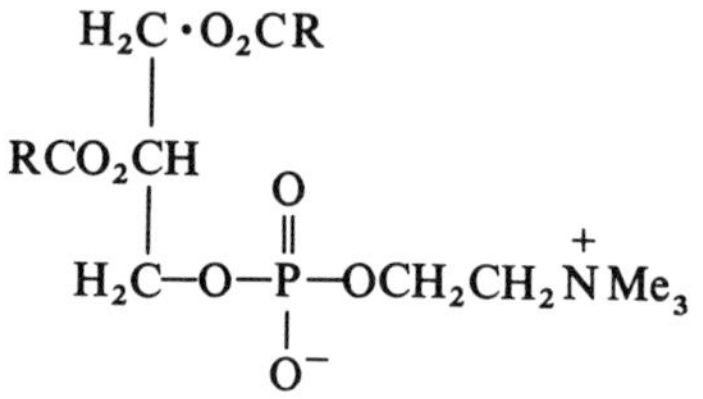

(1)

For proton resonance investigations, Seiter and Chan[5,6] have developed a theoretical approach based on Anderson's stochastic theory of resonance linewidths to calculate alkyl-chain bandshapes. The theory is successful in reproducing the main features of the observed bandshape in lecithin bilayers using parameters based on *gauche*$^+$–*gauche*$^-$ conformation changes ('kink' formation). In addition it is shown that the observed intensity loss (~50%) for $CH_3$ groups is a consequence of their restricted motion and non-averaged dipolar couplings. In a later paper[13] Feigenson and Chan show that the previously reported field-strength dependence of the alkyl-chain bandshape is a consequence of the chemical shifts between different groups.

The non-averaged dipolar coupling in lamellar phases can be removed by using aligned samples positioned in the probe so that the angle between the l.c. director and the magnetic field is equal to the magic angle. Samulski and co-workers[14] have found $T_2$ values of up to 15 ms for potassium oleate lamellar-phase samples as opposed to values of 100 $\mu$s observed for powder samples. They suggest that this can be exploited for the measurement of self-diffusion coefficients by the pulsed field-gradient technique.

Proton relaxation measurements for the alkyl-chain region in the different l.c. phases of potassium dodecanoate–water have been reported by Charvolin and Rigny.[15] The report includes a detailed investigation of the shape of the FID, and values of $T_1$, $T_{1\rho}$, and $T_{1D}$ for the lamellar, cubic, and hexagonal

[13] G. W. Feigenson and S. I. Chan, *J. Amer. Chem. Soc.*, 1974, **96**, 1312.
[14] E. T. Samulski, B. A. Smith, and C. G. Wade, *Chem. Phys. Letters*, 1973, **20**, 167.
[15] J. Charvolin and P. Rigny, *J. Chem. Phys.*, 1973, **58**, 3999.

phases are given. It is concluded that at high frequencies $T_1$ is determined by translational diffusion, and that chain deformation occurs on a time-scale of $<3.5 \times 10^{-6}$ s. Assuming that the self-diffusion coefficient of $1 \times 10^{-10}$ m$^2$ s$^{-1}$ measured for the cubic phase also holds for the other phases, as implied by the authors, a jump distance of 0.8 nm is calculated for the estimated translation diffusion $\tau_c$ value of $10^{-9}$ s. This seems rather large, and it is possible that a significant contribution to $T_1$ does in fact arise from chain rotations about the molecular long axis. Measurements of $T_1$ values in bilayers formed from deuteriated lipids are required for an unambiguous conclusion to be made. For proton relaxation in lecithin bilayers, Feigenson and Chan[13] have demonstrated that spin diffusion is rapid and that relaxation occurs *via* common heat sinks. These will be the alkyl groups with rotations closest to the n.m.r. frequency, and the conclusion that they correspond to the terminal methyl and 2—4 mobile $CH_2$ groups seems surprising. The overall rotation of the whole molecule about the long axis is unlikely to have a correlation time ($\tau_{\parallel}$) longer than $10^{-10}$ s, the value typically found for spin-labelled lipids. The alkyl groups closest to the head group are likely to have the longest (closest to the n.m.r. frequency) values of $\tau_{\parallel}$, and these could be expected to provide the $T_1$ mechanism, rather than the suggested torsional oscillations.

In a further study Ming-chu Hsu and Chan have reported[16] the effects of valinomycin (a depsipeptide) on the proton resonance of di-$C_{14}$- and di-$C_{16}$-lecithin bilayers. No changes in the alkyl-chain FID or the gel–l.c. transition temperature occurred for the di-$C_{16}$-lecithin bilayers, whereas a 10% increase in the time constant of the FID and a drop in the gel–l.c. transition temperature were observed for the di-$C_{14}$-lecithin system. These remarkable differences indicate that valinomycin interacts with di-$C_{16}$-lecithin in the region of the polar head groups and penetrates into the alkyl-chain regions of the di-$C_{14}$ bilayer. The suggestion that this difference in behaviour can be attributed to the increased stability of $C_{16}$ bilayers compared with $C_{14}$ bilayers seems speculative, and n.m.r. results on a series of mixed bilayers containing both di-$C_{16}$- and di-$C_{14}$-lecithin would be of value in assessing its validity.

**B. Water Studies.**—Lyotropic liquid crystals contain regions of different bulk magnetic susceptibility, and this gives rise to local magnetic field inhomogeneities that broaden CW spectra of water, decreasing the water FID owing to translational diffusion of water molecules from one region to another. The effects on $T_2$ measurements were thought to be removed by the use of a CPMG pulse sequence; Packer[17] has estimated that the effects of diffusion through the field gradients arising from heterogeneity in muscle tissue (expected to be similar to that in l.c. phases) on water transverse relaxation is negligible. However, Glasel and Lee[18] have pointed out that the CPMG

16 Ming-chu Hsu and S. I. Chan, *Biochemistry*, 1973, **12**, 3872.
17 K. J. Packer, *J. Magn. Resonance*, 1973, **9**, 438.
18 J. A. Glasel and K. H. Lee, *J. Amer. Chem. Soc.*, 1974, **96**, 970.

sequence only removes the effects of linear field gradients, and the non-linear gradients could give a large contribution to $T_2$ measurements. They report deuteron n.m.r. measurements on monodisperse glass-sphere–$D_2O$ dispersions, the size of the spheres being varied over the range 20—660 $\mu$m. CW linewidths show the existence of a large susceptibility broadening. Deuteron $T_1$ and $T_2$ values are very different, and while this is explicable in terms of a fast exchange between free water and water restricted at the glass surface, the authors point out that large non-linear field gradients would also result in unequal $T_1$ and $T_2$ values. An estimation of the maximum size of the susceptibility effects could be obtained from measurements at different frequencies. With glass-sphere–water dispersions this could be done using proton resonance for $H_2O$ at different frequencies or by making proton and deuteron measurements on a sample containing HOD, and it is unfortunate that these measurements were not made.*

The anisotropy of motion in the water layers is far smaller than that in the alkyl-chain regions, but it is sufficient to give a quadrupole splitting for water deuterons in $D_2O$. A dipolar doublet can also be observed for $H_2O$, but proton exchange is usually fast enough to make its observation difficult. For this reason measurements of deuteron splitting are more common, and Finer[19] has discussed in detail the motional processes that lead to reduction of these splittings. He demonstrated how splittings and relaxation rates (linewidths) can be analysed to give $\tau_c$ value for different types of bound water and how an estimation of the binding energy can be obtained.

Many reports of deuteron spectra for $D_2O$ in l.c. phases mention the occurrence of a sharp central line in addition to the quadrupolar doublet. This is often taken to indicate the presence of a second phase with isotropic water motion, but Wennerström *et al.*[20] have now demonstrated that in the lamellar (D) phase of sodium octanoate–octanoic acid–$D_2O$ the additional line is due to double-quantum transitions. Calculated changes in linewidth and amplitude of the central peak were in excellent agreement with experimental observations. In related papers from the same group the effects of chemical exchange on deuteron lineshapes have been investigated.[21,22] The authors demonstrate the presence of chemical exchange of deuterons between decanol OD groups and $D_2O$ in a lithium octanoate–decanol–$D_2O$ lamellar phase. Equations are presented that describe the main features of the spectra, although a quantitative fit between calculated and observed spectra was not obtained. Data are also given on the variation of water splittings with

[19] E. G. Finer, *J.C.S. Faraday II*, 1973, **69**, 1590.
[20] H. Wennerström, N.-O. Persson, and B. Lindman, *J. Magn. Resonance*, 1974, **13**, 348.
[21] N.-O. Persson, H. Wennerström, and B. Lindman, *Acta Chem. Scand.*, 1973, **27**, 1667.
[22] G. Lindblom, N.-O. Persson, and B. Lindman, Proceedings of the VIth International Conference on Surface Active Agents, Sept. 1972, Carl Hanser Verlag, München, 1973, p. 939.

* $T_2$ Values measured for protons and deuterons in a glass-granule–water dispersion show no measurable frequency dependence, indicating that susceptibility effects on $T_2$ are indeed small (T. E. Bull and G. J. T. Tiddy, unpublished results).

temperature and with the nature of the counter-ion. The changes are complex, and depend on the octanoate–decanol ratio.

A number of papers have been published on phospholipid–water l.c. phases. Finer and Darke[23] reported deuteron measurements on $D_2O$ in lamellar phases of egg lecithin, egg phosphatidylethanolamine, and ox-brain sodium phosphatidylserine. The spectra of some freshly prepared samples gave only broad singlets, which changed to doublet + singlet after several weeks. The initial broad singlet was attributed to rapid translational diffusion between small crystallites with different orientations. The crystallites slowly grow larger with time, eventually becoming large enough that diffusion between them is sufficiently slow on the n.m.r. time-scale for the observation of a splitting. However, the intensity of the residual sharp line was independent of added water concentration for some samples, and this may be due to double-quantum transitions as indicated by Wennerström *et al.*[20] rather than the explanation given by the authors. By applying the analysis[19] referred to above, Finer and Darke were able to distinguish two, three, and four different types of water for phosphatidylethanolamine, lecithin, and sodium phosphatidylserine respectively. These corresponded to the inner hydration shell, weakly bound water, and trapped water, exchange between them being rapid on the n.m.r. time-scale. The authors recognize that the analysis does not strictly apply for sodium phosphatidylserine because of competition between the counter-ion and the head group for bound water at low water concentrations. In a less extensive study, Persson and co-workers[24] have reported changes in deuteron splittings for $D_2O$ in lecithin lamellar phase on addition of different salts, cholesterol, and sodium cholate. No change in splitting was observed when the phase structure changed from lamellar to hexagonal on increasing sodium cholate concentration.

Two further studies on proton relaxation of $H_2O$ in lecithin lamellar phases have been reported.[25,26] Gottlieb and co-workers[25] have given high-resolution linewidths and $T_1$ and $T_2$ values for the egg lecithin lamellar phase. The half-height width increases rapidly below ~10% water, and $T_2$ values were dependent on the CPMG pulse spacing ($t_{cp}$). The dependence of the spin–lattice relaxation on water concentration was a smooth curve with no sharp breaks. The results were interpreted to indicate the existence of two types of water, slow exchange between them resulting in the observed dependence of $T_2$ on $t_{cp}$. Unfortunately the authors ignore the effects of susceptibility broadening on the linewidth measurements. They also neglect the possibility that the alkyl resonances may give a contribution to the observed transverse relaxation at short pulse spacings, giving rise to the changes in $T_2$. Transverse relaxation decays with time constants of ~0.5 s can be obtained for alkyl-

23 E. G. Finer and A. Darke, *Chem. and Phys. Lipids*, 1974, **12**, 1.
24 N.-O. Persson, G. Lindblom, B. Lindman, and G. Arvidson, *Chem. and Phys. Lipids*, 1974, **12**, 261.
25 A. M. Gottlieb, P. T. Inglefield, and Y. Lange, *Biochim. Biophys. Acta*, 1973, **307**, 444.
26 H. Prestegard and A. Wilkinson, *Biochim. Biophys. Acta*, 1974, **345**, 439.

chain resonances at small values of $t_{cp}$. Until these points have been considered, the conclusions should be treated with caution. Similar results have been reported for di-$C_{16}$-lecithin lamellar-phase samples.[26] Proton $T_1$ and $T_2$ values for two samples with different compositions were given as a function of temperature. An increase in $T_1$ at the gel–l.c. transition temperature demonstrated the greater mobility of water in the l.c. phase. The authors also observed a dependence of $T_2$ on $t_{cp}$ above the gel–l.c. transition temperature, and again attributed this to slow chemical exchange between different water species. Since the hydrocarbon chain $T_2$ becomes much longer at this phase transition, this paper is subject to the same criticism as the previous one.

A rather unusual line-broadening effect has been noted[27] for the $^2H$ spectrum of the sodium dodecyl sulphate–decanol–sodium sulphate–$D_2O$ hexagonal phase commonly used for orientation of solutes in high-resolution n.m.r. studies.* Broadening of either or both of the deuteron doublet lines could be caused by changing the position of the sample tube with respect to the receiver coil. The authors suggest that this effect arises from the presence of a concentration gradient along the sample. A range of splittings with slightly different chemical shifts results in the observed broadening.

Tobacco mosaic virus forms a hexagonal phase with water, having the virus molecules arranged in a two-dimensional hexagonal lattice with water in the intervening channels. Blinc and co-workers[28] have reported water proton $T_1$, $T_2$, and self-diffusion measurements, together with deuteron CW spectra, for this phase. No deuteron splittings were observed, and proton measurements indicated that the fraction of water molecules bound to the virus was less than 1%.

**C. Counter-ion Studies.**—The lipids often contain charged groups in a lyotropic l.c. phase, and inorganic salts may be present in the system. These ions exist in the aqueous regions, usually in association with the lipid head groups. This results in an anisotropy in the environment, which gives rise to quadrupolar splittings for ions with magnetic nuclei having $I \geqslant 1$. From the magnitude of the splittings, information on the average orientation of the ion, and hence the ion binding to the surface, can be deduced. Lindblom and co-workers have reported[22,29] quadrupole splittings for a number of different cations in the lamellar phases of sodium octanoate–decanol or octanoic acid–water. The most complete results have been obtained for $^{23}Na$ splittings, and these have a complex dependence on temperature, composition, and amphiphile head group. It was also reported that the order parameter measured for powder samples was smaller than the value measured for aligned samples.†

27 L. A. McLachlan, D. S. F. Natusch, and R. H. Newman, *J. Magn. Resonance*, 1973, **10**, 34.
28 R. Blinc, J. Pirš, K. Easwaran, and L. Ehrenberg, *Biophys. J.*, 1973, **13**, 622.
29 G. Lindblom and B. Lindman, *Mol. Crystals Liquid Crystals*, 1973, **22**, 45.

* The phase structure of the material should be classified as smectic, not nematic.

† Further experiments have shown that this was due to water loss during the preparation of the aligned samples (G. J. T. Tiddy and B. Lindman, unpublished results).

Sodium splittings in the hexagonal phase are smaller than those in the adjacent lamellar phase whereas those for the reversed hexagonal phase are larger. Other counter-ions studied in similar systems[22,29,30] were $^{7}Li^{+}$, $^{39}K^{+}$, $^{85}Rb^{+}$, $^{87}Rb^{+}$, and $^{133}Cs^{+}$, and all except $^{85}Rb^{+}$ gave appropriate splittings. Data were also presented on $Br^{-}$ and $Cl^{-}$ splittings in l.c. phases formed from cationic surfactants. Sodium splittings in lecithin–cholesterol–water lamellar phases with added sodium chloride have also been reported[24] and show a complex relationship between sodium splittings and cholesterol content.

It is possible to detect the existence of two-phase systems by n.m.r. if the two phases concerned have distinctly different spectra, with slow exchange of the magnetic nuclei between the two phases. It has been suggested that the tetragonal (C) phase in sodium octanoate–decanol–water is an emulsion of micellar solution and lamellar (D) phase. Because large $^{23}Na$ splittings were observed for C-phase and not for adjacent D-phase samples, the emulsion structure was eliminated and the tetragonal rod structure was thought[22,29] to be more likely.*

In a detailed study of ammonium and methylammonium ions in a hexagonal phase consisting of dodecyl sulphate anions–decanol–sodium sulphate–water, using both $^{1}H$ and $^{2}H$ n.m.r., Reeves and Tracy[31] were able to obtain the sign and degree of orientation for the different ions. Data were presented for one composition, and further studies concerning the dependence of orientation on temperature and phase composition should enable a detailed model of counter-ion binding to be developed for this case.

**D. Vesicle Studies.**—Vesicles are formed by sonication of a lamellar phase + water dispersion; they consist of a spherical lipid bilayer (radius ~12 nm) surrounding a water core. An important unresolved problem is the magnitude of the change induced in bilayer properties on vesicle formation. Molecular motion would be expected to be a sensitive measure of this, and n.m.r. is the obvious technique for the investigation of changes. The dramatic narrowing of the linewidth on vesicle formation has not previously been subjected to a detailed theoretical analysis, but in the past year several papers[5—7] have been devoted to this topic. Seiter and Chan,[5,6] using the theory referred to above together with a $\tau_c$ value for vesicle rotation, calculate that the linewidths of vesicles with unaltered alkyl-chain mobility should be several hundred hertz. Since the observed value is ~5 Hz they conclude that the local motions of alkyl chains in vesicles are much more rapid than in unsonicated bilayers. Chan *et al.* have also reviewed[6] previous evidence for a difference between vesicles and continuous bilayers, including the effects of solution viscosity and

[30] G. Lindblom and B. Lindman, ref. 22, p. 939.
[31] L. W. Reeves and A. S. Tracy, *J. Amer. Chem. Soc.*, 1974, **96**, 365.

* Further measurements of deuteron and $^{23}Na$ resonances in this system showed that large sodium splittings are present in the 'D' phase, and that the emulsion structure for the 'C' phase is correct (G. J. T. Tiddy and B. Lindman, unpublished results).

different vesicle radii on the linewidths. However, Finer[7] developed similar equations based on the theory for narrowing of resonance lines by specialized motions, and calculated the observed vesicle linewidths assuming no alteration in chain mobility on vesicle formation. He was also able to account for the effects of increasing solution viscosity. It is not immediately obvious which of the treatments is in error, and further experimentation (perhaps on simpler vesicle systems) will be required to resolve the problem.

Spin–lattice relaxation rates, particularly of $^{13}C$ nuclei, are more susceptible to interpretation than linewidths. Gent and Prestegard[32] have compared $^{13}C$ $T_1$ values for alkyl chains in continuous bilayers and vesicles containing egg lecithin and a $^{13}C$-enriched phosphatidylethanolamine. The activation energies for spin–lattice relaxation are the same in both cases, although, except for the α-carbon, the vesicle $^{13}C$ relaxation times are *ca.* 50% greater than those in the continuous bilayers. $^{13}C$ Relaxation studies have also been used to determine the position and motions of spin-labelled stearic acid derivatives in vesicles by measuring the changes in relaxation rates caused by their presence.[33] It was concluded that the stearic acid carboxy-group was positioned in the vicinity of the lecithin phosphate group, and the average position of the nitroxide group was also determined. Conclusions concerning the relative amplitudes of chain motions perpendicular to the long axis of the molecule for spin-labelled and normal alkyl chains were also made. It is difficult to assess the general validity of these, since the spin label probably causes alterations in the conformations of adjacent alkyl chains. Natural-abundance CW $^{13}C$ spectra of lecithin bilayers at temperatures above and below the gel–l.c. transition have been obtained by Urbina and Waugh[34] using proton enhanced spectroscopy. This involves the use of heteronuclear pulsed double resonance to obtain proton–carbon cross-polarization, giving intensity enhancement and line narrowing.

The difficulties in determining $\tau_c$ values from proton relaxation measurements in vesicles have been discussed by McLaughlin *et al.*[35] They report proton $T_1$ measurements at two frequencies for the chemically shifted resonances in di-$C_{16}$-lecithin. It is tentatively concluded that rotation of the whole molecule has a $\tau_c$ value of $7 \times 10^{-10}$ s and that the alkyl groups also have a $\tau_c$ value substantially longer than $10^{-9}$ s. Horwitz and co-workers[36] have outlined their previous relaxation-rate measurements on lecithin vesicles and have presented a detailed model to describe the alkyl-chain motions. Metcalfe *et al.* have also summarized[37] their previous papers describing the use of $^{13}C$ and $^1H$ relaxation rates and linewidths to obtain the

32 M. P. N. Gent and J. H. Prestegard, *Biochem. Biophys. Res. Comm.*, 1974, **58**, 549.
33 P. E. Godici and F. R. Landsberger, *Biochemistry*, 1974, **13**, 362.
34 J. Urbina and J. S. Waugh, *Ann. New York Acad. Sci.*, 1973, **222**, 733.
35 A. C. McLaughlin, F. Podo, and J. K. Blasie, *Biochim. Biophys. Acta*, 1973, **330**, 109.
36 A. F. Horwitz, M. P. Klein, D. M. Michaelson, and S. J. Kohler, *Ann. New York Acad. Sci.*, 1973, **222**, 468.
37 J. C. Metcalfe, N. J. M. Birdsall, and A. G. Lee, *Ann. New York Acad. Sci.*, 1973, **222**, 460.

motional gradient and self-diffusion coefficients for lipids in bilayers. These results must be considered in the light of the deuteron n.m.r. studies[1—4] and theoretical papers[5—7] referred to above. In particular the estimation of self-diffusion coefficients requires a number of assumptions that are not easy to verify.

The difficulties in interpretation of relaxation times in vesicles may be overcome by using results for liquid alkane derivatives. Levine and co-workers[38—41] have outlined a model for the calculation of correlation functions for alkyl chains and have applied the model to determine rotational diffusion coefficients for different bonds in various alkane derivatives.

The use of vesicles as model systems for biological membranes to study the interactions of drugs continues to be popular. Increases in the linewidths of various phospholipids caused by incorporation of antibiotics in the bilayer were interpreted in terms of reduced alkyl-chain motion.[42] A similar study of the interaction between chlorophentermine and lecithin has also been reported.[43] Haque and co-workers[44] observed two resonances for the ring protons of 1,1-dichloro-2,2-bis-(*p*-chlorophenyl)ethane (an insecticide) in vesicles formed from several different phospholipids. These were attributed to the occurrence of two different lipid phases (gel and l.c.) because of the dependence of peak intensities on temperature and phospholipid concentration.

It has been known for some time that separate signals from molecules inside and outside vesicles can be observed by adding shift reagents to sonicated dispersions. Kostelnik and Castellano[45] have reported that separate choline methyl signals can be resolved at 250 MHz without shift reagents. The magnitude of the chemical shift difference can be explained by the vesicles having a prolate or oblate spheroid shape. Fernández and Cerbón have used the effect of paramagnetic shift reagents to investigate further the interactions of local anaesthetics with vesicles.[46] Tetracane displaced the paramagnetic ions from the bilayer surface, thus removing the splitting of the choline resonance, whereas other local anaesthetics did not. The observation of two choline resonances is due to the fact that ion diffusion across the bilayer is slow. Ionophores can complex with ions and increase their transport across membranes. Fernández and co-workers[47] have demonstrated this effect for a particular antibiotic by showing that the signal due to non-complexed 'inside'


[38] Y. K. Levine, N. J. M. Birdsall, A. G. Lee, J. C. Metcalfe, P. Partington, and G. C. K. Roberts, *J. Chem. Phys.*, 1974, **60**, 2890.
[39] Y. K. Levine, P Partington, and G. C. K. Roberts, *Mol. Phys.*, 1973, **25**, 497.
[40] Y. K. Levine, *J. Magn. Resonance*, 1973, **11**, 421.
[41] N. J. M. Birdsall, A. G. Lee, Y. K. Levine, J. C. Metcalfe, P. Partington, and G. C. K. Roberts, *J.C.S. Chem. Comm.*, 1973, 757.
[42] J. M. Padfield and I. W. Kellaway, *J. Pharm. Sci.*, 1973, **62**, 1621.
[43] J. K. Seydel and O. Wassermann, *Naunyn-Schmiedeberg's Arch. exp. Pathol. Pharmakol.*, 1973, **279**, 207.
[44] R. Haque, I. J. Tinsley, and S. E. Randall, *Biochemistry*, 1973, **12**, 2346.
[45] R. J. Kostelnik and S. M. Castellano, *J. Magn. Resonance*, 1973, **9**, 291.
[46] M. S. Fernández and J. Cerbón, *Biochim. Biophys. Acta*, 1973, **298**, 8.
[47] M. S. Fernández, H. Célis, and M. Montal, *Biochim. Biophys. Acta*, 1973, **323**, 600.

choline groups is not observed if praseodymium ions are added to vesicles containing the antibiotic. However, a note of caution should be added concerning the interpretation of experiments of this type. Hauser and Barratt[48] have shown that whereas two choline resonances are observed on addition of $NdCl_3$ to sonicated dispersions of di-$C_{14}$- and di-$C_{16}$-lecithins, single, broad resonances are observed for di-$C_{10}$- and di-$C_{12}$-lecithins. Additional results from gel chromatography and electron microscopy show that vesicles formed from the shorter-chain lecithins are unstable and re-form multi-lamellar aggregates, giving the single, broad choline resonance. The stability of di-$C_{14}$-lecithin vesicles has also been investigated in a detailed study by Prestegard and Fellmeth.[49] They used changes in the integrated intensity of the high-resolution resonances to measure the kinetics of vesicle coalescence and found a sharp maximum in the rate constant near the gel–l.c. transition temperature.

Although the use of vesicles formed from phospholipids other than lecithin is fairly common, studies on the characterization of these are rather rare. Hauser and Phillips[50] have reported n.m.r. studies of phosphatidylserine vesicles, together with data on their characterization using other techniques. The resonances are sharper than those of lecithin vesicles, and differential broadening is observed on increasing the ionic strength of the medium. The mixing behaviour in co-sonicated phosphatidylglycerol and lecithin was studied by Michaelson *et al.*[36,51] using a paramagnetic shift reagent and line broadening by $Mn^{2+}$ ions. They found that the vesicle bilayer was asymmetric, and that phosphatidylglycerol preferentially occupied sites in the outer monolayer. Finer and Phillips[52] have also investigated mixtures of lecithins and glycerides. The lecithin peaks were broadened by the addition of glycerides, and the results were used in conjunction with surface-balance data to determine the important criteria for the formation of mixed monolayers. Cholesterol is an important constituent of many biological membranes. In vesicles containing both lecithin and cholesterol, Ching-Hsien Huang and co-workers have reported the dependence of the inner and outer choline shifts and peak areas on the lecithin/cholesterol ratio in the presence of praseodymium ions.[53] They conclude that for cholesterol levels greater than 30% the cholesterol preferentially occupies the inner sites.

The use of molecular probes to obtain information concerning different environments of a disperse system is a useful technique, provided that the probe does not disturb the environment. Saito and co-workers[54] have reported deuteron n.m.r. linewidths for a hexadeuteriodecanoic acid [$CD_3$-$(CH_2)_9CD_2CO_2D$] probe in sonicated and unsonicated egg lecithin dis-

[48] H. Hauser and M. D. Barratt, *Biochem. Biophys. Res. Comm.*, 1973, **53**, 399.
[49] J. H. Prestegard and B. Fellmeth, *Biochemistry*, 1974, **13**, 1122.
[50] H. Hauser and M. C. Phillips, *J. Biol. Chem.*, 1973, **248**, 8585.
[51] D. M. Michaelson, A. F. Horwitz, and M. P. Klein, *Biochemistry*, 1973, **12**, 2637.
[52] E. G. Finer and M. C. Phillips, *Chem. and Phys. Lipids*, 1973, **10**, 237.
[53] Ching-Hsien Huang, J. P. Sipe, S. T. Chow, and R. B. Martin, *Proc. Nat. Acad. Sci. U.S.A.*, 1974, **71**, 359.
[54] H. Saito, S. Schreier-Muccillo, and I. C. P. Smith, *F.E.B.S. Letters*, 1973, **33**, 281.

persions. A linewidth of ~17 Hz was observed for the vesicles, and this increased on addition of cholesterol. From the linewidth a rotational $\tau_c$ value of $10^{-9}$ s was deduced. Nogrady[55] has reported the use of methyl 12-dimethylsilastearate [$Me(CH_2)_5SiMe_2(CH_2)_{10}CO_2Me$] as a probe for the alkyl-chain regions of lecithin bilayers. Only disturbances altering the lecithin alkyl-chain regions changed the $SiMe_2$ peak width.

Lecithin dispersions are two-phase mixtures. On addition of a micelle forming surfactant, the phospholipid can be solubilized into the surfactant micelles, resulting in the formation of a single phase. Dennis and Owens[56] have utilized the different linewidths of lecithin in a dispersion and in triton X-100 micelles to determine the limit of lecithin solubility in the micellar phase. In a further publication[57] the effect of the lecithin gel–l.c. transition on the solubility is reported, together with an approximate phase diagram for the system.

## 4 Thermotropic Liquid Crystals.

Because thermotropic l.c. phases are only formed by rather complex organic molecules containing protons in different groups with different mobilities (*e.g.* benzene rings, alkyl chains), their relaxation behaviour is complex and is still not fully understood. Samulski and co-workers[58] have re-evaluated previous data[59] on the inter- and intra-molecular contributions to proton spin–lattice relaxation in *p*-azoxyanisole (PAA). The magnitudes of the two contributions were approximately the same in both the nematic and isotropic liquid phases. Correlation times for rotational and translational motions were also derived. These rotational $\tau_c$ values were used by Wiszniewska *et al.*,[60] together with measured deuteron $T_1$ values on $PAA(CD_3)_2$, to obtain a value of $147 \pm 8$ kHz for the quadrupole coupling constant along the C—D bond. This value was in agreement with values measured for other compounds. Values of the relaxation times of the dipolar energy ($T_{1D}$) in PAA have been reported by Sharp *et al.*[61] The relaxation rates were faster than expected from the data of Samulski *et al.*,[59] and it was concluded that an additional frequency-independent relaxation mechanism was present. Further calculations indicated that the additional process could be proton–nitrogen cross-relaxation. From the similarity of the activation energy for spin–lattice relaxation in the nematic phase of *N*-(*p*-ethoxybenzylidene)-*p*-butylaniline (EBBA) to the value for the methoxy-compound (MBBA) it was concluded[62]

55 T. Nogrady, *Canad. J. Biochem.*, 1973, **51**, 1129.
56 E. A. Dennis and J. M. Owens, *J. Supramol. Structure*, 1973, **1**, 165.
57 A. A. Ribeiro and E. A. Dennis, *Biochim. Biophys. Acta*, 1973, **332**, 26.
58 E. T. Samulski, C. R. Dybowski, and C. G. Wade, *Mol. Crystals Liquid Crystals*, 1973, **22**, 309.
59 E. T. Samulski, C. R. Dybowski, and C. G. Wade, *Phys. Rev. Letters*, 1972, **29**, 340; *ibid.*, p. 1050.
60 M. Wiszniewska, R. Y. Dong, E. Tomchuk, and E. Block, *Canad. J. Chem.*, 1974, **52**, 2294.
61 A. R. Sharp, W. F. Forbes, and M. M. Pintar, *J. Chem. Phys.*, 1973, **59**, 460.
62 S. D. Goren, C. Korn, S. B. Marks, and R. Potashnik, *Chem. Phys. Letters*, 1974, **24**, 249.

that the relaxation mechanism was translational diffusion, since this mechanism is thought to apply for MBBA.

The compound ethyl [*p*-(*p*-methoxybenzylidene)amino]cinnamate (EMC) (2) forms two solid phases, two smectic phases (A and B), and a nematic phase

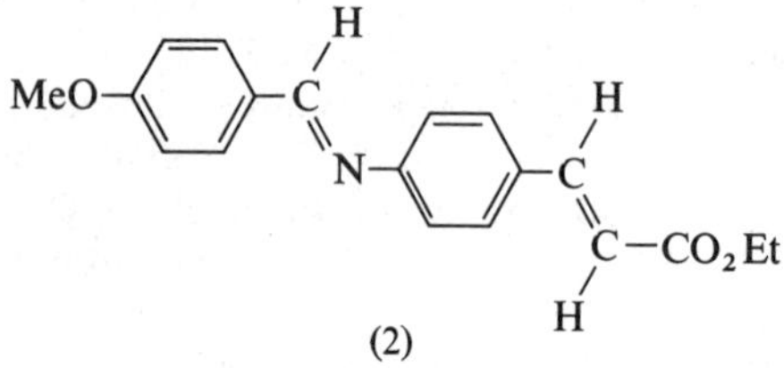


(2)

before melting. Tarr and Fuller have reported[63] $T_1$ and second-moment data for EMC as a function of temperature. The value of $T_1$ decreases with increasing temperature, and discontinuities are observed at phase boundaries. In the second-moment data the discontinuities are more marked. An interesting observation was that cooling an aligned nematic sample resulted in the formation of an aligned smectic A phase which, after rotation, reoriented so that the director was aligned along the magnetic field direction. No such behaviour was observed for the smectic B phase, and the authors indicate that the presence of impurities may cause this re-alignment (see below). Dong and co-workers have also reported $T_1$ values for EMC.[64] Their results are similar to those of Tarr and Fuller for the nematic phase, but are lower for the smectic A and B phases. Also, they observed an additional liquid-like resonance for the smectic phases. These discrepancies are difficult to reconcile, and since the EMC was not recrystallized before use, the presence of a significant concentration of impurities cannot be neglected.

The proton resonance lineshapes and FID's of thermotropic l.c. phases are determined by intramolecular dipolar coupling which, in turn, depends on the degree of order ($S$) and orientation of the molecule. Wise *et al.*[65] have used this property to determine the orientation of molecules in the smectic C phase. This phase consists of layers of molecules with the molecular axes tilted at an angle to the axis of the lamellae. Aligned smectic C phase samples were prepared from terephthal-bis-4-(4-n-butylaniline) (TBBA) and 4,4′-bis-(heptyloxy)azoxybenzene (HOAB). Calculations were made to determine the orientation for maximum FID with different tilt angles, and in practice this was found to be a good experimental technique for the determination of the tilt angle. Lineshape calculations were also reported, and a tilt angle of 45° was measured for HOAB, while that of TBBA was dependent on temperature. Dong and co-workers[66] have also examined the smectic C phase of HOAB,

[63] C. E. Tarr and A. M. Fuller, *Mol. Crystals Liquid Crystals*, 1973, **22**, 123.
[64] R. Y. Dong, A. Wiszniewska, E. Tomchuk, and E. Block, *Chem. Phys. Letters*, 1974, **25**, 299.
[65] R. A. Wise, D. H. Smith, and J. W. Doane, *Phys. Rev. (A)*, 1973, **7**, 1366.
[66] R. Y. Dong, M. Wiszniewska, E. Tomchuk, and E. Block, *J. Chem. Phys.*, 1973, **59**, 6266.

comparing the $T_2$ values (measured using $\pi/2$–$\pi/4$ pulse sequences) with those of the smectic A phase of MBBA. No dependence of $T_2$ on orientation was observed for MBBA, while a maximum at $\sim 75°$ was observed for HOAB.

Meiboom and Luz have reported an extended study of lineshapes for a range of different l.c. phases, and for dichloromethane dissolved in these phases.[67,68] Results are given for seven thermotropic compounds, chosen to cover smectic A, B (uniaxial and biaxial), and C phases. The n.m.r. spectra were sensitive to the uniaxial or biaxial nature of the phase, and tilt angles determined for the smectic C phase were in agreement with those of Wise *et al.*[65] Measurements were made on samples aligned by cooling from the isotropic or nematic phases in the presence of a large magnetic field. The authors noted that in fresh samples the director of the smectic phase did not realign along the magnetic field, but this did occur for aged samples where decomposition was suspected.[68]

In a graph of order parameter (measured from the lineshape) *versus* temperature a discontinuity is usually observed at a first-order phase transition. Measured values of the order parameter for a series of 4-n-alkoxybenzylidene-4′-phenylazoaniline were in good agreement with theoretical predictions from McMillan theory and showed discontinuities at the smectic-A–nematic phase transition, except for the ethoxy/derivative.[69] For this compound a second-order phase transition was predicted, and the observations were consistent with this.

From a study of lineshape and second moments of three related acetoxy- and/or methoxy-substituted thermotropic compounds, the acetoxy-group was shown to undergo hindered rotation in the nematic phase.[70] The order parameter in the nematic phase was also evaluated for these compounds.

The self-diffusion coefficient ($D$) in l.c. phases is a property of considerable interest, but its measurement by the usual pulse technique is precluded because of the short FID due to large intramolecular dipolar coupling. This can be eliminated by the use of aligned samples oriented at the magic angle. Doane and Parker[71] have used this technique to measure $D$ for the smectic A phase of 4-n-butoxybenzylidene-4′-n-octylaniline. Small molecules dissolved in l.c. phases can have a long FID and can be used to probe the anisotropy in diffusion properties. Murphy and co-workers[72] have measured the coefficients for self-diffusion ($D_{\parallel}$ and $D_{\perp}$) of TMS along and perpendicular to the director in smectic A and B phases. The values were similar in both phases, and $D_{\perp}$ was an order of magnitude larger than $D_{\parallel}$. Because the nematic phase director is

[67] S. Meiboom and Z. Luz, *Mol. Crystals Liquid Crystals*, 1973, **22**, 143.
[68] Z. Luz and S. Meiboom, *J. Chem. Phys.*, 1973, **59**, 275.
[69] J. W. Doane, R. S. Parke, B. Cvikl, D. L. Johnson, and D. L. Fishel, *Phys. Rev. Letters*, 1972, **28**, 1694.
[70] M. N. Avadhanlu and C. R. K. Murty, *Mol. Crystals Liquid Crystals*, 1973, **20**, 221.
[71] J. W. Doane and R. S. Parker, Proceedings of the XVIIth Congress Ampere, Turku, Aug. 1972, ed. V. Hovi, North Holland Publishing Co., Amsterdam–London, 1973, p. 410.
[72] J. A. Murphy, J. W. Doane, Y. Y. Hsu, and D. L. Fishel, *Mol. Crystals Liquid Crytals*, 1973, **22**, 133.

aligned along the magnetic field direction it was only possible to measure $D_{\parallel}$ for this phase, and the value observed was close to the average value $[\frac{1}{3}(D_{\parallel} + 2D_{\perp})]$ for the smectic phases. To overcome this problem Kruger and Spiesecke[73] constructed special field-gradient coils to apply field gradients parallel and perpendicular to the main magnetic field. They measured the anisotropy of **D** for several small molecules in nematic phases, and found values of $D_{\parallel}/D_{\perp}$ in the range 1.0—1.25.

The isotropic liquids formed by thermotropic compounds often exhibit pre-transitional properties. As an example of this Haywood and Packer[74] have reported measurements of $D$ for two nematogens, where the values were dependent on sample history, the variation being up to $\pm 50\%$.

## 5 Micelles

Micelles are aggregates consisting of 10 or more (usually $\geqslant 50$) surfactant monomers that form in water above a critical surfactant concentration (the c.m.c.). They give high-resolution spectra, indicating that $\tau_c < 10^{-10}$ s for molecular motions. In an attempt to determine details of molecular motions Roberts and Chachaty[75] have measured $^{13}$C $T_1$ values for sodium dodecyl sulphate (SDS) micelles. The $T_1$ values for C-1—C-9 were all similar, and only the values for C-10—C-12 were significantly larger. An important observation was that addition of salt caused an increase in the $T_1$ value for C-1, the value becoming larger than those for C-2—C-9. Correlation times of $6 \times 10^{-11}$ s (no salt) and $3 \times 10^{-11}$ s (0.3 mol $l^{-1}$ NaCl) were calculated for C-1. The authors suggest that the increased motion of the head group caused by the addition of salt is due to dehydration of the Stern layer by the added counterions. A more plausible explanation is that addition of salt reduces electrical repulsion between negatively charged head groups, and this results in increased motion. Alexandre and co-workers[76] have reported proton and $^{13}$C relaxation measurements for a number of different micellar solutions. For SDS at low concentrations the proton $T_1$ is dependent on frequency, and from this a $\tau_c$ value of $2 \times 10^{-9}$ s was calculated. This was attributed to translational diffusion around the micelle. However, this Reporter has used the $\tau_c$ value deduced from $^{13}$C measurements[75] to calculate proton relaxation rates, and values close to those measured experimentally are determined. Thus the contribution of molecular rotation to proton relaxation cannot be ignored. At high SDS concentrations an additional relaxation contribution to $T_{1\rho}$ with an estimated $\tau_c$ of $0.9 \times 10^{-5}$ s was observed. This was attributed to the formation of elliptical micelles with consequently slower diffusion around the micelle. Relaxation data of $^{13}$C in sodium decanoate similar to the SDS $^{13}$C data were also reported. Williams and co-workers[77] have also published $^{13}$C

73 G. J. Krüger and H. Spiesecke, *Z. Naturforsch.*, 1973, **28a**, 964.
74 R. J. Hayward and K. J. Packer, *Mol. Phys.*, 1973, **26**, 1533.
75 R. T. Roberts and C. Chachaty, *Chem. Phys. Letters*, 1973, **22**, 348.
76 M. Alexandre, C. Fouchet, and P. Rigny, *J. Chim. phys.*, 1973, **70**, 1073.
77 E. Williams, B. Sears, A. Allerhand, and E. H. Cordes, *J. Amer. Chem. Soc.*, 1973, **95**, 4871.

relaxation rates for micelles formed by cationic surfactants, and again the values were qualitatively similar to those of SDS, the formation of rod-shaped micelles having only a small effect.

Fluorocarbon surfactants have similar properties to hydrocarbon surfactants. Henriksson and Ödberg[78] have reported an increase in $^{19}F$ relaxation rates for two fluorocarbon surfactants on micelle formation. No difference between $H_2O$ and $D_2O$ solutions was observed for n-$C_7F_{15}CO_2Na$ micelles, but the relaxation of n-$C_3F_7CO_2H$ micelles was 10% faster in $H_2O$ than in $D_2O$. This was thought to be due to water–fluorocarbon intermolecular dipolar coupling for the $C_4$ compound, and the absence of the effect for the $C_8$ compound indicated that no water penetrated the $C_8$ micelle.

Podo and co-workers[79] have used proton chemical shifts and relaxation rates to study micelles formed from *p*-(1,1,3,3-tetramethyl)butylphenoxy-(polyoxyethylene)ethanols (Tritons). The $T_1$ values indicated that ethylene oxide group mobility increased away from the hydrocarbon chain. Also, in agreement with previous studies, water penetration into the hydrocarbon region was concluded to be small. The polyoxyethylene resonance was particularly sensitive to the environment, giving a broader resonance in the micelle than was observed for the monomer or neat liquid because of the distribution of $OCH_2$ chemical shifts. A similar result has been noted for a cationic surfactant with a phenyl group close to the head group.[80] One phenyl resonance was observed for the monomer in water while chemically shifted resonances were observed for the micelle. Solubilization of aromatic molecules can cause chemical shifts of groups in the environment of the solubilizate. Menger and co-workers[81] have measured the chemical shifts of glyco-deoxycholate (a bile-salt steroid surfactant) in the presence of solubilized naphthalene, and determined that the site of solubilization was the micelle core.

Proton chemical shifts and relaxation rates for water in aqueous solutions of non-ionic and zwitterionic surfactants of different chain lengths were measured by Walker.[82] The results were used to deduce changes in water structure caused by the presence of surfactants. Changes in $^{13}C$ chemical shifts as a function of concentration have been measured for sodium octanoate.[83] A break in the curve is observed at the c.m.c., and the magnitude of the shift on micelle formation depends on the position of the carbon. Comparison of the shifts with those of model compounds showed no evidence for the penetration of water into the micelle.

When micelle solutions have compositions adjacent to phase boundaries with l.c. phases, the micelles can become very large ($\sim 10^3$ monomers), and the micellar solution has some properties similar to those of the adjacent l.c.

78 U. Henriksson and L. Ödberg, *J. Colloid Interface Sci.*, 1974, **46**, 212.
79 F. Podo, A. Ray, and G. Nemethy, *J. Amer. Chem. Soc.*, 1973, **95**, 6164.
80 R. E. Atkinson, G. E. Clint, and T. Walker, *J. Colloid Interface Sci.*, 1974, **46**, 32.
81 F. M. Menger, J. U. Rhee, and L. Mandell, *J.C.S. Chem. Comm.*, 1973, 918.
82 T. Walker, *J. Colloid Interface Sci.*, 1973, **45**, 372.
83 T. Drakenberg and B. Lindman, *J. Colloid Interface Sci.*, 1973, **44**, 184.

phase. High-resolution n.m.r. linewidths, and relaxation times have been reported for two such micellar systems, both containing two surfactants.[84,85] The linewidths increased as the composition approached that of the phase boundary, while no concurrent changes in $T_1$ were detected. Increases in rheological properties (viscosity and viscoelasticity) were also observed. The changes were attributed to the formation of large cylindrical micelles in addition to the normal spherical micelles.

The state of counter-ions around micelles formed from charged surfactants has been the subject of a number of reports. Oakes[86] has used water relaxation enhancement due to the presence of $Mn^{2+}$ ions to determine increases of 30% and 100% for $Mn^{2+}$ rotation close to SDS micelles and in lamellar phases, respectively. Lindblom and co-workers[22,87,88] have reported an extended study of bromine relaxation in cationic micellar solutions. The $^{81}Br^-$ relaxation rate was increased by the presence of micelles, and the mechanism was thought to involve motion of the water of hydration. Smaller relative changes were observed for $^{35}Cl$ relaxation in the presence of micelles, possibly due to the difference in ionic radii.[87] The formation of rod-like micelles (at the second c.m.c.) increased the $^{81}Br^-$ relaxation rate.[88] At high surfactant concentrations, solubilization of small molecules in the head-group region of the micelle reduced the $^{81}Br$ relaxation rate owing to desorption of counter-ions, whereas solubilization in the micellar interior caused no change. Robb and Smith observed increases in relaxation rates of cations in the presence of anionic micelles.[89] Again the relaxation mechanism was thought to involve water motions causing distortion of the ion hydration layer.

## 6 Reversed Micelles

Reversed micelles are formed in non-polar solvents and consist of surfactant molecules aggregated so that the polar heads are grouped in a core surrounded by the alkyl chains and non-polar solvent. Water is often present in the core and is a requirement for the formation of the reversed micelles in many cases. El Seoud and co-workers have used $^1H$ n.m.r. to study the solubilization of small molecules by dodecylammonium propionate (DAP) in several non-polar solvents.[90,91] Chemical shift changes of DAP were used to demonstrate that the site of solubilization was the micelle core.[90] The solubilizate chemical shifts were used[91] to determine binding constants ($K_{ms}$) for the binding of the

84 C. A. Barker, D. Saul, G. J. T. Tiddy, B. A. Wheeler, and E. Willis, *J.C.S. Faraday I*, 1974, **70**, 154.
85 D. Saul, G. J. T. Tiddy, B. A. Wheeler, P. A. Wheeler, and E. Willis, *J.C.S. Faraday I*, 1974, **70**, 163.
86 J. Oakes, *J.C.S. Faraday I*, 1973, **69**, 1321.
87 G. Lindblom and B. Lindman, *J. Phys. Chem.*, 1973, **77**, 2531.
88 G. Lindblom, B. Lindman, and L. Mandell, *J. Colloid Interface Sci.*, 1973, **42**, 400.
89 I. D. Robb and R. Smith, *J.C.S. Faraday I*, 1974, **70**, 287.
90 O. A. El Seoud, E. J. Fendler, and J. H. Fendler, *J.C.S. Faraday I*, 1974, **70**, 450.
91 O. A. El Seoud, E. J. Fendler, and J. H. Fendler, *J.C.S. Faraday I*, 1974, **70**, 459.

solubilizate to the reversed micelle. It is difficult to understand why different values of $K_{ms}$ are calculated for the same compound from the data for different protons in the same molecule. The differences are much larger than experimental error, and suggest to this Reporter that one (or more) of the assumptions used in deriving the basic equations is invalid in this case. Solubilizate linewidth changes were used to determine the molecular exchange rate between the micelle and bulk solutions. The values must be treated with caution since the origin of the line broadening was not clearly proven.

Other studies of reversed micelles were concerned with lecithins. Haque and co-workers[92] have used chemical shift changes to determine the binding site of insecticides in lecithin reversed micelles, and also deduced binding constants from the data. A similar study of indole and skatole in lecithin micelles has also been reported.[93]

## 7 Microemulsions and Emulsions

Microemulsions are single-phase systems and consist of water (oil) drops of ~10 nm diameter dispersed in a continuous oil (water) phase. Hansen[94] has reported proton linewidths and chemical shifts of microemulsions formed in the potassium oleate–hexanol–water–benzene system. These indicated that oleate molecules are located at the water–oil interface of the water drops, and head-group mobility was restricted. No restriction of hexanol motion was observed. Relaxation rates ($T_1$ and $T_2$) for $D_2O$ decreased with increasing drop size. It was concluded that water existed in two states, a monolayer (~0.1 nm thick) of water bound to interfacial surfactant head groups, and bulk water, with fast exchange between the two types. Shah[95] has reported a similar study on the potassium oleate–hexanol–hexadecane–water system, additionally including the oil-in-water microemulsion and the transition region between two types of microemulsion. For the two microemulsion regions the results were similar to those of Hansen. Large chemical shift and broadening effects were observed in the transition region where l.c. phases formed. Since the hydrocarbon-chain resonance widths were far less than those of other l.c. phases, the systems were probably real emulsions containing l.c. phases. The chemical shift variations attributed to changes in water structure could also be due to bulk susceptibility effects. Frank and co-workers[96] have measured chemical shifts in one- and two-phase samples of the aerosol OT–electrolyte–octane–water system. The water chemical shifts and changes in peak areas indicated the occurrence of a reorganization in the aerosol OT micelle, leading to separation of additional water.

Above a certain temperature (the cloud point) the solubility of non-ionic

[92] R. Hague, I. J. Tinsley, and J. K. McCrady, *Pesticide Biochem. and Physiol.*, 1973, **3**, 73.
[93] T. M. Bray, J. A. Magnuson, and J. R. Carlson, *J. Biol. Chem.*, 1974, **249**, 914.
[94] J. R. Hansen, *J. Phys. Chem.*, 1974, **78**, 256.
[95] D. O. Shah, *Ann. New York Acad. Sci.*, 1973, **204,** 125.
[96] S. G. Frank, Y.-H. Shaw, and N. C. Li, *J. Phys. Chem.*, 1973, **77**, 238.

surfactants in water decreases sharply, and a surfactant-rich phase is formed. Ribeiro and Dennis[97] have studied the process by high-resolution n.m.r. for Triton X-100 and report an intensity decrease of Triton X-100 resonances just below the cloud point. This is attributed to broadening of surfactant resonances due to the formation of large micelles.

[97] A. A. Ribeiro and E. A. Dennis, *Chem. and Phys. Lipids*, 1974, **12**, 31.

# 8
# The Solid State

BY P. S. ALLEN AND W. DERBYSHIRE

## 1 General Introduction

In contrast to Volume 3, the current volume contains both spectral and relaxation studies of the solid state within a single chapter. In the two previous volumes, spectral and relaxation studies were dealt with in Chapters 9 and 3, respectively. One further difference from last year's survey is the inclusion of surface phenomena, to be found currently in Section 7. The organization of this chapter resembles closely that of Chapter 9 in previous volumes and is as follows. Non-metallic diamagnetic materials are dealt with first. However, in order to accommodate the differing features of dipolar and quadrupolar interactions, these materials are divided into those studied by means of spin-$\frac{1}{2}$ nuclei on the one hand, and, on the other, those studied by means of nuclei of spin $> \frac{1}{2}$. Secondly, paramagnetic materials have also been covered, but they too have been divided into non-metallic and metallic so as to separate different dominating interactions. Ordered magnetic materials, a grouping which encompasses ferromagnets, antiferromagnets, and ferrimagnets, are dealt with next, preceding the penaltimate section on experimental techniques. The final section is devoted to surface phenomena, a topic not covered for two years.

The aim of this Report is to be comprehensive without claiming to be completely exhaustive. Only papers published in main-stream languages are included. Furthermore, dissertations have been purposely excluded. It is unfortunate, but also quite possible, that a few papers may have been omitted because of an inappropriate title or because they appeared in an unusual journal. For these omissions we apologize to both authors and readers alike.

During the preparation of this chapter the programmes of three broadly-based international conferences have been available, namely the First Specialised Colloque Ampère held in Krakow, Poland, in August 1973; the 4th International N.M.R. Conference held in Bombay, India, in January 1974, and the 18th Ampère Congress held in Nottingham, U.K., in September 1974. Whilst the Bombay Conference catered for a wide spectrum of nuclear resonance interest and therefore included the solid state, the two Ampère meetings were more specifically oriented towards solids, and as a consequence their proceedings[1,2] contain 330 papers of relevance to the scope of this chapter. These conference papers are not referred to individually here, but they cannot be neglected by the serious student of resonance affairs. The papers submitted to the Bombay conference are appearing individually in the *Journal of Magnetic Resonance*, and the invited talks presented there will be published by IUPAC.

Finally, it only remains to reiterate the note of caution sounded last year over the problem of units. It is a fact of life that workers in different areas use different units. For this reason it is necessary for the reader to be alert to changes in units and ambiguities in reference points or reference substances.

## 2 Half-integral Spins in Diamagnetic Solids

**A. Introduction.**—The behaviour of nuclei of spin $\frac{1}{2}$ in diamagnetic solids is usually dominated by the internuclear dipole–dipole interaction. The effect that this interaction has on the nuclear resonance observables reflects both the motional and structural properties of the solid under investigation. As a result, its exploitation forms the basis of a large proportion of published nuclear resonance work. Papers which deal with the theoretical problems rather than individual materials are covered in subsections 2B and 2C whereas those whose prime interest is a specific material are covered in subsections 2D, 2E, and 2F.

The removal of dipolar broadening effects in diamagnetic solids enables the weaker chemical shielding interaction to be observed and measured. A great deal of effort continues to be devoted to the pursuance of chemical shielding effects in solids by sample spinning, by multi-pulse techniques, or by double-resonance techniques when a rare spin species is of interest. Work in this area is described in subsection 2G.

**B. Theory of the Resonance Spectrum.**—The characteristic lineshapes of n.m.r. have long been recognized to be the Gaussian and the Lorentzian. The former is often representative of absorption lines in solids, since it describes a distribution of static dipolar fields. The motionally narrowed lines of very mobile solids on the other hand usually resemble the latter rather closely. A general theory of bandshape which includes both these limiting cases was first proposed by Kubo and Tomita,[3] who related the absorption line to the Fourier transform of the autocorrelation function of the magnetization. A more recent formulation, which calculates the bandshape in terms of a memory function, which itself is related to the autocorrelation function, has been extended by Parker and Lado.[4] By using the qualitative form of the bandshape, together

[1] Proceedings of the First Specialized Colloque Ampère, ed. J. W. Hennel, Inst. Nuclear Physics, Krakow, 1973.
[2] Proceedings of the 18th Ampère Congress, ed. P. S. Allen, E. R. Andrew, and C. A. Bates, Nottingham, 1974.
[3] R. Kubo and K. Tomita, *J. Phys. Soc. Japan*, 1954, **9**, 888.
[4] G. W. Parker and F. Lado, *Phys. Rev.* (*B*), 1973, **8**, 3081.

with its first few moments, their bandshape evaluation represents a compromise between the qualitative approach and a full microscopic calculation of the proper spin autocorrelation functions. This method is equally valid for broad lines in solids and strongly narrowed lines in liquids. The specific examples discussed are pair bandshapes in rigid solids and during motional narrowing, the exchange-narrowed bandshape in paramagnetic $MnF_2$, and the perennial $CaF_2$ free induction decay (FID). The FID and $CaF_2$ are also the subjects of a further paper by Parker and Lado.[5] By making use of recent moment calculations, they re-examine the convergence of two different modified moment expansions for the FID.

The nuclear resonance spectrum of reorienting nuclear pairs has also been treated in detail by Andrew.[6] He considers $\theta°$ flips of the nuclear pairs between two directions in which the pairs spend unequal residence times. Spectra are computed as a function of $\theta$ and the residence-time ratio. The conformational order–disorder phenomenon in solid cyclobutane is quoted as an example of the application of this treatment. The echo response of nuclear spin pairs to a $90°–\tau–90°$ orthogonal pulse sequence has been investigated both experimentally[7] and theoretically[8] by Boden and co-workers. The variation of the echo amplitude as a function of the square of the pulse separation enables one to determine the strength of the interpair dipolar interaction. The response to an orthogonal $90°–\tau–90°$ sequence has also been considered by Demco and Simplaceanu.[9] A similar study of three-spin triangles has been carried out by Moskvich *et al.*[10] Janzen,[11] on the other hand, has discussed measurement of the local field from the adiabatic rapid passage (ARP) signal shape.

An evaluation of the spin density matrix for slowly reorienting molecules in zero field has been made by Lynden-Bell.[12] From this evaluation it is possible to calculate the resonance bandshape for any value of the rotational diffusion constant, provided that for the system in question the molecular angular velocity correlation time is less than the spin precession period. Shmueli *et al.* have derived a correction to the intramolecular second moment for the librational motion of the component atoms.[13] Their corrections are made for both powders and single crystals, and are applied to the example of anthracene. The extraction of experimental values for the second moment from FID data is the subject of a paper by Parker.[14]


[5] G. W. Parker and F. Lado, *Phys. Rev. (B)*, 1974, **9**, 22.
[6] E. R. Andrew, *J. Magn. Resonance*, 1973, **9**, 108.
[7] N. Boden and M. Mortimer, *Chem. Phys. Letters*, 1973, **21**, 538.
[8] N. Boden, Y. K. Levine, M. Mortimer, and R. T. Squires, *Phys. Letters*, 1974, **46A**, 329.
[9] D. Demco and V. Simplaceanu, *Rev. Roumaine Phys.*, 1973, **18**, 623.
[10] Y. N. Moskvich, N. A. Sergeev, and G. I. Dotsenko, *Soviet Phys. Solid State*, 1974, **15**, 1912.
[11] W. R. Janzen, *J. Magn. Resonance*, 1973, **12**, 307.
[12] R. M. Lynden-Bell, *Mol. Phys.*, 1973, **26**, 979.
[13] U. Shmueli, M. Polak, and M. Sheinblatt, *J. Chem. Phys.*, 1973, **59**, 4535.
[14] G. W. Parker, *J. Chem. Phys.*, 1973, **58**, 3274.

The study of the absorption bandshape of highly polarized nuclear spin systems represents a recent departure for n.m.r. spectroscopists. Abragam and co-workers[15] have considered a homonuclear system and made a theoretical study of the bandshape as a function of polarization. Their predictions were then tested and confirmed on $CaF_2$ polarized up to 90%. The general expressions for the spectral moments, corrected for paramagnetic impurities, were used for the absolute calibration of the fluorine polarization.

**C. Nuclear Spin–Lattice Relaxation Theory.**—Nuclear relaxation studies are, by and large, a more sensitive and informative indicator of the validity of a particular model of a system than straightforward analysis of the resonance spectrum. This opinion is borne out by the wealth of published data on diamagnetic systems covered in section 2E. The theoretical analyses of relaxation published during the last year vary, as always, in their level of sophistication and application.

On a more general plane, Emid[16] has derived a master equation for the diagonal part of the density matrix in the context of a semi-classical description of spin–lattice relaxation and in the basis of the irreducible representations of the symmetry group of the Hamiltonian. This derivation is generally valid, obtained without recourse to a spin-temperature assumption, equivalent to the master equation for the populations, and is claimed to be preferable to the master equation for the full density matrix. The actual characteristics of the spin correlation functions, however, are the subject of a computer simulation by Hermkes.[17] Soda and Chihara,[18] restricting themselves to exponential correlation functions, have used the symmetry properties of second-rank spherical harmonics to derive a general relation between the relative magnitudes of the power spectra, *viz.*

$$J^{(0)}(\omega):J^{(1)}(\omega):J^{(2)}(\omega) = 6:1:4$$

which holds exactly in the weak collision regime and which is independent of the mode of molecular motion giving rise to that spectrum.

The spin–lattice relaxation rate of a system undergoing three-fold reorientation has been shown[19] to depend strongly on the angle between its $C_3$ axis and the principal direction of the relaxation interaction for dipolar, quadrupolar, and anisotropic chemical-shielding mechanisms. Extensions to the three-fold jump process have been considered by Anderson,[20] who includes an asymmetry to the three-fold potential, and by Versmold,[21] who compounds it with the rotational diffusion of the parent molecule as a whole.

[15] A. Abragam, M. Chapellier, J. F. Jacquinot, and M. Goldman, *J. Magn. Resonance*, 1973, **10**, 322.
[16] S. Emid, *Physica*, 1973, **70**, 616.
[17] H. Hermkes, *Phys. Letters*, 1973, **46**, 241.
[18] G. Soda and H. Chihara, *J. Phys. Soc. Japan*, 1974, **36**, 954.
[19] P. S. Allen, *J. Phys. (C)*, 1973, **6**, 3174.
[20] J. E. Anderson, *J. Magn. Resonance*, 1973, **11**, 398.
[21] H. Versmold, *J. Chem. Phys.*, 1973, **58**, 5649.

Isotropic overall reorientation of a molecule has been treated together with multiple internal rotations by Levine *et al.*[22]

Wennerström[23] presents a new approach to spin diffusion, in which he maintains that the total spin of a homonuclear system can reach internal equilibrium through coupling with the lattice. He obtains the circumstances under which the spin–lattice relaxation can be expected to be governed by a single time constant.

Jacquinot and Goldman[24] have developed a formalism of spin–lattice relaxation in a nuclear spin system subjected to r.f. irradiation either close to or far from resonance. When the irradiation is close to resonance a single spin temperature exists in the rotating frame, and the form of the expression for the relaxation rate depends on the relative magnitudes of $\tau_c$ and $T_2$. However, if the r.f. field is far from resonance, a single expression governs the relaxation rate, independent of the relative values of $\tau_c$ and $T_2$. The situation corresponding to irradiation in the form of intense resonant pulses has been treated by Pines and Waugh,[25] whereas the situation corresponding to relaxation caused by electronic spins rather than other nuclear spins is the subject of a paper by Devreax *et al.*[26]

**D. Structural Studies.**—The anisotropy of the dipolar interaction, as a function of the angle between the static magnetic field and the internuclear vector, can cause either a featureless broadening, or a well-defined structure of the nuclear resonance signal. This depends on whether the distribution of nuclear spins is random or regimented into fairly simple well-organized molecular groups, *e.g.* the proton pairs of water molecules of crystallization. Measurement of this anisotropy in the well-regimented cases frequently leads to structural information about the molecular groups, in particular their internuclear separations.

A good example of this is the observation by Scheller and Lippold[27] of the proton doublet separation as a function of orientation for the water of crystallization in a single crystal of $Ba(ClO_4)_2,3H_2O$. Their data are consistent with a proton separation of 0.158–0.160 nm and a structure of space group $P6_3$. Similar studies[28] on a single crystal of urea at 77 K, made with the intention of circumventing the thermal motion which affected the previous work[29] done at 300 K, have obtained smaller N—H bond lengths than the experiments at higher temperatures. Both sets of nuclear resonance data are compared with the results of neutron-diffraction work.

An interesting extension of this behaviour to the electron–nuclear dipole–


22 Y. K. Levine, P. Partington, and G. C. K. Roberts, *Mol. Phys.*, 1973, **25**, 3, 497.
23 H. Wennerström, *J. Magn. Resonance*, 1973, **11**, 219.
24 J. F. Jacquinot and M. Goldman, *Phys. Rev.* (*B*), 1973, **8**, 1944.
25 A. Pines and J. S. Waugh, *Phys. Letters*, 1974, **47A**, 337.
26 F. Devreax, J. P. Boucher, and M. Nechtschein, *J. Phys.* (*France*), 1974, **35**, 271.
27 D. Scheller and B. Lippold, *Z. phys. Chem.* (*Leipzig*), 1973, **253**, 105.
28 Z. M. ElSaffar, E. M. Peterson, and D. E. O'Reilly, *J. Chem. Phys.*, 1973, **59**, 1557.
29 E. R. Andrew and D. Hyndman, *Discuss. Faraday Soc.*, 1955, **19**, 195.

dipole interaction has been invoked[30] in a single-crystal study of $CdF_2$ with 10% $ErF_2$. Weak satellite $^{19}F$ resonance lines were attributed to those fluorines neighbouring $Er^{3+}$ ions. The anisotropy of these satellites gave the strength of the electron–nuclear dipolar interaction, and an analysis of the intensity ratios supported the idea that $Er^{3+}$ ions are distributed randomly throughout the lattice.

Relaxation data coupled with *X*-ray and thermal measurements have revealed[31] the possible existence of two amorphous phases and two crystalline phases of solid 2,3-dimethylbutane within the temperature range 80–160 K. At 80 K the crystalline phase is orthorhombic, with cell dimensions 1.251, 1.013, and 0.474 nm. In bis(cyclo-octatetraene)iron on the other hand, the overall bandshape shows[32] that there are two component lines, one broad and the other narrow. The temperature dependence of their relative amplitudes gives evidence for dynamic tautomerism in the solid state of this compound.

**E. Motional Studies (Classical Regime).**—The majority of motional studies make use of the temperature dependence of nuclear relaxation rates. This is because the relaxation rates can be conveniently expressed in terms of a correlation time for the motion, which itself is usually assumed to be thermally activated. The activation energy so obtainable then provides a useful characterization of the motional mobility. The temperature dependence of the bandshape, though providing only a coarse monitor of the motion over a restricted temperature range, is still occasionally employed to this end. No differentiation between these two techniques is made below, where the papers are organized in order of increasing complexity of molecular motion.

*Two-spin Reorientation.* Bruno and Pintar[33] have made both spectral and relaxational analyses of the reorientational jumps and diffusion of proton pairs in hexagonal ice. At $-24$ °C they estimate that the correlation time for diffusion is $(37 \pm 6)$ $\mu$s. The effects of the reorientation of $NH_2$ groups on the proton second moment in benzidine[34] and in *p*-aminobenzoic acid[35] have been observed by Banerjee *et al.*

*Three-spin Reorientation.* The investigation of the three-fold reorientation of methyl groups and ammine groups provides the basis for a large number of publications on molecular motion. These[35—59] are listed in Tables 1 and 2, where the quoted activation energies for $C_3$ reorientation are noted. In a number of these materials additional motion is observed at the higher temperatures, and these facts are also contained in the tables.

30 E. Banks, M. Greenblatt, and B. R. McGarvey, *J. Chem. Phys.*, 1973, **58** 4787.
31 K. J. Anderton and J. P. Llewellyn, *J.C.S. Faraday II*, 1973, **69,** 1249.
32 G. Allegra, A. Colombo, and E. R. Mognaschi, *Gazzetta*, 1972, **102,** 1060.
33 G. F. Bruno and M. M. Pintar, *J. Chem. Phys.*, 1973, **58,** 5344.
34 A. K. Banerjee, V. D. Agrawal, and R. C. Gupta, *Z. phys. Chem. (Frankfurt)*, 1973, **83,** 7.
35 A. K. Banerjee, P. Agrawal, and R. C. Gupta, *J. Prakt. Chem.*, 1973, **315,** 251.

**Table 1** *Compounds containing methyl groups for which three-fold reorientation has been studied*

| *Compound* | $E_a$/kJ mol$^{-1}$ | *Other Motions Studied* | *Reference* |
|---|---|---|---|
| $MeCO_2D$ | 3.22 | | 36 |
| MeCOMe | 5.69 | | 36 |
| MeOH | — | | 37 |
| $MePCl_2$ | 9.4 | | 38 |
| $MePSCl_2$ | 9.8 | | 38 |
| MeSCN | 10.0 | | 38 |
| $MeCD_2I$ | 13.4 | | 38 |
| $MeCD_2Br$ | 15.1 | | 38 |
| MeNCS | 2.38 | | 38 |
| MeCOCl | 6.1 | | 38 |
| $MeNH_3Cl$ | 18.6 | $NH_3$ reorientation; cation reorientation | 39 |
| $Cl_3CCMe_2Cl$ | 10.5 | Isotropic tumbling and diffusion | 40 |
| MeC(NOH)C(NOH)Me | | Whole molecule rotation | 41 |
| MeSOMe | 17.7 | | 42 |
| *m*-$(MeO)_2C_6H_4$ | 7.9 | | 43 |
| *p*-$(MeO)_2C_6H_4$ | 15.1 | | 43 |
| Decamethonium | 20.9 | Reorientation of $Me_3N$ | 44 |
| $Me_3N^+$—$Al^-Me_3$ | 12.1 on Al<br>55.6 on N | Relative motion of the moieties | 45 |
| $Me_3N^+$—$Ga^-Me_3$ | 12.1 on Ga<br>~54 on N | Relative motion of the moieties | 45 |
| $Me_4NCl$ | | Diffusion of cations | 46 |
| $Me_4NBr$ | ~27 | Cation tumbling | 47 |
| $Me_4NNO_3$ | 28.5 | Cation reorientation | 48 |
| $Me_4NClO_3$ | 20.9 | | 48 |
| $Me_4NClO_4$ | 19.2 | | 48 |
| $(Me_4N)_2SO_4$ | 26.4 | Cation reorientation | 48 |
| $(Me_4N)_2CO_3$ | 27.2 | Cation reorientation | 48 |
| $Me_6N_3B_3$ | 20.9 | | 49 |
| $Me_3N_3B_3Cl_3$ | 18.8 | | 49 |
| $MeC_5H_4N$ | 0.063 | | 50 |
| $(MeCH_2)_2NH$ | 12.1 | | 51 |

[36] J. A. Ripmeester and B. A. Dunell, *Canad. J. Chem.*, 1974, **52**, 191.
[37] S. K. Garg and D. W. Davidson, *J. Chem. Phys.*, 1973, **58**, 1898.
[38] A. Kumar and C. S. Johnson jun., *J. Chem. Phys.*, 1974, **60**, 137.
[39] S. Albert and J. A. Ripmeester, *J. Chem. Phys.*, 1973, **58**, 541.
[40] S. Albert and J. A. Ripmeester, *J. Chem. Phys.*, 1973, **59**, 1069.
[41] S. C. Agarwal, V. D. Agrawal, and R. C. Gupta, *Z. phys. Chem.* (*Frankfurt*), 1973, **83**, 1.
[42] J. D. Cutnell and W. Venable, *J. Chem. Phys.*, 1974, **60**, 3795.
[43] J. D. Cutnell and L. Verduin, *J. Chem. Phys.*, 1973, **59**, 258.
[44] J. Dufourcq and B. Lemanceau, *J. Chim. phys.*, 1973, **70**, 1129.
[45] T. T. Ang and B. A. Dunell, *J.C.S. Faraday II*, 1974, **70**, 17.
[46] S. Albert and J. A. Ripmeester, *J. Chem. Phys.*, 1973, **58**, 2870.
[47] M. Pollak and M. Sheinblatt, *J. Magn. Resonance*, 1973, **12**, 261.
[48] M. Mahajan and B. D. Nageswara Rao, *J. Phys.* (*C*), 1974, **7**, 995.
[49] J. Gallier, M. A. Chassonneau, and J. Meinnel, *Compt. rend.*, 1973, **276**, *C*, 1231.
[50] A. E. Zweers, H. B. Brom, and W. J. Huiskamp, *Phys. Letters*, 1974, **47A**, 347.
[51] P. S. Allen, A. W. K. Khanzada, and C. A. McDowell, *Mol. Phys.*, 1973, **25**, 1273.

**Table 1** *continued*

| *Compound* | $E_a$/kJ mol$^{-1}$ | *Other Motions Studied* | *Reference* |
|---|---|---|---|
| $(MeCH_2)_2NH$–$6.8D_2O$ clathrate | 9.79 | | 51 |
| $MeCNH_2$ | 13.4 | | 52 |
| $Me_3CNH_2$–$9.75D_2O$ clathrate | ~7 | t-Butyl and molecular re-orientation | 52 |
| $Me_2NH$ | 11.3 | | 53 |
| $Me_2NH$–deuterate clathrate | 9.46? | Molecular tumbling | 53 |

**Table 2** *Studies of the three-fold reorientation of ammine groups*

| *Compound* | $E_a$/kJ mol$^{-1}$ | *Other Motions Studied* | *Reference* |
|---|---|---|---|
| L-alanine | 38.9 | | 54, 55 |
| L-aspartic acid | 38.1 | | 54 |
| L-glutamine | 33.9 | | 54 |
| α-glycine | 28.9 | | 54 |
| L-isoleucine | 46.0 | | 54 |
| DL-leucine | 49.8 | | 54 |
| DL-methionine | 46.0 | | 54 |
| DL-serine | 39.7 | | 54 |
| L-valine | 35.1 | | 54 |
| *trans*-$[Co(NH_3)_4Cl_2]Cl$ | 4.18 | Whole ion rotation | 56 |
| *cis*-$[Co(NH_3)_4Cl_2]Cl$ | 0.8 | Whole ion libration | 56 |
| *trans*-$[Co(NH_3)_4Br_2]Br$ | 3.97 | Whole ion rotation | 57 |
| *trans*-$[Co(NH_3)_4Br_2]Cl$ | 1.05 | Whole ion rotation | 57 |
| $[Co(NH_3)_4CO_3]_2SO_4,H_2O$ | 7.9 and 18.0 | | 58 |
| *cis*-$[Co(NH_3)_4(H_2O)_2]_2[SO_4]_3$ | 3.8 | 180° flips of $H_2O$ | 59 |

The non-exponential magnetization recovery often observed in triangular groups and due to cross-correlations of the intramethyl dipole–dipole interactions[60] is the specific interest of investigations in dimethoxybenzene,[43] dimethyl sulphone,[42] and a single crystal of trifluoromethyl silver acetate.[61] Both inter-methyl and intra-methyl group cross-correlations have been observed by Burnett and Muller[62] in ethane and its partially deuteriated analogue.

52 P. S. Allen, A. W. K. Khanzada, and C. A. McDowell, *J. Chem. Phys.*, 1973, **59**, 470.
53 A. W. K. Khanzada, C. A. McDowell, and P. Raghunathan, *J. Chem. Phys.*, 1974, **60**, 3025.
54 E. R. Andrew, W. S. Hinshaw, M. G. Hutchins, and P. C. Canepa, *Chem. Phys. Letters*, 1974, **26**, 50.
55 S. C. Agarwal, P. Agarwal, and R. C. Gupta, *J. prakt. Chem.* 1973, **315**, 443.
56 S. E. Ulrich and B. A. Dunell, *J.C.S. Faraday II*, 1973, **69**, 1609.
57 S. E. Ulrich and B. A. Dunell, *J. Magn. Resonance*, 1974, **13**, 137.
58 B. A. Dunell, M. D. Pachal, and S. E. Ulrich, *Canad. J. Chem.*, 1973, **51**, 1107.
59 B. A. Dunell, J. F. Nattrass, and S. E. Ulrich, *Canad. J. Chem.*, 1974, **52**, 900.
60 R. L. Hilt and P. S. Hubbard, *Phys. Rev. (A)*, 1964, **134**, 392.
61 M. Mehring and H. Raber, *J. Chem. Phys.*, 1973, **59**, 1116.
62 L. J. Burnett and B. H. Muller, *Chem. Phys. Letters*, 1973, **18**, 553.

Other work on three-spin systems will be found in the next section, under quantum-tunnelling effects.

*Four-spin Reorientation.* Solid $SnH_4$ exhibits[63] a great freedom of motion, both reorientation and translational diffusion occurring at low temperatures. The commonest four-spin group $NH_4$ has been observed in $(NH_4)_2GeF_2$ to show anomalous $T_1$ behaviour below 15 K,[64] and in $NH_4PF_6$ to be highly mobile at all temperatures above 77 K.[65] Additional work on the $NH_4^+$ ion will be quoted below (p. 262), and in the subsection on quantum-tunnelling effects.

*Six-spin Reorientation.* The six-spin group, in its octahedral configuration and involving $^{19}F$ nuclei, is a highly symmetric and mobile rotor, as evidenced by work on the $PF_6^-$ ion. Gutowsky and co-workers[65] have studied $NH_4PF_6$ in addition to the alkali-metal hexafluorophosphates.[66,67] The $AsF_6^-$ ion and the $SiF_6^-$ ion display a similar freedom of movement in the alkali-metal hexafluoroarsenates[68] and in hydrated fluorosilicates,[69] respectively.

An exploitation of $^{19}F$ relaxation data to explain the motional behaviour of the planar configuration of six spins in hexafluorobenzene has been made by Boden *et al.*[70]

*Miscellaneous Molecular Reorientation.* Both $^1H$ and $^{19}F$ n.m.r. have been employed to study the structure and molecular dynamics of trifluorobenzene. This compound has been observed[71] to exhibit a phase transition at 200 K, which is attributed to 120° rotational jumps about the triad axis with an activation energy of 42.7 kJ mol$^{-1}$. Acrylonitrile also exhibits a number of phase transitions in the solid state. In the phase between 164 K and its melting point at 191 K, it undergoes binary reorientation around an axis defined by the nitrogen atom and the middle of the vinyl double-bond. The activation energy for this motion is 17.6 kJ mol$^{-1}$.[72] Reorientation about the chain axis has been investigated in n-alkylammonium chlorides by Tsau and Gilson[73] and in n-$C_{19}H_{40}$ by Dehl.[74]

*The Reorientation of Guest Molecules in Clathrates.* The majority of clathrate studies appearing this year involve the deuterate clathrates. However, Gregoire *et al.*[75] have resisted the trend and conducted a spectral analysis of $CH_3OH$, $HCO_2H$, and $CH_3CN$ in hydroquinone clathrates in order to ascertain the dynamics of the trapped molecules. Of the deuterate clathrates,

63 L. Numela, M. Hirvonen, and U. Lahteenmaki, *Phys. Letters*, 1972, **39A**, 323.
64 E. E. Ylinen, J. E. Tuohi, and L. K. E. Niemela, *Chem. Phys. Letters*, 1974, **24**, 447.
65 S. Albert and H. S. Gutowsky, *J. Chem. Phys.*, 1973, **59**, 3585.
66 H. S. Gutowsky and S. Albert, *J. Chem. Phys.*, 1973, **58**, 5446.
67 L. Niemela and M. Komu, *Ann. Acad. Sci. Fennicae*, 1973, **A403**, 3.
68 L. Niemela, *Ann. Univ. Turku*, 1973, **163**, 99.
69 K. Muthukrishnan and J. Ramakrishna, *J. Chem. Phys.*, 1973, **59**, 5571.
70 N. Boden, P. P. Davis, C. H. Stam, and G. A. Wesselink, *Mol. Phys.*, 1973, **25**, 87.
71 N. Boden and M. Gibb, *Mol. Phys.*, 1974, **27**, 1359.
72 B. Ban and C. Chachaty, *Canad. J. Chem.*, 1973, **51**, 3889.
73 J. Tsau and D. F. R. Gilson, *Canad. J. Chem.*, 1973, **51**, 1990.
74 R. E. Dehl, *J. Chem. Phys.*, 1974, **60**, 339.
75 P. Gregoire, J. Gallier, and J. Meinnel, *J. Chim. phys.*, 1973, **70**, 1247.

Hayward and Packer[76] have observed ethylene oxide, tetrahydrofuran, and triethylenediamine as guest molecules. Each of these molecules appears to undergo reorientation about at least one axis over part of the temperature range from 77 K to the decomposition point. In order to remove ambiguities between naturally surrounded and enclathrated molecules, McDowell and his collaborators have adopted the practice of presenting together the data from a molecule embedded in its parent substance on the one hand and encaged in its clathrate deuterate on the other. Data in this form have been published for t-butylamine,[52] diethylamine,[51] and dimethylamine.[53]

*Motion in Plastic Crystals.* The motion of globular molecules, which usually form solids exhibiting a plastic crystal phase below their melting points, continues to evoke interest. Molecular diffusion in the five organic compounds triethylenediamine, 3-azabicyclononane, norbornane, norbornylene, and norbornadiene has been quantified by Folland *et al.*[77] In a study[78] of pressure dependence, the same group show, by monitoring the Zeeman relaxation in the laboratory frame, that overall molecular reorientation is only weakly affected by pressure. However, they discovered from studies of slow motion, using rotating-frame relaxation rates, that self-diffusion in the plastic phase depends strongly on pressure. Such studies were undertaken for six compounds, *viz.* hexamethylethane, hexamethyldisilane, cyclohexane, plus the last three mentioned under reference 77. Other plastic crystals subjected to recent investigations are sulpholan,[79] quinuclidine,[80] and $Cl_3CCMe_2Cl$.[40]

*Motional Studies of Ferroelectrics, and Phase Transitions.* In the KDP family of ferroelectric substances, an exhaustive enquiry by Dalal *et al.*[81] on the ammonium arsenates, using both e.s.r. and n.m.r., supports the conclusion that the motion of the $NH_4^+$ ion plays a dominant role in the antiferromagnetic phase transition. However, this is not the only motion present in ammonium arsenate. Grosescu,[82] by measuring the temperature dependence of $T_1$ and $T_{1D}$, claims to have detected the low-frequency reorientations of the $H_2AsO_4$ groups. A spectral analysis of substances in this family has also been made by Nicholson and Soest.[83]

The hydrogen-bonded ferroelectrics colmanite [$K_4Fe(CN)_6.3H_2O$] and triglycine sulphate,[84] together with dicalcium strontium propionate,[85] have been the subject of relaxation studies near their critical points. Special attention was paid to the anomalous part of the relaxation rate near the critical

[76] R. J. Hayward and K. J. Packer, *Mol. Phys.*, 1973, **25**, 1443.
[77] R. Folland, R. L. Jackson, J. H. Strange, and A. V. Chadwick, *J. Phys. and Chem. Solids*, 1973, **34**, 1713.
[78] R. Folland, S. M. Ross, and J. H. Strange, *Mol. Phys.* 1973, **26**, 27.
[79] D. F. R. Gilson and P. P. Saviotti, *J.C.S. Faraday II*, 1974, **70**, 1.
[80] F. Fried and C. Brot, *Mol. Phys.*, 1974, **27**, 1701.
[81] N. S. Dalal, C. A. McDowell, and R. Srinivasan, *J. Chem. Phys.*, 1974, **60**, 3787.
[82] R. Grosescu, *Chem. Phys. Letters*, 1973, **21**, 80.
[83] J. Y. Nicholson tert. and J. F. Soest, *J. Chem. Phys.*, 1974, **60**, 715.
[84] G. Brosowski, W. Buchheit, D. Muller, and J. Peterson, *Phys. Status Solidi*, 1974, **62**, 93.
[85] T. Hikita, K. Sakata, and I. Tatsuzaki, *J. Phys. Soc. Japan*, 1973, **34**, 1248.

temperature in order to evaluate the critical fluctuations of the electric dipole system.

*Diffusion.* Van Steenwinkel[86] has measured the $^{19}F$ relaxation times $T_1$, $T_{1D}$, and $T_2$ as functions of temperature in single crystals of $CaF_2$ and NaF. These measurements have enabled him to determine the activation energy for vacancy diffusion, and the formation energy of defect pairs.

**F. Quantum Effects.**—*Quantum Effects in Molecular Rotors.* Simple symmetric molecular groups exhibit an energy splitting of their torsional energy states, which is often given the name of tunnelling splitting. This splitting arises because of the overlap of the partially localized spatial wavefunctions, and separates each torsional level into different nuclear spin sub-states. At low temperatures the nuclear resonance observables are sensitive to this tunnelling splitting, and are no longer necessarily consistent with the semi-classical predictions that were realistic at higher temperatures.

For the simplest two-spin system Hardy and Berlinsky[87] explain the anomalous bandshape in ordered ortho-hydrogen and para-deuterium in terms of a pumping process brought about by ortho–para conversion. A theory of ortho–para conversion is the subject of a companion paper.[88]

That model three-spin system, the methyl group, occurs in many more materials, and therefore receives much more attention, than the hydrogen molecule. Johnson[89] has extended the bandshape analysis to the proton pair in the $CH_2D$ system. He finds that when the tunnelling splitting is much greater than the dipolar energy the bandshape does not tend to the same limit as classical fast rotation, as does the unadulterated triangle. Together with Cobb,[90] he shows that for a significant difference to occur between the spectrum of a pure $C_3$ system and one with a $C_6$ component, the $C_6$ term needs to be unrealistically high.

The effects of tunnelling on the relaxation of triangular rotors have been observed in the methylbenzenes[91] and, by Von Schutz *et al.*,[92] in methylnaphthalene. Both systems display extra minima over and above BPP predictions but the respective authors conflict in their explanations. Resonances in the spin–lattice relaxation of methyl protons have been observed by Clough *et al.*[93] when the Larmor frequency of unpaired electrons in the system is equal to the sum of the tunnelling frequency and the nuclear Larmor frequency.

The possibility that the effective tunnelling splitting is temperature-dependent has been considered recently.[94] This has bearing on the temperature

86 R. Van Steenwinkel, *Z. Naturforsch.*, 1974, **29a,** 278.
87 W. N. Hardy and A. J. Berlinsky, *Phys. Rev.* (*B*), 1973, **8,** 4996.
88 A. J. Berlinsky and W. N. Hardy, *Phys. Rev.* (*B*), 1973, **8,** 5013.
89 C. S. Johnson jun., *J. Chem. Phys.*, 1973, **59,** 623.
90 T. B. Cobb and C. S. Johnson jun., *J. Chem. Phys.*, 1973, **59,** 4478.
91 P. S. Allen and A. J. Snell, *J. Phys.* (*C*), 1973, **6,** 3478.
92 J. U. Von Schutz, W. Gutter, and H. C. Wolf, *Z. Naturforsch.*, 1973, **28a,** 69.
93 S. Clough, W. S. Hinshaw, and T. Hobson, *Phys. Rev. Letters*, 1973, **31,** 1375.
94 P. S. Allen, *J. Phys.* (*C*), 1974, **7,** L22.

dependence of nuclear resonance bandshapes in solids[95] as well as sideband frequencies in e.s.r.[96]

The four-spin tetrahedron is yet another system which exhibits tunnelling. Evidence for the existence of nuclear spin symmetry states in solid silane is presented by Jones.[97] The ground state bandshape has been evaluated, with particular reference to $NH_4Br$, $NH_4I$, $NH_4SCN$, $(NH_4)_2CrO_4$, and $NH_4NO_3$, by Ikeda and McDowell.[98] The temperature dependence of relaxation phenomena has been reported by Guttler and Von Schutz,[99] who make use of their data to estimate the tunnelling splitting.

*Resonance in* $^3He$. Notwithstanding the fact that some are concerned with the liquid state, papers published on $^3He$ are included here because quantum effects play a dominant role in determining the nuclear resonance properties of that system. In a rather fundamental paper,[100] Leggett postulates a novel ordering property for liquid $^3He$ below 3 mK. He calls this phenomenon 'spontaneously broken spin–orbit symmetry' and uses it to explain shifts in the resonance frequency of $^3He$ nuclei. Ahonen *et al.*[101] have since noted that the bulk liquid resonance properties are not observed in superfluid $^3He$ phases in the presence of finely divided Pt powder, *i.e.* a metal closely resembling a weak electronic paramagnet. No explanation is offered.

The bulk polarization of solid $^3He$ was measured by Johnson *et al.*,[102] whereas Sullivan and Chapellier[103] report the spin–lattice relaxation rates in the b.c.c. structure. For Larmor frequencies that are large compared with exchange frequencies, the temperature-independent relaxation rates were observed between 0.45 and 0.85 K.

In $^3He$–$^4He$ mixtures, the $^3He$ nuclear resonance has been used to determine the roton energy gap from spin diffusion coefficients[104] and the nature of the critical singularity in the spin–spin relaxation time.[105]

**G. High Resolution in Solids.**—A rapidly growing area of nuclear resonance is its application to high resolution in solids. This application was one of the principal subjects of the First Specialized Colloque Ampère,[1] and the conference began with a review of the field by Andrew,[106] who covered all the current methods of achieving a high resolution in the solid state. Of these methods, three feature regularly in the published literature. They are (i) the

95 C. S. Johnson and C. Mottley, *Chem. Phys. Letters*, 1973, **22**, 430.
96 S. Clough and J. Hill, *J. Phys. (C)*, 1974, **7**, L20.
97 E. P. Jones, *Phys. Letters*, 1973, **43A**, 19.
98 R. Ikeda and C. A. McDowell, *Mol. Phys.*, 1973, **25**, 1217.
99 W. Guttler and J. U. Von Schutz, *Chem. Phys. Letters*, 1973, **20**, 133.
100 A. J. Leggett, *J. Phys. (C)*, 1973, **6**, 3187.
101 A. I. Ahonen, M. T. Haikala, and M. Krusius, *Phys. Letters*, 1974, **47A**, 215.
102 R. T. Johnson, D. N. Paulson, R. P. Gifford, and J. C. Wheatley, *J. Low Temp. Phys.*, 1973, **10**, 35.
103 N. S. Sullivan and M. Chapellier, *J. Phys. (C)*, 1974, **7**, L195.
104 K. Fukuda, Y. Hirayoshi, and A. Hirai, *Phys. Letters*, 1974, **47A**, 377.
105 S. Saito, *Phys. Letters*, 1973, **43A**, 241, 242.
106 E. R. Andrew, Proc. First Specialized Colloque Ampère, ed. J. W. Hennel, Inst. of Nuclear Physics, Krakow, 1973, p. 1.

magic-angle rapid sample rotation method, (ii) the multiple-pulse methods, and (iii) the double-resonance methods on rare spins. The following subsections cover each of these methods in turn.

In spite of all the current effort to narrow the resonance line from solids, Terao and Hashi[107] describe a 'new' method of measuring chemical shifts in solids from an unnarrowed line. The essence of this technique, which they call the 'first moment method', is the precise determination of the first moment of a broad resonance. It is claimed to be simple and accurate for a single chemical-shift tensor, but if more than one tensor contributes to the spectrum it gives only an average, and the technique cannot resolve the contributions. Terao and Hashi measure the average axially symmetric proton chemical-shift tensor components in $KH_2PO_4$ relative to liquid $C_6H_{12}$, the $^{31}P$ chemical shift components in $KH_2PO_4$ with reference to $H_3PO_4$, and (in $CaF_2$) the $^{19}F$ chemical-shift tensor relative to $CF_3CO_2H$ liquid.

*Magic-angle Rotation Method.* First it is worth mentioning that a review of the work conducted over a period of years at the Friedrich Schiller University, Jena, has been presented by Schnabel.[108] Secondly, the individual papers using this technique and published in the past year report (i) the resolution of the doublet and quartet structure, and thus the interaction constants characteristic of an $AB_3$ type of spectrum, in polycrystalline $P_4S_3$;[109] (ii) the characterization of the slow motional modes in some polycrystalline amides;[110] and (iii) the $^1H$ and $^{19}F$ chemical shifts, which permit the investigation of the monomer ratio and conformational structures, in several elastomers.[111,112]

*Multiple-pulse Methods.* The development of multiple-pulse cycles over recent years has led to the observation of progressively narrower resonance lines in solids. The price paid for increasing the resolution has been the increasing complexity of the pulse cycles. The symmetry requirements of the one-cycle propagator associated with the design of line-narrowing pulse cycles have been discussed by Wang.[113] He confines himself to an explicit calculation of the FID and dipolar echo modulation connected with the symmetrized WAHUHA cycle[114] and to a pair of spins.

The factors limiting the reduction in bandwidth which have been the subject of recent attention are the finite pulse width, $B_1$ inhomogeneity, off-resonance effects, r.f. phasing errors, and residual dipolar effects. Rhim *et al.*[115] consider the nature of those interactions which are important in determining the residual linewidth, with particular emphasis on the Waugh four-pulse cycle[114]

107 T. Terao and T. Hashi, *J. Phys. Soc. Japan*, 1974, **36,** 989.
108 B. Schnabel, *Wiss. Z. Friedrich-Schiller Univ. Jena, Math.-Nat. R.*, 1973, **22,** 335.
109 E. R. Andrew, W. S. Hinshaw, and A. Jasinski, *Chem. Phys. Letters*, 1974, **24,** 399.
110 H. Pivcova, D. Doskocilova, Z. Veksli, and B. Schneider, *J. Magn. Resonance*, 1974, **14,** 182.
111 H. Benoit and M. Rabii, *Chem. Phys. Letters*, 1973, **21,** 466.
112 M. Rabii, *Compt. rend.*, 1973, **277,** *B*, 269.
113 C. H. Wang, *J. Chem. Phys.*, 1973, **59,** 225.
114 J. S. Waugh, L. M. Huber, and U. Haeberlen, *Phys. Rev. Letters*, 1968, **20,** 180.
115 W. K. Rhim, D. D. Elleman, and R. W. Vaughan, *J. Chem. Phys.*, 1973, **59,** 3740.

and the eight-pulse cycle of Rhim *et al.*[116] They show that $B_1$ inhomogeneity causes residual broadening both directly and indirectly, due to cross-coupling with the other Hamiltonian terms of the spin system. Furthermore, Rhim and his co-workers maintain that their eight-pulse cycle is much less dependent on the stability of the r.f. transmitter than the Waugh four-pulse cycle. In a later paper[117] the same group present a systematic method of dealing with the effects of pulse errors by incorporating a number of pulse imperfections (phase errors, length errors, r.f. inhomogeneity, and phase transient effects) into the average Hamiltonian formulation. By this means they calculate the coupling terms of these pulse imperfections with other components of the Hamiltonian, such as dipolar and off-resonance terms. This approach enables them to explain many of the features of spectra obtained with an improperly tuned spectrometer.

Mansfield and Haeberlen[118] confine their attention to the various coherent phase errors encountered in this type of experiment. They present a general scheme for self-correcting these errors, which greatly facilitates the phase alignment by reducing the sensitivity of the multiple-pulse response to phase control settings and drifts. This scheme involves phase-modulating the r.f. carrier, and it automatically compensates for the systematic phase errors of four independent r.f. pulses; at the same time it removes phase 'glitch' errors.

By means of a four-pulse cycle, multiple-pulse sequence, Raber *et al.*[119] have measured (from powder samples) the chemical-shift anisotropies of carboxyl protons in squaric acid and butynedioic acid. The data, which show that the anisotropies cover the whole range of proton chemical shifts observed in liquids, were referred to TMS *via* $CH_3CO_2Ag$ and $C_6H_6$. Van Hecke *et al.*,[120] on the other hand, have worked with a single-crystal sample, *i.e.* anhydrous $\alpha$-oxalic acid. The WAHUHA four-pulse cycle was again used, and adamantane was employed as a reference sample. Their proton chemical shielding tensors indicate approximate axial symmetry of the chemical shift about the nominal hydrogen-bond direction, with the largest shielding occurring when the external field is parallel to the hydrogen bond.

A more extensive investigation of proton chemical-shielding tensors has been undertaken by Haeberlen and Kohlschutter.[121] They have observed proton multi-pulse spectra in powder samples of anhydrous oxalic acid; oxalic acid dihydrate; malonic, succinic, phthalic, maleic, and fumaric acids; succinic, phthalic, and maleic acid anhydrides; and ferrocene. The work was initiated with the four-pulse WAHUHA cycle but transferred later to the cycle

[116] W. K. Rhim, D. D. Elleman, and R. W. Vaughan, *J. Chem. Phys.*, 1973, **59**, 566.
[117] W. K. Rhim, D. D. Elleman, L. B. Schreiber, and R. W. Vaughan, *J. Chem. Phys.*, 1974, **60**, 4595.
[118] P. Mansfield and U. Haeberlen, *Z. Naturforsch.*, 1973, **28a**, 1081.
[119] H. Raber, G. Brunger, and M. Mehring, *Chem. Phys. Letters*, 1973, **23**, 400.
[120] P. Van Hecke, J. C. Weaver, B. L. Neff, and J. S. Waugh, *J. Chem. Phys.*, 1974, **60**, 1668.
[121] U. Haeberlen and U. Kohlschutter, *Chem. Phys.*, 1973, **2**, 76.

introduced by Mansfield and Haeberlen in ref. 118. All the results obtained could be represented by axially symmetric shielding tensors, and $\Delta\sigma = \sigma_{\parallel} - \sigma_{\perp}$ was found to be positive and comparatively large for carboxyl protons (~ +18 p.p.m.), smaller but with a quite large variation for aliphatic protons (+6 to +10 p.p.m.), below resolution (2 p.p.m.) for olefinic protons, and negative for aromatic protons (~ −5 p.p.m.).

$^{19}F$ chemical-shielding tensors have been measured in polycrystalline $BeF_2$[122] and some polycrystalline hexafluorophosphates[123] by Sears, who employs the 'adiabatic pulse' method of Kunitomo.[124] The $BeF_2$ result is used to complement recent multiple-pulse data on a series of cubic difluorides by Vaughan *et al.*,[125] who found a remarkably good correlation between the $^{19}F$ chemical shifts and the electronegativities of the cations, using Gordy's[126] original electronegativity scale. Similar correlations were sought in the hexafluorophosphate data.

A thorough $^{19}F$ n.m.r. analysis of potassium tetrafluorophthalate is the subject of a recent paper by Griffin *et al.*[127] Using a single crystal and the four-pulse cycle, they have evaluated its chemical-shielding tensors and crystal structure. The most shielded elements of the tensors lie perpendicular to the plane of the carbon ring and of the two in-plane components; the more shielded is along the C—F bond. The data are also interesting in that they allow one to examine the '*ortho*' effect of Karplus and Das.[128] The fluorines in $C_6F_4(CO_2K)_2$ do not fit quantitatively into the 'additive' picture of the chemical shifts displayed by the fluorobenzenes, upon which was based the reasoning of Karplus and Das. More theoretical work is obviously required before a satisfactory understanding of fluorine chemical shifts is obtained.

*Double-resonance Observation of Rare Spins.* The high-resolution n.m.r. observation of isotropically or chemically dilute nuclear spins by means of double-resonance techniques that involve an abundant spin species, I, as well as the rare species of interest, S, has recently been the subject of two substantial papers.[129,130] The method of Pines *et al.*[129] invokes the direct observation of the rare spins, S, after repeatedly transferring polarization to them from the more abundant I spins. The gain in power sensitivity as compared with the conventional observation of rare spins approaches:

$$N_I I(I+1)\gamma_I^2 / N_S S(S+1)\gamma_S^2$$

which is $\sim 10^3$ for $S = {}^{13}C$ and $I = {}^{1}H$. High resolution of the S spins is then

122 R. E. J. Sears, *J. Chem. Phys.*, 1973, **59**, 5213.
123 R. E. J. Sears, *J. Chem. Phys.*, 1973, **59**, 973.
124 M. Kunitomo, *J. Phys. Soc. Japan*, 1971, **30**, 1059.
125 R. W. Vaughan, D. D. Elleman, W. K. Rhim, and L. M. Stacey, *J. Chem. Phys.*, 1972, **57**, 5388.
126 W. Gordy and W. J. O. Thomas, *J. Chem. Phys.*, 1956, **24**, 439.
127 R. G. Griffin, H. N. Yeung, M. D. LaPrade, and J. S. Waugh, *J. Chem. Phys.*, 1973, **59**, 777.
128 M. Karplus and T. P. Das, *J. Chem. Phys.*, 1961, **34**, 1663.
129 A. Pines, M. G. Gibby, and J. S. Waugh, *J. Chem. Phys.*, 1973, **59**, 569.
130 P. K. Grannell, P. Mansfield, and M. A. B. Whittaker, *Phys. Rev. (B)*, 1973, **8**, 4149.

obtained by decoupling the abundant I spins with a continuous spin-locking r.f. field. This paper also presents a didactic review of direct and indirect methods of observing rare nuclear spins. Grannell *et al.*,[130] on the other hand, pursue the indirect observation of a rare spin species through its effect on the signal of the abundant spin species. They describe a high-resolution double-resonance scheme which facilitates the measurement of the chemical-shift structure and the intrinsic bandshape of low-abundance spins in solids. In addition, they compare quantitatively the sensitivities of the direct and indirect methods. As an example of the application of their method they quote results on $^{13}C$ in PTFE, solid adamantane, and solid methylurea.

The $^{13}C$ chemical-shielding tensors obtained from a single crystal of durene[131] show that the well-known tendency of the isotropic shift to be smaller for aromatic than for aliphatic carbons is ascribable largely to the very small shielding which occurs for the former when the applied field is in the plane of the ring and tangential to it at the C atom in question. The methyl carbons have much less anisotropic shielding than the ring carbons, and it is further suggested that non-bonded interactions in the crystal play an unimportant role in determining the shielding tensors. Chang *et al.*[132] make use of the 'direct' method to obtain the first full chemical-shielding tensor of $^{13}C$ in a carboxy-group. Their experiments were performed on a single crystal of ammonium *d*-tartrate, and they found that the $^{13}C$ nucleus is most shielded when $B_0$ is roughly perpendicular to the $^{13}COO^-$ plane, and least shielded when $B_0$ is along the C—C bond. Chemical-shift anisotropy powder patterns for $^{13}C$-enriched benzene adsorbed on charcoal and silica-gel, and for toluene adsorbed on charcoal, were found[133] to be very similar to those for the pure solids. They also reveal molecules in various states of molecular motion.

Other nuclei to have their high-resolution spectra displayed by double-resonance techniques in the solid state are $^{19}F$[134] and $^{15}N$.[135] Yannoni[134] establishes the feasibility of doing double resonance on a spin species that has been made 'rare' by dilution rather than by natural abundance. An erratum to this paper is given in ref. 136. The $^{15}N$ resonance, enhanced from protons,[135] shows no anisotropy in its chemical shielding in $(NH_4)_2SO_4$ and glycine. In $NH_4NO_3$, however, the shielding tensor of the nitrate-ion nitrogen displays a perceptible departure even from axial symmetry.

**H. N.M.R. Image Formation.**—An exciting new technique using n.m.r. has emerged during the past year. When fully developed this technique should enable one to determine accurately the spatial distribution of one type of spin species in a heterogeneous sample.


131 B. Pansak, A. Pines, and J. S. Waugh, *J. Chem. Phys.*, 1973, **59**, 591.
132 J. J. Chang, R. G. Griffin, and A. Pines, *J. Chem. Phys.*, 1974, **60**, 2561.
133 S. Kaplan, H. A. Resing, and J. S. Waugh, *J. Chem. Phys.*, 1973, **59**, 5681.
134 C. S. Yannoni, *J. Chem. Phys.*, 1973, **58**, 1773.
135 M. G. Gibby, R. G. Griffin, A. Pines, and J. S. Waugh, *Chem. Phys. Letters*, 1972, **17**, 80.
136 C. S. Yannoni, *J. Chem. Phys.*, 1973, **59**, 566.

It has long been known that a uniform field gradient leads to an inhomogeneously broadened resonance line, which is a projection of the sample spin density onto a line in the direction of the gradient. Exploration of this idea by Lauterbur[137] has led to the production of two-dimensional spin-density maps, when the lineshape was measured for many different gradient directions and the image reconstructed from all the data. This method of image formation has been improved and simplified by Hinshaw,[138] who has introduced a time dependence into both the magnitude and direction of the field gradients. Hinshaw points out that, in addition to spin mapping, such observables as $T_1$ can be simultaneously measured for different regions of the sample using his method.

Mansfield and Grannell,[139,140] using multiple-pulse line-narrowing sequences on a layered sample of macroscopic dimensions in an applied field gradient, were able to observe diffraction effects in the transient nuclear signal. When Fourier-transformed, the transient nuclear response gave rise to structure in which the number of layers was clearly resolved. Neither the multiple-pulse resolution nor the available magnitudes of field gradients are sufficient at the moment to extend this technique to a microscopic scale.

## 3 Nuclei of Spin $> \frac{1}{2}$ in Diamagnetic Solids

**A. Introduction.**—Nuclei of spin $I > \frac{1}{2}$ have an electric quadrupole moment in addition to a magnetic dipole moment. This section is restricted to the situation where the interaction between the nuclear electric quadrupole moment and the electric field gradient is less than that between the nuclear magnetic dipole and the externally applied magnetic field. This division is a consequence of the prime interest of this volume being nuclear resonance. However, Scott[141—143] has reviewed papers concerned with quadrupole effects published in the calendar years 1970, 1971, and 1972 without applying this restriction, and has provided 90, 117, and 136 references, respectively. In addition, the proceedings at the First International Symposium on Nuclear Quadrupole Resonance have been published[144] and those of the Second Symposium are in the press. These proceedings provide a useful indication of the types of study currently in progress. The quadrupolar interaction can provide a relaxation mechanism, and information on the quadrupole interaction can be derived from relaxation studies on the fluid phases. Such investigations are not reviewed in this chapter, devoted to the solid state, nor are those involving liquid crystals. For such studies the reader is referred to

137 P. C. Lauterbur, *Nature*, 1973, **242**, 190.
138 W. S. Hinshaw, *Phys. Letters*, 1974, **48A**, 87.
139 P. Mansfield and P. K. Grannell, *J. Phys.* (*C*), 1973, **6**, 422.
140 P. Mansfield, P. K. Grannell, A. N. Garroway, and D. C. Stalker, Proc. First Specialized Colloque Ampère, ed. J. W. Hennel, Inst. Nuclear Physics, Krakow, 1973, p. 16.
141 T. A. Scott, *Magn. Resonance Rev.*, 1972, **1**, 103.
142 T. A. Scott, *Magn. Resonance Rev.*, 1973, **2**, 69.
143 T. A. Scott, *Magn. Resonance Rev.*, 1973, **2**, 221.
144 J. A. S. Smith, 'Advances in Nuclear Quadrupole Resonance', Heyden, New York, 1974.

the relevant chapters of this volume. Furthermore, as this particular section is restricted to diamagnetic homogeneous insulators, quadrupolar effects in conducting, para- and ferro-magnetic systems and in adsorbed species will be discussed in other sections of this chapter. Pseudo-quadrupole effects will not be discussed in detail; one example will suffice. Moskvich *et al.*[145] observed a $^{1}H$ 'quadrupolar echo' in a three-spin system, sulphamic acid ($NH_3SO_3$), with only dipolar interactions present, the three protons being equivalent to single particles of spin-$\frac{3}{2}$ and spin-$\frac{1}{2}$.

For the heavier atoms the distortion of the electron cloud by the electric field gradient becomes more and more significant, and calculated electric field gradients due to external charge distributions have to be scaled by the Sternheimer antishielding factor. Stout and Garrett[146] have made a perturbation calculation of the induced field gradient showing that the Sternheimer shielding contains a significant amount of anisotropy, which can lead to errors of the order of 25%. Whilst the authors applied their calculation to gas molecules, it is clearly relevant to the solid state, but their method of calculation would not seem to be adequate for the extraction of reliable numerical values for the solid. Ray *et al.*[147] have discussed the influence of self-consistency effects on calculations of the antishielding factor for positive ions, selecting $Rb^+$ and $Cs^+$ for detailed study, and showing that neglect leads to errors of $\frac{1}{2}$%.

In a paper involving a nucleus with $I > \frac{1}{2}$ but with no direct interest in quadrupolar effects, Jacquinot *et al.*[148] produced a $^{43}Ca$ sample polarized to $80 \pm 15\%$. The sample chosen was $CaF_2$ with a $^{43}Ca$ concentration of 0.13%. A three-step process was involved: dynamic polarization of $^{19}F$ nuclei, followed by adiabatic demagnetization in the rotating frame, and then r.f. thermal mixing with $^{43}Ca$, again in the rotating frame.

**B. $^{2}H$ N.M.R.**—Deuteron magnetic resonance remains a popular activity. A high degree of accuracy is readily attainable as quadrupolar interactions are in the range 0—250 kHz, whereas the nuclear resonance frequency is typically 8—10 MHz, and dipolar linewidths are *ca.* 1 kHz. A paper typical of single-crystal studies is that by Berglund and Tegenfeldt,[149] who examined partially deuteriated sodium hydrogen oxalate, $NaHC_2O_4,H_2O$, with the intention of testing *ab initio* MO–LCAO–SCF calculations. At 25 °C they observed that the $D_2O$ molecules were not undergoing the 180° flips customarily observed in hydrates at this temperature. Two quadrupole coupling constants, $\chi = e^2qQ/h$, were obtained (235.2 and 228.5 kHz), with asymmetry factors $\eta$ of 0.90 and 0.13, respectively; these results are typical of the ac-

145 Yu. N. Moskvich, N. A. Sergeev, and G. I. Dotsenko, *Soviet Phys. Solid State*, 1974, **15**, 1912.
146 E. W. Stout, jun., and B. B. Garrett, *J. Chem. Phys.*, 1973, **59**, 3604.
147 S. N. Ray, T. Lee, T. P. Das, and R. M. Sternheimer, *Phys. Rev.*, 1974, **9**, 1108.
148 J. F. Jacquinot, W. Th. Wenckebach, M. Goldman, and A. Abragam, *Phys. Rev. Letters*, 1974, **32**, 1096.
149 B. Berglund and J. Tegenfeldt, *Mol. Phys.*, 1973, **26**, 633.

curacy of the technique. The principal directions of the major components deviated by 0.9° from the reported OH bond directions. In a subsequent paper[150] the same authors used i.r. and $^2H$ n.m.r. to examine lithium formate monohydrate, $LiHCOO,D_2O$, at 25 °C, obtaining nuclear quadrupole coupling constants (NQCC's) of 198.7 and 231.3 kHz, corresponding to H-bond lengths differing by 0.021 nm. The corresponding $\eta$ values were 0.060 and 0.097 and the major principal directions made angles of 0.59° and 0.94° with the OH bond directions. No evidence for a 180° flip motion was obtained up to 70 °C, where dehydration occurred, although Muthukrishnan and Ramakrishna[151] had attributed a reduction in $^1H$ second moment at 50 °C to this mechanism. Weiss and Wust[152] examined the $^1H$ and $^2H$ n.m.r. in single crystals of $Na_2ZnCl_4,3H_2O$ and $Na_2ZnCl_4,3D_2O$, obtaining Pake-type curves for the $^1H$ resonance. The $^2H$ n.m.r. demonstrated that the water molecules undergo a rapid 180° flip motion at 22 °C, giving an averaged electric field gradient (e.f.g.) tensor with a NQCC of 132 kHz and asymmetry factor 0.754. By −22 °C the flip motion was ineffective, and separate NQCC's of 239 and 236 kHz were obtained, with asymmetry factors of 0.118 and 0.125. It was observed that the temperature at which the 180° flip motion becomes effective was proportional to the height of the energy barrier, in this case 53.1 kJ $mol^{-1}$. Hutton and Pedersen[153] reported that the use of the Volkoff method[154] can lead to ambiguities, which can be removed by analysis of the powder spectrum or by rotation about a non-symmetry axis. The authors used their own work on $CaSO_4,2D_2O$[155] as an example.

Although measurements on single crystals are accurate and provide directional information, they are time-consuming, and single crystals may not always be available. Values of $\chi$ and $\eta$ can readily be obtained from the powder bandshape, but with reduced accuracy. In this category of experiment, Wei and Maeland[156] studied the H-bonded aluminium hydroxide systems gibbsite, bayerite, and nordstrandite, the oxyhydroxides diaspore and boehmite, and anodic oxide film. The $^2H$ n.m.r. spectra of the deuteriated analogues typically contained two regions. The first, a narrow central line from $D_2O$ molecules undergoing a rapid isotropic motion, could be removed under vacuum. The width of the second feature, due to structural OD, could be related to hydrogen-bond length using the Chiba relationship.[157] Hydrogen bonding results in a motion that is slow on the n.m.r. timescale; spectra were practically unchanged between −50 and −100 °C. For boehmite the measurements were consistent with one *X*-ray determination, which gave a hydrogen-bond length of 0.270 nm, and not with a second giving 0.247 nm;

150 B. Berglund, J. Lindgren, and J. Tegenfeldt, *J. Mol. Structure*, 1974, **21**, 135.
151 K. Muthukrishnan and J. Ramakrishna, *Acta Cryst.*, 1973, **B29**, 375.
152 A. Weiss and A. Wust, *Z. Naturforsch.*, 1973, **28**, 1919.
153 G. Hutton and B. Pedersen, *J. Magn. Resonance*, 1974, **13**, 119.
154 G. M. Volkoff, H. E. Petch, and D. W. Smellie, *Canad. J. Phys.*, 1952, **30**, 270.
155 G. Hutton and B. Pedersen, *J. Phys. and Chem. Solids*, 1969, **30**, 235.
156 I. Y. Wei and A. J. Maeland, *J. Chem. Phys.*, 1974, **60**, 3718.
157 G. Soda and T. Chiba, *J. Phys. Soc. Japan*, 1969, **26**, 717.

for diaspore the measurements were consistent with *X*-ray data, but the hydrogen-bond was found to be bent. For bayerite the measurements were inconsistent with the *X*-ray structure, and anodic $Al_2O_3$ gave a spectrum different to that from single crystals. Scheie *et al.*[158] used $^1H$ $T_1$, $T_{1\rho}$, and second-moment determinations, together with CW $^2H$ n.m.r. to investigate molecular reorientation in the plastic phase of acetylene; the results suggested the occurrence of a 180° flip process and not spherical or cubical reorientations. At 143 K the NQCC was 183 kHz, slightly less than the gas-phase value of 200 kHz. The appearance of a central signal was tentatively ascribed to a double-quantum transition. O'Reilly *et al.*[159] performed a comprehensive n.m.r. investigation on the three solid phases of $Bu^tCl$ from 90 K to the melting point at 249 K, involving $T_1$ ($^1H$), $T_{1\rho}$($^1H$), $^2H$ CW dispersion mode, and $^{35}Cl$ n.q.r. measurements. The $^2H$ studies confirmed the $^1H$ $T_1$ relaxation studies, monitoring firstly the onset of methyl-group rotation, then of rotation of the t-butyl group about the C—Cl bond, and finally of rapid translational diffusion. Each transition results in a characteristic change in powder pattern.

Niemela and his colleagues have published a series of papers using $^2H$ spin–lattice relaxation measurements on rapidly rotating groups to obtain correlation times, activation energies, and NQCC's (assuming zero asymmetry). In $SnD_4$[160] spin–lattice relaxation was observed between 68 and 129 K on both sides of a phase transition at 102.6 K; the NQCC is $66 \pm 5$ kHz. In $(ND_4)_2GeF_6$,[161,162] $^1H$, $^2H$, and $^{19}F$ $T_1$ values were obtained between 40 and 300 K for $^2H$ and between 4 and 400 K for $^1H$ and $^{19}F$. Correlation times were obtained for $ND_4$ and $GeF_6$ group rotations, the respective energy barriers to rotation being 4.2 and 56 kJ $mol^{-1}$. Below 50 K the $NH_4$ group underwent quantum-mechanical tunnelling.

To supplement earlier *ab initio* calculations, Lindgren and Tegenfeldt[163] compared NQCC's obtained for $D_2O$ with the O—H stretching frequency, a relationship developed by Salem.[164] Mokarram and Ragle[165] compared the e.f.g. at the $^2H$ site with the force constant for hydrides of the first- and second-row elements, obtaining a very good linear correlation but unfortunately having to use calculated e.f.g. values. Goren[166] noted that previous calculations[167,168] had indicated that $\chi$ was not very sensitive to hydrogen-bond length and, citing neutron-diffraction data, suggested that bond bending was significant. Using a minimal SCF basis set, Goren varied

158 C. E. Scheie, E. M. Peterson, and D. E. O'Reilly, *J. Chem. Phys.*, 1973, **59**, 2758.
159 D. E. O'Reilly, E. M. Peterson, C. E. Scheie, and E. Seyfarth, *J. Chem. Phys.*, 1973, **59**, 3576.
160 L. Niemela and J. Makela, *Phys. Letters*, 1973, **43A**, 343.
161 J. Tuohi, E. E. Ylinen, and L. K. E. Niemela, *Chem. Phys. Letters*, 1974, **28**, 35.
162 J. Tuohi and E. E. Ylinen, *Proc. Ann. Conf. Finn. Phys. Soc.*, 1974, **70**, No. 4, p. 31.
163 J. Lindgren and J. Tegenfeldt, *J. Mol. Structure*, 1974, **20**, 335.
164 L. Salem, *J. Chem. Phys.*, 1963, **38**, 1227.
165 M. Mokarram and J. L. Ragle, *J. Chem. Phys.*, 1973, **59**, 2770.
166 S. D. Goren, *J. Chem. Phys.*, 1974, **60**, 1892.
167 M. Weissmann, *J. Chem. Phys.*, 1966, **44**, 422.
168 S. D. Gornostansky and C. W. Kern, *J. Chem. Phys.*, 1971, **55**, 3254.

bond lengths and angles, finding agreement up to 30°. The asymmetry factor will be less than 0.1 if the hydrogen-bond bending is a few degrees. Bersohn[169] calculated $\eta$ factors in aromatic hydrocarbons using simple wavefunctions. He concluded that the source of a non-zero asymmetry arose from an inequivalence between carbon $2p_\sigma$ and $2p_\pi$ orbitals, with the necessity of employing different screening constants. Bearfield *et al.*,[170] using a localized orbital model, calculated NQCC's in formaldehyde, formic acid, and the formate ion, whilst Dixon *et al.*[171] calculated $\chi(^2H)$ for methane using LCAO–MO–SCF wavefunctions with optimized C—H bond length and H orbital exponent. In both papers comparison was made with experiment. The current availability of $^2H$ NQCC's in a range of compounds is proving to be very useful in stimulating and testing detailed quantum-mechanical calculations.

The properties of the $D_2$ molecule at low temperatures are of considerable interest because of the insight they provide into various fundamental quantum-mechanical processes. Scopazzi and White[172] reported spin–lattice relaxation data for $D_2$. The sample prepared was 95% in the para-form, but this value was reduced to 93% by conversion into the ortho-form. Below 2.7 K only the para-form ($I = 1$) was observed, the $T_1$ value for the ortho-form ($I = 2$) being too long for it to be detected, although above 2.7 K both isomers were observed. As in ortho-$H_2$, relaxation was produced by inelastic scattering of librons by the nuclear moments. Hardy and Berlinsky[173] were interested in the anomalous $^1H$ and $^2H$ bandshapes observed in solid ortho-hydrogen and para-deuterium. In the $^1H$ resonance the intensity of the low-frequency line in a Pake doublet was enhanced relative to the high-frequency component, which could even be negative, corresponding to an emission of energy. Two Pake doublets were observed in the $^2H$ resonance; the proposed explanation invoked a pumping process due to ortho–para conversion, the theory of the conversion and its effect on n.m.r. bandshapes having been developed by the same authors.[174] Weinhaus *et al.*[175] reported $T_1$ and $T_2$ in h.c.p. $D_2$ for the $I = 1$ and $I = 2$ spin systems below 5 K. Relaxation in the $I = 1$ system was a two-stage process, but that for $I = 2$ was single, the relaxation processes of the different spin systems being coupled by cross-relaxation. The temperature dependence of the second moment of the $I = 1$ species was observed below 8 K. In a subsequent paper[176] $T_2$ measurements were extended to 17 K and rotating-frame measurements made between 4 and 13 K. Relaxation was attributed to thermally activated diffusion. The temperature dependence of the relative isomer concentrations was also determined.


169 R. Bersohn, *Mol. Phys.*, 1974, **27**, 605.
170 D. W. Bearfield, M. A. Robb, and J. A. S. Smith, *J.C.S. Faraday II*, 1974, **70**, 920.
171 M. Dixon, T. A. Claxton, and R. E. Overill, *J. Magn. Resonance*, 1973, **12**, 193.
172 C. Scopazzi and D. White, *J. Chem. Phys.*, 1974, **60**, 323.
173 W. N. Hardy and A. J. Berlinsky, *Phys. Rev.* (*B*), 1973, **8**, 4996.
174 A. J. Berlinsky and W. N. Hardy, *Phys. Rev.* (*B*), 1973, **8**, 5013.
175 F. Weinhaus, H. Meyer, and S. M. Meyers, *Phys. Rev.* (*B*), 1973, **7**, 2948.
176 F. Weinhaus, H. Meyer, and S. M. Meyers, *Phys. Rev.* (*B*), 1973, **7**, 2960.

Edmonds and his colleagues[177,178] continue to exploit their sensitive spin-mixing technique, reviewed in previous volumes, to extract information on the quadrupolar interaction in various systems. For systems and temperatures where relaxation rates are favourable this is likely to remain a powerful technique. In deuteriated acetamide, $CH_3COND_2$, at 77 K,[177] they obtained $\chi$ and $\eta$ values of 196.42 kHz and 0.161, respectively, at one $^2H$ site, and 194.17 kHz and 0.173, respectively, at the other. They observed fine structure due to magnetic and pseudo-magnetic dipole interactions.

**C. First-row Elements.**—Edmonds and Summers[178] used the same technique to determine $\chi(^{14}N)$ for 18 crystalline amino-acids at 77 K, whilst Hunt and Mackay[179] reported $^{14}N$ and $^2H$ NQCC's in 19 amino-acids. Deuteriation caused small changes in the $^{14}N$ interaction. Mixed transitions of the form $\nu_N \pm \nu_D$ were also observed. In a subsequent paper Hunt[180] considered the symmetry of the e.f.g. tensor with reference to the tetrahedral $^{14}N^+$ site in amino-acids, and with particular interest in the discrepancies between the observed asymmetry factor and that predicted on the basis of simple argument. Blinc *et al.*[181] applied the pulsed double-resonance technique (equivalent in many ways to that of Edmonds) to the determination of the $^{17}O$ quadrupolar interaction in 10%-enriched $KH_2PO_4$. The temperatures selected, 90 and 77 K, were such that observations were made in the para- and ferro-electric phases. In the paraelectric phase oxygen sites were equivalent, $\chi$ = 5.16 MHz and $\eta$ = 0.55, whereas in the ferroelectric phase the degeneracy was removed, the NQCC's being 5.96 and 4.85 MHz and corresponding asymmetry factors 0.72 and 0.18.

**D. Second-row Elements.**—Hughes[182] considered the spin–lattice relaxation due to the quadrupolar interaction. On assuming that the lattice motions are independent of magnetic field, the application of the principle of time invariance implies that the transition probabilities $W_1$ and $W_2$ are invariant to the sign of the magnetic field. If the magnetic field does affect the lattice $W_1$ and $W_2$ are still invariant, but the relaxation tensor is not necessarily real. Hence it was concluded that the anomalous orientation dependence of $W_2/W_1$ in $^{23}NaNO_3$ is not attributable to a failure in the principle of time-reversal invariance. Yagi and Tatsuzaki[183] determined the $^{23}Na$ e.f.g. tensor in $\gamma$-irradiated $NaNO_2$. The $\eta$ term exhibited a temperature dependence, which in turn exhibited a dose dependence, equivalent to a shift to lower temperature. The e.f.g. tensor was related to the long-range order parameter in the ferroelectric material. Very precise temperature control was required to identify the

177 D. T. Edmonds, M. J. Hunt, and A. L. Mackay, *J. Magn. Resonance*, 1973, **11**, 77.
178 D. T. Edmonds and C. P. Summers, *J. Magn. Resonance*, 1973, **12**, 134.
179 M. J. Hunt and A. L. Mackay, *J. Magn. Resonance*, 1974, **15**, 402.
180 M. J. Hunt, *J. Magn. Resonance*, 1974, **15**, 113.
181 R. Blinc, J. Seliger, R. Osredkar, and T. Prelesnik, *Chem. Phys. Letters*, 1973, **23**, 486.
182 D. G. Hughes, *J. Phys.* (*C*), 1973, **6**, 3797.
183 T. Yagi and I. Tatsuzaki, *J. Phys. Soc. Japan*, 1973, **35**, 1675.

temperature at which $\nu$ became zero. Fitzgerald and Casabella[184] examined the effect of radiation on the $^{23}Na$ quadrupolar splitting in the non-ferroelectric phases of Rochelle salt, supplementing earlier work.[185] Irradiation at 30 °C decreased the spontaneous polarization and the ferroelectric temperature range. No bias field was produced; otherwise, effects were similar to radiation at 0 °C. Irradiation at 77 K was followed by annealing effects as the temperature was increased, a bias field on entry to the ferroelectric region, and additional damage on entry to the upper non-ferroelectric phase. Taki and Satoh[186] used pulsed double resonance to determine the $^{23}Na$ e.f.g. in solid solutions of NaI and NaCl. In a later paper[187] work was extended to a single crystal and to solutions of NaCl and NaBr with the bromine concentration at 0.125 mole %. The origin of the e.f.g. was discussed in some detail. Meyer *et al.*[188] examined a related system, $Ag^+$-doped NaCl. The $Ag^+$ ions break the inversion symmetry at $Na^+$ nearest neighbours, although distant nuclei remain sites of inversion symmetry. Hence an oscillating electric field induces $\Delta m = \pm 2$ transitions in near but not distant neighbours. By saturation of the magnetic resonance of distant neighbours, transitions $2\nu_0 \pm \nu_Q$ could be obtained, to give a $\nu_Q$ value of 12.5 kHz with asymmetry factor 0.28. Thus the prediction that the first shell has a low quadrupolar interaction was confirmed, and such signals would be lost under the main line in a normal double-resonance experiment. The electric field experiment was performed at 77 K and a cross-relaxation bottleneck observed. Speight and Jeffrey[189] observed $^{14}N$ and $^{35}Cl$ spin–lattice relaxation in $NH_4Cl$, finding a discontinuity for $^{14}N$ and a cusp for $^{35}Cl$ at the $\lambda$-transition at 175 K. The $^{14}N$ relaxation was considered to be dipolar in origin in the neighbourhood of the $T_1$ minimum, but that for $^{35}Cl$ was always quadrupolar. O'Reilly *et al.*[190] examined the $^{35}Cl$ spin–spin relaxation in solid and liquid $CCl_4$. $T_2$ values were continuous between liquid and f.c.c. solid, and there was no abrupt change in intensity. Over long periods, spectra were broadened by conversion into the rhombohedral form.

Von Meerwall and his colleagues have used n.m.r. of nuclei of spin $I = \frac{3}{2}$ in the quadrupolar regime where the quadrupolar energy is greater than the Zeeman energy, observing transitions between the $m = \pm\frac{1}{2}$ levels. They developed expressions from which the quadrupolar coupling and chemical-shift anisotropy could be obtained from measurements on single crystals. The technique was tested using the $^{35}Cl$ n.m.r. in *p*-dichlorobenzene.[191]

184 M. E. Fitzgerald and P. A. Casabella, *Phys. Rev.* (*B*), 1973, **8**, 3076.
185 M. E. Fitzgerald and P. A. Casabella, *Phys. Rev.* (*B*), 1973, **7**, 2193.
186 T. Taki and M. Satoh, *J. Phys. Soc. Japan*, 1974, **36**, 312.
187 T. Taki, T. Kanashito, T. Ohno, and M. Satoh, *J. Phys. Soc. Japan*, 1974, **36**, 470.
188 W. J. Meyer, D. V. Long, and C. P. Slichter, *Phys. Rev.* (*B*) 1973, **8**, 1924.
189 P. A. Speight and K. R. Jeffrey, *J. Magn. Resonance*, 1973, **10**, 195.
190 D. E. O'Reilly, E. M. Peterson, and C. E. Scheie, *J. Chem. Phys.*, 1974, **60**, 1603.
191 R. B. Creel, E. von Meerwall, C. F. Griffin, and R. G. Barnes, *J. Chem. Phys.*, 1973, **58**, 4930.

**E. Other Nuclei.**—Von Meerwall *et al.*[192] extended their analysis of n.m.r. in the quadrupolar regime to polycrystalline samples, developing expressions for the extraction of $\chi$, $\eta$, and the chemical-shift anisotropy from the powder band-shape. The technique was tested on $^{81}$Br in *p*-dibromobenzene, but a more complete study on this compound, published later,[193] used both $^{79}$Br and $^{81}$Br nuclei. Hatanaka *et al.*[194] applied the pulsed double-resonance technique to extract the $^{39}$K quadrupole interactions in $KH_2AsO_4$ at 77 K and at 4.2 K below the transition temperature of 96 K. At 77 K a single interaction was observed, with $\chi = 0.837$ MHz and $\eta = 0.493$, the major principal direction being along the crystalline *y*-axis. At 4.2 K two symmetry-related interactions were observed, $\chi$ was 0.829 MHz and $\eta$ was 0.444. Some evidence for As signals was obtained. Andrew *et al.*[195] measured the $^{63}$Cu, $^{65}$Cu, and $^{81}$Br spin–lattice relaxation rates in polycrystalline cuprous halides between 78 and 300 K. The $T_1$ behaviour was describable in terms of the Van Kranendonk temperature function involving a two-phonon relaxation process. The ratio of the $^{63}$Cu and $^{65}$Cu relaxation rates was consistent with published nuclear quadrupole moments. Jeffrey *et al.*[196] determined the $^{79}$Br and $^{81}$Br NQCC's in $NH_4Br$ on either side of the order–disorder transition at 234.5 K. Above 234.5 K, single lines were obtained, but below that temperature there were six lines for each isotope. The data were interpreted in terms of movement of the $Br^-$ ions relative to the $NH_4^+$ ions. Cowgill and Norberg[197] measured the $^{83}$Kr chemical shift and $T_1$ values in solid and liquid krypton from 40 K to the critical point at 209 K, observing a density dependence which was compared with those found for $^{21}$Ne and $^{131}$Xe. The relaxation was dominated by quadrupolar effects modulated by atomic diffusion and thermal phonons. Bailey and Story[198] determined the $^{14}$N and $^{87}$Rb quadrupolar interactions in single crystals of ammonium aluminium alum $NH_4Al(SO_4)_2,12H_2O$ and in $RbAl(SO_4)_2,12H_2O$, yielding values of 1.968 and 12.9 MHz. The $^{14}$N $\eta$ value was zero and $\chi$ ($^{87}$Rb) was only extracted from the central $+\frac{1}{2}$ to $-\frac{1}{2}$ transition, the ratio of the NQCC's giving an antishielding factor of $-4$ for $^{14}$N in the $NH_4$ ion. Kasahara and Tatsuzaki[199] also extracted the $^{87}$Rb e.f.g. from the second-order quadrupolar shifts, this time in single crystals of $RbDSO_4$ between $-100$ and $+18$ °C. Above $-22$ °C in the paraelectric phase two inequivalent sites were detected, but below $-22$ °C in the ferroelectric phase each produced two further non-equivalent sites, and gave a splitting with a

192 E. von Meerwall, R. B. Creel, C. F. Griffin, and S. L. Segal, *J. Chem. Phys.*, 1973, **59**, 5350.
193 C. F. Griffin, R. B. Creel, J. Worth, and E. von Meerwall, *J. Phys. and Chem. Solids*, 1974, **35**, 71.
194 H. Hatanaka, T. Terao, K. Morimoto, and T. Hashi, *J. Phys. Soc. Japan*, 1974, **36**, 1207.
195 E. R. Andrew, W. S. Hinshaw, and R. S. Tiffen, *J. Phys. (C)*, 1973, **6**, 2217.
196 K. R. Jeffrey, A. G. Brown, and R. L. Armstrong, *Phys. Rev. (B)*, 1973, **8**, 3071.
197 D. F. Cowgill and R. E. Norberg, *Phys. Rev. (B)*, 1973, **8**, 4966.
198 W. C. Bailey and H. S. Story, *J. Chem. Phys.*, 1974, **60**, 1952.
199 M. Kasahara and I. Tatsuzaki, *J. Phys. Soc. Japan*, 1974, **36**, 786.

temperature dependence of the form $(T_c - T)^\beta$, where $0.28 \leqslant \beta \leqslant 0.29$. No evidence was obtained for the thermal hysteresis previously reported. Blinc *et al.*[200] described a method of determining the fast quadrupolar relaxation of a non-resonant spin involving rotating- and dipolar-frame $^1H$ relaxations. The temperature dependence of the $^{127}I$ spin–lattice relaxation was obtained in $Ag_2H_3IO_6$ and $(NH_4)_2H_3IO_6$ in situations where the ($^{127}I$, $^1H$) cross-relaxation was fast and those where it was slow compared with the $^{127}I$ relaxation.

## 4 Systems with Unpaired Electrons in the Absence of Magnetic Ordering

**A. Non-metallic Paramagnetic Systems.**—The relaxation of $^{19}F$ in $CaF_2$, doped with rare-earth ions to form paramagnetic centres, has been investigated by Lin and Hartmann.[201] They first eliminated the fluorine–fluorine spin-diffusion in the long $\tau_c$ limit, by operating off-resonance and at a $B_1$ value which satisfied the magic-angle condition. They were then able to study the $^{19}F$ relaxation by its interaction with the paramagnetic centres alone. When $\tau_c$ is short, spin-diffusion of orders higher than the first becomes important due to multiple-spin flip processes. In this latter regime Lin and Hartmann have therefore studied the effective field dependence of the higher-order spin-diffusion constant.

With an eye to obtaining high proton polarizations, Lichti and Stapleton[202] have observed proton relaxation rates and proton polarization enhancements in ytterbium-doped yttrium hydroxide. The protons were dynamically polarized above their thermal equilibrium values by mechanically rotating the sample in a spin refrigerator. The temperature and field dependences of the proton–lattice relaxation rate are shown to agree with theory. However, the dependence of this rate on the $Yb^{3+}$ ion concentration was found to differ from that expected on theoretical grounds.

Skjaeveland *et al.*[203] have measured the crystal-field splitting of $Fe^{2+}$ ions in $FeSiF_6,6H_2O$ to be $(12.2 \pm 1.0)$ cm$^{-1}$. They obtained this value from the exponential temperature dependence of the proton $T_1$ and by invoking a crystal-field distortion of magnitude 0.54 cm$^{-1}$, in line with e.s.r. results.

**B. Metals**—Kaplan and Evans[204] present a procedure for calculating the temperature dependence of the FID in a metal. They take account of the coupling between the spin–lattice interaction, which is assumed to be of the Korringa form, and the magnetic dipole–dipole interaction, which is made to encompass all dipolar-like mechanisms (such as the pseudo-dipolar part of the indirect exchange interaction). A study of the influence of the skin-effect on the

200 R. Blinc, J. Pirs, and S. Zumar, *Phys. Rev.* (*B*), 1973, **8**, 15.
201 N. A. Lin and S. R. Hartmann, *Phys. Rev.* (*B*) 1973, **8**, 4079.
202 R. L. Lichti and H. J. Stapleton, *Phys. Rev.* (*B*), 1973, **8**, 4134.
203 S. M. Skjaeveland, I. Jvare, and D. P. Tunstall, *Phys. Letters*, 1974, **47A**, 121.
204 J. I. Kaplan and W. A. B. Evans, *Phys. Rev.* (*B*), 1974, **9**, 3670.

shape of spin-echo signals in bulk metals has been made by Kanert, Kotzur, and Mehring.[205—207] They calculate the amplitude and phase of the spin-echo signal after a $\beta_1$–$\tau$–$\beta_2$ ($\alpha$) pulse sequence, where $\beta$ represents the pulse lengths and $\alpha$ their relative phases, for half-integral spins from $\frac{3}{2}$ to $\frac{9}{2}$. The calculations were tested on copper foils, copper doped with tin, and aluminium.

Because of the low electronic contribution to the spin–lattice relaxation in beryllium metal, Chabre[208] has been able to observe both the dipolar and quadrupolar contributions, which arise respectively because of beryllium atomic motions and the field gradient fluctuations coming from vacancy and impurity motions. The dipolar contribution suggests a self-diffusion activation energy of 1.35 eV, whereas a migration energy of 0.65 eV is indicated by the quadrupolar contribution. Beryllium is also the subject of a multiple-pulse high-resolution experiment,[209] which obtains $(-10.0 \pm 3)$ p.p.m. for the $^9Be$ Knight shift. The same authors obtain $(1636 \pm 3)$ p.p.m. for the Knight shift of $^{27}Al$. The reference substances were aqueous solutions of $BeCl_2$ and $AlCl_3$, respectively. Using the alternative high-speed magic-angle sample-rotation technique, Andrew and Hinshaw[210] obtained $(1640 \pm 1)$ p.p.m. for the $^{27}Al$ shift relative to the same reference. In another paper, Andrew and Hinshaw[211] employed the same sample-spinning technique to evaluate an indirect scalar coupling constant of $(230 \pm 10)$ Hz for nearest neighbours in metallic copper. For next-nearest neighbours they found the coupling constant to be negative and *ca.* 100 Hz. A theoretical calculation of the Knight shift in copper, together with the other noble metals silver and gold, is given by Abati *et al.*[212] Their agreement with experiment for copper is quite good, but falls off for Ag and Au.

The electron–phonon enhancement of the electronic specific heat and the effective Coulomb enhancement of the spin susceptibility are taken into account by Ganguly[213] in calculating the nuclear spin–lattice relaxation rates in transition metals. The vanadium $T_1$ obtained experimentally was found to be consistent with a Knight shift which is almost entirely due to orbital paramagnetism. In addition, the inclusion of the above-mentioned corrections was found to improve the agreement between theory and experiment for both Nb and Pt.

The superconducting state is the subject of papers by Williamson and MacLaughlin and by Kobayashi *et al.* Williamson and MacLaughlin[214] concentrate on pure indium, in addition to indium successively alloyed with small

205 D. Kotzur, O. Kanert, and M. Mehring, Proc. XVII Congress Ampère, ed. V. T. Houi, Turku, 1973, p. 322.
206 D. Kotzur, M. Mehring, and O. Kanert, *Z. Naturforsch.*, 1973, **28a**, 1607.
207 O. Kanert, D. Kotzur, and M. Mehring, *Solid State Comm.*, 1973, **13**, 1017.
208 Y. Chabre, *J. Phys. (F)*, 1974, **4**, 626.
209 M. Mehring and H. Raber, *Solid State Comm.*, 1973, **13**, 1637.
210 E. R. Andrew, W. S. Hinshaw, and R. S. Tiffen, *Phys. Letters*, 1973, **46A**, 57.
211 E. R. Andrew and W. S. Hinshaw, *Phys. Letters*, 1973, **43A**, 113.
212 S. Abati, E. Borchi, P. Coppi, and S de Gennaro, *Phys. Letters*, 1973, **46A**, 255.
213 B. N. Ganguly, *Phys. Rev. (B)*, 1973, **8**, 1055.
214 J. D. Williamson and D. E. MacLaughlin, *Phys. Rev. (B)*, 1973, **8**, 125.

amounts of Ga, Cd, Tl, Hg, and Pb. They observed that for a fixed $T_c/T$ ($T_c$ being the critical temperature) the relaxation rate decreases with increasing impurity concentration, thus indicating that the anisotropy of the energy gap decreases with decreasing mean free path. Their data also support the notion of an r.m.s. energy-gap inhomogeneity of about 2%. Kobayashi *et al.*[215] focus their attention on the properties of fine superconducting particles with diameters of less than 50 nm. In particular they investigate the effect on the nuclear resonance observables of the discreteness of the electronic energy levels. Electronic quantum effects in small metal particles are also discussed in a review paper by Knight.[216] In order to preserve charge neutrality in particles with diameters <10 nm, there is a strict conservation of electron number. This leads at low temperatures to significant deviations from the properties of bulk materials and to distinguishability between particles with odd and even numbers of electrons.

**C. Alloys.**—Work on impurities in metals is included in this subsection as well as genuine alloy systems

The presence of small amounts of impurity atoms in metallic lattices causes measurable perturbations on the resonance properties of the host nuclei. Walstedt and Walker[217] calculate the effect on the host bandshape of randomly placed magnetic impurities in a metallic lattice. The properties of the impurities themselves have been considered (*a*) by Spanjaard and Hartmann-Boutron,[218] who discuss the observability of nuclear orientation n.m.r. from a paramagnetic impurity in a diamagnetic lattice; (*b*) by Terakura and Kanamori,[219] who take the opposite case and calculate the electronic structure of a non-transition element impurity in a transition metal; and (*c*) by Gotze and Schlottmann,[220] who calculate the longitudinal impurity susceptibility in dilute magnetic systems.

The genuine intermetallic compounds and alloy systems which have been covered in the literature over the past year are summarized in Table 3.[221—237]

215 S. Kobayashi, T. Takahashi, and W. Sasaki, *J. Phys. Soc. Japan*, 1974, **36**, 714.
216 W. D. Knight, *J. Vac.-Sci. Technol.*, 1973, **10**, 705.
217 R. E. Walstedt and L. R. Walker, *Phys. Rev.* (*B*) 1974, **9**, 4857.
218 D. Spanjaard and F. Hartmann-Boutron, *J. Phys.* (*F*), 1973, **3**, 1178.
219 K. Terakura and J. Kanamori, *J. Phys. Soc. Japan*, 1973, **34**, 1520.
220 W. Gotze and P. Schlottmann, *J. Low Temp. Phys.*, 1973, **12**, 149.
221 T. K. Halstead, W. U. Benesh, R. D. Gulliver jun., and R. A. Huggins, *J. Chem. Phys.*, 1973, **58**, 3531.
222 L. E. Drain, *Phys. Rev.* (*B*) 1973, **8**, 3628.
223 V. Niculescu, I. Pop, and M. Rosenberg, *Acta Phys. Polonica*, 1973, **A43**, 279.
224 V. Niculescu, I. Pop, and M. Rosenberg, *J. Phys. and Chem. Solids*, 1973, **34**, 961.
225 M. Coldea, M. Crisan, V. Niculescu, and I. Pop, *Phys. Status Solidi*, 1973, **56**, 749.
226 A. J. Arko, F. Y. Fradin, and M. B. Brodsky, *Phys. Rev.* (*B*), 1973, **8**, 4104.
227 K. Terakura, N. Yokoyama, S. Kohzuki, and K. Asayama, *J. Phys. Soc. Japan*, 1974, **36**, 130.
228 J. Buttet and K. Lauger, *Phys. Rev.* (*B*), 1974, **9**, 4643.
229 P. Panissod, *J. Phys.* (*F*), 1974, **4**, 484.
230 L. B. Welsh and C. W. Chu, *Phys. Rev.* (*B*), 1973, **8**, 1026.
231 H. Alloul and P. Bernier, *J. Phys.* (*F*), 1974, **4**, 870.
232 D. C. Lo, D. V. Lang, J. B. Boyce, and C. P. Slichter, *Phys. Rev.* (*B*), 1973, **8**, 1026.

**Table 3** *Studies of nuclei in alloy systems that have been undertaken recently*

| *Probe Nucleus* | *Alloy Systems* | *Reference* |
|---|---|---|
| $^{7}Li$ | $\gamma$-$Li_{1+x}V_3O_{8+\lambda}$, with $0<x<0.12$ and $-0.1<\lambda<0$ | 221 |
| $^{27}Al$ | AlCu,AlAg,AlGa,AlGe,AlMg | 222 |
| $^{27}Al$ | $Ce_{1-x}R_xAl_3$ (R = Y, La, Gd, Tb, or Er) | 223,224 |
| $^{27}Al$ | $Cr_2Al_{11}$ | 225 |
| $^{27}Al$ | $U_{1-x}Pu_xAl_2$ | 226 |
| $^{45}Sc$ | PdSc,PtSc | 227 |
| $^{51}V$ | PdV | 227 |
| $^{51}V$ | $V_3Si$ | 228 |
| $^{51}V$ | VCo,VMo,VRh,VTa,VW,VRe,VOs,VIr | 229 |
| $^{51}V$ | VRu | 230 |
| $^{63}Cu$ | AlCu | 222 |
| $^{63}Cu$ | CuMn,CuCr | 231 |
| $^{63}Cu$ | CuNi,CuCo | 232,233 |
| $^{71}Ga$ | AlGa | 222 |
| $^{71}Ga$ | $Au_{1-x}Pd_xGa_2$ | 234 |
| $^{71}Ga$ | $AuGa_2$,$Au_{0.95}Pd_{0.05}Ga_2$,$AuIn_2$ | 235 |
| $^{119}Sn$ | PtSn | 236,237 |
| $^{195}Pt$ | PtSi,PtGe,PtSn,$Pt_2Si$,$Pt_2Ge$ | 236,237 |

The probe nuclei range from $^{7}Li$ to $^{195}Pt$ and the purposes of the experiments are also diverse.

**D. Semiconductors.**—Ternary semiconductors of the type $I_B$–III–VI, which have attracted interest recently because of their nonlinear optical properties, have been investigated using magnetic resonance by Speiss *et al.*[238] They observed the $^{27}Al$, $^{63}Cu$, $^{69}Ga$, and $^{115}In$ resonances in a series of these systems in order to determine the respective nuclear quadrupole coupling constants. The quadrupole coupling constants were then discussed in connection with the *X*-ray structural data. Measurements of the $^{115}In$ Knight shift in *n*-type InSb at low temperatures are presented by Miranda *et al.*[239] These measurements were obtained as a function of magnetic field near to the quantum limit for several different conduction-electron concentrations. The hyperfine contribution to the shift exhibited quantum oscillations as the field was varied, thus reflecting the oscillations in the conduction electron density of states at the Fermi surface.

The semi-metallic actinide compounds are represented by Lam and Fradin's

233 D. V. Lang, D. C. Lo, J. B. Boyce, and C. P. Slichter, *Phys. Rev.* (*B*), 1974, **9**, 3077.
234 H. T. Weaver, J. E. Schirber, and A. Narath, *Phys. Rev.* (*B*), 1973, **8**, 5443.
235 W. W. Warren, R. W. Shaw, A. Merith, F. Di Salvo, A. R. Storm, and J. H. Wernick, *Phys. Rev.* (*B*), 1973, **7**, 1247.
236 J. P. van Dyke and H. T. Weaver, *Solid State Comm.*, 1973, **12**, 905.
237 H. T. Weaver, R. K. Quinn, R. J. Baughman, and R. C. Kramer, *J. Chem. Phys.*, 1973, **59**, 4961.
238 H. W. Speiss, U. Haeberlen, G. Brandt, A. Rauber, and J. Schneider, *Phys. Status Solidi*, 1974, **62B**, 183.
239 G. A. Miranda, J. A. McNeill, and W. G. Clark, *Phys. Rev.* (*B*), 1974, **9**, 495.

paper on NpP.[240] The $^{31}P$ Knight shift is shown to follow a Curie–Weiss behaviour and its spin–lattice relaxation rate is Korringa-like between 175 and 360 K. However, at lower temperatures (~120 K) a complex ordering takes place.

**E. Metal Hydrides.**—Proton mobility in the metal hydrides has long been a subject of investigation by nuclear resonance methods. Dickens *et al.*[241] obtained an activation energy of 16 kJ mol$^{-1}$ for the process in hydrogen tungsten bronzes $H_xWO_3$, where $0 < x < 0.6$. In uranium hydride, however, Cinader *et al.*[242] have observed a much more highly hindered diffusion process, *i.e.* $E_a = 80.5$ kJ mol$^{-1}$. The relaxation rate in this hydride was found to be the sum of three contributions, namely (*a*) time-dependent nuclear dipole–dipole interaction caused by hydrogen diffusion, (*b*) the contact interaction between the conduction electrons and the protons, causing Korringa relaxation, and (*c*) the Rudderman–Kittel interaction between the protons and the localized *f*-electrons *via* the conduction electrons. Separation of the three contributions from each other is made using their different temperature dependence.

The titanium and zirconium hydrides, where the protons occupy tetrahedral sites of the f.c.c. or tetragonal lattice to form a $CaF_2$ structure, have been used by Nakamura[243] to test several bandshape formulations in dipolar solids.

The investigation of the phase diagrams in hydrides has been made by Arons *et al.*[244] in the V–H and V–D systems. They were able to use the quadrupolar structure of the $^{51}V$ and $^{2}H$ resonances, since the quadrupole coupling displayed different magnitude and symmetry in each of the $\alpha,\beta$, and $\delta$ phases. An independent resonance study of the structure of $\beta$-$V_2D$ has also been made by Nakamura.[245]

Another hydride exploited by Nakamura[246] is the Nb–D system. He finds that the order–disorder ($\beta \rightarrow \alpha$) phase-transition temperature depends on the presence of small amounts of interstitial impurities, as well as on the concentration of deuterons. A tentative phase diagram for Nb–D is also proposed.

**F. One-dimensional Metals.**—The class of materials known as 'one-dimensional metals', which exhibit conduction in one dimension, are currently of considerable interest. Included in this class of compounds is the TCNQ group of materials. Devreux and Nechtschein[247] have shown that the proton relaxation times $T_{1D}$ and $T_{1\rho}$ are typical thermally activated processes in the $(Ph_3AsCH_3)^+(TCNQ)_2^-$ salt. This they attribute to the creation and annihilation of excitons. Alizon *et al.*,[248] on the other hand, have observed the

240 D. J. Lam and F. Y. Fradin, *Phys. Rev.* (*B*), 1974, **9**, 238.
241 P. G. Dickens, D. J. Murphy, and T. K. Halstead, *J. Solid State Chem.*, 1973, **6**, 370.
242 G. Cinader, M. Peretz, D. Zamir, and Z. Hadari, *Phys. Rev.* (*B*), 1973, **8**, 4063.
243 K. Nakamura, *J. Magn. Resonance*, 1974, **14**, 31.
244 R. R. Arons, H. G. Bohn, and H. Lutgemeier, *J. Phys. and Chem. Solids*, 1974, **35**, 207.
245 K. Nakamura, *Bull. Chem. Soc. Japan*, 1973, **46**, 2588.
246 K. Nakamura, *Bull. Chem. Soc. Japan*, 1973, **46**, 2028.
247 F. Devreux and M. Nechtschein, *Phys. Letters*, 1973, **46A**, 49.
248 J. Alizon, J. P. Blanc, J. Gallice, and H. Robert, *Compt. rend.*, 1973, **276**, *B*, 623.

temperature dependence of the proton polarization in order to clarify the nature of the electron–proton coupling, and to detect the phase transition in the alkali-metal salts of TCNQ.

## 5 Ordered Magnetic Materials

**A. Ferromagnetic Materials.**—N.m.r. in cobalt, the ferromagnetic metal in which nuclear resonance in the ordered internal hyperfine field was first reported in 1960, continues to reveal important information. Bailey, Creagh, and Wilson[249] have observed the 'domain' n.m.r. signal from h.c.p. cobalt metal. Their results show that the 'domain' resonance is from nuclei at the edges of the domains, and that such nuclei have spin–lattice relaxation rates which are a factor of 13 less than those at the centre of domain walls. An application of fast-passage techniques to ferromagnetic materials has been made by Butler.[250] He analyses how the spectra from cobalt metal and nearest-neighbour impurities in b.c.c. iron metal are distorted by anisotropic hyperfine fields and by large values of the r.f. excitation.

Nuclear spin–lattice relaxation of $^{59}$Co has been studied as a function of the external magnetic field both in nickel[251] and in multidomain iron.[252] In nickel, Streever and Caplan claim to have minimized non-nuclear relaxation effects and to have obtained actual nuclear relaxation rates in order to test the consistency of the $T_1T$ product. In multidomain iron, the nuclear orientation n.m.r. technique was employed on those nuclei in the bulk of the domains. It was again found that the relaxation rates of nuclei in the domains of a multi-domain sample were greater than those of bulk domain nuclei in single-domain samples. No theoretical explanation of this fact seems to agree with experiment at the moment.

The transverse nuclear relaxation of $^{59}$Co in cobalt metal and of $^{151}$Eu in the ferromagnetic insulator EuO have been observed by Barak *et al.*[253] as a function of the external field. Non-exponential spin-echo decays were seen in both cases and the relaxation rates were each found to be field-dependent. Their analysis of the results confirms the important role of the Suhl–Nakamura interaction in the spin–spin relaxation process.

The understanding of nuclear resonance in ferromagnetic insulators is to some extent simpler than in metallic ferromagnets, because of the absence of conduction-electron effects. In EuO,[254] the $^{115}$Eu spin–lattice relaxation is found to be dominated by the dipolar-induced two-magnon process at temperatures below 14 K, but by the exchange-scattering-enhanced three-magnon process at higher temperatures. In $GdCl_3$, Hessler[255] has utilized the $^{35}$Cl nuclear resonance to obtain a critical-point study of $GdCl_3$. From the

249 S. G. Bailey, D. G. Creagh, and G. V. H. Wilson, *Phys. Letters*, 1973, **44A**, 229.
250 M. A. Butler, *Phys. Rev.* (*B*) 1973, **8**, 5122.
251 R. L. Streever, *Phys. Rev.* (*B*), 1973, **8**, 3241.
252 R. Kieser, N. Kaplan, and B. G. Turrel, *Phys. Rev.* (*B*), 1974, **9**, 2165.
253 J. Barak, I. Siegelstein, A. Gabai, and N. Kaplan, *Phys. Rev.* (*B*), 1973, **8**, 5282.
254 J. Barak, A. Gabai, and N. Kaplan, *Phys. Rev.* (*B*), 1974, **8**, 4914.
255 J. A. Hessler, *Phys. Rev.* (*B*), 1973, **8**, 3151.

internal field splitting of the $^{35}Cl$ n.q.r. spectrum, it was possible to deduce that the internal field at the chlorine sites is parallel to the $C_3$ symmetry axis and antiparallel to the magnetization. Near to the Curie temperature of 2.214 ±0.0016) K, the critical exponent was evaluated to be (0.3904 ± 0.006).

The alloy systems upon which nuclear resonance work has been published in the past year include FeCr;[256] $FeBe_2$,[257] HoGd, HoTb, and HoDy;[258] and $YFe_2$, $YFe_3$ and $Y_2Fe_{17}$.[259] The nuclear resonance of Ru as an impurity in iron, cobalt, and nickel hosts[260] has enabled a value of (1.121 ±0.002) to be obtained for the ratio of the nuclear moments of the two Ru isotopes $^{101}Ru$ and $^{99}Ru$. Using the Mössbauer evaluation of the $^{99}Ru$ moment[261] therefore, $\mu_{101} = (0.698 \pm 0.024)\mu_N$ is obtained.

**B. Antiferromagnetic Materials.**—Dynamic nuclear polarization and nuclear relaxation experiments[262] in the nearly one-dimensional Heisenberg system Tanol (2,2,6,6-tetramethyl-4-piperidinol-1-oxyl) have enabled values to be obtained for the electron-spin frequency correlation function at both the electronic and nuclear Larmor frequencies. In addition, it is found that the inter-chain couplings in directions perpendicular to the chain axis are about two orders of magnitude weaker than the intra-chain exchange. Another quasi one-dimensional exchange-coupled system is $Me_4NMnCl_3$. The temperature and frequency dependence of its proton spin–lattice relaxation rate has been investigated in the paramagnetic state[263] in order to study its electron-spin dynamics. Spin diffusion in the paramagnetic state of a second antiferromagnet, $RbMnF_3$, has been discussed in a letter by Tucker,[264] who compares his own previous calculations with experimental results. The planar antiferromagnet $K_2MnF_4$ was investigated *via* its $^{19}F$ nuclear resonance bandwidth by Bucci and Guidi.[265] By making use of the two types of fluorine nuclei (axial and planar) they were able to evaluate the influence of the interaction anisotropy on the magnetic fluctuations.

Antiferromagnetic alloys are not well represented in this year's literature; however, Kohara and Asayama[266] have observed the $^{55}Mn$ resonance in a number of α-Mn alloys. They report the internal field resonance values at various crystalline sites and in various concentrations of the alloys α-MnFe and α-MnCr.

256 E. F. Mendis, *J. Phys.* (*F*), 1974, **4**, L59.
257 B. Lemius, *J. Phys. Soc. Japan*, 1973, **35**, 1785.
258 I. S. Mackenzie, M. A. H. McCausland, and A. R. Wagg. *J. Phys.* (*F*), 1974, **4**, 315.
259 A. Oppelt and K. H. J. Bushchow, *J. Phys.* (*F*), 1973, **3**, L212.
260 J. J. Murphy, T. J. Burch, and J. I. Budnick, *J. Phys. Soc. Japan*, 1974, **36**, 634.
261 O. C. Kistner, *Phys. Rev.*, 1966, **144**, 1022.
262 J. P. Boucher, F. Ferrieu, and M. Nechtshein, *Phys. Rev.* (*B*) 1974, **9**, 3871.
263 D. Hone, C. Scherer, and F. Borsa, *Phys. Rev.* (*B*), 1974, **9**, 965.
264 J. W. Tucker, *J. Phys.* (*C*), 1974, **7**, L16.
265 C. Bucci and G. Guidi, *Phys. Rev.* (*B*), 1974, **9**, 3053.
266 T. Kohara and K. Asayama, *J. Phys. Soc. Japan*, 1974, **34**, 1105.

**C. Ferrimagnetic Materials.**—The zero-field spectrum of $^{167}$Er in erbium iron garnet shows the existence of two types of rare-earth site.[267] Hyperfine fields of 574 T and 287 T are found, corresponding to the two sites, together with respective quadrupole coupling constants of 17 MHz and 1 MHz. In addition, the spin–lattice relaxation rates were measured for both sites and interpreted in terms of relaxation of the rare-earth ion. In the manganese ferrites,[268] Davis and Searle have concentrated on the spin–spin relaxation of $^{55}$Mn and found that the Suhl–Nakamura interaction is the predominant mechanism. By means of this interaction they are also able to explain the observed formation of multiple echoes.

## 6 New Apparatus and Techniques

Discussion will be brief, mostly in tabular form, and will be restricted to developments relevant to studies reviewed in this chapter, namely solids and inhomogeneous systems. Some developments in technique and analysis are discussed in other sections, where the system studied is deemed to be of great interest.

**A. Techniques.**—Janzen[269—271] produced three papers concerned with the ARP technique. In the first he considered the ARP signal shape and passage conditions, showing that the inequality $dB/dt \ll \gamma B_1^2$ should be greater for solids than for liquids, and that the effect of $T_{1\rho}$ should also be considered. In the second paper he examined the use of the dispersion signal decay to determine $T_1$ values. The use of the maximum absorption signal can lead to spurious $T_1$ values because of non-normal saturation. Tests were made using potassium caproate and lithium stearate. In the third paper the ARP technique was applied to $CaF_2$. Stepisnik and Slak[272] described a method of applying saturation to determine dipolar spin–lattice relaxation, using the Provotorov equations to determine the thermal contact between Zeeman and dipolar subsystems. The technique is useful for long relaxation times when $T_{1D}/T_1$ is not too small; it was tested on $NaH_3(SeO_3)_2$ at 192 K. Demco *et al.*[273,274] described a single-scan method of determining $T_1$ using a 'flip-flop' sequence $(90_x–\tau^1–90_{-x}–\tau–)_n$, with $\tau^1 < T_2$ and $T_2{}^*$, and $\tau \leqslant T_1$. Murday[275] described a method of measuring field gradients by the effect on the FID, considering errors due to non-orthogonality of axes, baseline errors, and being off-resonance.

267 R. L. Streever and P. J. Caplan, *Phys. Rev.* (*B*), 1973, **8**, 3138.
268 J. H. Davis and C. W. Searle, *Phys. Rev.* (*B*), 1974, **9**, 323.
269 W. R. Janzen, *J. Magn. Resonance*, 1973, **12**, 71.
270 W. R. Janzen, *J. Magn. Resonance*, 1973, **12**, 276.
271 W. R. Janzen, *J. Magn. Resonance*, 1973, **12**, 307.
272 J. Stepisnik and J. Slak, *J. Magn. Resonance*, 1973, **12**, 148.
273 D. E. Demco, V. Simplaceanu, and I. Ursu, *J. Magn. Resonance*, 1974, **15**, 166.
274 D. E. Demco and V. Simplaceanu, *Rev. Roumaine Phys.*, 1973, **18**, 623.
275 J. S. Murday, *J. Magn. Resonance*, 1973, **10**, 111.

**B. Analysis.**—Alexander *et al.*[276] considered the effects of molecular jumps in an ordered environment, coupled with rotational diffusion of that environment. For systems that are sufficiently symmetric, central lines will appear in powder spectra even at slow jump rates. Schoenberger *et al.*[277] demonstrated experimentally and theoretically that non-random magnetic field inhomogeneities can produce bandshapes similar to those due to chemical-shift or Knight-shift anisotropy. Hughes[278] considered the errors in second moments arising from non-observation of the wings of Gaussian and Lorentzian curves. Moore and Yalcin[279] discussed the experimental determination of relaxation rates in the presence of noise, whilst Wardell and McBrierty,[280] in a paper aimed at the analysis of FID's from polymer samples, considered the confidence limits for resolution of two and three components, selecting polymers to give specified ratios of $T_2$ values. They used a Weibull function $\exp[-(t/T_2)^E]$, where $E = 1$ for a Lorentzian curve and 2 for a Gaussian.

**C. Sources of Error.**—Bull[281] experimentally and numerically considered errors in $T_2$ determinations arising from inhomogeneities in the static and r.f. fields. Even quite small r.f. inhomogeneities produce significant error, particularly when exchange effects are present. Assink[282] considered the effect of r.f. field inhomogeneities on $T_1$ determination, advocating the use of a $180_y$–$\tau$–$90_{-y}$ sequence. Hughes and Lindblom[283] considered baseline errors arising from the use of CPMG pulse sequences due to phase transients associated with the 180° pulses. The conclusions were similar to those of Vold *et al.*[284] Pines and Ellett[285] warned against the use of the assumption that $B_{1(\mathrm{eff})} = B_1\tau_p/t_p$ in multiple-pulse and time-sharing experiments, on the basis that this is equivalent to discarding the sidebands in a Fourier transform. Hiltbrand *et al.*[286,287] considered the 'displacement' of the Larmor frequency as a result of a non-resonant r.f. field.

**D. Apparatus.**—New apparatus is concisely summarized in Table 4.[288—308]

276 S. Alexander, A. Baram, and Z. Luz, *Mol. Phys.*, 1974, **27**, 441.
277 J. Schoenberger, R. B. Creel, B. K. Lunde, and R. G. Barnes, *J. Magn. Resonance*, 1973, **11**, 20.
278 D. G. Hughes, *J. Magn. Resonance*, 1973, **11**, 83.
279 W. S. Moore and T. Yalcin, *J. Magn. Resonance*, 1973, **11**, 50.
280 G. E. Wardell and V. J. McBrierty, *Proc. Roy. Irish. Acad.*, 1973, **73A**, 63.
281 T. E. Bull, *Rev. Sci. Instr.*, 1974, **45**, 232.
282 R. A. Assink, *J. Magn. Resonance*, 1973, **12**, 88.
283 D. G. Hughes and G. Lindblom, *J. Magn. Resonance*, 1974, **13**, 142.
284 R. L. Vold, R. R. Vold, and H. E. Simon, *J. Magn. Resonance*, 1973, **11**, 283.
285 A. Pines and J. D. Ellett jun., *J. Amer. Chem. Soc.*, 1973, **95**, 4437.
286 E. Hiltbrand, B. Borcard, R. Sechehaye, C. Piron, and G. J. Bene, *Compt. rend.*, 1974, **278B**, 243.
287 E. Hiltbrand, R. Sechehaye, B. Borcard, and G. J. Bene, *Compt. rend.*, 1973, **277B**, 531.
288 D. R. Torgeson, *Rev. Sci. Instr.*, 1973, **44**, 982.
289 M. Rakos, O. Bartko, and M. Jelsinia, *Czech J. Phys.*, 1973, **23**, 394.
290 A. J. Nanassy, *Rev. Sci. Instr.*, 1973, **44**, 1633.
291 T. K. Davies and B. C. Manning, *Rev. Sci. Instr.*, 1973, **44**, 1513.

**Table 4**

| *Authors* | *Comments* | *Ref.* |
|---|---|---|
| *CW spectrometers* | | |
| Torgeson | Versatile d.c. amplifier to interface signal averager to field or frequency scanner | 288 |
| Rakos *et al.* | $M_2$ determination using analogue circuit and optical curve follower | 289 |
| Nanassy | Hermetic sealing of r.f. probe | 290 |
| Davies and Manning | Pre-amplifier using cascaded dual FET's (aimed at FT spectrometers, but valid for CW) | 291 |
| Hawk *et al.* | Broad-band time-sharing spectrometer for field stabilization between 0.4 and 2.3 T. | 292 |
| Richards *et al.* | CW millidegree thermometer based on Robinson circuit | 293 |
| *Pulse Spectrometers* | | |
| Aalto *et al.* | Pulsed n.m.r. spectrometer suitable for Pt spin thermometry and $T_1$ determinations | 294 |
| Hensel and Smith | Pulsed n.m.r. thermometer for use below 0.1 K | 295 |
| Narath and Barham | High-power 2—200 MHz pulsed spectrometer | 296 |
| Khoi and Veillet | U.h.f. spin-echo spectrometer, 170—1700 MHz | 297 |
| Dorn *et al.* | Dual r.f. gate | 298 |
| Taylor *et al.* | Flexible modular pulse programmer | 299 |
| Grannell *et al.* | Low-noise FET analogue switch μs to s with tuned receiver coil and signal pre-amplifier | 300 |
| El-Hanany | Air-cored-transformer-coupled matching stage. Passive quenching device to give good recovery | 301 |
| Clark and McNeil | Optimization criteria for series resonant circuit, 1—100 MHz. Active damping circuit | 302 |
| Lowe and Engelsberg | Fast-recovery probe using delay line | 303 |
| Vanderhart | Electrical feed-through capacitance for use in high-pressure probes | 304 |
| Ladjouze and Benoit | Discussion of filling factor for two-coil probe | 305 |
| Gerig *et al.* | Automatic data collection schemes: interfaced computer | 306 |
| Wright and Rogers | Interfacing of minicomputer for two-pulse experiments | 307 |
| *Magnets* | | |
| Dworschak *et al.* | Capacitative discharge unit for production of pulsed magnetic field of 44 T for 1 ms | 308 |

[292] R. M. Hawk, R. R. Sharp, and J. W. Tolan, *Rev. Sci. Instr.*, 1974, **45,** 96.
[293] M. G. Richards, P. S. Tofts, and P. R. Turner, *Cryogenics*, 1973, **13,** 182.
[294] M. I. Aalto, H. K. Collan, R. G. Gylling, and K. O. Nores, *Rev. Sci. Instr.*, 1973, **44,** 1075.
[295] P. Hensel and D. T. Smith, *Cryogenics*, 1974, 121.
[296] A. Narath and D. C. Barham, *Rev. Sci. Instr.*, 1974, **45,** 100.
[297] La Dhang Khoi and P. Veillet, *Rev. Sci. Instr.*, 1974, **45,** 759.
[298] H. C. Dorn, S. J. Uurtamo, and G. E. Maciel, *Rev. Sci. Instr.*, 1973, **44,** 128.
[299] D. G. Taylor, S. Booth, and P. S. Allen, *J. Phys.* (*E*), 1974, **7,** 105.
[300] P. K. Grannell, M. J. Orchard, P. Mansfield, A. N. Garroway, and D. C. Stalker, *J. Phys.* (*E*), 1973, **6,** 1201.
[301] U. El-Hanany, *Rev. Sci. Instr.*, 1973, **44,** 1069.
[302] W. G. Clark and J. A. McNeil, *Rev. Sci. Instr.*, 1973, **44,** 844.
[303] I. J. Lowe and M. Engelsberg, *Rev. Sci. Instr.*, 1974, **45,** 631.
[304] D. L. Vanderhart, *Rev. Sci. Instr.*, 1974, **45,** 111.
[305] H. Ladjouze and H. Benoit, *Rev. Sci. Instr.*, 1973, **44,** 1780.

## 7 Surface Phenomena

**A. Introduction.**—As this subject was not reviewed in Volume 3, the period covered here will be the two years to June 1974. N.m.r. investigations of surface phenomena are in general concerned with structures and mobilities of molecules adjacent to and probably 'bound' to surfaces. Similar interactions occur and similar mechanisms operate in the binding of small molecules to macromolecules. Hence this section is very much related to its equivalent section in the Macromolecules chapter, and for a more complete coverage the two sections should be read in parallel. To avoid duplication of material this section will not describe adsorption onto macromolecules, exchange resins, or the properties of solvent molecules in macromolecular solutions. Coverage will be restricted to zeolites and adsorption onto surfaces such as silica, titania, alumina, *etc.*

A number of reviews have been published during this period. Resing[309] has reviewed n.m.r. studies of water in heterogeneous systems, selecting zeolites for most of his examples. In a related paper with Murday[310] the present state of the art of relaxation in zeolite systems was discussed. The prime interest was the determination of the correlation times for guest molecules from measurements of relaxation times and diffusion coefficients. Pfeifer[311] has published a more general review entitled 'N.M.R. and Relaxation of Molecules Adsorbed on Solids'. Boddenberg[312] has published a brief introductory article describing the use of n.m.r. to study surface mobility, taking benzene on silica and methylated silica as an example, and finding that some 5% of a monolayer is very tightly bound. Derouane *et al.*[313] reviewed the use of n.m.r. in investigations of adsorption and heterogeneous catalysis, covering 236 papers in the period to 1971 on the adsorption of water, ammonia, formic and acetic acids, alcohols, hydrocarbons, and various miscellaneous molecules, onto silica gel, alumina, and other simple oxides, hydrides, and various mixed oxides. Woessner and Snowden[314] discussed $^1H$ and $^2H$ n.m.r. in a wide range of systems, mostly of water–polymer type, making comparison with clay–water systems. Using $T_2$ and $T_3$ measurements (where $T_3$ characterizes the signal echoes following a $90°–\tau–90°$ pulse sequence), they inferred that the tetrahedral model was more appropriate than rotational diffusion. There was some discussion of the use of the $T_3$ measurement (based on the 'solid echo') in such systems. Rakvin and Herak[315] used a method of moments for the calculation of the relationship between linewidth and surface concentration.

[306] J. T. Gerig, G. B. Matson, and A. D. Stock, *J. Magn. Resonance*, 1973, **12**, 48.
[307] D. A. Wright and M. T. Rogers, *Rev. Sci. Instr.*, 1973, **44**, 1189.
[308] G. Dworschak, F. Haberey, P. Hildebrand, E. Kneller, and D. Schreiber, *Rev. Sci. Instr.*, 1974, **45**, 243.
[309] H. A. Resing, *Adv. Mol. Relaxation Processes*, 1972, **3**, 199.
[310] H. A. Resing and J. S. Murday, *Adv. in Chem. Series*, 1973, **121**, 414.
[311] H. Pfeifer, *N.M.R. Basic Principles and Progress*, 1972, **7**, 53.
[312] B. Boddenberg, *Ind. chim. belge*, 1973, **38**, 368.
[313] G. Derouane, J. Fraissard, J. J. Fripiat, and W. E. E. Stone, *Catalysis. Rev.*, 1972, **7**, 121.
[314] D. E. Woessner and B. S. Snowden jun., *Ann. New York Acad. Sci.*, 1973, **204**, 113.
[315] B. Rakvin and N. J. Herak, *J. Magn. Resonance*, 1974, **13**, 94.

Assuming a random arrangement and dipolar interactions, and neglecting exchange interactions, Gaussian and truncated Lorentzian lines were calculated for cubic, hexagonal, and two-dimensional lattices. Derouane[316] provided a brief analysis of OH group motion at the surface of oxides, and of methods of detection of activated complexes (transient species) including i.r., n.m.r., e.s.r., and dielectric studies. Resing[317] discussed methods of determining the kinetics of surface reactions from n.m.r. relaxation times, demonstrating that the dissociation of the benzene–surface complex on charcoal is unimolecular, and considering the reactions of water with surface hydroxyls in keratin and zeolite X.

**B. Adsorption on Graphite and Charcoal.**—Riehl and Fisher[318] reported the spin–lattice and spin–spin relaxation of hydrogen adsorbed on graphite for between 0.03 and 0.33 statistical monolayer coverage, at between 12 and 28 K. The observation of a temperature-independent multi-component spin–lattice relaxation was attributed to adsorption at different sites. Relaxation times increased with coverage in a manner analogous to the density dependence of a gas. The $T_2$ temperature dependence gave an activation energy of 418 kJ mol$^{-1}$. A slow decrease in signal height observed at temperatures below 30 K represents a catalysed ortho–para conversion process. The different temperature dependences of $T_1$ and $T_2$ indicate the occurrence of highly anisotropic molecular reorientation. In an earlier paper, Riehl and Koch[319] recorded the single-component $T_1$ and $T_2$ relaxation of methane on graphitized carbon black at between 0.02 and 0.9 monolayer coverages, and between 4 and 100 K. The methane behaved as a 'gas' phase at high and a 'solid' phase at low temperatures. The energy barrier for escape from the surface was found to be 12.6 kJ mol$^{-1}$ and for translation along the surface 290–330 J mol$^{-1}$. The lifetime at 100 K was 1 $\mu$s. $T_1$ exceeded $T_2$, again implying an oriented gas. A detailed discussion of relaxation was presented; spin–rotation was considered and rejected, and dipolar relaxation favoured. At high coverages two $T_1$ minima were observed, the high-temperature one being due to the intermolecular interaction. Dubault[320] used dynamic nuclear polarization to investigate interactions between adsorbed fluids and paramagnetic centres on charcoal, studying $C_6H_6$, $C_6F_6$, perfluorobut-2-ene, octafluorocyclobutane, and $CCl_2FCF_2Cl$.

**C. Adsorption on Silica.**—Silica remains a popular adsorbent. Morariu and Mills[321—324] have published a series of papers concerned with various aspects

316 E. G. Derouane, *Bull. Classe Sci.*, 1972, **5**, 377.
317 H. A. Resing, *J. Phys. Chem.*, 1974, **78**, 1279.
318 J. W. Riehl and C. J. Fisher, *J. Chem. Phys.*, 1973, **59**, 4336.
319 J. W. Riehl and K. Koch, *J. Chem. Phys.*, 1972, **57**, 2199.
320 M. Dubault, A. P. Legrand, J. Uebersfeld, J. M. Perrot, M. Roques, and M. Bastick, *Carbon*, 1973, **11**, 363.
321 V. V. Morariu and R. Mills, *Z. phys. Chem.* (*Frankfurt*), 1972, **78**, 298.
322 V. V. Morariu and R. Mills, *Z. phys. Chem* (*Frankfurt*), 1972, **79**, 1.
323 V. V. Morariu and R. Mills, *Z. phys. Chem.* (*Frankfurt*), 1973, **85**, 38.
324 V. V. Morariu and R. Mills, *Z. phys. Chem.* (*Frankfurt*), 1973, **83**, 41.

of water adsorption, investigating the $^{1}H$ CW linewidths and diffusion coefficients on silicas of different acidity as a function of coverage, and finding evidence of exchange effects. In the later papers they examined the influence of various cations, finding that $^{1}H$ exchange was still significant on treatment with $K^{+}$ and $Ba^{2+}$, but that $Mg^{2+}$ had an inhibiting effect. In the most recent work they examined the exchange effect in more detail, estimating the adsorption-site binding energy and the enthalpy of diffusion. $D_2O$ was preferentially adsorbed onto high-energy sites. Pearson and Derbyshire[325] investigated the dependence of linewidth on temperature and coverage in a number of silicas of different porosity and pore dimensions, with particular interest in the behaviour on freezing and the amount of non-freezable water. In a related study Belfort *et al.*[326] examined the relaxation behaviour of water on various porous (Vycor) glasses as a function of coverage between $-91$ and $+92$ °C. Freezing was only observed in the glasses with largest pore diameter. Boddenberg *et al.*[327] determined the surface diffusion and relaxation times of benzene adsorbed on modified silica surfaces as a function of impurity content, pore texture, pretreatment, *etc.* Coverage plots allowed $T_1$ values to be obtained for separate layers. Methylation increased $T_1$ by a greater factor than $T_2$, and at low but not high coverages increased the diffusion coefficient. In the hydroxylated sample, $D$ was maximum at one monolayer coverage. The paper was an extension of earlier work.[328] Gay[329] reported the $^{13}C$ n.m.r. of $CO_2$, $C_2H_4$, $C_2H_4Cl_2$, $C_2H_5OH$, $CH_3CHO$,$(CH_3)_2O$, $(C_2H_5)_2O$, cyclo-$C_6H_{12}$, and *cis*- and *trans*-$C_4H_8$, adsorbed on a high-area silica gel at 0.2 to 1 monolayer coverage. The largest change in chemical shift occurred for the carbonyl peak in acetone. The change in shielding, $\Delta\sigma$, was a function of the outgassing treatment, $\Delta\sigma$ being reduced by the severity of the treatment, presumably owing to a reduction in surface OH concentration. Miyamoto and Cantow[330] studied the adsorption of poly(methyl methacrylate), PMMA, on silica gel from $CHCl_3$ solution, with particular interest in conformational changes and the effect of polar additives. The disappearance of the spectrum showed that the lengths of free loops between binding points were small. Methanol competed successfully with PMMA for the silica binding sites.

In a paper of more fundamental interest, Glasel and Lee[331] used $^{2}H$ relaxation to investigate the occurrence of large magnetic field inhomogeneities in a system comprising coated glass spheres packed in $D_2O$. Surprisingly, relaxation was a function only of surface area and not of surface nature. $T_1$ and $T_2$ were single-component exponential functions, and the measured $T_2$ values exhibited no pulse-spacing dependence. Spin–lattice and spin–spin relaxation

325 R. T. Pearson and W. Derbyshire, *J. Colloid Interface Sci.*, 1973, **46**, 232.
326 G. Belfort, J. Scherfig, and D. O. Seevers, *J. Colloid Interface Sci.*, 1974, **47**, 106.
327 B. Boddenberg, R. Haul, and G. Oppermann, *Adv. Mol. Relaxation Processes*, 1972, **3**, 61.
328 B. Boddenberg, R. Haul, and G. Oppermann, *J. Colloid Interface Sci.*, 1972, **38**, 210.
329 I. D. Gay, *J. Phys. Chem.*, 1974, **78**, 38.
330 T. Miyamoto and H. J. Cantow, *Makromol. Chem.*, 1972, **162**, 43.
331 J. A. Glasel and K. H. Lee, *J. Amer. Chem. Soc.*, 1974, **96**, 970.

rates exhibited different dependences upon the surface area, *i.e.* the radii of the glass spheres. On assuming quadrupolar relaxation and a single 'bound' phase with a single correlation time, a value of that correlation time could be extracted using the expressions of Carver and Richards,[332] and hence $T_b$, the relaxation time of the bound phase (either $T_1$ or $T_2$), could be obtained. Using these $T_b$ values, $P_b$, the population of the bound phase, was found to be negligible. Hence to explain the relaxation behaviour, large field gradients were invoked, and it was concluded that water molecules could not diffuse sufficiently rapidly to produce an averaging on an n.m.r. timescale, a conclusion at variance with that of Packer,[333] who considered skeletal muscle systems. A weak point in the analysis is the assumption that the 'bound' phase is single and may be represented by one correlation time.

**D. Adsorption on Aluminas and other Surfaces.**—Palmer[334] studied the spin–lattice relaxation and equilibrium of $Li^+$ ions in LiCl solutions in contact with alumina, thereby extending the treatment of Headley *et al.*[335] Magnetic dipolar and electric quadrupolar contributions to relaxation were found to be approximately equal. The data were analysed using a two-phase model, and the population and relaxation rate of the 'bound' phase, $P_b$ and $1/T_b$, were examined as a function of surface coverage. There was some discussion of the mechanism for surface relaxation. Kiviat and Petrakis[336] investigated the surface acidity of aluminas modified by transition metals, using absorbed pyridine, which is both a Lewis and a Brønsted base. Cobalt- and nickel-impregnated aluminas contained only Lewis-acid surface sites whereas those impregnated with Mo contained additional Brønsted-acid sites. Kline *et al.*[337] considered the effect of the penetration of water into non-stoicheiometric $\beta$-alumina $Na_2O,11Al_2O_3$ on the $^{23}Na$ quadrupole interaction by observing the $+\frac{1}{2} \leftrightarrow -\frac{1}{2}$ transition from $-170$ to $+250$ °C. Several $^{27}Al$ quadrupole interactions were observed which did not exhibit a dependence upon water content. Reference was made to similar observations by Gabuda *et al.*[338,339] and Genser[340] on $^{23}Na$ and $^{27}Al$ NQCC's in zeolites.

Headley[341] examined $^1H$ and $^7Li$ relaxation in Appalachian petroleum reservoir rocks containing a LiCl solution. $^1H$ n.m.r. was found to be a useful indicator of rock permeability. Variations in relaxation rate between samples are possibly due to paramagnetic impurities. Relaxation of both $^7Li$ and $^1H$

332 J. P. Carver and R. E. Richards, *J. Magn. Resonance*, 1972, **6**, 89.
333 K. J. Packer, *J. Magn. Resonance*, 1973, **9**, 438.
334 G. M. Palmer, *J. Magn. Resonance*, 1973, **11**, 388.
335 L. C. Headley, W. E. Wallace, jun., and P. Waldstein, *J. Magn. Resonance*, 1971, **5**, 168.
336 F. E. Kiviat and L. Petrakis, *J. Phys. Chem.*, 1973, **77**, 1232.
337 D. Kline, H. S. Story, and W. L. Roth, *J. Chem. Phys.*, 1972, **57**, 5180.
338 S. P. Gabuda, I. V. Grinchenko, and V. N. Scherbakov, *Soviet Phys. Solid State*, 1969, **11**, 663.
339 V. N. Scherbakov, S. P. Gabuda, and S. A. Seryshev, *Soviet Phys. Solid State*, 1971, **12**, 2477.
340 E. E. Genser, *J. Chem. Phys.*, 1971, **54**, 4612.
341 L. C. Headley, *Nature Phys. Sci.*, 1973, **242**, 87.

was non-exponential, but in a porcelain sample with uniform pore size, single-component relaxation was observed. Using a two-site model, reasonable agreement was obtained for $^7Li$ and $^1H$ estimations of the bound fraction. In a related paper, Headley[342] examined $^{23}Na$ relaxation in a porous material containing NaCl solution. Relaxation was single exponential, the signal amplitude was proportional to pore volume, but $T_2$ showed a pulse-spacing dependence. The dependence of $T_1$ upon surface area was normal but that of $T_2$ anomalous. Hecht and Geissler[343] examined the $^1H$ n.m.r. of absorbed water in a synthetic montmorillonite, $Mg_{5.1}Na_{0.7}Si_8F_4O_{20}Na_{1.1}$. Below 200 K, $M_2$ was obtained as a function of angle, and $T_1$ determined at different frequencies between 2 and 112 K. Evidence was obtained for the occurrence of $H_3O^+$ and $OH^-$. Above 200 K, molecular rotation occurred, and two species of water were observed, giving rise to a narrow central line and a dipolar split doublet, both temperature-dependent. The exchange rate between $H_2O$ and $D_2O$ gave a diffusion coefficient of $10^{-14}$ cm$^2$ s$^{-1}$ at 300 K. In a later paper the same authors[344] prepared samples comprising one and two layers of water. The narrow central line was attributed to random jumps in an ice-like structure. The sample suffered an ageing process. $T_1$ measurements on the one-layer sample were not repeatable one year later. Spin–spin relaxation was interpreted in a manner similar to that used by Goldman and Shen[345] on $LaF_3$, involving exchange between two-spin systems, one diffusing quickly and the other slowly. Whitney and Gay[346] reported the 100 MHz $^1H$ n.m.r. of $C_2H_6$ and $C_4H_{10}$ adsorbed on a sample of ZnO having low surface area at a low coverage. Linewidths were 30—120 Hz, and there was no change in chemical shift relative to the free molecules. In olefins the lines were wider and shifted downfield by 0.6 and 0.8 p.p.m.

**E. Zeolites.**—Of the systems examined in this section, zeolites are the most intensively studied. Freude *et al.*[347] used CW and relaxation techniques to investigate the arrangement and thermal mobility of structural hydroxy-groups in decationized zeolites, finding a H—Al length of 0.23 nm and two types of hydroxyls, with different correlation times and thermal stabilities. In a later paper[348] the same group examined hydroxy-group mobility on decationized zeolites of types A, X, and Y and in ion-exchanged zeolites, below the pretreatment temperature, using bandwidths and solid echoes. The jump frequency was increased by a factor of 60 on pyridine adsorption, considerably greater than that observed for toluene. Other relevant papers are those of Fenzke,[349]

[342] L. C. Headley, *J. Appl. Phys.*, 1973, **44**, 3118.
[343] A. M. Hecht and E. Geissler, *Bull. Soc. Franc. Mineral Crist.*, 1972, **95**, 291.
[344] A. M. Hecht and E. Geissler, *J. Colloid Interface Sci.*, 1973, **44**, 1.
[345] M. Goldman and L. Shen, *Phys. Rev.*, 1966, **144**, 321.
[346] A. G. Whitney and I. D. Gay, *J. Catalysis*, 1972, **25**, 176.
[347] D. Freude, D. Muller, and H. Schmiedel, *Z. phys. Chem.* (*Leipzig*), 1972, **250**, 345.
[348] D. Freude, W. Oehme, H. Schmiedel, and B. Staudte, *J. Catalysis*, 1974, **32**, 137.
[349] D. Fenzke, D. Freude, D. Muller, and H. Schmiedel, *Phys. Status Solidi* (*B*), 1972, **50**, 209.

Mestdagh,[350] Freude and Muller,[351] and Freude *et al.*[352] Karger *et al.*[353] measured the diffusion of water in NaY zeolites, extending the work of Riedel.[354] The timescale of the experiment was such that water molecules diffused over several crystallites. The observed value of $D$ was greater than that of pure water for filling factors, $\theta$, just less than 1. The enthalpy of diffusion, also a function of $\theta$, was less than the desorption energy. Geschke *et al.*[355] recorded the $T_1^H$ values of water in NaPtY zeolite to obtain information on the distribution of Pt (1% by weight) as a function of the thermal treatment. Renner and Karger[356] observed a non-equilibrium absorption process in a degassed zeolite, the 5A zeolite–butane system. Another relevant paper is that of Gutsze *et al.*[357] Michel and Robiger[358] reported the $T_1^H$ and $T_2^H$ values of butene in synthetic zeolites of A and Y types as a function of temperature and pore-filling factor. In A zeolites a high-temperature $T_1$ minimum was obtained at 150 °C whereas in the Y zeolites the minimum $T_1$ (at 16 MHz) occurred between $-30$ and $-50$ °C. Hauser and Hoch[359] recorded $^1H$ and $^7Li$ $T_1$ values in zeolitic germanates over wide temperature ranges and water contents. The presence of intracrystalline water did not affect cation diffusion. A $T_1^H$ minimum observed in the ammonium germanate sample in both hydrated and dehydrated samples was attributed to $NH_4$ group rotation. Weaver[360] measured $^3He$ $T_1$ and $T_2$ relaxation times for helium adsorbed on two types of zeolites of pore diameters 0.3 and 1 nm as a function of pressure between 1.2 and 4 K. The helium was more tightly bound in the 1 nm sample, where the lineshape was Lorentzian, as opposed to the Gaussian shape observed in the 0.3 nm zeolite. Geschke[361] recorded the $^{13}C$ n.m.r. of benzene absorbed in a NaY zeolite with $\theta = 0.92$. The spectrum shift of 45 p.p.m. was explained using Karplus–Pople theory.[362] Deininger and Michel[363] reported the $^{13}C$ n.m.r. of toluenes and n-butenes adsorbed on NaX and NaY zeolites, again reporting shifts relative to the pure liquid. In the same sequence of papers Michel[364] also reported the natural-abundance $^{13}C$ spectra of the hydrocarbons $CH_2{=}CHCH_2CH_3$, $CH_3CH{=}CHCH_3$, and $CH_2{=}CHCH_2CH_2CH_2CH_3$, recording chemical shifts as a function of pore-filling

[350] M. Mestdagh, W. E. Stone, and J. J. Fripiat, *J. Phys. Chem.*, 1972, **76**, 1220.
[351] D. Freude and D. Muller, *Exp. Tech. Phys.*, 1972, **20**, 379.
[352] D. Freude, A. A. Pripylov, and H. Schmiedel, *Phys. Status Solidi*, (*B*), 1973, **57**, K73.
[353] J. Karger, H. Pfeifer, E. Riedel, and H. Winkler, *J. Colloid Interface Sci.*, 1973, **44**, 187.
[354] E. Riedel, *Z. phys. Chem.* (*Leipzig*), 1973, **252**, 161.
[355] D. Geschke, H. Winkler, and G. Wendt, *Z. phys. Chem.* (*Leipzig*), 1973, **252**, 235.
[356] E. Renner and J. Karger, *Z. phys. Chem.* (*Leipzig*), 1973, **253**, 12.
[357] A. Gutsze, D. Deininger, H. Pfeifer, G. Finger, W. Schirmer, and H. Stach, *Z. phys. Chem.* (*Leipzig*), 1972, **249**, 383.
[358] D. Michel and V. Robiger, *Z. phys. Chem.* (*Leipzig*), 1973, **252**, 312.
[359] E. Hauser and M. J. R. Hoch, *J. Magn. Resonance*, 1973, **10**, 211.
[360] H. T. Weaver, *J. Phys. and Chem. Solids*, 1973, **34**, 421.
[361] D. Geschke, *Z. phys. Chem.* (*Leipzig*), 1972, **249**, 125.
[362] M. Karplus and J. A. Pople, *J. Chem. Phys.*, 1963, **38**, 2003.
[363] D. Deininger and D. Michel, *Wiss. Z. Karl Marx Univ. Leipzig* (*Math. Naturwiss. R.*) 1973, **22**, H6.
[364] D. Michel, *Z. phys. chem.* (*Leipzig*), 1972, **252**, 263.

factor and the nature of the cations. Michel has reviewed the area,[365] summarizing $^{1}H$ and $^{13}C$ n.m.r. spectra and $T_1$ values of various butene isomers sorbed on NaX and NaY zeolites. The $^{13}C$ spectra were the much the more informative, yielding resolved spectra, whereas the $^{1}H$ spectrum typically consisted of two broad lines. The NOE was reported for $=CH_2$ and $—CH_2—$ groups.

Lechert,[366,367] using a simplified point charge-multipole model, calculated e.f.g. tensors at sodium sites in a zeolite Linde 13x, whilst Hennig and Lechert obtained[368] two $^{23}Na$ quadrupole interactions in a dehydrated powder of a 4A zeolite; NQCC's were 6.0 and 2.8—3.0 MHz, with $\eta = 0$ and 0.74. Freude *et al.*[369] reported a CW and relaxation study of Tl in a zeolite. The chemical shift was investigated as a function of temperature and water content.

365 D. Michel, *Surface Sci.*, 1974, **42,** 453.
366 H. Lechert, *Ber. Bunsengesellschaft. phys. Chem.*, 1973, **77,** 697.
367 H. Lechert, *Adv. in Chem. Series*, 1973, **121,** 74.
368 H. J. Hennig and H. Lechert, *Z. phys. Chem.* (*Frankfurt*), 1973, **87,** 104.
369 D. Freude, A. Hauser, H. Pankau. and H. Schmiedel, *Z. phys. Chem.* (*Leipzig*), 1972, **251,** 13

# 9

# Medium Effects

BY M. I. FOREMAN

## 1 Introduction

The best approach to any discussion of medium effects on n.m.r. chemical shifts is still *via* the equation (1) of Buckingham, Schaefer, and Schneider,[1] whereby the contribution of the medium to the nuclear shielding is expressed by $\Delta\sigma$,

$$\Delta\sigma = \sigma_b + \sigma_a + \sigma_w + \sigma_E + \sigma_c \qquad (1)$$

where $\sigma_b$ is the bulk susceptibility effect, $\sigma_a$ takes account of the diamagnetic anisotropy of the solvent molecules, $\sigma_w$ is the van der Waals contribution, and $\sigma_E$ the 'reaction field' term. The final parameter, $\sigma_c$, encompasses the effect of specific interactions between the molecules present in the system.[2]

## 2 Non-specific Medium Effects

One of the major problems in n.m.r. studies of such relatively weak interactions is the difficulty of finding an inert internal reference standard, which in principle removes the $\sigma_b$ term from further consideration. For proton studies in aqueous solution the sodium salt of 5,5-dimethyl-2-silapentanesulphonate (DSS) is often used. Recently, however, it has been shown that this material interacts quite strongly with purines, purine nucleosides, and purine nucleotides.[3] For non-aqueous systems, TMS is the almost universally adopted standard. It is, however, becoming increasingly clear that TMS cannot be used for detailed studies without due care. Jameson and Jamieson[4] have studied the temperature dependence of the proton shift of TMS, along with the $^{19}$F shift of the two common $^{19}$F reference standards, hexafluorobenzene and dibromotetrafluorobenzene. The temperature-independent shift of $^{129}$Xe gas extrapolated to zero density was used as a base reference, and a calibration table is presented by which due correction may be made for the temperature variability of the above three compounds. In addition, Bacon and Maciel[5] have reported a lengthy and detailed discussion of the effect of solvent on the nuclear shieldings of TMS and cyclohexane. For TMS, it has been shown that

[1] A. D. Buckingham, T. Schaefer, and W. G. Schneider, *J. Chem. Phys.*, 1960, **32,** 1227.
[2] T. Schaefer and W. G. Schneider, *J. Chem. Phys.*, 1960, **32,** 1218, 1224.
[3] D. H. Live and S. I. Chan, *Org. Magn. Resonance*, 1973, **5,** 275.
[4] A. K. Jameson and C. K. Jamieson, *J. Amer. Chem. Soc.*, 1973, **95,** 8559.
[5] M. Bacon and G. E. Maciel, *J. Amer. Chem. Soc.*, 1973, **95,** 2413.

the $^{13}C$ resonance is some three times more sensitive to solvent than the proton resonance, whilst the $^{29}Si$ resonance is affected to about the same extent as for $^{1}H$.[6] Generally, the $^{13}C$ shift of cyclohexane is less sensitive to solvent; however, Bacon and Maciel[5] point out that the $^{13}C$ resonance of cyclohexane in hydrocarbon solvents is rather more sensitive to a change in the solvent carbon skeleton than that of TMS, particularly if both aromatic and aliphatic hydrocarbon solvents are being considered. Evidence is also presented which may support the view that TMS has a very slight tendency to complex with aromatic solvents whereas cyclohexane does not. For example, the protons of TMS are subject to a greater reduction in shielding on going from non-aromatic to aromatic solvents than those of cyclohexane. Bacon and Maciel[5] have further treated their data by an extension of the Factor Analysis approach, first applied in this context by Malinowski and Weiner and reviewed in some detail in Volume 2 of this series. As expected for solutes of this type, the solvent shifts induced by the halogenobenzenes and halogenocyclohexanes are dependent on two factors, one of which is the neighbour anisotropy term $\sigma_a$. Attempts to calculate the magnitude of the second factor in terms of van der Waals dispersion forces were not, however, successful. Some minor contributions from a third effect were also discerned, particularly for the solvents hexafluorobenzene and carbon disulphide. This may be due to weak dipole–induced dipole interactions between solvent and solute.

Abraham and co-workers have applied a not dissimilar approach to a study of the dependence of $^{19}F$ shifts on solvent.[7] Measurements on nineteen types of $^{19}F$ resonance in non-polar solvents and the gas phase are presented, including a conversion table which allows fluorine shifts quoted with respect to a variety of common $^{19}F$ standards to be referenced to $CFCl_3$. It is generally held that the $\sigma_w$ term dominates the solvent dependence of $^{19}F$ shifts in non-polar solvents. Abraham *et al.* have accordingly concentrated on this term, which they express as

$$\sigma_w = S^v S^u \tag{2}$$

$S^u$ is a measure of the ease with which the $^{19}F$ resonance of a given solute may be perturbed by non-polar solvents. Using $CF_3CCl_3$ as the standard compound, $S^v$ for any solvent is then the gas-to-solution shift for $CF_3CCl_3$ in that solvent. By a least-squares treatment, the best-fit $S^u$ and $S^v$ values are obtained, which provide good agreement between observed and calculated shifts for all cases except those of the perfluoroalkanes in hydrocarbon solvents. The discrepancy is attributed to a specific solute–solvent interaction. In terms of current understanding of van der Waals interactions, the solvent factor $S^v$ should correlate with some function of the solvent refractive index, although there is a variety of theoretical approaches to the calculation of the $\sigma_w$

[6] M. Bacon, G. E. Maciel, W. K. Musker, and R. Scholl, *J. Amer. Chem. Soc.*, 1971, **93**, 2537.
[7] R. J. Abraham, D. F. Wileman, and G. R. Bedford, *J.C.S. Perkin II*, 1973, 1027.

term.[8] In fact, a plot of $S^v$ *versus* $(\eta^2-1)/(2\eta^2+1)$ gave[9] the anticipated linear correlation; the line did not, however, pass through the origin, or gas-phase point. This effect cannot be due to neglect of the solvent anisotropy term, since the deviations from the gas-phase point are a function of the solute. As with Bacon and Maciel's work,[5] therefore, the conclusion is that current methods for calculation of the $\sigma_w$ term are inadequate. Far more compelling is the correlation between the solute number, $S^u$, and the degree of electronic screening of the nucleus, measured as $\Sigma_i Z_i/r_i^2$, where $Z_i$ is the number of electrons in the $i$-th atom a distance $r_i$ from the nucleus of interest (Figure 1).

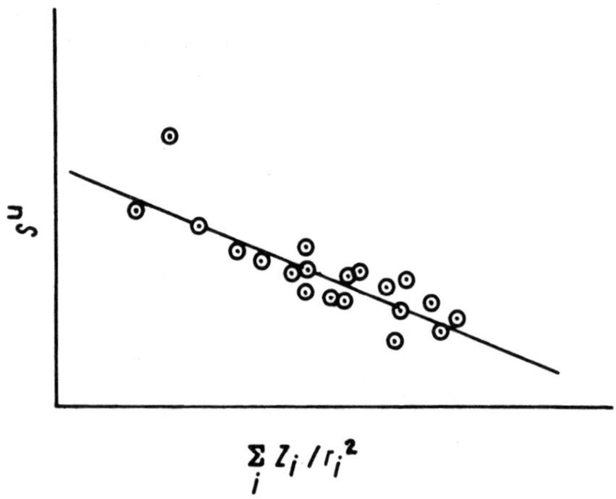


**Figure 1** *Solute numbers* ($S^u$) versus *the electronic screening term* $\Sigma_i Z_i/r_i^2$ *for a series of fluoroalkane solutes*
(Reproduced by permission from *J. Chem. Phys.*, 1960, **32,** 1218)

Subsequently, this approach was extended to include the effects of anisotropic and polar solvents on $^{19}F$ shifts.[10] For the anisotropic solvents, the shifts may be expressed by equation (3). For a series of non-polar solutes, for

$$\Delta\delta_{\text{gas-solution}} = S^u S^v + \sigma_a \qquad (3)$$

which $S^u$ values have previously been determined,[7] a plot of $S^u$ against the chemical shift will therefore be linear, with slope $S^v$ and intercept $\sigma_a$. In this way, $\sigma_a$ for benzene was found to be $-0.25$ p.p.m., and for carbon disulphide $+0.5$ p.p.m.[10] For polar solutes the van der Waals term is again found to be dominant; there is, however, an additional factor which could not be accounted for in terms of the anticipated reaction field effect. The reaction field in this context does not, in fact, appear to be a significant factor. Specifically,

8 V. P. Senthilnathan and S. Singh, *Chem. Phys. Letters*, 1973, **21,** 147.
9 B. B. Howard, B. Linder, and M. T. Emerson, *J. Chem. Phys.*, 1962, **36,** 485.
10 R. J. Abraham and D. F. Wileman, *J.C.S. Perkin II*, 1973, 1521.

whereas the solvent dielectric constant does seem to be related to the observed solvent shifts, the solute dipole moment does not. Abraham *et al.*[10] point out that this is in direct contradiction to current understanding of the effects of polar solvents on $^{19}$F shifts.

Wilbur and Jonas[11] have studied the dependence of $^{19}$F shifts of benzotrifluoride and *p*-chlorobenzotrifluoride on solvent, pressure, and temperature, using modified n.m.r. apparatus capable of moderate resolution (2—4 Hz at 60 MHz) at pressures of up to 5 kbar. As before,[10] the $\sigma_E$ term seems to be of little consequence, and $\sigma_w$ dominates the shift. Since[12] $\sigma_w \propto r^{-6}$, a high-frequency shift occurs as the pressure is increased, due to a net reduction in the intermolecular distances. One other report of pressure-dependent chemical shifts has appeared,[13] in this case for the proton resonance of chloroform in a series of substituted benzenes. The behaviour of the shifts may be seen from Figure 2. For the pure liquid, the $\sigma_E$ term and any molecular self-association

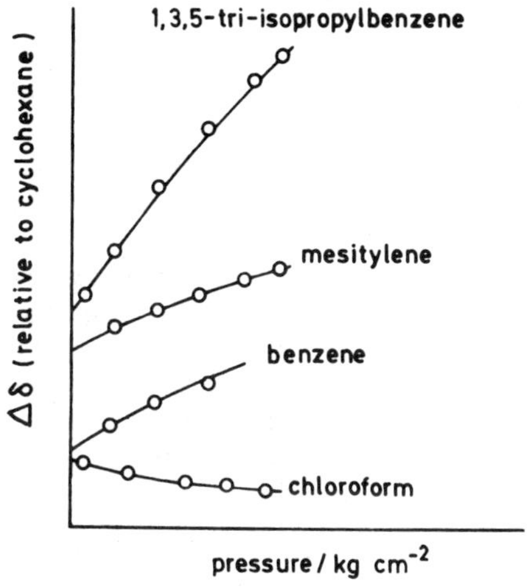


**Figure 2** *Plot of the proton chemical shift of chloroform as a function of pressure in methylbenzene solvents and in the pure liquid*
(Reproduced by permission from *J. Amer. Chem. Soc.*, 1974, **96**, 1935)

would increase with pressure, in addition to the $\sigma_w$ term. For the aromatic solvents, it is felt that the low-frequency shift is due to enhanced solute–solvent association, the chloroform proton becoming, therefore, progressively shielded by close proximity to the aromatic rings.

A slightly different type of $^{19}$F solvent-effect study has been reported by Dayal and Taft,[14] in which the effect of the solvent on the transmission of

[11] D. J. Wilbur and J. Jonas, *J. Magn. Resonance*, 1973, **10**, 279.
[12] H. J. Bernstein and F. H. A. Rummens, *J. Chem. Phys.*, 1965, **43**, 2971.
[13] H. Yamada, T. Ishihara, and T. Kinugasa, *J. Amer. Chem. Soc.*, 1974, **96**, 1935.
[14] S. K. Dayal and R. W. Taft, *J. Amer. Chem. Soc.*, 1973, **95**, 5595.

substituent effects in compounds of general structure (1) is described. In the present instance, G represents an uncharged group. If $P$ is the effect of the group X on the $^{19}F$ shift,

$$P = \rho_I\sigma_I + \rho_R\sigma_R \quad (4)$$

where $\sigma_I$ and $\sigma_R$ are the substituent inductive and resonance effect parameters, and $\rho_I$ and $\rho_R$ are the respective susceptibilities of the shift to the substituent

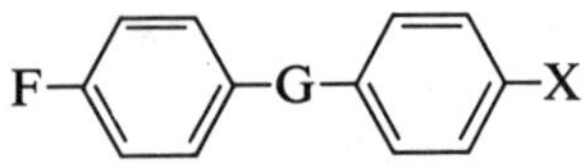

(1)

properties. From an extensive study it is apparent that transmission of the substituent shifts occurs *via* the $\pi$-electron system, and that the susceptibility of the $^{19}F$ chemical shift to the $\sigma_I$ term is much more dependent upon the solvent, specifically the solvent polarity, than is that for the substituent resonance effect.

Temperature-dependent $^{13}C$ shifts of benzene, cyclohexane, pyridine, cyclohexene, neopentane, hexane, and heptane have been reported for the liquid state over the range +10 to +70 °C.[15] Qualitatively, the temperature dependence for $^{13}C$ nuclei in a ring is less than that for those in a chain, whether the ring is aromatic or not. Similar observations have been reported earlier, the effect being attributed to hydrogen-bonding. For most of the solvents studied in this work, such an explanation is clearly inappropriate.[15] There has also been a suggestion[16] that solvation entropy effects may also contribute to the solvent shift. The $^{19}F$ shifts of certain *p*-fluorophenylcarbonium ions in fluorosulphonic acid at −30 °C fall into three distinct groups depending on whether the ion is acyclic, cyclic, or bicyclic. This may reflect the negative entropy effect experienced by each group on becoming co-ordinated to the solvent.

## 3 Effects due to Specific Interactions with the Medium

The involvement of specific interactions of a solute with its environment is usually invoked, sometimes perhaps too readily, whenever the solvent shifts cannot be rationalized in terms of one or other theory for the estimation of the individual non-specific contributions to the shift given by equation (1). In some instances the dividing line between the two types of effect is by no means clear. There is evidence, for example, that a degree of short-range order is present in alkane solutions of aromatic compounds,[17] and other weak hydrophobic interactions have been detected by n.m.r. studies.[18] These effects are not easily classified in the terms of equation (1). 'Collision com-

[15] W. M. Litchman and D. R. McLaughlin, *Chem. Phys. Letters*, 1973, **22**, 424.
[16] D. G. Farnum and D. S. Patton, *J. Amer. Chem. Soc.*, 1973, **95**, 7728.
[17] M. Lamotte and J. Joussot-Dubien, *Chem. Phys.*, 1973, **2**, 245.
[18] C. Jolicoeur, E. Firkins, and P. Bernier, *J. Phys. Chem.*, 1974, **78**, 851.

plexes', which have been invoked[19] to explain the marked solvent dependence of the enantiotropic thienyl protons of (2), are another case in point. In some instances, even where the interactions involved are quite strong, care must be exercised in deciding the source of the effect. For example, it has been argued[20] that the shift to high frequency for 1-methoxynaphthalene protons on being progressively added to a solution of butyl-lithium is evidence for an

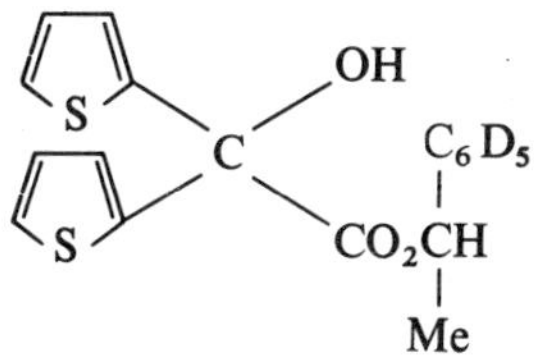

(2)

association between these molecules in solution. Subsequently, it has been pointed out that this behaviour simply reflects the concentration-dependent chemical shifts of the naphthalene derivatives arising from self-association.[21]

**A. The ASIS Effect.**—Traditionally one of the more popular fields of study, the ASIS effect has received very little attention within this reporting period. There have been qualitative accounts of benzene-induced shifts in various metal complexes.[22,23] Some authors persist in discussion of the solute–solvent interaction in terms of a specific complex having a 1:1 stoicheiometry.[22] That the interaction is dipolar in origin for polar solutes is given further support by a reported correlation between the observed benzene solvent shift and the solute dipole moment for a series of structurally similar solutes.[24] A related correlation has also been reported for shifts in hexafluorobenzene.[25]

**B. Hydrogen-bonded Systems.**—Studies of hydrogen-bonding have generally followed the pattern of previous years. $^{13}C$ shieldings have been reported for the phenol–acetone system in carbon tetrachloride,[26] from which it would seem that the $^{13}C{=}O$ resonance is quite sensitive to hydrogen-bonding interactions, but that the $^{13}C$—OH signal is rather less so. This kind of qualitative understanding of the effect of hydrogen-bonding on nuclear resonances can

[19] B. Ostman, *Tetrahedron*, 1973, **29**, 2115.
[20] B. M. Graybill and D. A. Shirley, *J. Org. Chem.*, 1962, **27**, 1939.
[21] R. A. Ellison and F. N. Kotsonis, *Tetrahedron*, 1973, **29**, 805.
[22] G. E. Manoussakis and C. A. Tsipis, *Z. anorg. Chem.*, 1973, **88**, 398.
[23] H. A. Meinema, A. Mackor, and J. G. Noltes, *J. Organometallic Chem.*, 1974, **70**, 79.
[24] R. S. Armstrong, M. J. Aroney, R. K. Duffin, H. J. Stootman, and R. J. W. Le Fèvre, *J.C.S. Perkin II*, 1973, 1272.
[25] R. S. Armstrong, M. J. Aroney, R. K. Duffin, H. J. Stootman, and R. J. W. Le Fèvre, *J.C.S. Perkin II*, 1973, 1362.
[26] T. K. Nakishima, D. D. Traficante, and G. E. Maciel, *J. Phys. Chem.*, 1974, **78**, 124.

prove quite useful in understanding solution interactions. In one example, the $^{1}H$ and $^{13}C$ resonances of the $CONH_2$ group of nicotinamide in solution show low-frequency shifts on dilution; this is not consistent with the removal of diamagnetic shielding which would occur if stacked aggregates were present. It is, however, consistent with a reduction in inter-amide hydrogen-bonding,[27] and provides an easy insight into the nature of the species in the system.

In another context, the effect of dilution on the shielding of hydrogen-bonded nuclei continues to be used to distinguish inter- from intra-molecularly hydrogen-bonded systems.[28–31] There has, however, been a degree of controversy relating to hydrogen-bonded interactions involving cyclopropane. It has been argued that an intramolecular hydrogen bond is formed in cyclopropanols, possibly of the kind shown in structure (3).[32] More recently

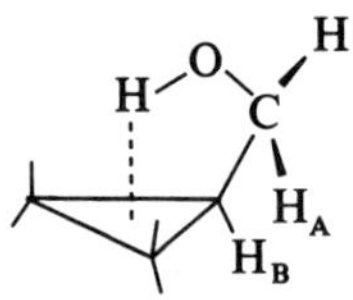


(3)

it has been pointed out that the solvents pyridine and DMSO should compete effectively with such a bond, and that, since the conformation of (3) is sterically unfavourable, $J(H_A,H_B)$ should therefore show a degree of solvent dependence.[33] Such a solvent dependence has in fact been shown for similar systems,[34] but is not observed in the present case.[33] Other workers have, however, reported[35] infinite-dilution shifts for ethanol, cyclopropylmethanol, and 2-cyclopropyl-ethanol. The shift for the cyclopropylmethanol is 0.23 p.p.m. to high frequency of the others, which would seem to indicate that intramolecular bonding is present.

The suggestion has been made that it may be possible to correlate solvent shifts due to hydrogen bonding with solute structure.[36] For a fairly representative group of compounds, methyl groups $\beta$ to hydroxyl show solvent shifts which are a function of the dihedral angle $\theta$ between these two groups. The relationships (5) and (6) were derived from a study of 28 compounds for which

$$\Delta\delta[\mathrm{CDCl_3}\text{–pyridine}] = 0.11 + 0.1 \cos\theta\ (\pm 0.02)\ \text{p.p.m.} \qquad (5)$$

27 B. Birdsall, J. Feeney, and P. Partington, *J.C.S. Perkin II*, 1973, 2145.
28 Y. Murakami and J. Sunamoto, *J.C.S. Perkin II*, 1973, 1231.
29 N. N. Shapet'ko, *Org. Magn. Resonance*, 1973, **5**, 215.
30 V. Folli, D. Jarossi, and F. Taddei, *J.C.S. Perkin II*, 1973, 848.
31 E. Kiehlmann, B. C. Menon, and N. McGillivary, *Canad. J. Chem.*, 1973, **51**, 3177.
32 L. Joris, P. von R. Schleyer, and R. Gleiter, *J. Amer. Chem. Soc.*, 1968, **90**, 327.
33 J. L. Pierre and R. Perraud, *Tetrahedron Letters*, 1973, 1935.
34 J. L. Pierre, R. Perraud, and P. Arnaud, *Bull. Soc. chim. France*, 1970, 4459.
35 M. Oki and H. Iwamura, *Tetrahedron Letters*, 1973, 4003.
36 B. P. Hatton, C. C. Howard, and R. A. W. Johnstone, *J.C.S. Chem. Comm.*, 1974, 744.

$$\Delta\delta[(CD_3)_2SO\text{–pyridine}] = 0.17+0.146\cos\theta\ (\pm 0.03)\ \text{p.p.m.} \qquad (6)$$

$0° < \theta < 180°$, and are thought also to hold for methyl groups *ortho* to phenolic hydroxyl.[36] The correlations do not hold in cases where a carboxy-group is close to the methyl, nor, probably, where there is hindered rotation about the C—OH bond.[36] The correlation is wholly empirical, and too few structures have been studied to justify its use; the preliminary results do, however, merit further consideration.

One other useful technique employing hydrogen-bonding shifts is exemplified by studies of the conformations of medium and large molecules in solution. The cyclic peptides *cyclo*-(L-X-D-Phe-L-Pro)$_2$, where X = Ala, Orn, or His, adopt conformations in solution for which only the peptide proton of the

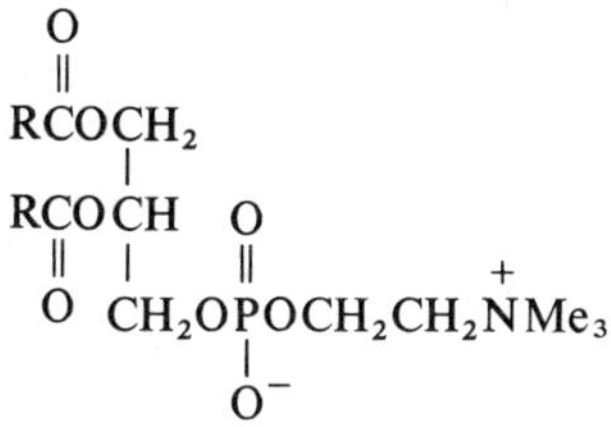

(4)

D-Phe residue is 'exposed' to the solvent. This is evident from the observation that only this proton undergoes a change in shift on going from strong to weak proton-acceptor solvents.[37] The binding of water to egg yolk lecithin (4) has been studied in a similar way.[38] Addition of water shifts the resonance of

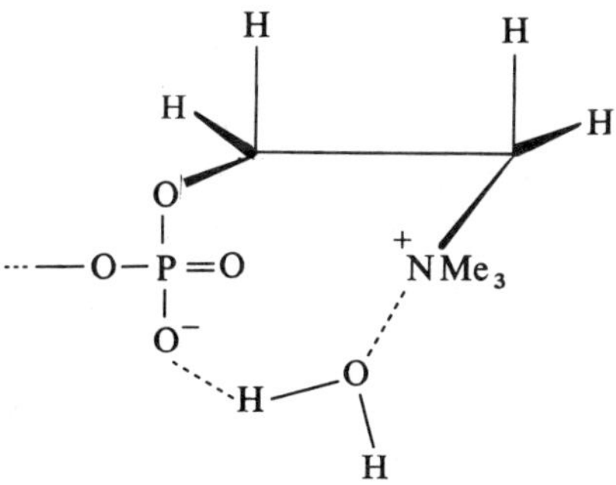

**Figure 3** *Binding of water to choline nitrogen and the phosphate group of egg yolk lecithin*

protons close to the choline nitrogen to low frequency; those close to phosphate move in the opposite direction. This is thought to represent the type of

[37] K. D. Kopple, A. Go, T. J. Schamper, and C. S. Wilcox, *J. Amer. Chem. Soc.*, 1973, **95,** 6090.
[38] J. H. Shaw, L. S. Kan, and N. C. Li, *J. Magn. Resonance*, 1973, **12,** 209.

binding illustrated in Figure 3. The water proton resonance shifts progressively to low frequency as water is added to the system up to six moles of water per mole of lecithin; thereafter the signal shifts to high frequency.[38] This seems to represent progressive binding of water to the lipid, followed by water self-association. The value of deuterium resonance studies to elucidate the interactions of $^{2}H_2O$ with macromolecules generally, using lecithin as a model system, has been described by Finer.[39]

There have been relatively few reports of the effects of hydrogen-bonding on coupling constants. The ($^{15}N$,$^{13}C$) coupling of aqueous formamide has been found to show a small variation in proportion to the amount of water present,[40] correcting an earlier report.[41] The behaviour of the $^{13}C$ shift on addition of water is also reported: the signal moves to low frequency initially, (up to 0.3 mole fraction of water); thereafter the direction is reversed.[40] This behaviour is qualitatively similar to that of the $^{13}CO_2H$ resonance of aqueous acetic acid.[42] Ng and Wong[43] have reported solvent-dependent geminal (H, H) couplings of $\alpha$-chloroacrylonitrile. In solvents that are $\pi$-electron donors, evidence is presented for hydrogen-bonded structures such as (5).

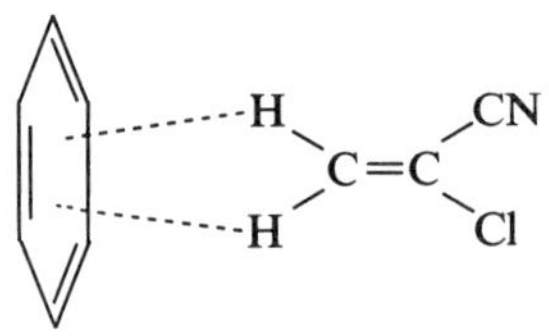


(5)

Morishima, Yonezawa, and co-workers have extended their study of the nature of the hydrogen bond using stable free-radical 'probes' in solution,[44] in this case using $\alpha,\gamma$-bisdiphenylene-$\beta$-phenylallyl (6) as the proton acceptor. An indication of the shift contribution due to the diamagnetic ring current of (6) was obtained from similar studies involving the parent closed-shell hydrocarbon. Shifts induced in proton donor substrates are found to be in proportion to the donor strength. $^{13}C$ Shifts are reported for $^{13}CHCl_3$ and PhC≡$^{13}$CH in carbon tetrachloride; the shifts observed are in the opposite sense to those of the adjacent protons, which is in accord with a spin polarization mechanism for the transmission of spin density.[44]

Slightly more quantitative approaches towards an understanding of hydrogen-bonding effects have appeared. Ng *et al.*[45] have obtained estimates of the

[39] E. G. Finer, *J.C.S. Faraday II*, 1973, **69,** 1590.
[40] J. F. Hinton, K. H. Ladner, and W. E. Stewart, *J. Magn. Resonance*, 1973, **12,** 90.
[41] J. F. Hinton and K. H. Ladner, *J. Magn. Resonance*, 1972, **6,** 586.
[42] G. E. Maciel and D. D. Traficante, *J. Amer. Chem. Soc.*, 1966, **88,** 220.
[43] K. F. Wong and S. Ng, *J. Magn. Resonance*, 1974, **13,** 239.
[44] I. Morishima, K. Toyoda, K. Yoshikawa, and T. Yonezawa, *J. Amer. Chem. Soc.*, 1973, **95,** 8627.
[45] K. F. Wong, T. S. Pang, and S. Ng, *J.C.S. Chem. Comm.*, 1974, 55.

limiting hydrogen-bond shift $\Delta\delta_c$ using equation (7), which, for a series of

$$\frac{1}{\Delta\delta_{obs}} = \frac{1}{\Delta\delta_c} + \frac{1}{\Delta\delta_c K[\text{acceptor}]} \qquad (7)$$

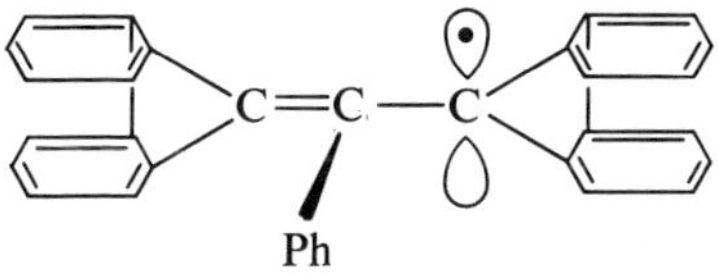

(6)

closely similar bases with a given proton donor, correlate quite well with appropriate enthalpies for hydrogen-bond formation ($\Delta H$). Slejko and Drago have pursued their work on this topic.[46] These authors are critical of many reported estimates of limiting hydrogen-bond shifts.[46—48] In this work, linear correlations are reported between $\Delta\delta_c$ values and $\Delta H$ for quinuclidine, tetrahydrothiophen, or l-phospha-2,6,7-trioxa-4-ethylbicyclo[2,2,2]octane for a series of proton donors, but such correlations do not obtain for acetonitrile, THF, or diethyl ether. However, in almost all cases, $\Delta H$ values do appear to be linearly related to $\alpha_{\parallel}$, where $\alpha_{\parallel}$ is the polarizability along the donor X—H bond.[46]

The enthalpy of formation of cholesterol dimers in chloroform has been estimated[49] from n.m.r. shift measurements to be $-10.8$ kJ mol$^{-1}$; that for cholesterol binding to the solvent is $-3.8$ kJ mol$^{-1}$. Chloroform hydrogen-bonding to thioethers has, over the temperature range $-50$ to $+50$ °C, been studied in a similar way.[50] Litchman[51] has measured pyridine $^{13}$C shifts in a wide range of solvents; the shifts appear to be expressible as a linear combination of terms corresponding to interactions of types (7) and (8).

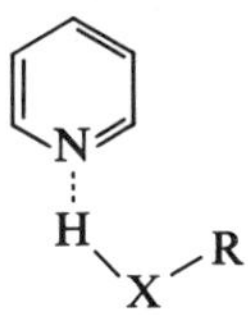

(7)

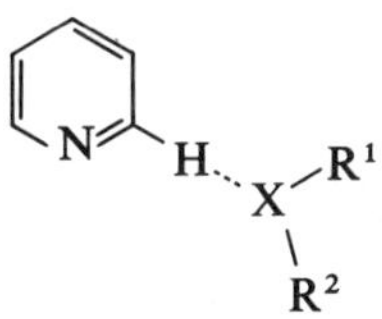

(8)

Perhaps the most rigorous quantitative study reported to date relates to the self-association of t-butyl alcohol in hexadecane.[52] It is generally true that n.m.r. chemical shift measurements are not readily amenable to unambiguous

[46] F. L. Slejko and R. S. Drago, *J. Amer. Chem. Soc.*, 1973, **95**, 6935.
[47] F. L. Slejko and R. S. Drago, *Inorg. Chem.*, 1973, **12**, 176.
[48] F. L. Slejko, R. S. Drago, and D. G. Brown, *J. Amer. Chem. Soc.*, 1972, **94**, 9210.
[49] J. J. Feher, L. D. Wright, D. B. McCormick, *J. Phys. Chem.*, 1974, **78**, 250.
[50] K. W. Jolley, L. M. Hughes, and I. D. Watson, *Austral. J. Chem.*, 1974, **27**, 287.
[51] W. M. Litchman, *J. Magn. Resonance*, 1974, **14**, 286.
[52] E. E. Tucker and E. D. Becker, *J. Phys. Chem.*, 1973, **14**, 1783.

analysis when several associated species are present, all undergoing fast exchange. In this work, n.m.r. results are complemented by vapour-pressure and i.r. measurements, which identify a linear trimer as the predominant species at low concentrations, rather than the anticipated linear dimer. This is demonstrated from the n.m.r. shift data using a Lippert plot.[53]

**C. Aqueous Systems.**—Although the behaviour of aqueous solutions is wholly dominated by hydrogen-bonded interactions, the importance of such systems justifies their inclusion in a separate section, although numerous additional references to aqueous solutions occur throughout this chapter. Theoretical approaches to calculation of the water proton shielding are generally unsatisfactory; there has, however, been a report of the successful prediction of water proton shielding for the linear perpendicular dimer.[54] There have been two fairly detailed studies of binary solvent systems involving water. Kingston and Symons[55] have observed the hydroxy proton shift in a number of binary solvent mixtures over the entire mole fraction range at various temperatures. At low concentrations of water, two factors contribute to the observed shift, the strength of the hydrogen bonds formed and the number of free, or non-hydrogen-bonded hydroxy-groups. For example, the behaviour of the hydroxyl resonance in acetone and triethylamine solutions is rather similar. In the former the bonding is rather weak but there are virtually no non-bonded hydroxy-groups. In the latter the bonding is much stronger, but there are fewer hydrogen bonds formed. In the mid-concentration range it is evident that water–water binding occurs well before all of the potential solute binding sites have been accommodated. At high water concentrations, where the water structure becomes quite extensive, the data are discussed in terms of competition between acid–base-type interactions and effects of water structuring. A somewhat different approach is taken by Marciacq-Rousselot and Lucas.[56] These authors have adopted the standpoint that hard-sphere solutes enhance the binding between water molecules below 4 °C but decrease the binding above this temperature.[57] In an attempt to find experimental support for this proposition, the water proton resonance in aliphatic alcohol solutions has been studied as a function of temperature and concentration. The authors make the somewhat doubtful assumption that the effect of alcohol molecules on the water structure is that expected of a hard-sphere solute. Nevertheless, evidence is presented which suggests that alcohols do have a structure-enhancing effect at low temperatures, becoming structure breakers as the temperature is increased. The actual temperature at which the transition occurs is unclear, however.

One very interesting development relates to measurements of the water proton resonance in liquid water at low temperature, down, in fact, to the

[53] E. Lippert, *Ber. Bunsengesellschaft phys. Chem.*, 1963, **67**, 267.
[54] M. Jaszuński and A. J. Sadlej, *Theor. Chim. Acta*, 1973, **30**, 257.
[55] B. Kingston and M. C. R. Symons, *J.C.S. Faraday II*, 1973, **7**, 978.
[56] M.-M. Marciacq-Rousselot and M. Lucas, *J. Phys. Chem.*, 1973, **77**, 1056.
[57] M. Lucas, *J. Phys. Chem.*, 1972, **76**, 4030.

homogeneous nucleation point at about −40 °C. Angell, Shuppert, and Tucker[58] have measured the resonance between 0 and −38 °C, both for an emulsion in a suitable hydrocarbon and for water in glass capillaries. In both systems ice nucleation is inhibited and the water remains in the supercooled liquid state. Under these conditions, below 0 °C, the chemical shift is a non-linear function of temperature, in marked contrast to the behaviour above 25 °C. The shift moves increasingly to high frequency as the temperature decreases. This is thought to represent a co-operative increase either in the hydrogen-bond strength or in the rate at which such bonds are formed.

With regard to aqueous systems at low temperatures, there have been a number of studies relating to the observation of an anomalously narrow water proton resonance which is observed in the spectrum of frozen aqueous solutions of macromolecules,[59—62] in particular, proteins. It is supposed that this narrow line, having an intensity proportional to the protein concentration, represents water of hydration of the protein. Ramirez, Cavanaugh, and Purcell[63] have studied this behaviour in some detail. It is apparent from studies of binary systems involving low molecular weight solutes that the narrow line appears on freezing the system, decreases in intensity as the temperature is lowered, and finally disappears at a temperature characteristic of the solute. It is argued that, as the temperature decreases, ice crystals form progressively throughout the system; the narrow signal corresponds therefore to water remaining in the liquid state. A similar situation holds true for solutions of amino-acids and polypeptides. The situation with regard to proteins is, however, quite different. Here, the narrow signal does not decrease with temperature, but persists down to the lowest temperatures accessible. For bovine serum albumin (BSA) the signal does, however, decrease if the pH of the system is lowered, although this should result in an increase in hydration of the protein. For the ternary system water–BSA–KF a narrow $^{19}F$ signal is observed only if BSA is present, the intensity decreasing with temperature. The authors suggest that the KF is dissolved in 'non-freezable' water which is maintained even at low temperature as a distinct thermodynamic liquid phase by the protein.[63]

**D. Ionic Solutions.**—This section is largely concerned with ion binding to suitable substrates or to the solvent. In this context, interest has been shown in studies of $^{205}Tl^+$,[64,65] possibly because the ion is a useful substitute for $K^+$ in studies of biological systems. $^{205}Tl^+$ chemical shifts at infinite dilution in

58 C. A. Angell, J. Shuppert, and J. C. Tucker, *J. Phys. Chem.*, 1973, **77**, 3092.
59 I. D. Kuntz, T. S. Brassfield, G. D. Law, and G. V. Purcell, *Science*, 1969, **163**, 1329.
60 I. D. Kuntz and T. S. Brassfield, *Arch. Biochem. Biophys.*, 1971, **142**, 660.
61 I. D. Kuntz, *J. Amer. Chem. Soc.*, 1971, **93**, 516.
62 J. P. White, I. D. Kuntz, and C. R. Cantor, *J. Mol. Biol.*, 1972, **64** 511.
63 J. E. Ramirez, J. R. Cavanaugh, and J. M. Purcell, *J. Phys. Chem.*, 1974, **78**, 807.
64 H. S. Gutowsky and B. R. McGarvey, *Phys. Rev.*, 1953, **91**, 81.
65 R. Freeman, R. P. H. Gasser, R. E. Richards, and D. N. Wheeler, *Mol. Phys.*, 1959, **2**, 75.

water, formamide, DMF, methanol, pyridine, pyrrolidine, and n-butylamine have been shown to depend on both the solvent and the anion.[66] It appears that the more strongly interacting solvents cause the greatest high-frequency shifts. $^{205}Tl^{+}$ shieldings for $D_2O$ and $H_2O$ solutions have been compared; the isotope shift can be shown to be dependent on the anion and to vary markedly with concentration.[67] As in the previous study,[66] it was also shown that co-ordinating species cause very pronounced high-frequency shifts, sometimes of the order of several hundred p.p.m.[67] Detailed theoretical calculations have been presented[68,69] for the proton and $^{17}O$ chemical shifts of water co-ordinated to metal ions. $Nd^{III}$ binding to amino- and carboxylic acids has been measured in aqueous solution;[70] the results are in good accord with those derived from potentiometric studies, despite the use of DSS as an internal reference standard. Fuhr and Rabenstein have also used $^{13}C$ n.m.r. to locate specific binding sites for $Cd^{2+}$, $Zn^{2+}$, $Pb^{2+}$, and $Hg^{2+}$ within the tripeptide glutathione.[71] These binding sites show a high degree of ion specificity; $Hg^{2+}$, for example, binds only to the sulphydryl group up to a $Hg^{2+}$:glutathione ratio of 0.5.[71]

Several authors report studies of ion solvation at low temperatures, such that exchange between the ion co-ordination sphere and the bulk solvent becomes slow. For solutions of $AlCl_3$ in ethanol at sufficiently low temperatures, several hydroxyl proton resonances had been observed and attributed to polymeric species involving the cation and the solvent only.[72,73] Recently, Stockton and Martin[74,75] have removed this multiplicity by replacing $Cl^-$ by $ClO_4^-$, which suggests that the $Cl^-$ anion is involved in the co-ordination sphere whereas $ClO_4^-$ is not. This has also been confirmed by a $^{13}C$ study of ethanol co-ordinated to $Mg^{2+}$ and $Al^{3+}$.[75] The $^{13}C$ shieldings of the co-ordinated ethanol appear to be dominated by electrostatic perturbations due to the ion. Studies of the ethanol exchange rates estimated from $^{13}C$ measurements are reported to agree with those from proton studies, which suggests that, in these cases, the behaviour of the hydroxyl proton resonance does reflect whole-molecule exchange. Other low-temperature studies of $Mg^{2+}$ ions in water–acetone mixtures at $-95$ °C show complex proton spectra, indicating that 'hybrid' co-ordination spheres are formed, containing acetone and water simultaneously.[76]

Eaton and Chua have studied the strength of solvent binding in the second

[66] J. J. Dechter and J. I. Zink, *J.C.S. Chem. Comm.*, 1974, 96.
[67] S. O. Chan and L. W. Reeves, *J. Amer. Chem. Soc.*, 1974, **96**, 404.
[68] J. W. Akitt, *J.C.S. Dalton*, 1974, 175.
[69] J. Sadlej and A. J. Sadlej, *J. Magn. Resonance*, 1974, **14**, 97.
[70] A. D. Sherry, C. Yoshida, E. R. Birnbaum, and D. W. Darnall, *J. Amer. Chem. Soc.*, 1973, **95**, 3011.
[71] B. J. Fuhr and D. L. Rabenstein, *J. Amer. Chem. Soc.*, 1973, **95**, 6944.
[72] H. Grasdalen, *J. Magn. Resonance*, 1971, **5**, 84.
[73] H. Grasdalen, *J. Magn. Resonance*, 1972, **6**, 336.
[74] J. S. Martin and G. W. Stockton, *J. Magn. Resonance*, 1973, **12**, 218.
[75] G. W. Stockton and J. S. Martin, *Canad. J. Chem.*, 1974, **52**, 744.
[76] F. Toma, M. Villemin, and J. M. Thiéry, *J. Phys. Chem.*, 1973, **77**, 1294.

co-ordination sphere of $Co^{II}$ pyrazolylborate.[77] Estimates of the equilibrium constants have been determined for the process

$$XS + S' \rightleftharpoons XS' + S \tag{8}$$

where X is the $Co^{II}$ pyrazolylborate complex, S is carbon tetrachloride, the reference solvent, and S′ is a second co-ordinating species such as acetone or benzene. Values of $K$ were obtained by studying the effect of the addition of species S′ on the *ortho* proton resonance of aniline in a mixture of the cobalt complex and aniline in carbon tetrachloride. This provides a measure of the extent to which a given solvent (S′) will displace aniline from the second co-ordination sphere of the complex. On this basis it seems that the binding

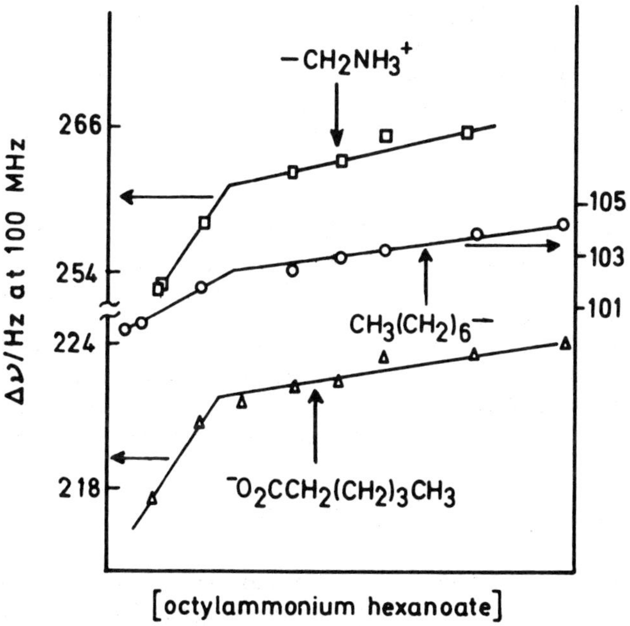


**Figure 4** *Selected proton shifts of octylammonium hexanoate at* 30 °C *as a function of concentration in benzene solution*
(Reproduced by permission from *J. Chem. Phys.*, 1973, **77**, 1432)

strength is governed almost entirely by van der Waals forces; hydrogen-bonding of the incoming molecule to the pyrazolyl groups seems to be of little importance.

Ion-pairing interactions of sodium ion in propylene carbonate solution have been studied by $^{23}Na$ spectroscopy.[78] For the iodide and thiocyanate

[77] D. R. Eaton and K. L. Chua, *Canad. J. Chem.*, 1973, **51**, 4137.
[78] M. S. Greenberg, D. M. Wied, and A. I. Popov, *Spectrochim. Acta*, 1973, **29A**, 1927.

salts the shifts are strongly concentration dependent, which suggests that ion-pairing is occurring despite the high dielectric constant of the solvent.

Generally, n.m.r. studies of ionic systems are carried out at moderate to high dilution. One report has, however, appeared of the proton shift behaviour of water in ionic solutions of such high concentration that 'bulk' water effectively does not exist in the system.[79] In this instance the observed change in the water proton chemical shift with temperature can no longer be attributed to disruption of water–water hydrogen bonds but is presumed to be due to disruption of water–anion hydrogen bonds within the cation co-ordination sphere. One other novel study[80] involves binary mixtures of organic solvents and tetra-alkylammonium tetra-alkylborides. The latter species are generally liquid at normal temperatures, and can only interact with other molecules *via* dipolar and dispersion interactions. In the presence of benzene, the cationic fragment shows large low-frequency shifts, whereas the anion shifts to high frequency. This suggests some form of non-random interaction whereby the benzene has a net orientation which produces the observed shift.[81] The enthalpy of the interaction appears to be small and negative. The data cannot, however, be rationalized in terms of traditional understanding of 'collision complexes'.

**E. Micellar Aggregation.**—Several examples have appeared of the utility of n.m.r. chemical shift measurements in determining critical micelle concentrations (c.m.c.). Figure 4 illustrates selected proton shifts for octylammonium hexanoate in benzene solution at 30 °C plotted against concentration.[82] The sharp discontinuity observed in this and similar plots is a sensitive indication of the concentration at which micelle formation occurs.[82] Above the c.m.c. the concentration of monomeric material is given by equation (9), where $\delta_M$ and $\delta_m$ are the shifts of the appropriate nucleus in the micelle (M) and monomer

$$[m] = \frac{[\text{surfactant}]_{\text{total}} \cdot (\delta_M - \delta_{obs})}{\delta_M - \delta_m} \quad (9)$$

(m) respectively. Substituting this value into equation (10) then yields an estimate of the 'equilibrium constant' ($K$) for the aggregation process, and also of $n$, the average number of monomers in the micelle. This approach, using $^1$H n.m.r., has been applied to systems of 'reversed' micelles (surfactants in non-aqueous systems),[82,83] and to aqueous solutions of sodium n-octanoate using $^{13}$C spectroscopy.[84]

79 E. J. Sare, C. T. Moynihan, and C. A. Angell, *J. Phys. Chem.*, 1973, **15**, 1869.
80 W. T. Ford and D. J. Hart, *J. Amer. Chem. Soc.*, 1974, **96**, 3261.
81 R. P. Taylor and I. D. Kuntz, *J. Amer. Chem. Soc.*, 1970, **92**, 4813.
82 E. J. Fendler, J. H. Fendler, R. T. Medary, and O. A. El Seoud, *J. Phys. Chem.*, 1973, **77**, 1432.
83 O. A. El Seoud, E. J. Fendler, J. H. Fendler, and R. T. Medary, *J. Phys. Chem.*, 1973, **77**, 1876.
84 T. Drakenberg and B. Lindman, *J. Colloid Interface Sci.*, 1973, **44**, 184.

$$\log\{[\text{surfactant}]_{\text{total}} - [\text{m}]\} = \log(nK) + n\log[\text{m}] \qquad (10)$$

Qualitative information may be deduced by a variety of techniques. Addition of DMSO to reversed micelles of dodecylammonium propionate (DAP) in benzene, chloroform, or dichloroethane markedly perturbs the $NH_3^+$ proton resonance.[85] This strongly suggests that the DMSO molecule is incorporated into the centre of the micelle, probably by hydrogen-bonding to the polar group. The equilibrium constants for incorporation of such additives into the micelle have been estimated from studies of proton shifts as a function of DAP concentration.[86] In another example,[87] the proton spectra of *p*-t-octylphenoxy(polyethoxy)ethanols (9) (i) in the monomeric form at high dilutions, (ii) as the pure liquid, and (iii) in the aggregated micellar state have been compared. In the final case the resonances due to the ethoxy-groups are to low frequency relative to the monomeric and pure-liquid states. In fact, there is a steady gradation of chemical shift along the surfactant chains within the

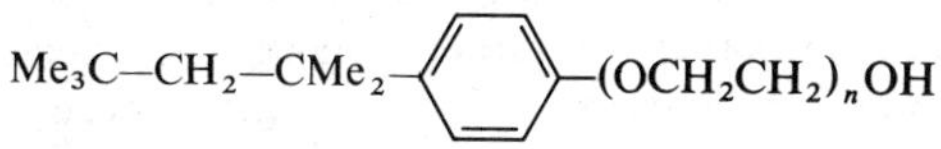


(9)

micelles, which appear, from nuclear relaxation measurements, to have a liquid-like interior.[87]

The effect of non-ionic,[88] zwitterionic,[88] and anionic[89] surfactants on the proton resonance of water has been studied. The evidence suggests that water molecules may penetrate to some degree into the micelle interior.[88] As an example of complications which may arise in these studies, measurements have been made of the binding of 1,2-dimethoxyethane (DME) to micelles of sodium diethyl n-butylmalonate.[90] It is found that increase in DME concentration in the system causes the protons of the $CH_3$ and $CH_2$ groups of DME to become progressively more shielded, in the absence of oxygen. If oxygen is introduced into the system, the DME signals rapidly revert to the normal values. Throughout the whole process, there is no detectable change in the surfactant micelles, and it is therefore tentatively supposed[54] that there is very weak, oxygen-sensitive binding of DME to the micelle surfaces.

## 4 Shift Reagents

The term 'shift reagent' is by now almost synonymous with the paramagnetic lanthanide chelates, particularly those of $Eu^{III}$ and $Pr^{III}$. Other types of shift

85 O. A. El Seoud, E. J. Fendler, and J. H. Fendler, *J.C.S. Faraday I*, 1974, **70**, 450.
86 O. A. El Seoud, E. J. Fendler, and J. H. Fendler, *J.C.S. Faraday I*, 1974, **70**, 459.
87 F. Podo, A. Ray, and G. Némethy, *J. Amer. Chem. Soc.*, 1973, **95**, 6164.
88 T. Walker, *J. Colloid Interface Sci.*, 1973, **45**, 372.
89 J. Clifford and B. A. Pethica, *Trans. Faraday Soc.*, 1964, **60**, 1433.
90 H. E. Zangg and R. S. Egan, *J. Phys. Chem.*, 1973, **77**, 1038.

reagent continue to be used, including $BPh_4^-$, which has been found to perturb the proton resonances of anilinium cations by up to 2.5 p.p.m.,[91] the diamagnetic/paramagnetic $Co^{II}$ mesoporphyrin IX dimethyl esters,[92] and $Co^{II}$ tetraphenylporphines.[93] Other, less common, shift reagents are certain $U^{IV}$ chelates [particularly the complex with 1,1,1-trifluoro-4-phenylbutane-2,4-dione, which is reported to effect proportionally greater perturbations of the pyridine proton resonances in carbon tetrachloride than $Eu(tmhd)_3$[94]] and $TiCl_4$.[95] This latter reagent is favoured because of its cheapness and ready availability; the induced shifts are, however, generally much smaller than for the common lanthanide reagents.

As has been said, the majority of reported studies relate to the use of complexes of $Eu^{III}$ or $Pr^{III}$. Typical examples of the application of these species are in spectral simplification,[96,97] assignment of molecular configurations,[98—100] analysis of epimer[101] and *cis–trans* isomer ratios,[102] and studies of molecular conformations in solution.[103] Most such work is concerned with $^1H$ n.m.r., and to a lesser extent $^{13}C$ n.m.r., with one report of $^{19}F$ shift perturbations induced in (10) by $Eu(fod)_3$.[104] Studies are generally carried out in solution; reports have, however, appeared of shift-reagent effects in the nematic phase[105] and in a high polymer.[106]

The basic requirement for effective use of a shift reagent is that the sub-

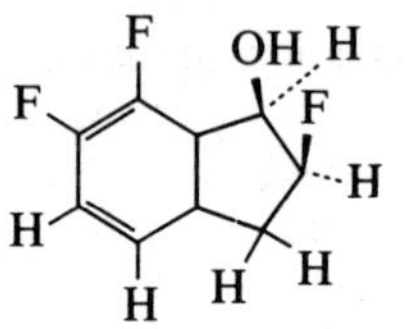


(10)

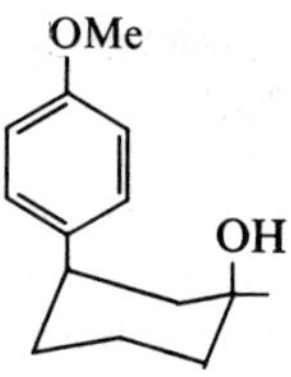


(11)

[91] G. P. Schiemenz, *Org. Magn. Resonance*, 1973, **5**, 257.
[92] C. D. Barry, H. A. O. Hill, B. E. Mann, P. J. Sadler, and R. J. P. Williams, *J. Amer. Chem. Soc.*, 1973, **95**, 4545.
[93] M. Gonedard, F. Gaudemer, and A. Gaudemer, *Tetrahedron Letters*, 1973, 2257.
[94] G. Folcher, J. Paris, P. Plurien, P. Rigny, and E. Soulié, *J.C.S. Chem. Comm.*, 1974, 3.
[95] A. K. Bose, P. R. Srinivasan, and G. Trainor, *J. Amer. Chem. Soc.*, 1974, **96**, 3670.
[96] J. W. Paschal and P. W. Rabideau, *J. Amer. Chem. Soc.*, 1974, **96**, 272.
[97] K. Tori, I. Horibe, Y. Tamura, and H. Tada, *J.C.S. Chem. Comm.*, 1973, 620.
[98] J. Paasivirta, *Suomen Kem.*, 1973, **46**, 162.
[99] S. D. Gero, D. Horton, A. M. Sepulchre, and J. D. Wander, *Tetrahedron*, 1973, **29**, 2963.
[100] F. Cavagna and A. Groebel, *Org. Magn. Resonance*, 1973, **5**, 133.
[101] R. Enríquez, J. Taboada, I. Salazar, and E. Diaz, *Org. Magn. Resonance*, 1973, **5**, 291.
[102] G. Montaudo, V. Librando, S. Caccamese, and P. Maravigna, *J. Amer. Chem. Soc.*, 1973, **95**, 6365.
[103] P. E. Young, V. Madison, and E. R. Blout, *J. Amer. Chem. Soc.*, 1973, **95**, 6142.
[104] G. M. Brooke and R. S. Matthews, *Tetrahedron Letters*, 1973, 3469.
[105] I. M. Armitage, E. E. Burnell, M. B. Dunn, L. D. Hall, and R. B. Malcolm, *J. Magn. Resonance*, 1974, **13**, 167.
[106] S. Amija, I. Andro, and R. Chûjô, *Polymer. J.*, 1973, **4**, 385.

strate should bind to the reagent to form a labile seven- or eight-co-ordinate complex. It is therefore of interest to know which substrates are likely to form sufficiently strong complexes with a given shift reagent to be amenable to this type of study. The more common usage of shift reagents with hydroxy-compounds has been extended to include nitriles,[107,108] pyrazolines,[109] azo-compounds,[110] azoles,[111] and sulphydryl[112] and organophosphorus compounds.[113] One interesting point has arisen with regard to the binding of $Eu(fod)_3$ to the oxygen atom of methyl ether groups. Generally, such binding is believed to be very slight, as for example in the case of (11), where binding is almost exclusively to the hydroxy-group.[114] Nevertheless, in an earlier study of the reserpine proton spectrum in the presence of $Eu(fod)_3$ it was reported[115] that the most perturbed resonance was in fact that of the *p*-methoxy-group of the trimethoxybenzoyl ring, and that this was by implication the preferred binding site, despite the presence of a number of other potential sites within the molecule. It now seems that three adjacent methoxy-groups present a stronger site for binding than, say, an ester carbonyl, possibly owing to bidentate binding to the shift reagent.[116] A detailed discussion of $Eu(fod)_3$ binding to compounds containing bivalent sulphur gives further support to the belief that the strength of substrate binding to the shift reagent is largely determined by the polarizability of the substrate binding site.[117] With respect to the type of substrate suitable for a shift-reagent study, there has been a timely report by Shoffner[118] which serves to remove a long-standing misapprehension. It had been thought that organic acids and phenols bring about decomposition of lanthanide chelates generally, and that such species could not therefore be studied by this technique.[119] This has been shown to be incorrect, at least for the case of $Eu(fod)_3$, and shifted spectra have been reported for n-hexanoic acid and 2,4-xylenol.[118]

The next step in the discussion is to identify the nature of the effect that gives rise to the observed shifts. Two terms are generally held to be responsible, and the overall chemical shift change $\Delta\delta_{i,\mathrm{obs}}$ may be written as

$$\Delta\delta_{i,\mathrm{obs}} = \Delta\delta_{i,\mathrm{contact}} + \Delta\delta_{i,\mathrm{dipolar}} \qquad (11)$$

where $\Delta\delta_{i,\mathrm{contact}}$ arises from transmission of electron spin density through the molecular bonds and $\Delta\delta_{i,\mathrm{dipolar}}$ is a 'through space' dipolar term. The latter is

107 J. A. Young, J. G. Grasselli, and W. M. Ritchey, *J. Magn. Resonance*, 1974, **14**, 194.
108 R. E. Davis, M. R. Willcott, tert., R. E. Lenkinski, W. von E. Doering, and L. Birladeanu, *J. Amer. Chem. Soc.*, 1973, **95**, 6846.
109 M. Franck-Neumann and M. Sedrati, *Org. Magn. Resonance*, 1973, **5**, 217.
110 S. R. Wilson and R. B. Turner, *J.C.S. Chem. Comm.*, 1973, 557.
111 J. Elguero, C. Marzin, and J. D. Roberts, *J. Org. Chem.*, 1974, **39**, 375.
112 H. Yanagawa, T. Kato, and Y. Kitahara, *Tetrahedron Letters*, 1973, 2137.
113 F. S. Mandel, R. H. Cox, and R. C. Taylor, *J. Magn. Resonance*, 1974, **14**, 235.
114 B. L. Shapiro, M. D. Johnston, jun., and M. J. Shapiro, *J. Org. Chem.*, 1974, **39**, 796.
115 G. E. Wright and T. Y. Tang Wei, *J. Pharm. Sci.*, 1972, **61**, 299.
116 G. E. Wright and T. Y. Tang Wei, *Tetrahedron*, 1973, **29**, 3775.
117 T. C. Morrill, R. J. Opitz, and R. Mozzer, *Tetrahedron Letters*, 1973, **38**, 3715.
118 J. P. Shoffner, *J. Amer. Chem. Soc.*, 1974, **96**, 1599.
119 J. K. M. Sanders and D. H. Williams, *Chem. Comm.*, 1970, 422.

widely calculated on the basis of equation (12),[120] in which $r_i$ is the length

$$\Delta\delta_{i,\mathrm{dipolar}} \propto (3\cos^2\theta_i - 1)/r_i^3 \qquad (12)$$

of the vector joining nucleus $i$ to the paramagnetic centre, and $\theta_i$ is angle between this vector and the principal magnetic susceptibility axis. The validity of the use of this equation in the present context was discussed in an earlier volume of these Reports[121] and will be referred to again below. Possibly the major reason for the interest in lanthanide shift reagents is the widespread belief that contributions from the contact term are negligible; the observed shift is then wholly dipolar in origin and a potential source of information about molecular geometries. An increasing number of reports have, however, appeared which suggest that neglect of the contact term is not valid. In Figure 5 for example, the alternating magnitude of the induced $^{13}C$ shift along the carbon chain suggests a major contribution from the contact term.[107] Similarly, shifts induced in the $^{31}P$ resonance of organophosphorus compounds are subject to a contact interaction to varying degrees.[113] With

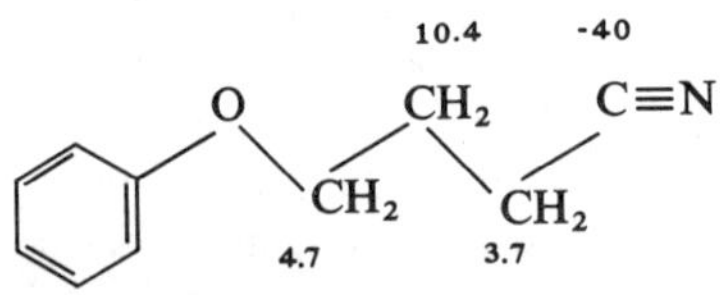


**Figure 5** $^{13}C$ *Shift gradients (from a plot*[107] *of induced shift* versus *shift reagent:substrate ratio) induced by* Eu(fod)$_3$

trimethyl phosphate the result is striking; the induced shifts for Pr(fod)$_3$ and Pr(tmhd)$_3$ are in opposite directions. Possibly, stronger binding of the substrate to the former reagent increases the involvement of the contact term sufficiently to reverse the direction of the shift.[113] Further evidence may be obtained by comparing the shifts induced in a given substrate by two shift reagents that are likely to form shift reagent–substrate complexes having very similar geometries. If the induced shifts for a given nuclear species in both cases are of a purely dipolar origin, the shift ratio should be constant irrespective of the particular nucleus measured. For acridine, quinoline, and isoquinoline with Eu(tmhd)$_3$ and Pr(fod)$_3$ the proton shift ratios are effectively constant, but those for the $^{13}C$ shifts vary widely.[122] Similar results for protons have been obtained from a study of polyglycodimethyl ethers.[123]

One further approach is to attempt to define a structure for the substrate–shift reagent complex which is in accord with the observed shifts, assuming these to be wholly dipolar in origin. Often the agreement between calculated

[120] H. M. McConnell and R. E. Robertson, *J. Chem. Phys.*, 1958, **20**, 1361.
[121] 'Nuclear Magnetic Resonance' ed. R. K. Harris (Specialist Periodical Reports), The Chemical Society, London, 1973, Vol. 2. Ch. 10.
[122] M. Hirayama and Y. Hanyu, *Bull. Chem. Soc. Japan*, 1973, **46**, 2687.
[123] A. M. Grotens, J. J. M. Backus, and E. de Boer, *Tetrahedron Letters*, 1973, 4343.

and observed shifts may be significantly improved if nuclei very close to the paramagnetic metal atom are removed from consideration. In this way, a contact effect has been inferred for the α-carbons of monofunctional keto- and hydroxy-steroids with $Pr(fod)_3$, and for the α- and β-carbons using $Eu(fod)_3$.[124] There are objections to all such approaches to estimate the involvement of the contact term; nevertheless there is a considerable degree of unanimity in results from many different sources. Firstly, it seems that $^{13}C$ shifts are more susceptible to a contact shift term than are the protons in any given case. Secondly, $Eu^{III}$ reagents are likely to produce a greater contact term than $Pr^{III}$ species; this is particularly so for $Eu(fod)_3$. Reuben[125] has in fact reported calculations which suggest that, for the lanthanide reagents, the contact term is likely to become progressively more important along the series Yb < Tm < Dy < Tb < Er < Ho < Nd < Eu. Several authors have in fact stressed the usefulness of $Yb^{III}$ shift reagents which, based on the experimental criteria discussed above, produce quite large shifts which do seem to be wholly dipolar in origin.[113,125—127]

One quite serious objection remains with regard to most of the above work, this being the almost universal neglect of what might loosely be called the 'diamagnetic term'. Equation (11) should in fact be

$$\Delta\delta_{i,\mathrm{obs}} = \Delta\delta_{i,\mathrm{dipolar}} + \Delta\delta_{i,\mathrm{contact}} + \Delta\delta_{i,\mathrm{diamag}} \qquad (13)$$

where $\Delta\delta_{i,\mathrm{diamag}}$ in practice encompasses all those factors which contribute to the shifts of the ligand nuclei other than those arising from the paramagnetism of the metal atom. The importance of this term has been stressed by Tori and Yoshimura.[128] In a study of $^1H$ and $^{13}C$ shifts of γ-picoline *N*-oxide, *p*-toluidine, *p*-cresol, and γ-picoline in chloroform solution a check was made by observing the shifts induced by the diamagnetic $La(fod)_3$ reagent. These shifts were in many cases a significant fraction of those induced by $Pr(fod)_3$ and $Eu(fod)_3$, and in a few instances appropriate correction actually reversed the sign of the observed paramagnetic shift.[128] Here the diamagnetic contribution may be unusually large. The α-geminal proton coupling at C-3 of camphor is known to vary in proportion to the amount of $Eu(fod)_3$ or $Eu(tmhd)_3$ present;[129] the effect persists if the diamagnetic complexes $La(tmhd)_3$ or $Lu(tmhd)_3$ are used,[130] without, however, there being any change in the substrate chemical shifts. Nevertheless, since it is experimentally very simple to check that the $\Delta\delta_{i,\mathrm{diamag}}$ term is effectively zero in any given case, such a check should be carried out as a routine part of any shift-reagent investigation.[130] Further support for this view comes from the observation of

124 J. W. ApSimon, H. Beierbeck, and J. K. Saunders, *Canad. J. Chem.*, 1973, **51**, 3874.
125 J. Reuben, *J. Magn. Resonance*, 1973, **11**, 103.
126 D. J. Chadwick and D. H. Williams, *J.C.S. Chem. Comm.*, 1974, 128.
127 K. L. Williamson, D. R. Clutter, R. Emch, M. Alexander, A. E. Burroughs, C. Chua, and M. E. Bogel, *J. Amer. Chem. Soc.*, 1974, **96**, 1471.
128 K. Tori and Y. Yoshimura, *Tetrahedron Letters*, 1973, 3127.
129 B. L. Shapiro, M. D. Johnston, jun., and R. L. Towns, *J. Amer. Chem. Soc.*, 1972, **94**, 4381.
130 J. J. Uebel, C. Pacheco, and R. M. Wing, *Tetrahedron Letters*, 1973, 4383.

an appreciable diamagnetic correction required for the carbonyl carbon of certain ketones with $Pr(tmhd)_3$.[126] To conclude this section, two papers are reported which present a novel approach. A correlation has been observed between $\Delta\delta_{obs}$ and $\Delta\Delta E_R$, the change in energy resulting from changes in non-bonded repulsion interactions involved in forming the complex, for aliphatic amine[131] and alcohol systems[132] with lanthanide shift reagents. It is argued that $\Delta\delta_{obs}$ is a measure of the steric strain energy involved in forming the complex. Such a correlation, if it exists, may well be fortuitous, and the work illustrates the dangers of taking empirical correlations too seriously. In any event such steric effects are wholly taken into account by the $\Delta\delta_{i,diamag}$ term of equation (13).

There have been numerous attempts to identify the molecular geometry which best fits the observed dipolar shifts, within the limitations of the above discussion, generally in an effort to locate the position of the metal atom relative to the substrate. Chadwick and Williams,[126] perhaps alone amongst workers publishing such papers appearing within the present reporting period, have made proper correction of the $\Delta\delta_{i,diamag}$ term in studies of shift reagent–ketone geometries. From this work, involving $Eu^{III}$ and $Pr^{III}$ complexes, the metal atom is believed to be above the plane of the $\rangle$C=O bonds, and by implication, therefore, bound to the oxygen lone-pair electrons. A similar result has been reported from a study of $Eu(tmhd)_3$ and $Pr(tmhd)_3$ with *exo*-2-methoxy-3-aza-4-keto-7,8-benzobicyclo[4,2,1]nonene.[133] To locate the metal atom position, a detailed *X*-ray crystallographic determination was made of the substrate. The shifts induced by each shift reagent were then used to calculate a set of $R_{i,j}$ values where

$$R_{i,j} = \Delta_{i,obs}/\Delta_{j,obs} \qquad (14)$$

This procedure effectively removed the proportionality constant of equation (12) from further consideration. The metal atom position was then located, using the *X*-ray data for the substrate geometry, effectively by an iterative least-squares technique assuming a rigid shift reagent–substrate complex. The metal atom in each case was found to be above the plane of the amide group, at an angle to the C—O bond direction (Figure 6).[133] In this example the metal atom positions were found to be rather sensitive to quite small changes in the substrate geometry. A significant difference in the calculated metal atom positions was found if 'ideal' bond lengths and angles were used in the calculation, rather than those determined from the crystal data.[133] If this observation proves to be generally true, it would suggest that *X*-ray studies may be of limited value for such calculations, since changes in bond lengths and angles of this order of magnitude may obtain on going from the crystal

131 Y. Sasaki, H. Kawaki, and Y. Okazaki, *Chem. and Pharm. Bull.* (*Japan*), 1973, **21**, 917.
132 Y. Sasaki, H. Kawaki, and Y. Okazaki, *Chem. and Pharm. Bull.* (*Japan*), 1973, **21**, 2488.
133 H. L. Ammon, P. H. Mazzocchi, W. J. Kopecky, jun., H. J. Tamburin, and P. H. Watts, jun., *J. Amer. Chem. Soc.*, 1973, **95**, 1968.

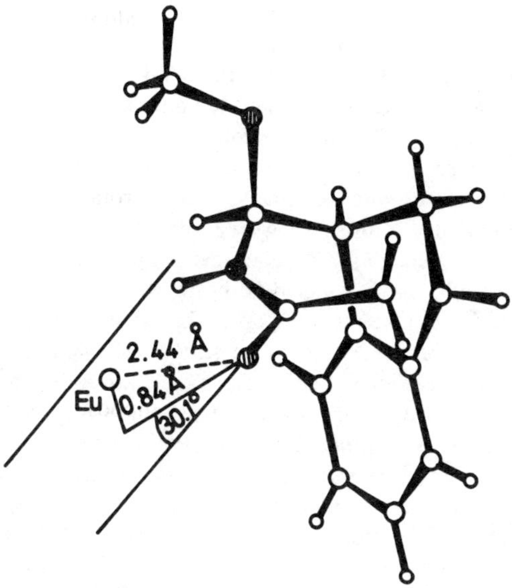


**Figure 6** *Structure of the complex of* exo-2-*methoxy*-3-*aza*-4-*keto*-7,8-*benzobicyclo*[4,2,1]*nonene with* Eu(tmhd)$_3$, *estimated from dipolar shift measurements*

(Reproduced by permission from *J. Amer. Chem. Soc.*, 1973, **95**, 1968)133

to the liquid, with the removal of the crystal packing forces. Details of an alternative computational approach for determining the 'best-fit' structure for the complex are given by Montaudo *et al.*,[134] including a sub-routine by which the peak assignments may be changed and checked in cases where such assignments are in doubt. Conformational changes of the complex may also be taken into account. Structures determined for $\alpha,\beta$-unsaturated aldehydes, ketones, and amides by this approach appear to be in good agreement with independent estimates.[134]

A somewhat different method has been described by Wing and co-workers,[135] which is specifically designed to take account of motional averaging of the substrate. In this technique the position of the substrate is varied relative to the fixed metal atom, rather than the opposite approach, which is more widely used. Here the advantage is that substrate rotamer averaging can readily be taken into account by the use of appropriate co-ordinate transformations, the population weightings being included as adjustable parameters. Non-axial symmetry and non-alignment of the molecular symmetry axis with the principal magnetic axis may also be included in the computation.

134 G. Montaudo, S. Caccamese, V. Librando, and P. Maravigna, *Tetrahedron*, 1973, **29**, 3915.

135 R. M. Wing, J. J. Uebel, and K. K. Anderson *J. Amer. Chem. Soc.*, 1973, **95**, 6046.

Studies using this technique are reported[135] for a number of organic compounds, all assumed to have 1:1 stoicheiometry and without correction for the diamagnetic or contact contribution. Nevertheless the predicted rotamer populations seem reasonable. Furthermore the advantage of the inclusion of rotamer averaging is particularly evident for one case where the assumption of complex rigidity leads to incorrect proton assignments.

The frequently successful use of the McConnell–Robertson equation[120] in shift reagent studies, despite the many, seemingly arbitrary, assumptions involved, has puzzled a number of workers in this field. For example, the assumption that the principal molecular axis is coincident with the principal magnetic susceptibility axis, and that both lie along the shift reagent–substrate bond direction, seems quite untenable on the basis of the crystallographic studies reviewed earlier.[121] Wing *et al.*[135] allowed for the possibility that the above two axes were not coincident; calculation then showed that the best fit for the structure–dipolar shift correlation did indeed correspond to collinear axes. It has been pointed out that rapid rotation of the substrate about the shift reagent–substrate bond will lead to effective coincidence of the two axes.[136] Crammer and Dubois[137] have, however, shown subsequently that a fairly high (> 29 kJ mol$^{-1}$) barrier to rotation exists in the Eu(tmhd)$_3$–(3-picoline)$_2$ complex, and since the substrate in this case is not unduly bulky, such restricted rotation may be quite general for these systems. A perhaps more reasonable approach, therefore, supposes that the substrate may be incorporated into the complex in a number of different ways; rapid exchange between these systems would then give rise to an apparent complex with effectively axial symmetry.[138]

A sufficient number of detailed studies of the stoicheiometry of the shift reagent–substrate interaction have now appeared to demonstrate quite clearly that assumption of a 1:1 stoicheiometry is no longer valid. Generally, both 1:1 and 1:2 complexes *at least* coexist in the solutions, the relative amounts being determined by the appropriate binding constants. An additional complication arises whenever self-association of the shift reagent also occurs. From ebulliometric measurements it has been shown that Eu$^{III}$, Gd$^{III}$, Tb$^{III}$, Er$^{III}$, Ho$^{III}$, and Tm$^{III}$ chelates with the 3-t-butylhydroxymethylene-(+)-camphorato (bhmc) ligand are monomeric at 76.5 °C in the range 0—0.2 mol l$^{-1}$, whilst those of Pr$^{III}$, Nd$^{III}$, and Sm$^{III}$ show marked aggregation above about 0.02 mol l$^{-1}$.[139] These results are also valid at 5.5 °C, from cryoscopic measurements.[139] Results of this type should not be regarded as being a general property for all shift reagents of a particular metal, for, although association of Pr(fod)$_3$ has been observed in $CCl_4$ by vapour-pressure osmometry, which is in accord with the above study, Eu(fod)$_3$ has

[136] J. M. Briggs, G. P. Moss, E. W. Randall, and K. D. Sales, *J.C.S. Chem. Comm.*, 1972, 1180.
[137] R. E. Crammer and R. Dubois, *J.C.S. Chem. Comm.*, 1973, 936.
[138] W. De W. Horrocks, jun., *J. Amer. Chem. Soc.*, 1974, **96**, 3022.
[139] R. G. Denning, F. J. C. Rossotti, and P. J. Sellars, *J.C.S. Chem. Comm.*, 1973, 581.

also been shown to aggregate extensively in the same system,[140] and furthermore, the degree of aggregation is a function of the extent of hydration of the shift reagent. Self-association of $Pr(fod)_3$ has also been detected in benzene solution.[141]

Possibly the first really convincing evidence that 1:1 and 1:2 complexes exist together in shift reagent–substrate systems was presented by Shapiro and Johnston.[142] Following this work, a number of workers have reported similar conclusions. Reuben[143] has discussed the effect that the simultaneous existence of such complexes would have on the shape of plots of $\Delta\delta_{obs}$ *versus* the shift reagent–substrate ratio, and has made the point that the observation of linear plots does not allow any conclusion to be drawn with regard to the complex stoicheiometry. The only conclusion that may in fact be drawn is that an insufficient range of concentrations has been studied. Reuben has obtained evidence for 1:1 and 1:2 complexation of $Eu(fod)_3$ with acetone, DMSO, propan-2-ol and $\beta$-picoline.[143] In addition it was shown that simulated shifts based on a 1:1 and 1:2 complex system can in fact be analysed quite satisfactorily assuming only 1:1 complexation; the equilibrium constant so determined, however, bears no relation whatever to those used in generating the data.[143]

More direct evidence has been obtained from low-temperature studies where the ligand exchange is slow, although the relevance of results so obtained to the situation at higher temperatures may be disputed. The stoicheiometry of the principal complex in a variety of systems has been studied in this way at temperatures ranging from $-30$ to $-90$ °C in $CCl_4$–$CDCl_3$–$C_6H_6$ (1.5:1.8:1.0 v/v) mixtures;[144] numerous instances of a 1:2 complex are observed. From ebulliometric studies it seems that 1:1 adduct formation alone occurs for $Gd(bhmc)_3$ and $Ho(bhmc)_3$ with butan-1-ol, and for $Eu(bhmc)_3$ and $Ho(bhmc)_3$ with (+)-$\alpha$-methylbenzylamine.[139] In contrast, c.d. studies of $Pr(tmhd)_3$ complexes with optically active simple amines have been reported.[145] In these experiments, the first Cotton effect is observed at *ca.* 315 nm and a second at 290 nm, of opposite sign, for all of the cases studied. Further, the value of $[\theta]$ was found to increase rapidly up to a shift reagent:substrate ratio of 0.5, which suggests that the predominant species in these instances is the 1:2 complex. In only one report, however, have such approaches been combined with an n.m.r. study. In a study[140] of $Eu(fod)_3$ and $Pr(fod)_3$ binding to methyl dimethylcarbamate, $Me_2NCO_2Me$, in $CCl_4$ solution, shift reagent self-association was evaluated in advance by vapour-pressure osmometry. These equilibria were then taken into account

140 A. H. Bruder, S. R. Tanny, H. A. Rockefeller, and C. S. Springer, jun., *Inorg. Chem.*, 1974, **13**, 880.
141 R. Porter, T. J. Marks, and D. F. Shriver, *J. Amer. Chem. Soc.*, 1973, **95**, 3548.
142 B. L. Shapiro and M. D. Johnston, jun., *J. Amer. Chem. Soc.*, 1972, **94**, 8185.
143 J. Reuben, *J. Amer. Chem. Soc.*, 1973, **95**, 3534.
144 D. F. Evans and M. Wyatt, *J.C.S. Chem. Comm.*, 1973, 339.
145 G. N. Mitchell and F. I. Carroll, *J. Amer. Chem. Soc.*, 1973, **95**, 7912.

when analysing the n.m.r. shift data. The equilibrium constants shown in Scheme 1 were thereby obtained

$$SR + S \rightleftharpoons SR{-}S \qquad K = 1600 \pm 600 \text{ l mol} - 1$$

$$SR{-}S + S \rightleftharpoons SR{-}S_2 \qquad K = 107 \pm 10 \text{ l mol} - 1$$

$$SR + SR \rightleftharpoons (SR)_2 \qquad K = 367 \text{ l mol} - 1$$

$$(SR)_2 + SR \rightleftharpoons (SR)_3 \qquad K \approx 12 \text{ l mol} - 1$$

**Scheme 1** *Binding constants for the system; methyl dimethylcarbamate* (S)–*anhydrous* Eu(fod)$_3$, (SR)–*carbon tetrachloride; at* 32 °C[140]

In this study the limiting shifts for the complexes are also reported, and it is perhaps worth noting that, in contrast to the systems studied by Shapiro and Johnston,[142] here $\Delta\delta_{1:2} > \Delta\delta_{1:1}$.[140] Too little information is presently available to permit generalizations to be made about the relative magnitudes of these terms, and it is certainly unwise to assume their relative sizes in any given situation.

Further evidence has been presented to indicate that binding substrate molecules to shift reagents may modify the position of any conformational equilibrium process which may be occurring. The equilibrium between structures (12a) and (12b) is thought to be modified by the presence of Eu(fod)$_3$

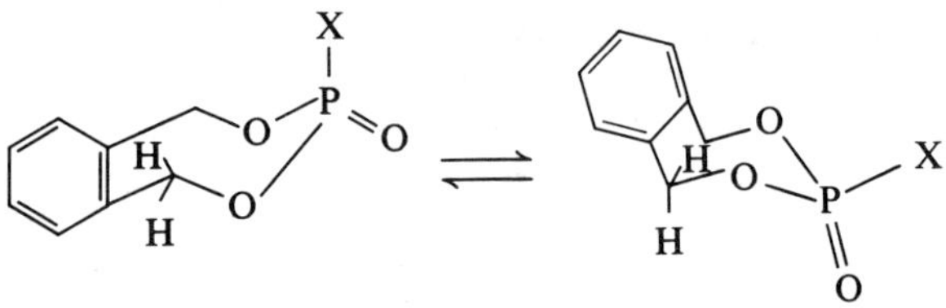

(12a) (12b)

since the $^3J$(P,H) coupling constants depend upon the relative amount of shift reagent present.[146] However, it is known that spin density may be transferred to the $^{31}$P nucleus in similar systems, and if this were the case here it could account for changes in these coupling constants. Changes in the coupling constants of 3,3-dimethylbutan-1-ol induced by Eu(tmhd)$_3$ have also been observed, and were attributed to a change in conformational equilibrium.[127] Interconversion of (13a) and (13b) is also affected by Eu(tmhd)$_3$, favouring (13b). The solvent dependence of the equilibrium is also considerably modified.[147]

For piperidine derivatives, an attempt has been made to determine, from a shift-reagent study, whether the conformation having an axial lone pair of

[146] T. Sato and K. Goto, *J.C.S. Chem. Comm.*, 1973, 494.
[147] A. Tangerman and B. Zwanenburg, *Tetrahedron Letters*, 1973, 5195.

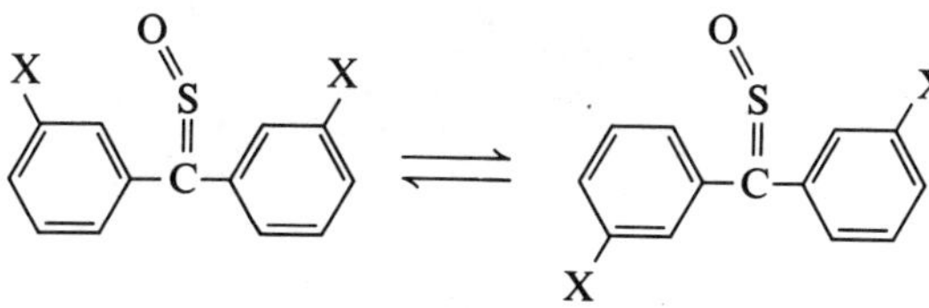

(13a) (13b)

electrons is favoured over that having an equatorial orientation.[148–150] Based on the assumption that the shift reagent, in this case $Co(acac)_2$, binds both conformers equally strongly, it was proposed that the axial conformer is the favoured one. Katritzky and co-workers[151] have contested this conclusion, and argue that the assumption of equal binding for the two conformers is invalid. In an alternative approach, based on Scheme 2, the possibility that different binding constants exist for the two conformers is explicitly recognized. Using this more rigorous approach the opposite conclusion is drawn, namely that the equatorial conformation is the more favourable.

Other applications of shift reagents to conformational studies include a detailed study of $^1H$ and $^{13}C$ n.m.r. of eighteen isomeric six-carbon aliphatic

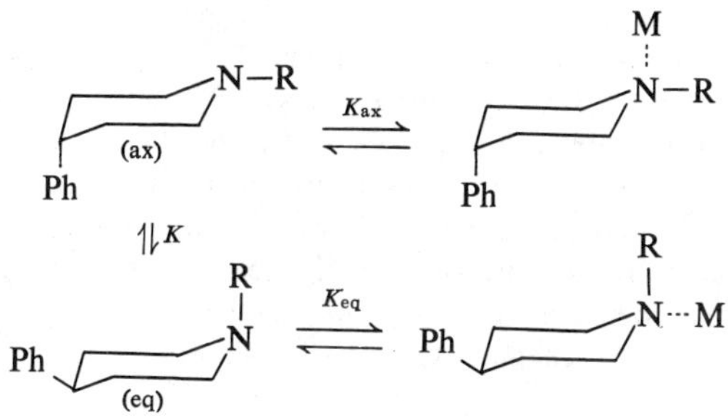


**Scheme 2**

alcohols.[127] The average solution conformation derived using the simple McConnell–Robertson equation from the $^{13}C$ data is in every case the same as that from the proton data. Each individual alcohol is discussed in detail. It is also pointed out that the fitting of the shift data to the dipolar equation in terms of the correlation coefficient is in some instances quite insensitive to a relatively large degree of rotation about the carbon–oxygen bond.[127]

There has also been a particularly interesting study of rotation about the

148 T. Yonezawa, I. Morishima, and Y. Ohmori, *J. Amer. Chem. Soc.*, 1970, **92**, 1267.
149 I. Morishima, K. Okada, M. Ohashi, and T. Yonezawa, *Chem. Comm.*, 1971, 33.
150 I. Morishima, K. Okada, T. Yonezawa, and K. Goto, *J. Amer. Chem. Soc.*, 1971, **93**, 3922.
151 I. D. Blackburne, A. R. Katritzky, and Y. Takeuchi, *J. Amer. Chem. Soc.*, 1974, **96**, 682.

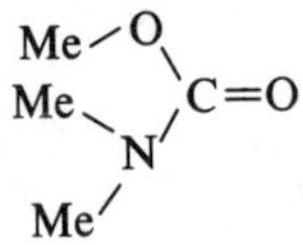

(14)

C—N bond of (14).[152] Bandshape analysis of the *N*-methyl proton signal is impractical in this case since the shift difference between the two methyl groups in the absence of rotational exchange is not more than 0.030 p.p.m. in $CDCl_3$ at −23 °C. The signal separation can, however, be increased to the desired extent using a suitable shift reagent. A computational approach to the total bandshape analysis as a function of the relative concentration of shift reagent is described. This yields pre-exchange lifetimes and thereby the apparent first-order rate constants for the process ($k_{app}$). It was found that $k_{app}$ is proportional to the substrate saturation fraction ($S$), so that $k_{free}$, the rate constant for methyl exchange in the absence of shift reagent, may be determined by extrapolating the plot of $k_{app}$ *versus* $S$ to $S = 0$.

Possibly the most profitable application of shift reagents has been in the study of systems of chiral molecules. Following the initial report[153] by Whitesides and Lewis that lanthanide shift reagents formed from a chiral diketonate ligand may be used to distinguish the proton resonances of enantiomers in a racemic mixture, there has been a great deal of interest in the nature and potential applications of this effect. The effectiveness of a number of chiral shift reagents has been evaluated in terms of their ability to separate the proton resonances of a wide variety of racemates in this way.[154] Details of the synthesis of each reagent are given; it also seems that most reagents of this type are decomposed by acid and by competing ligands such as α-diketones or dioximes. Based on the magnitude of the differential shifts induced in the

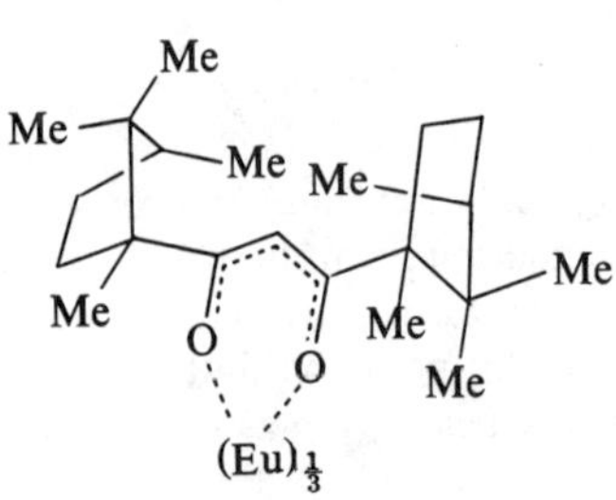

(15)

[152] S. R. Tanny, M. Pickering, and C. S. Springer, jun., *J. Amer. Chem. Soc.*, 1973, **95**, 6227.
[153] G. M. Whitesides and D. W. Lewis, *J. Amer. Chem. Soc.*, 1970, **92**, 6979.
[154] M. D. McCreary, D. W. Lewis, D. L. Wernick, and G. M. Whitesides, *J. Amer. Chem. Soc.*, 1974, **96**, 1038.

enantiomers studied ($\Delta\Delta\delta_{obs}$) and also on the ease with which the reagent may be synthesized, it has been concluded that tris-[*d*,*d*-dicampholylmethanato] $Eu^{III}$ (15) is the best single reagent for studies of this kind.[154] In a similar study, Goering *et al.*[155] have examined reagents having the general structure (16). These reagents seem to induce slightly smaller $\Delta\Delta\delta_{obs}$ values than those discussed above,[154] but may be more easily synthesized. The effect of the shift reagent:substrate ratio on both $\Delta\Delta\delta_{obs}$ and $\Delta\delta_{obs}$ is discussed in some detail in this work.[155] It is apparent that the complex stoicheiometries vary from system to system, but in many cases it appears that 1:1 and 1:2 complexes are present simultaneously. Variation in the relative amounts of the two species is thought to explain the otherwise extraordinary behaviour of the $\Delta\Delta\delta_{obs}$ values in certain cases. For example, $\Delta\Delta\delta_{obs}$ for the *ortho*-protons of 2-phenylbutan-2-ol with $Eu(hfbc)_3$ initially increases in proportion to the relative amount of shift reagent up to a 0.5 molar ratio, decreases to zero as the molar ratio is further increased to 1.2, and again increases thereafter but with reversed sign. It has been suggested, on the basis of these observations,

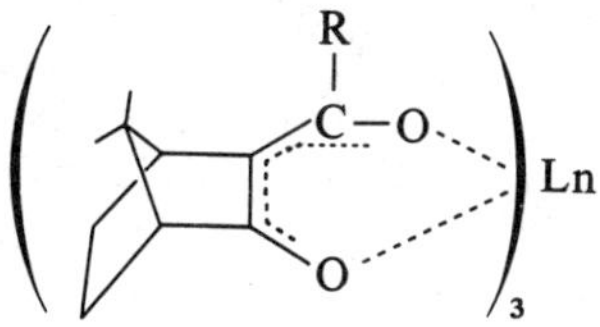


(16) R = $CF_3$ is Ln $(facam)_3$
R = $C_3F_7$ is Ln $(hfbc)_3$

that structural features of the shift reagent–substrate complex are the major factor which determines the $\Delta\Delta\delta_{obs}$ values in any given instance, rather than differences in the enantiomer binding constants to the shift reagent.

The simplest application of chiral shift reagents is in the estimation of optical purity. If two resonances, one from each enantiomer, can be clearly resolved, it is then possible to obtain directly the composition of the mixture by integration. For compounds having quite complex spectra, this may only become possible at high shift reagent:substrate ratios, where line-broadening effects may become objectionable. Selective spin decoupling has been used to alleviate this difficulty,[156] although NOE effects may then become a problem.

Another application of chiral shift reagents is related to their ability to resolve the resonances of anisochronous nuclei in a *meso*-compound. As an example, 1,2-dichloro-1,2-difluoroethane exists as the *d*-, *l*- and *meso*-forms. Resolution of the $^{19}F$ resonances of the *d*- and *l*-forms can be effected by observing the spectrum in a chiral solvent such as *l*-bornyl acetate.[157] With certain chiral shift reagents the two $^{19}F$ resonances of the *meso*-form can also be resolved.[157] The resonances of the *meso*-form are, of course, readily

155 H. L. Goering, J. N. Eikenberry, G. S. Koermer, and C. J. Lattimer, *J. Amer. Chem. Soc.*, 1974, **96**, 1493.
156 O. Červinka, P. Malon, and P. Trska, *Coll. Czech. Chem. Comm.*, 1973, **38**, 3299.
157 R. J. Abraham and T. M. Siverns, *Org. Magn. Resonance*, 1973, **5**, 253.

assigned, since a vicinal coupling will be apparent in this case, whilst the *d*- and *l*-forms will each give rise to a single line. The dimethyl ester of 3,3,4,4-tetramethylcyclobutane-1,2-dicarboxylic acid (17) may be assigned as *cis* (*meso*) or *trans* (*dl*) on the basis of the observation or otherwise of vicinal ($H_A$,$H_B$) coupling in the shifted spectrum produced by Eu(facam)$_3$.[158] Similar behaviour is observed for the H-5 and H-4 protons of 4,5-dihydroxy-4,5-dihydroaldrin (18).[159] For nuclei at a prochiral centre of an alcohol, as for

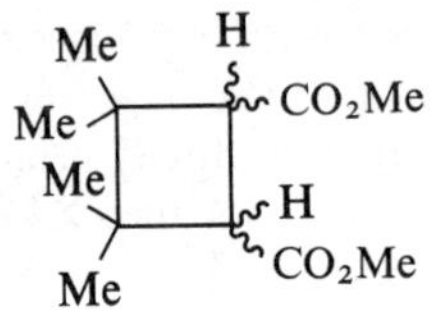


(17)

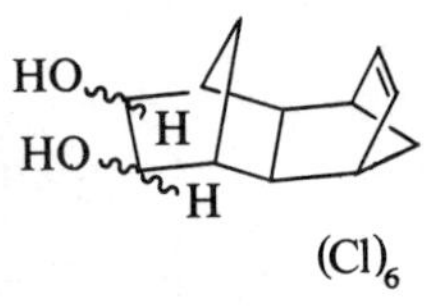


(18)

example the methylene protons of benzyl alcohol, resolution of the signals may be effected by forming the ester with *l*-camphanic acid. The addition of an achiral reagent such as Eu(fod)$_3$ may then be used to induce very marked shift differences between the enantiotopic protons at the prochiral centre.[160] The magnitude of the shift differences is quite large, and may reflect the possibility of bidentate binding of the substrate in the complex.[160] The technique is useful in studies of α-deuterio-alcohols, and in general it is found that the greater shift is induced in the *pro-R* proton, which may prove to be a useful diagnostic result.[160] More direct assignments for α-deuterio-alcohols are, however, possible using[161] tris-[(3-heptafluoropropylhydroxymethylene)-*d*-camphorato]Eu$^{III}$, where again the greater $\Delta\delta_{obs}$ values are recorded for the *pro-R* proton. A preliminary report suggests that the same may hold true for primary amines, in which case this approach may prove a useful method for the assignment of absolute configurations.[160]

158 G. L. Goe, *J. Org. Chem.*, 1974, **38**, 4285.
159 J. D. McKinney, H. B. Matthews, and N. K. Wilson, *Tetrahedron Letters*, 1973, 1895.
160 H. Gerlach and B. Zagalak, *J.C.S. Chem. Comm.*, 1973, 274.
161 C. J. Reich, G. R. Sullivan, and H. S. Mosher, *Tetrahedron Letters*, 1973, 1503.

# Author Index